THE CORRESPONDENCE OF ISAAC NEWTON

VOLUME VII

1718–1727

Kneller's portrait of Newton, painted in 1720 for Varignon; see Letter 1348. Reproduced by courtesy of Lord Egremont and the Petworth Estate. *Frontispiece*

THE CORRESPONDENCE OF
ISAAC NEWTON

VOLUME VII

1718–1727

EDITED BY

A. RUPERT HALL

AND

LAURA TILLING

PUBLISHED FOR THE ROYAL SOCIETY

CAMBRIDGE UNIVERSITY PRESS

CAMBRIDGE

LONDON · NEW YORK · MELBOURNE

CAMBRIDGE UNIVERSITY PRESS
Cambridge, New York, Melbourne, Madrid, Cape Town, Singapore, São Paulo, Delhi

Cambridge University Press
The Edinburgh Building, Cambridge CB2 8RU, UK

Published in the United States of America by Cambridge University Press, New York

www.cambridge.org
Information on this title: www.cambridge.org/9780521087230

First published 1977
This digitally printed version 2008

A catalogue record for this publication is available from the British Library

Library of Congress Cataloguing in Publication data

Newton, Sir Isaac, 1642–1727.
Correspondence.
Vols. 1–3 edited by H. W. Turnbull
Vol. 4 edited by J. F. Scott.
Vols. 5– edited by A. R. Hall and L. Tilling.
Includes bibliographical references.
Contents: – v. 1. 1661–1675. – v. 2. 1676–1687. –
v. 3. 1688–1694. [etc.]
Vol. 7 has index.
1. Newton, Sir Isaac, 1642–1727. 2. Physicists – Correspondence.
3. Physicists – Biography. I. Turnbull, Herbert Westren, 1885– ed.
QC16.N7A4   509′.2′4 [B]   59–65134

ISBN 978-0-521-08723-0 hardback
ISBN 978-0-521-08597-7 paperback

CONTENTS

THE CORRESPONDENCE

CONTENTS

vi

UNDATED CORRESPONDENCE

CONTENTS

LIST OF PLATES

PREFACE

The letters in this, the seventh and final volume of *The Correspondence of Isaac Newton*, are divided into two quite distinct groups. The first group begins with the remaining letters of the main chronological sequence written during the closing years of Newton's life, 1718–27, and then proceeds to those few letters to which we have been unable to assign a date with any certainty; these latter are arranged according to the scheme described on p. 357. The second group of letters, placed in Appendix I, contains corrections and additions to the letters printed in the earlier volumes of the *Correspondence*. For a note on the numeration used in Appendix I, and a general account of the correspondence it includes, see pp. 384–7. A genealogical table is added as Appendix II to help the reader through the intricacies of Newton's family tree. Throughout this volume in transcribing letters we have adhered to the conventions used in earlier volumes.

As always we are indebted to a number of colleagues and friends for their invaluable advice and assistance in the preparation of this volume. Many of those who assisted us in previous volumes have helped us again in this.

We must particularly mention Mme Bilodeau, of the École Pratique des Hautes Études, Paris, who provided useful bibliographical information; Professor J. O. Fleckenstein for so kindly sending us copies of a large portion of the Varignon–Bernoulli correspondence, which he is preparing for publication as part of the Bernoulli Edition; and Mr P. E. Spargo, Mr Julius Heimann and Dr Ivo Schneider for help in locating documents. The librarians and archivists of the following institutions who have under their care the documents which we have transcribed and the printed works to which we have had to refer have been as generous with their time and advice as ever: Babson College; the University Library, Basel; the Bibliothèque Nationale, Paris; King's College Library, Trinity College Library, and the University Library, Cambridge; the Scottish Record Office, Edinburgh; the Bibliothèque Publique et Universitaire de Genève; the Jewish National and University Library, Jerusalem; the Brotherton Library, Leeds; the British Library, Goldsmiths' Hall, London University Library and the Public Record Office, London; the Burndy Library, Norwalk, Connecticut; the Bodleian Library and Christ Church College Library, Oxford; Hale Observatories Library, Pasadena; Stanford University Libraries; and the Yale University Library. We should also like to thank Viscount Lymington for courteous assistance and for allowing us to transcribe documents from his collections. We owe our gratitude to the anonymous noble owner of an outstanding library for a similar permission, and also to both Lord Egremont and the Petworth Estate, and to Miss M. G. Williams and the National Trust, for allowing the portrait of the frontispiece to be rephotographed and reproduced here.

Our greatest debt of gratitude is to Dr D. T. Whiteside, who has helped us in every way in our struggle to interpret the maze of manuscripts which Newton left behind him, and who has been unstinting in his advice, even though himself heavily engaged in preparing *The Mathematical Papers of Isaac Newton*, companion volumes to our own.

Finally our thanks go to Mrs Frances Couch for her care in preparing the typescript of this volume, and to the Imperial College of Science and Technology, London University, for space in which to work.

A. RUPERT HALL
LAURA TILLING

SHORT TITLES AND ABBREVIATIONS

Bernoulli Edition	In preparation under the editorship of Professor J. O. Fleckenstein on behalf of the Naturforschenden Gesellschaft of Basel.
Bernoulli, *Opera Omnia*	*Johannis Bernoulli...Opera Omnia*, 4 vols. Lausanne and Geneva, 1742.
Brewster, *Memoirs*	Sir David Brewster, *Memoirs of the Life, Writings and Discoveries of Sir Isaac Newton.* Edinburgh, 1855; reprinted, Johnson, New York and London, 1965.
Cal. Treas. Books	William A. Shaw (ed.), *Calendar of Treasury Books preserved in the Public Record Office.* London H.M.S.O., 1904 etc.
Cal. Treas. Papers	J. Redlington (ed.), *Calendar of Treasury Papers preserved in Her Majesty's Public Record Office.* London, Longman & Co., 1868–89.
Charta Volans	The broadside, dated 29 July 1713, [N.S.], written anonymously by Leibniz; see Number 1009, vol. VI, pp. 15–21.
Cohen, *Introduction*	I. Bernard Cohen, *Introduction to Newton's 'Principia'.* Cambridge, 1971.
Commercium Epistolicum	*Commercium Epistolicum D.* Johannis Collins, *et aliorum de Analysi promota: jussu Societatis Regiæ in lucem editum.* London, 1712. Second edition, London, 1722.
Craig, *Mint*	Sir John Craig, *The Mint. A History of the London Mint from A.D. 287 to 1948.* Cambridge, 1953.
Craig, *Newton*	Sir John Craig, *Newton at the Mint.* Cambridge, 1946.
De Villamil	Richard de Villamil, *Newton: The Man.* London, n.d. [1931].
Des Maizeaux, *Recueil*	Pierre Des Maizeaux (ed.), *Recueil de Diverses Pièces sur la Philosophie, la Religion Naturelle, l'Histoire, les Mathematiques, etc., par Mrs Leibniz, Clarke, Newton et autres Auteurs célèbres.* Amsterdam, 1720, 1740. The page numbers of the second edition are referred to unless otherwise stated.
Edleston, *Correspondence*	Joseph Edleston, *Correspondence of Sir Isaac Newton and Professor Cotes including Letters of other Eminent Men.* London, 1851; reprinted, Cass, London, 1969.
Foster	C. W. Foster, 'Sir Isaac Newton's Family', *The Architectural and Archaeological Society of the County of Lincoln,* **39** (1928), 1–61.

Hall & Hall, *Oldenburg* A. Rupert Hall and Marie Boas Hall, *The Correspondence of Henry Oldenburg*. University of Wisconsin Press, 1965 onward: in progress.

Hall & Hall, *Unpublished Scientific Papers* A. Rupert Hall and Marie Boas Hall, *Unpublished Scientific Papers of Isaac Newton*. Cambridge, 1962.

Horsley, *Opera omnia* Samuel Horsley (ed.), *Isaaci Newtoni Opera quæ exstant omnia*, 5 vols. London, 1779–85.

Huygens, *Œuvres Complètes* *Œuvres Complètes de Christiaan Huygens publiées par la Société Hollandaise des Sciences*, 22 vols. La Haye, 1888–1950.

Keill, *Epistola ad Bernoulli* *Joannis Keill M.D. & R.S.S. in Academia Oxoniensi Astronomiæ Professoris. Epistola ad Virum Clarissimum Joannem Bernoulli in Academia Basiliensi Mathematum Professorem*. London, 1720.

Keynes MSS. Lord Keynes' collection of Newton's papers and correspondence at King's College Library Cambridge; also available on microfilm at the University Library, Cambridge.

Koyré and Cohen, *Principia* *Isaac Newton's Philosophiæ Naturalis Principia Mathematica. The Third Edition (1726) with variant readings assembled and edited by Alexandre Koyré and I. Bernard Cohen with the assistance of Anne Whitman*, 2 vols. Cambridge, 1972.

Maclaurin *An Account of Sir Isaac Newton's Philosophical Discoveries* in Four Books by Colin Maclaurin, A.M....Published from the Author's Manuscript Papers. By Patrick Murdoch, M.A. and F.R.S. London, 1748.

Mint Papers Newton's private file of papers concerning Mint business, sold at Sotheby's in 1936, now bound in three volumes in the Public Record Office (Mint/19, I–III).

More Louis Trenchard More, *Isaac Newton, a Biography*. New York and London, 1934.

Nichols John Nichols, *Illustrations of the Literary History of the Eighteenth Century*, vol. IV. London, 1822.

P.R.O. Manuscripts in the Public Record Office, London.

Raphson, *History of Fluxions* Joseph Raphson, *The History of Fluxions Shewing in a compendious manner The first Rise of, and various Improvements made in that Incomparable Method*. London, 1715. (For the Appendix of letters concerning the calculus controversy, added by Newton to a reprint of 1718, see vol. VI, p. 254, note (2).)

Recensio	'Recensio Libri Qui inscriptus est Commercium Epistolicum Collinii et aliorum, de Analysi Promota...' prefaced to the second edition of the *Commercium Epistolicum* (London, 1722), pp. 1–59, and also published separately in both an English and a French translation in 1715. (See vol. VI, p. 242, note (2).)
Rigaud, *Correspondence*	S. P. and S. J. Rigaud, *Correspondence of Scientific Men of the Seventeenth Century...in the Collections of the Earl of Macclesfield*, Oxford, 1841; reprinted, Georg Olms, Hildesheim, 1965.
Rigaud, *Essay*	S. P. Rigaud, *Historical essay on the first publication of Sir Isaac Newton's Principia*. Oxford, 1838.
Shaw	William A. Shaw, *Select Tracts and Documents Illustrative of English Monetary History, 1626–1730*. London, 1896.
Sotheby Catalogue	*Catalogue of the Newton Papers Sold by Order of the Viscount Lymington...which will be sold by Auction by Messrs. Sotheby and Co.* [on 13 and 14 July, 1936].
Taylor, *Contemplatio Philosophica*	Brook Taylor, *Contemplatio Philosophica: a posthumous work, of the late Brook Taylor,... To which is prefixed a Life of the Author by his Grandson, Sir William Young...with an Appendix containing sundry original Papers, Letters from the Count Raymond de Montmort, Lord Bolingbroke, Marcilly de Villette, Bernoulli &c.* London, 1793.
Taylor, *Methodus Incrementorum*	Brook Taylor, *Methodus Incrementorum directa et inversa*. London, 1715.
Turnor	Edmund Turnor, *Collections for the History of the Town and Soke of Grantham...*London, 1806.
U.L.C.	Manuscripts in the University Library, Cambridge, mostly from the Portsmouth Collection.
Wallis, *Opera Mathematica*	*Joannis Wallis S.T.D. Opera Mathematica*, 3 vols. Oxford, vol. I, 1695; vol. II, 1693; vol. III, 1699.
Whiteside, *Mathematical Papers*	D. T. Whiteside (ed.), *The Mathematical Papers of Isaac Newton*. Cambridge, 1967 onward: in progress.
Wollenschläger, *De Moivre*	Karl Wollenschläger, 'Der mathematische Briefwechsel zwischen Johann [I] Bernoulli und Abraham de Moivre', *Verhandlungen der Naturforschenden Gesellschaft in Basel*, **43** (1931–3), 151–317.

INTRODUCTION

After the creative power of his genius had deserted him, Newton retained to the very end of his extremely long life his characteristic clarity of thought. He suffered increasingly many of the physical discomforts of old age, and eventually illness forced him to move to the clearer air of Kensington in early 1725, and to curtail his business journeys into town. James Stirling, visiting him later that year, wrote, 'S[ir] Isaac Newton lives a little way of[f] in the country. I go frequently to see him, and find him extremely kind and serviceable in everything I desire but he is much failed and not able to do as he has done.'* From mid-1722 onwards, as Newton's health deteriorated, the Chair at meetings of the Royal Society was taken more and more frequently by one or other of the Deputy Presidents. The relative inactivity of the Mint meant that, although he apparently delegated few of his responsibilities to others, Newton's concerns there were now not onerous. Thus it is not surprising that in the last nine years of his life (the period covered by the present volume), and particularly from 1725 onwards, there was a decrease in Newton's output of letters; but those which he did write remain as lucid as ever. There is some indication that in writing to Pierre Varignon and Johann [I] Bernoulli Newton received considerable advice and encouragement from Abraham de Moivre, but he nevertheless finally penned the letters himself, and as late as February 1727, little over a month before his death, he was still able to write cogently to Thomas Mason concerning some metallurgical assays (Letter 1498).

Few of Newton's letters in this volume may justly be described as scientific. In his extensive exchanges with Varignon and Bernoulli about the calculus dispute he was preoccupied with historical arguments over priority and authorship, not with the mathematical methods involved, and from this correspondence we learn far more of the emotional characters of the protagonists than of the nature of their scientific intellects. In further correspondence with Varignon about the Paris edition of the *Traité d'Optique* (1722) we find virtually no discussion of the scientific content of the work; Newton confined himself in his letters to problems of administrative detail and left to Abraham de Moivre the business of communicating to Varignon minor corrections, although clearly he had discussed these privately with De Moivre. The extensive series of letters from Henry Pemberton to Newton about the preparation of the third edition of the *Principia* lacks all the richness of the Cotes–Newton correspondence concerning the second (see vol. v), but this is partly because the cor-

* Charles Tweedie, *James Stirling* (Oxford, 1922), p. 13.

rected copy of the second edition from which Pemberton worked is now lost, together with any letters Newton wrote to him. Pemberton did not have the mental calibre of Cotes, and his own amendments to the edition are usually trivial.

Newton's multitudinous drafts, and other evidence supplied by the writings of his contemporaries, indicate that his concern for the topics just mentioned—the calculus dispute and the new editions of the *Traité d'Optique* and the *Principia*—was far greater than his extant letters alone would show, and here, as in earlier volumes, we have had frequent recourse to ancillary material in order to maintain a balanced view of Newton's preoccupations during the closing years of his life. But in other areas his lack of interest seems to have been real. His correspondence as President of the Royal Society is purely formal in nature. As one of the panel selected by the Admiralty for assessing new methods of determining longitude at sea he was frequently called upon to make reports upon the methods proposed, but only two such reports are to be found amongst his papers for this period, and both largely reiterate comments made many years earlier. Although he received a number of letters from scientific colleagues at home and abroad, he apparently left most of these un-answered, and certainly showed no desire to build up extensive epistolary exchanges.

Newton's correspondence with Pierre Varignon is perhaps the most interest-ing and important sequence of letters in this volume. Varignon, as a member of the Académie Royale des Sciences and as a professor at both the Collège Mazarin and the Collège de France, held an important position in the scientific community of Paris. He acted as an intermediary between Newton and the Académie, of which he was an *associé étranger*, despatching to him regularly its *Mémoires* and the *Connoissance des Temps*. He maintained friendly relations with Newton while pushing the Paris edition of the *Traité d'Optique* through the press, albeit slowly; he also possessed a wide circle of acquaintances amongst continental scientists, and Newton could have used him as a means of ob-taining scientific news from Europe. He showed little inclination to do so, however, although Varignon no doubt took every opportunity to keep his friends well informed of Newton's activities.

In 1718 Varignon took it upon himself to act as mediator in the calculus dispute. Since Leibniz's death, Johann Bernoulli had become the chief con-tinental protagonist in the debate, despite his efforts to maintain anonymity; while he regarded his dispute with John Keill as a separate and private affair, nevertheless, in the eyes of the world, the argument was between the English and the continental mathematicians, with Newton and Bernoulli as their major representatives. Varignon had so far maintained at least a superficial

The *Charta Volans* had of course been Leibniz's response to the *Commercium Epistolicum*; Leibniz had later revealed (truthfully, but against Bernoulli's wishes) Bernoulli's authorship of the letter inserted in it, but despite the obvious realization of all the learned world that he *was* the author, and against all reason and honesty, Bernoulli tried to maintain his innocence. In June 1719 (Letter 1320) he was at last prompted, largely by Varignon, to reply to Newton to thank him for his gifts, to give a polite but downright denial of his authorship of the letter in the *Charta Volans*, and to complain that he had heard that he had been cited yet again as author of it in the Appendix to Raphson's *History of Fluxions*, which he had not yet seen. We learn from Bernoulli's correspondence with Varignon that he was annoyed on other counts too: he had discovered that it was Varignon, not Newton himself, who had suggested the gifts of the *Opticks*; but more aggravating still he had learnt from Rémond de Monmort what he had already suspected, that Newton was the moving force behind Keill's recent activities in the dispute. Keill had, of course, inveighed more strongly than had Newton against the 'Eminent Mathematician' in his 1714 'Answer' to the *Charta Volans* (see Letter 1053a, vol. VI, note (1)): Bernoulli had already replied to him in his 'Epistola pro Eminente Mathematico' of 1716 (see Letter 1196, vol. VI), published anonymously, but with its authorship so ill-disguised that there was never any real doubt about its source. Bernoulli was, of course, correct in assuming that Newton was deeply involved in the preparation of Keill's early papers; he would have been correct also had he accused Newton of being behind the publication of the Newton–Leibniz letters of 1715 and 1716 in the Appendix to Raphson's *History of Fluxions*. There is no doubt too that Newton was responsible for the 'Recensio' of the *Commercium Epistolicum* printed in the *Philosophical Transactions* in 1715. But it is also clear (although Bernoulli could not know this) that Newton, around 1719, was losing patience with Keill, and that he positively discouraged publication of Keill's 'Lettre à Bernoulli' published in the *Journal Literaire* for 1719 and his *Epistola ad Bernoulli* and its 'Additamentum' of 1720; he certainly was never drawn into the contemporary mathematical wrangles in which Bernoulli took so active a part. Considerable correspondence concerning these debates is extant, much of it very interesting; but since Newton was so little concerned in it, either privately or publicly, we touch on it only briefly in this volume. Rémond de Monmort was perhaps the chief trouble maker: it was he who told Bernoulli of Newton's influence on Keill; it was he who passed on to Bernoulli (and to Jakob Hermann and Nikolaus [II] Bernoulli in a modified form) Keill's 'challenge' problem concerning the path of a projectile in a resisting medium (see Letter 1303); and it was he who told Brook Taylor that he suspected Bernoulli of writing the 'Epistola pro Eminente Mathemcoati' and of being responsible for Nikolaus Kruse's 1718 defence of Bernoulli (see Letter 1307). His death in October 1719 put an end to these machinations, and also to a philosophical dispute with Taylor in which Monmort supported the Cartesian view of the universe against the Newtonians. Taylor and Monmort were friends despite their disagreements; but Taylor's attitude towards Bernoulli was not so amiable. Bernoulli, in his 'Epistola pro Eminente Mathematico', had charged Taylor with plagiary, and Taylor had angrily retaliated with his 'Apologia contra Bernoullium' in the *Philosophical Transactions* for 1719 (see Letter 1316) attacking Bernoulli on a more general front. Both Nikolaus [II] Bernoulli and Johann Burchard Mencke replied on Bernoulli's behalf.

It was Keill's 'challenge' problem to Bernoulli that probably instigated the next step. Bernoulli gave the general solution (but no demonstration) in his 'Responsio' printed in the *Acta Eruditorum* for May 1719, and there went so far as to cast aspersions on Keill's morals, calling him 'a certain individual of Scottish race who has become no less notorious among his own people for his immorality than odious everywhere to foreigners'; he also once again brought up the whole vexed question of Newton's mistake in Book II, Proposition 10 of the first edition of the *Principia*, and the extent to which Bernoulli had pointed out other errors in the *Principia* before the second edition went to press. Newton clearly saw this paper, for in his reply (Letter 1329a) to Bernoulli's letter, he was at pains to refer to the correspondence between himself and Roger Cotes during preparations for the second edition in order to prove that no corrections were made as a result of information received from Bernoulli, *except* the correction to Book II, Proposition 10.

Newton accepted Bernoulli's denial of authorship of the Letter in the *Charta Volans*; in his second letter to Newton (Letter 1332) Bernoulli repeated it—but in a much more tentative form: 'I am not certain of what kind that letter addressed to Mr. Leibniz is of which you speak, which is dated 7 June 1713 [N.S.]. I do not remember having written to him myself that day, yet I would not deny it altogether, since I have not kept copies of all the letters I have written.' But just as Newton's annoyance had been rekindled by Bernoulli's 'Responsio' to Keill, so Bernoulli had further reason to be cold towards Newton—first, he had (wrongly) understood Newton to say that the Appendix to Raphson's book had been reprinted, and second, he had heard that he, Bernoulli, had been expelled from the Royal Society, and demanded an explanation.

Newton's draft reply (Letter 1333) seems unwarrantably severe; if Newton suspected Bernoulli of not being entirely honest in his answer, he must have known full well that he himself was equally guilty of equivocation, and his terse denial of having any part in the publication of Raphson's book is downright falsehood. Apparently, however, this reply was never sent; and at this period, Abraham de Moivre, who had previously carried letters back and forth between Newton and Varignon, and himself maintained a correspondence with Varignon, now became increasingly important as Newton's spokesman and mentor (see Letter 1334). De Moivre reported to Varignon Newton's reactions on reading Bernoulli's letter (Letter 1332), and Varignon, although his information was confidential, passed it on to Bernoulli. Apparently on first seeing the letter Newton had reacted favourably, but on a second reading his anger had been stirred, and he adamantly refused to reply.

The dispute had reached an impasse: on the surface there was still no real argument between Bernoulli and Newton, since Bernoulli continued to deny authorship of the letter in the *Charta Volans*, and also insisted that when he had communicated certain challenge problems to Leibniz long ago he had not intended that they should be posed to the English; and Newton continued to deny any part in the publication of Raphson's *History of Fluxions* and any influence over Keill. Neither therefore felt the need to offer an explanation—far less an apology—to the other. Meanwhile irritations grew; Bernoulli hinted that he was privy to certain information from Monmort which could put Newton's actions in a bad light; the publication of Pierre Des Maizeaux's *Recueil* and of Keill's strongly worded *Epistola ad Bernoulli* in 1720 did nothing to improve matters. Newton was speaking the truth when he said he had had virtually no influence on the latter publication, but he probably could have acted more effectively in the case of the *Recueil*, had he so wished. Des Maizeaux had sent him proofs in 1718 (see vol. VI), but Newton then took no action to prevent publication—probably it was only after he had received Bernoulli's protestations of innocence that he saw the need to delay the press (see Letter 1330), but nonetheless the book appeared, containing the 1715 and 1716 Leibniz–Newton correspondence via Conti, together with Newton's 'Observations' on it and the clear implication that Bernoulli was author of the notorious letter in the *Charta Volans*, with only a brief intimation by Des Maizeaux in the Preface that Bernoulli might not, in fact, be the author. When Bernoulli saw the *Recueil* he was incensed—this was the first time he had actually seen the Leibniz–Newton correspondence of 1715–16. In his angry reaction to Varignon (see Letter 1357, note (2)) he singled out three phrases Newton had used: 'homo novus', 'pretended mathematician' and 'Knight-errant'. Bernoulli claimed that Newton had used all three in description of himself. He was correct in respect of the first one, but Newton claimed that since he there referred to Bernoulli when he *was* a very young man, it could scarcely be called an insult; the second and third Newton used only to describe the anonymous writer of the letter in the *Charta Volans*, whom in 1715 he understandably believed to be Bernoulli, a misapprehension Des Maizeaux corrected in his Preface, admittedly somewhat obscurely. Bernoulli presumably felt that Newton should have influenced Des Maizeaux to stress his innocence more strongly.

Varignon now redoubled his efforts to bring about a reconciliation. He enlisted De Moivre's help in trying to persuade Newton to write an explanatory open letter; he himself put the whole matter before Bernoulli again, but to no avail. He almost despaired when he *did* receive another letter from Newton (Letter 1372) with a full and detailed explanation, but instructions that it should on no account be published; a reply which Bernoulli would not accept. At the end of August 1721 Keill died; De Moivre wrote to Varignon (see Letter 1374) 'la Mort de M. Keil...leve un grand obstacle à la paix'. Varignon, however, was still determined to bring about a positive settlement of the dispute between Newton and Bernoulli. In desperation he offered to draft on Newton's behalf an explanatory letter to be approved by him and then published; Newton eventually agreed to this plan and a draft was prepared (Letter 1390*a*) and sent to Bernoulli in May 1722. Bernoulli's reaction horrified Varignon: he accepted the explanation (although he would have preferred it to be more apologetic) but, astonishingly, wrote that he had decided *not* to publish it, out of consideration for Varignon. This, of course, put Varignon in an insupportable position; how could he go back to Newton and De Moivre and tell them that after all their trouble Bernoulli would not publish the letter? His death, in December 1722, spared him the embarrassment of doing so.

And with Varignon's death, at last, came the end of the dispute. No solution was reached, and no public explanation was made, but interest waned. With no intermediary to fan the flames of their anger, neither Newton nor Bernoulli could sustain their argument, and De Moivre seems to have shown no interest in taking over Varignon's rôle. There was one last letter from Bernoulli to Newton in January 1723 (Letter 1404) in which he pointed out to Newton a work by Nicolas Hartsoeker insulting (so he said) to both Bernoulli and Newton, and retrenched so far as to suggest that he *might* have written the letter in the *Charta Volans*, but certainly was not responsible for referring to himself as an 'Eminent Mathematician'. The second edition of the *Commercium Epistolicum* (1722) produced no further furore; the letter in the *Charta Volans* reprinted in its appendix was not attributed to Bernoulli (see Letter 1395).

The other major concern in Newton's correspondence with Varignon was with the publication of the Paris edition of the *Traité d'Optique* (1722). Newton had, of course, not many years previously, revised the *Opticks*, in particular the Queries appended to it, in preparation for the second English edition of 1717 (see Letter 1304, note (3)). The Latin edition of 1719 (in print in 1718) was simply a straightforward translation of this. When Varignon saw the Latin edition he commented briefly on the new Queries (Letter 1323), perhaps in an abortive attempt to draw Newton into an epistolary discussion of them. But whereas Newton's earlier reluctance to enter into metaphysical debate had overlain an intense interest in theories of matter for which we have strong evidence in the extensive manuscript revisions of the Queries to be found in Cambridge University Library (Add. 3970), there is no real proof that he maintained this interest once the 1717 edition had been published. Yet since his methodological position remained the same he felt a need to stress it, for in a reply (Letter 1329*b*) to a letter (now lost) from Fontenelle, Newton, speaking of the 1719 *Optice*, commented, 'Here I cultivate the experimental philosophy as that which is worthy to be called philosophy, and I consider hypothetical philosophy not as knowledge but by means of queries. And the matters I have added to this edition are of the latter type.' Clearly Newton shunned metaphysical argument as adamantly as ever.

Varignon's involvement with the second French edition of the *Traité d'Optique* began in April 1720, when he was asked to examine the work on behalf of the official censor; he not only wrote a formal *Approbation*, but himself took charge of this Paris edition. The first French edition had been printed at Amsterdam in 1720 from Pierre Coste's translation; Varignon used the same translation, incorporating corrections by De Moivre, who was clearly ideally suited to provide assistance, being closely acquainted with both Varignon and Newton, fluent in French and English, and well versed in science. It is difficult to say to what extent De Moivre's corrections were directed by Newton himself, but it is clear from Varignon's correspondence with Bernoulli and from autograph notes by De Moivre on Newton's draft letters, that De Moivre kept in close touch with both Varignon and Newton throughout the printing of the work. Newton himself recommended to Varignon (Letter 1353) that, 'You should pay very little attention to the corrections of others (should you receive any), in case they cause you trouble or delay. If any other points turn up, they will be looked into here and sent there [*to Paris*].' In August 1721 Coste himself wrote a bitter letter (Letter 1365) to Newton, protesting that Newton and Varignon had tactlessly paid very little attention to corrections submitted by Coste himself, and had not bothered to keep him informed of those of De Moivre, although they had promised to do so. The letter also indicates that Varignon had abdicated virtually all responsibility for corrections to the *Traité*, whilst Newton kept a fairly tight rein on what was going on, perhaps unwillingly, for he had written to Varignon (Letter 1363), 'I hope and pray that you will use your own judgement and freely make any alterations which seem necessary in the corrections sent to you. For Mr De Moivre submits everything to your judgement.'

In August 1721, shortly after printing started, it was interrupted by serious delays, largely the fault of the printer Montalant. However, in July 1722 (Letter 1396) Varignon was able to report that the edition was at last finished, complete with a *vignette* by the famous Swiss miniaturist, John Arlaud, produced from Newton's own sketch. Varignon commented that De Moivre's corrections were of greater help than Coste's, and that he himself only used his judgement when the two were at variance.

The very existence of Varignon's edition of the *Traité d'Optique* is indicative of the growing French interest in Newton's experimental optics.* These indications are reinforced by a letter addressed by Sebastien Truchet to Newton (Letter 1350) in which the writer explains how he has repeated many of Newton's optical experiments in front of a group of eminent scientists (including Varignon himself); Truchet's observations agreed with Newton's own and he admired Newton's experimental technique, but in common with a number of his countrymen, rejected his explanation and provided a completely different— and very obscure—interpretation of the formation of the spectrum. Newton apparently passed over in silence this solitary communication concerning optics.

Little need be said of Henry Pemberton's letters to Newton in the period October 1723 to February 1726 concerning the third edition of the *Principia*, although they make up in bulk for what they lack in interest, forming a large proportion of the printed matter in the second half of this volume. Pemberton lacked the intellectual brilliance of Cotes, and it is by no means clear why Newton chose him as editor for this new edition; professionally a physician, he seems to have been introduced to Newton in 1722 through Richard Mead, Newton's own doctor, but he was also acquainted with others of Newton's circle, notably John Keill and James Wilson, who held him in considerable esteem. Pemberton had begun in 1721 or 1722 to write *A view of Sir Isaac Newton's Philosophy* (finally published in London in 1728), when he had not met Newton and very much wanted to make his acquaintance. Much later, possibly after Newton's death, he planned to publish an English translation of the *Principia*, together with a commentary upon it, expanding and explaining Newton's demonstrations; the plan never came to fruition.†

Newton must have sent Pemberton a corrected printed copy of the second edition, together with several loose sheets of drafts of the longer corrections (this material has never been found, although earlier annotated copies of the *Principia* are still in existence); Pemberton sent Newton comments on this

---

* See A. Rupert Hall, 'Newton in France: a new view', *History of Science*, **13** (1975), 233–50.

† See I. Bernard Cohen, 'Pemberton's translation of Newton's *Principia*, with notes on Motte's translation', *Isis*, **54** (1963), 319–51.

annotated version as he worked through it, and simultaneously sent proof sheets from the third edition as printing progressed. Newton accepted a number of trivial corrections suggested by Pemberton—corrections to style, spelling and so on—but Pemberton's more radical suggestions he largely rejected.* (On one occasion Pemberton was so tediously prolix that we have departed from our usual rule of printing in full all letters addressed to Newton, and have omitted part of a long proof which Pemberton sent (see Letter 1486).)

Newton had planned a third edition of the *Principia* long before he submitted it to Pemberton at the end of 1723. The 'Addenda and Corrigenda' which Newton had sent Cotes in 1713 were clearly not so much corrections of the printer's errors in the second edition as afterthoughts by Newton.† In June 1714 De Moivre mentioned Newton's plans for a third edition to Johann Bernoulli (see vol. VI, p. 152); certainly by September 1719 the extent of the revision must have been considerable, since Newton then referred in a letter to Varignon (Letter 1329) to a copy he had already corrected with a view to producing a third edition. He had long ago considered supplementing the *Principia* with some of his work on the calculus, and later thought of annexing the *De Quadratura Curvarum* in an expanded form to a revised reprint of the second edition of the *Principia*; plans of this sort were still in his mind as late as 1718 (see Letter 1298). (The *De Analysi*, the *De Quadratura Curvarum*, the *Enumeratio Linearum tertii Ordinis* and the *Methodus Differentialis* were all appended to the 1723 Amsterdam reissue of the second edition, together with excerpts from his correspondence with Oldenburg, but this work was presumably printed without Newton's authorization, and it is certainly not mentioned in his correspondence.) Despite this apparent enthusiasm for republishing part of his mathematical work, Newton appears to have been unwilling to allow others to undertake the task on his behalf. He seems to have totally ignored two letters from James Wilson (Letters 1349 and 1355) encouraging Newton to put into print certain of his early mathematical manuscripts, copies of which Wilson himself had seen. (Presumably Wilson hoped to act as editor.)

But, however far Newton's revision of the *Principia* had proceeded by 1719, he was still willing to incorporate new observational data as they became available. There is direct evidence for this in the book itself; on p. 353 of the third edition some observations made by Desaguliers in July 1719 are inserted, and on p. 524 are observations of the comet of 1723. As late as 1724–5 Halley and Newton exchanged letters (Letters 1449, 1455 and 1460) in which new values for the motions of the 1680 comet are discussed. But there is evidence too

---

* For the preparation of the third edition of the *Principia* and Pemberton's part in it, see Cohen, *Introduction*, pp. 258–86, and Koyré and Cohen, *Principia*, pp. 827–47.

† See Edleston, *Correspondence*, pp. 160–5.

that Newton was anxious to see the edition complete; Pemberton complained of too little time to make all the alterations required, and Newton himself drafted prefaces to the third edition, the earliest of which is dated 1724, presumably an indication that he thought the edition was *then* almost complete.*

Apart from the correspondence concerning his quarrel with Bernoulli and the new editions of his works, there are virtually no letters from Newton on scientific matters in this volume. We have already mentioned that the letters he wrote as President of the Royal Society were purely formal in nature. Those he received in this capacity apparently remained unanswered, but a number of them are of considerable interest. 'sGravesande wrote in 1721 to tell Newton of Orffyraeus' perpetual motion machine (Letter 1364); in 1723 Philippe Naudé wrote from Berlin a laudatory but very unimaginative letter concerning Newton's mathematics (Letter 1405); in the same year, or perhaps a little later, J. C. F. von Hatzfeld, an obscure and probably eccentric scientist, complained to Newton of the rejection by the Royal Society of his letters on perpetual motion (Letter 1418); in 1724 Luigi Ferdinando Marsigli reported on his recent travels and publications (Letter 1424 and 1440). The latter, at least, was clearly offended by Newton's failure to reply, and it seems probable that Newton did not answer *any* of these letters. Nor did he reply to Desaguliers' letter complaining that he had been accused of performing an insufficient number of experimental demonstrations at meetings (Letter 1463); perhaps Newton passed this complaint on to another member of council.

Newton also continued to receive a number of letters as one of the Committee set up by the Admiralty to investigate proposed methods of finding the longitude at sea. The majority of these methods were nonsensical, and put forward by obscure men hoping for easy fame, and fortune in the shape of the £20 000 prize offered by the Board of Longitude for a sufficiently accurate method. In a number of cases the manuscripts describing the methods are no longer to be found, and all we know is the name of the inventor. In all, only two replies from Newton are extant in the period covered by the present volume, and in both he expresses his annoyance at the futility of the methods proposed, very much in the terms he had used earlier (see vol. VI, p. 197 and pp. 221–2). Newton clearly thought that the only feasible way to find the longitude at sea was to improve methods of making *astronomical* measurements aboard ship. Refinements in clock mechanisms could only result in *keeping* the longitude, not in finding it afresh (see Letter 1377). His second reply (Letter 1476) was more explicit: only two methods of finding the longitude would prove useful, one by the motion of the Moon and the other by the eclipses of the innermost satellite of Jupiter.

* See Koyré and Cohen, *Principia*, pp. 848–50.

In one or two cases, however, there is at least slight evidence of Newton taking the trouble to give helpful replies to letters he had received. Roger Cotes' cousin, Robert Smith, later author of a celebrated text on Newtonian optics, wrote at the end of 1718 (Letter 1310) asking Newton for comment on the posthumous publication he was preparing of Cotes' *Harmonia Mensurarum*, and nearly two years later sent Newton a completed copy enclosed with a letter (Letter 1343) which implied that he had received some communication from Newton, although this may have been merely verbal. Newton also helped another exponent of Newtonian science, Colin Maclaurin, later author of *An Account of Sir Isaac Newton's Philosophical Discoveries*, by providing letters of recommendation for his appointment as joint Professor of Mathematics at Edinburgh University (see Letters 1475, 1481 and 1482). Clearly it was advantageous to Newton's own reputation to forward the careers of these two young men; at the same time there is no indication amongst his extant papers to show that he was at all interested in the mathematical points which each raised with him.

In the same way there is nothing to show that Newton welcomed the news of Italian mathematical concerns that James Stirling, mathematician and ardent Newtonian, sent him in a letter written from Venice in August 1719. Little is known of Stirling's movements abroad; the Abbé Conti had apparently at some time suggested that Newton should contribute to Stirling's maintenance in Italy, so that Stirling could defend the Newtonian cause there; but in the event it was not Conti, or Newton, but Nicolas Tron, Venetian ambassador at the English court, who was responsible for Stirling's visit and it seems unlikely that Newton ever furnished financial support. There is no doubt, however, that Newton always believed Conti to be intrumental in bringing about Stirling's visit, and later interpreted this as yet another of his subterfuges.

For when the affair of the *Abrégé de Chronologie* blew up in 1725, Newton looked back over the years and saw as Janus-like most of Conti's actions with regard to himself. Newton gave his opinion of Conti in his account of the affair in *Philosophical Transactions* for 1725, and in an even more bitter draft of the same passage* he wrote,

When Senior Conti came first into England (which was in Spring 1715) he insinu[at]ed himself into my acquaintance...Soon after he began to be at work in engageing me in disputes & sometimes finding persons to defend me. And lately he found out one in Italy to oppose me whither he had sent one before to defend me. And now he has found out another in France to oppose me. But I hope that the divulging of this Chronological

* Jewish National and University Library, Jerusalem; Yahuda Collection, Newton MS. 27, fo. 20.

Index [*the* Abrégé], the sending of Mr Sterling into Italy to be ready to defend me, the disputes raised against me there by Senior Rizzetti, & the perpetual motion, will be the last efforts of the friends of Mr Leibnitz to embroil me.

Certainly Conti and his friends were highly culpable for the publication of the pirated *Abrégé*, but Newton himself was not entirely without blame. Newton had compiled the *Abrégé* (in English) for Caroline, Princess of Wales, and had agreed that Conti should have a copy too, but stipulated that it be kept secret. Conti in fact let several friends see it, and eventually it was translated into French and published by Nicolas Fréret, a notable French scholar and theologian. The printer, Guillaume Cavelier, wrote three times to Newton asking permission to publish, his first letter (Letter 1436) being sent in April 1724. Eventually Newton replied in May 1725 (Letter 1469) refusing permission; but printing was already under way, and it was too late to withdraw the publication. Newton's objection to the publication was two-fold: first, the *Abrégé* had never been intended as a finished work, and second Fréret's only aim in publishing it was to print at the same time his own ill-considered criticisms of it. Newton's chronology, based on astronomical as well as biblical fact, was seen as a menace to orthodox theological interpretation, as it shifted by many hundred years certain events in the Old Testament. Eventually in 1728 Newton's full-scale *Chronology of Ancient Kingdoms amended* was published.*

The extant documents provide a very incomplete record of Newton's activities at the Mint, so that it becomes difficult to compile a coherent and consistent account. The evidence that towards the end of his life he unburdened many of his responsibilities as Master of the Mint upon his Deputy, Francis Fauquier, or, after the latter's death in 1726, upon John Conduitt, is slight. True, in July 1725 Fauquier rendered accounts for the copper coinage to the Treasury on Newton's behalf,† and in June 1726 Newton asked Conduitt's help over the business of importing some gold into the Mint (Letter 1490); but these are isolated incidents. As late as the autumn of 1726 Newton himself was dealing with the problem of the appointment of a new Engraver. Nevertheless in these last nine years of Newton's life the affairs of the Mint seem not to have consumed so much of his time as before, probably because of its generally inactive state.

The recoinage at Edinburgh was over by 1709, but the problem of financing the Scottish Mint dragged on, and correspondence during the period 1718 to 1725 strongly indicates that Charles, Earl of Lauderdale, General of the Scottish Mint, was using his position, now virtually a sinecure, for personal financial gain. Against strong political pressures Newton continually tried to restrain the Treasury from over-financing the Edinburgh Mint. The business

* See Frank E. Manuel, *Isaac Newton Historian* (Cambridge, 1963), pp. 21–36.
† P.R.O., T/27, 24, p. 95.

of the Alva silver mine (see vol. VI) also dragged on until 1724, again because of difficulties over payments for services rendered.

At the Tower Mint domestic affairs continued to cause Newton trouble. The long-standing rivalry between Ordnance and Mint over their respective privileges at the Tower blew up again in 1720 into a dispute concerning the smithy, where work was apparently carried out for both Mint and Ordnance (see Letter 1341). The payment of the lower-paid Mint employees was a perpetual headache, and even a general review of salaries in 1722 did not put a stop to perpetual but usually justifiable petitions requesting further money. The prosecution of counterfeiters was a responsibility which had long ago devolved upon the solicitor to the Mint (see Letter 1296), but the more general question of ways and means of discouraging illegal coining remained one of Newton's concerns. In October 1720 he faced the problem of craftsmen who possessed presses for stamping out small articles which could easily be misapplied to coining. Newton sent to the Secretary of State a list of sixteen men all charged with 'Keeping Engines or presses in their houses Usefull for makeing of Counterfeit Money' (Letter 1347); all were acquitted on the grounds that the 'Engines' were essential to their trades. Newton felt that this was an encouragement to coiners, and in private he jotted down possible solutions to the problem (see Letter 1341). These recommendations (which he never passed on to the Treasury) were that possession of illegal presses should be treated as high treason; that it should be possible to destroy such presses without prosecution; and that certain artisans should be licensed to possess presses for trade purposes.

The difficulties of the bimetallic currency and the relative flow of gold and silver had not really been adequately resolved in 1717 when the Treasury, acting largely on a recommendation from Newton, had reduced the value of the guinea by sixpence (see Letter 1270, vol. VI). The Treasury had been so plagued by complaints over this reduction that they clearly began to have doubts about its effectiveness and asked Newton for a further report, printed in this volume as Number 1302. Newton, realizing the immense complexity of the monetary system and its dependence on foreign exchange rates, rightly pointed out that other contrary factors had masked the good effect of the reduction, which perhaps in any case should have been more radical.

By far the most important of Newton's concerns at the Mint during this period was the copper coinage. After a cessation of sixteen years, which left the country desperately short of copper money, coinage had started again in January 1718, using copper bars provided by private contractors. At the beginning of 1719 the Treasury held up the coinage again, presumably because the recently enhanced price of copper meant that the coins were being produced at a loss. Newton's suggestions concerning possible new, cheaper ways of manufacturing coin show that he had an intimate knowledge of the methods used in refining and coining copper (see Letter 1325).

In 1717, and again in 1719 (see Letter 1319) William Wood had offered to supply the Mint with copper bars at $17\frac{1}{2}d$ per lb, thus undercutting the tender of $18d$ per lb made by the successful contractors, Appleby and Hines. Later Wood took on responsibility for the coinage of copper for both Ireland and the West Indies. He proposed to carry out this coinage at a private Mint in Bristol, and the tentative acceptance of this proposal by the Treasury brought a spate of troubles (see Letter 1403). The Commissioners of Revenue in Ireland objected, saying that a private coinage would be prejudicial to trade; the Moneyers at the Tower represented that this was an encroachment upon the province of their activities; and Newton himself was very unhappy about the risks of counterfeiting involved, and suggested that Wood prepare blanks only, to be sent to the Tower Mint for stamping. Nevertheless the coinage went ahead at Bristol, with Newton appointed as official Comptroller. Further complaints from the Irish Parliament (see Letter 1430) forced the Treasury to accept Newton's proposal of special assays of Wood's coin, which proved satisfactory and so the coinage continued. (Curiously the coin for the West Indies did *not* prove fine enough, and that coinage was stopped.) However, eventually pressure on the Privy Council persuaded it to reduce the coinage from the intended 360 tons to 142 tons (see Letter 1434).

Little is revealed of Newton's personal affairs by the letters in this volume. A number of begging letters, many undated, from impecunious relatives and other obscure persons, should not be taken as proof of any great generosity of spirit on Newton's part; we do not know if the demands were met, and the benefits desired were in any case always extremely small. The most interesting aspect of these letters is the light they occasionally throw on the family relationships of the would-be recipients, and they have been of considerable help in the construction of the genealogical table which we give on p. 485. The financial help Newton gave towards the restoration of Colsterworth Church in 1725–6 must have been very welcome to the recipients, but the sums involved could hardly have been a burden on Newton's pocket. Far more important to Newton must have been his investments in South Sea Stock (see Letter 1342) although it is difficult to interpret the exact form of Newton's dealings with the Company, since it is not clear whether all the investments were made with his own money, or with balances he held for the Mint.

Newton left no will; after his death on 20 March 1727 O.S. the family estates— and the income of about £80 p.a. arising from them—passed to his heir, John Newton (see Appendix II, Newton's Genealogy, p. 486, note (18)), whilst his personal estate, estimated by John Conduitt at £32000, was divided between his eight surviving half-nephews and half-nieces, namely Thomas, Mary (or possibly Hannah; see p. 487, note (25)) and George Pilkington; Benjamin and Newton Smith and their sister Hannah Tompson; and Catherine Conduitt and Margaret Warner, the daughters of Robert Barton. (Documents relating to the winding up of Newton's estate are amongst the Keynes MSS., particularly MS. 127A.) But there are indications that he did give careful consideration to the distribution of his wealth amongst his relatives; before his death he gave away considerable estates and monies: lands worth £30 p.a. to John Newton of Woolsthorpe (see p. 486, note (15)); £100 to an unidentified Ayscough; £500 to Benjamin Smith; an estate to Catherine Conduitt, daughter of his niece; and, ten days before his death, an estate in Berkshire to the children of his dead nephew Robert Barton. To his godson, Isaac Warner, he gave a grant of £100 (see British Museum, Add. 5017*, fo. 73), apparently in return for services rendered by John Warner, Isaac's father, in administering Newton's property (see P.R.O., PROB/3/26/66; later Warner claimed that at his death Newton owed him £1699. 18s. 0d).

Shortly after Newton died, John Conduitt wrote to a number of his contemporaries asking for reminiscences of Newton's life with a view to compiling a biography. Conduitt, who had married Newton's niece Catherine Barton and resided (when in town) in Newton's house for several years before Newton's removal to Kensington in 1725, could justly claim an intimate acquaintanceship with him; he obtained further information from Fatio de Duillier, from William Derham, from Humphrey Newton, and from William Stukeley. This material (some of it now in the Keynes collection, and printed by Sir David Brewster and others) is largely eulogistic, and when we ignore all the familiar anecdotage associated with Newton's early years and look only at the close of his life—the concern of the present volume—we obtain an image of Newton as a grand old man, bearing ill-health with fortitude, gentle, kind and dignified, beloved of many.

Augustus De Morgan's *Essays on the Life and Work of Newton* (Chicago and London, 1914) were written in reaction against the adulation of Newton's character by both his contemporaries and his Victorian biographers, in particular Brewster. While De Morgan was mainly concerned to analyse Newton's character in youth and middle-age, an examination of the letters in the present volume seems to consort better with De Morgan's image of Newton than with Conduitt's. In the majority of his letters he shows a straightforward, business-like application to the matter in hand; but sometimes—in particular in his correspondence with Varignon and Bernoulli—we see him as a cantankerous old man, unpredictable in his reaction, whose memory is blurred and whose faculties are dulled. We may go further, and accuse Newton of prevarication, of falsehood even, in his dealings with Johann Bernoulli. With Coste he was tactless; with Cavelier dilatory; to the majority of his correspondents unresponsive. In not one of his extant letters does he show humour or affection, or forbearance of the failings of others, unless in his allusion to his grand-niece Kitty.

Was Newton's extreme old age the unhappy product of tensions that more modern biographers than Brewster and De Morgan have detected in Newton's psychological development? As editors of his letters, we are not called upon to speculate upon such a question. Perhaps it is enough to recall that in Newton's day few retired voluntarily from the (often delegated) cares of paid office; and that he could find no final escape from that endless, bitter quarrel which was not of his making. Newton's last years were neither peaceful nor relaxed nor free from pain. Burdensome tasks fell continuously upon him, which his conscience did not allow him wholly to neglect (the pathos of Letter 1459 is as evident as the ill-temper of its missing precursor) and he had little enough comfort from family and friends. Could Newton, could any mortal, confidently assign to himself in the eyes of posterity that grand place which subsequent history has granted him? Perhaps it is needless, in this period of Newton's life, to look beyond such obvious considerations as these for an understanding of the increasingly evident symptoms of harshness, lack of humour and want of imagination in Newton's letters.

THE CORRESPONDENCE

1296 CALVERLEY PINCKNEY TO NEWTON
6 AUGUST 1718
From the original in the Bodleian Library, Oxford[1]

Mr John Furly's Mercht. 3: Doors
below ye. Church in St. Martin's lane
Cannon Street 6°: Augt. 1718

Hond. Sr.

The Ballance due to me on my Bill is £175 : 6 : 10[2] For wch. sum Dr. Fauquier tells me he took Sr Rich[ar]ds[3] Receit (last week before he left ye. Towne) to you And on paymt. thereof I am to give him my Receit for Sr. Rich[ar]d wch. I hope will be tantam[oun]t to Sr. Rich[ar]d's Bill on you & my Receit thereon for sd. Ballance as you seem'd to advise this morning

Above I give you my addresse In case you sho'd have any Commands to or for

Hond. Sir.
Your most h[umble] & Obedt. Servt.
CALVERLEY PINCKNEY

For
Sr. Isaac Newton att his
House Goeing downe out of the south
Side of
Lester Feilds
St Martins street

NOTES
(1) Bodleian New College MS. 361, II, fo. 69.
(2) The writer, as solicitor to the Mint, had been responsible for the prosecution of forty-four criminals in all parts of the country, according to his bill of charges for the period 29 July 1715 to 25 March 1718. See Letter 1293, vol. VI.
(3) Sir Richard Sandford, Warden of the Mint.

1297 C. STANHOPE TO NEWTON
12 AUGUST 1718
From the clerical minute in the Public Record Office[1]
For the answer see Letter 1301

Charles Earl of Lauderdale Gen[era]l of the Mint in Scotland prays paym[en]t of the summe of £2400 for the Service of the said Mint for two Years from 1st Decr 1715 to 1st Decr 1717 at the rate of £1200 per annum pursuant to Act of parliament, and such further summs as shall be thought necessary for defraying the Charge & Expences of the Coynage in the said Mint[2]

12th August 1718, Referrd to Sr Isaac Newton Knt Master & Worker of his Ma[jesty]s Mint to consider and Report his Opinion

C STANHOPE

NOTES

(1) T/17, 4, p. 355.

(2) On 4 November Lowndes wrote to Lauderdale for a list of the Edinburgh Mint salaries, and on 17 November a Warrant to Lauderdale authorized the payment of salaries at the reduced rate of £930 p.a. from Christmas 1714, and allowed the Mint to coin money under its existing Indenture (but this it did not do). See P.R.O. T/17, 4, p. 424 and *Cal. Treas. Books* xxxii (Part ii), 1718, pp. 627–8; also vol. vi, p. 411, note (2).

1298 NEWTON TO VARIGNON
[29 AUGUST 1718]
From a holograph draft in the University Library, Cambridge[1]
For the answer see Letter 1304

Quod Acta Philosophica[2] a Secretario Academiæ vestræ edita ad me aliquoties misisti gratiæ meæ tibi debentur easque quam maximas tibi reddo, ut et pro exemplaribus Ephemeridum[3] omni rerum varietate refertos & pro secundo Exemplari Actorum anni 1714. Et oro ut gratias meas Dno Fontenel Academiæ vestræ Secretario etiam reddas eo quod in Elogio Dni Leibnitij honorifice de me locutus sit.[4] Nuper mihi narratum est tuo nomine te in controversia Leibnitiana nihil contra me fecisse eaque de re tibi etiam gratias ago.[5] Misi tibi nuper exemplar novæ Editionis Libri Opticarum Anglice,[6] & missurus sum exemplar Editionis Latinæ quam primum prodierit.[7] Jam quadraginta sunt anni et amplius ex quo commercium per literas circa res Philosophicas & mathematicas habere desij[8] ideoque de controversia illa Leibnitiana nihil tibi scribo. Eademque de causa, molestiam tibi amplius creare nollem mittendi annuatim ad me tua dona Philosophica. Nam et Senio confectus a rebus Philosophicis et mathematicis quantum per negotia Societatis Regalis licet, me abstineo, & quod D. Leibnitio respondi ægre effectu est consilijs Aulicorum. Dona tamen tua plurimi facio quatenus sunt amicitiæ tuæ testimonia.

Viro Celeberrimo Dn. Abbati de Varignon in Academia Scientiarum Parisijs

ISAACUS NEWTON salutem

2

Newton's holograph English draft of the same Letter

Sr

I have been long indebted my thanks to you for several presents of the annual Memoirs[2] of your Academy & now return them to you very heartily. I thank you also for the presents you made me of Almanacks[3] filled with all variety of things relating to time [and for the second copy of the *Memoires* for 1714][9]. And I beg the favour that you will be pleased to return my thanks to Mr Fontenell [secretary of your Academy][9] for the honourable mention wch he made of me in the Elogium upon Mr Leibnitz.[4] I was lately told as from your self that you had done nothing against me in all that dispute & I beleive it & thank you for your friendship.[5] [By[10] the contrivance of some of the Court of Hannover I was prevailed with to write an Answer to a Postscript of a Letter which Mr Leibnitz wrote, to Mr l'Abbe Conti that both might be shewed to the king.[11] But while the Commercium Epistolicum remains unshaken I see no need of my medling with that Controversy any further unless perhaps I should cause that Book to be reprinted, or write a short Preface to the Book De Quadratura Curvarum to shew that that Book (which has been accused of Plagiary) was in MS before Mr Leibnitz knew anything of the Differential Method.[12]] A few weeks ago I sent you a copy of a new edition of my English Opticks.[6] By the contrivance of some of the Court of Hannover I was prevailed with to write an Answer to the Postscript of a Letter written by Mr Leibnitz to Mr l'Abbe Conti, that both might be shewed to the King. I did it with reluctancy. And by the Letters wch Mr Leibnitz thereupon wrote to several at Court I found that he was at the bottom of the designe. It is now above 40 [years] since I left of all correspondence by Letters about Mathematicks & Philosophy,[8] & therefore I say nothing further to you about these matters.

NOTES

(1) Add. 3968 (42), fo. 613 v. The date is taken from Varignon's reply, Letter 1304. Newton's letter was sent enclosed within one from De Moivre to Varignon (see Letter 1304). We have placed Newton's English draft of the same letter (U.L.C., Add. 3968(42), fo. 613 r) immediately following, in lieu of an editorial translation.

For Pierre Varignon, see Letter 1024, note (1) (vol. vi, p. 43). The present letter marks the beginning of an extensive correspondence between Newton and Varignon concerning the calculus dispute and especially Johann Bernoulli's part in it. Varignon's original letters to Newton are in the Keynes MSS. in King's College Library, Cambridge; Newton's letters to Varignon are extant only in draft form, in the University Library, Cambridge. A number of letters are missing, but we know of their existence indirectly, largely from information contained in the correspondence between Varignon and Bernoulli (see the Bernoulli Edition).

(2) *Histoire et Mémoires de l'Académie Royale des Sciences*, Paris, probably the volume for 1716.

(3) The *Connoissance des Temps*, which began as a private venture in 1679; in 1702 it was

taken over by the Académie. It remained a straightforward Almanac until 1766, when it began to include astronomical memoirs and news.

(4) Fontenelle's 'Éloge' of Leibniz had appeared in the *Histoire de l'Académie Royale* for 1716, pp. 94–128; see also Letter 1305, note (3), p. 19. In it Fontenelle discussed the calculus dispute, concluding that Newton and Leibniz invented the calculus independently, and that Newton could claim the prior invention.

(5) Varignon had earlier written formal letters of praise to Newton (see Letters 1024 and 1115, vol. vi); he was not so sympathetic towards Keill (see vol. vi, p. 27, note (2)).

(6) *Opticks: or, A Treatise of the Reflections, Refractions, Inflections and Colours of Light*, 2nd edition (London, 1718), or the earlier issue of 1717. Newton in fact sent Varignon three copies. The Latin draft adds that Newton will also send a copy of the Latin edition as soon as it is published (see Letter 1300, note (3)).

(7) The remaining passage differs considerably from the English draft. We translate it as follows (using Newton's English words where possible):

It is now above forty years since I left off correspondence by letters about philosophical and mathematical matters, and so I write nothing to you about that Leibnizian controversy. For the same reason, I do not wish to give you the further trouble of sending me annually your philosophical gifts. For, now I am old, I eschew philosophical and mathematical matters, as far as the business of the Royal Society allows, and what I have with reluctancy answered to Mr Leibniz was done by the contrivance of the Court. Nevertheless, I attach the more importance to your gifts in that they are proof of your friendship.

Isaac Newton greets the most celebrated M. l'Abbé de Varignon in the Academy of Sciences of Paris.

(8) This is a gross exaggeration by Newton; but possibly he has in mind his more public correspondence with the continental mathematicians, which largely petered out after his letters to Leibniz in the 1670s.

(9) This phrase appears only in the Latin version.

(10) These sentences, enclosed in square backets by Newton to indicate his intention to omit them and partly repeated later in the English version, are not found in the Latin draft.

(11) Newton refers to Leibniz's letter to Conti (Letter 1170, vol. vi) his own reply (Letter 1187, vol. vi) and also to Leibniz's letters to the Baroness von Kilmansegge and Count Bothmar (Letter 1203, vol. vi, and notes).

(12) The *Commercium Epistolicum* was not reprinted until 1722. Newton had, however, in about 1715, planned to add a revised version of the *De Quadratura Curvarum* to a new edition of the *Principia*, together with a preface explaining this joint publication. A number of drafts of this preface are extant, containing interesting but not wholly consistent autobiographical statements relating to the calculus dispute. (See Cohen, *Introduction*, pp. 345–9, where some of these drafts are printed, and also U.L.C., Add. 3968(5), fos. 24–9.)

4

1299   C. STANHOPE TO NEWTON
7 OCTOBER 1718
From a clerical copy in the Public Record Office[1]
For the answer see Number 1302

Sir.

I am commanded by the Lords Comm[issione]rs of his Ma[jesty']s Treasury to desire you will please to lay before them any Remarks and Observations You may have made upon the State of the Gold and Silver Coins of this Kingdom since the Reduction of 6d. in the Guinea made in the last Session of Parliamt.[2]

7th. Octob. 1718

C STANHOPE

NOTES

(1) T/27, 22, p. 355.
(2) See Letters 1264 and 1270, vol. VI.

1300   NEWTON TO VARIGNON
13 OCTOBER 1718
From the holograph draft in King's College Library, Cambridge[1]
For the answer see Letter 1323

Celeberrimo Viro Dno Abbati Varignonio Isaacus Newton S.P.D.

Vir Celeberrime

Misi tibi nuper exemplaria quædam Optices meæ recusæ Anglice,[2] & jam mitto (uti per Epistolam promisi) Exemplaria quinque Latina,[3] unum pro te ipso, aliud pro Secretario vestro Dno Fontenell, tertium pro Dno Monmort, quartum pro Bibliotheca Academiæ vestræ & quintum pro quovis alio amico tuo qui res Opticas callet. Et veniam a te peto ob molestiam quam tibi exhibeo in earum distribu[tione.] Et spero quod Epistolam meam nuper accepisti.[4] Vale.

Dabam Londini
13 *Octob.* 1718 *st. v.*

Translation

Isaac Newton sends his greetings
to the celebrated Mr Varignon

Most famous Sir,

I have sent you recently some copies of my Opticks printed in English,[2] and now (as I promised in a letter) I send five Latin copies,[2] one for you yourself, another for your secretary Mr Fontenelle, a third for Mr Monmort, a fourth for the library of your Academy, and a fifth for any other friend of yours who understands optical matters. And I beg your pardon for the trouble which I cause you over their distribution. And I hope that you received my letter not long ago.[4] Farewell.

London,
13 *October* 1718, *O.S.*

NOTES

(1) Keynes MSS. 142(U); microfilm 1011. 26. The letter was not received by Varignon until 26 May 1719 (see Letter 1323).

(2) See Letter 1298.

(3) *Optice: sive de Reflexionibus, Refractionibus, Inflexionibus & Coloribus Lucis,* 2nd edition (London, 1719). This translation of the second English edition was clearly in print at the end of 1718.

(4) Varignon probably did not receive Newton's previous letter of 29 August 1718 (Letter 1298) until the beginning of November.

1301 NEWTON TO THE TREASURY
14 OCTOBER 1718
From the holograph draft in the Mint Papers[1]
Reply to Letter 1297

To the Rt Honble the Lords Comm[ission]ers
of his Ma[jes]ties Treasury

May it please your Lordps

In obedience to your Lordps Order of Reference upon the Memorial of the Rt Honble the Earl of Lauderdale General of his Ma[jesty']s Mint in Scotland for impresting upon account, two thousand four hundred pounds to his Lordp out of the Coynage Duty as due for the service of the said Mint for the two years ending at Christmas last, together with such further sums as your Lordps shall judge necessary for bearing the charge of coinage I humbly represent that in pursuance of the Acts for coinage made in the 18th and 25th years of Carol. II the moneys arising by those Acts in England have always been paid into the Exchequer quarterly & by the Coinage Act made 7mo Annæ the duty was continued & extended to all great Britain & every clause article & sentence in ye said Acts of 18 & 25 Car. II then in force was revived & continued in force & extended to all the duty in Great Britain, & by the Act of Union the Mint in Scotland was to be conformable to the Mint in England, & by the Act which passed in the first year of his present Majesty for continuing the former Acts for encouragement of coinage, your Lordps are em-

6

poured out of the moneys arising in great Britain by the said Act, to cause so much money to be issued out of the Exchequer to the Master of the Mint in England & to the General of the Mint in Scotland respectively by way of imprest & upon account, as shall be necessary for defraying the expenses of the said Mints; & that all the moneys issued out of the Exchequer to his Ma[jes]ties Mint in England are (pursuant to the Act 18 of Car. II) imprest in general words for the use & service of this Mint, so that they may be applied to the charge of coinage as well as to any other charges of the Mint.

The salaries of the Officers of the Mint in Scotland amount only to 980£ per annum whereof 50£ per annum is to cease at the death of the late Clerk of the bullion, & the charge of keeping the Offices in repair & the dwelling houses of the Officers wind & water tite [*sic*], can be but small; & therefore twelve hundred pounds per annum allowed to that Mint hitherto without including the charge of coinage, is too much. If to the 1200£ allready in their hands for one year, the summ of 1800£ more be added to make up the whole summ 3000£ for the three years ending at Christmas last, & the 1800£ be imprest for the use & service of that Mint in general, so that it may be applicable to all sorts of charges, & the Account be made up to Christmas last, the Ballance of the Account may be then certified to yr Lordps in order to be considered when the next moneys shall be imprest. And so on from time to time.

And considering that the Indenture of that Mint doth not impower the General to pay any moneys, I am humbly of opinion that there should be a Signe Manual sent to him impowering & directing him to pay the Salaries & other expences to whom they shall become due by the said Indenture, & to observe all the Rules & Precepts set down in the said Indenture (except the alterations made by Parliament,) untill a new Indenture shall be made. [2]

And that there may be no more complaints of the want of moneys to beare the charge of coinage, the said General may have notice to pay those charges in the first place, & let the deficiency (whenever there shall be any) fall upon the Salaries.

All which is most humbly submitted to Your Lordps great wisdome

Mint Office   ISAAC NEWTON
14 *Octob*. 1718.

NOTES

(1) III, fos. 107–8. There are antecedent drafts (with different salary figures) at fos. 102–3 and 105–6, the latter dated 7 October, as was the present draft before Newton altered the date.

(2) Newton's holograph drafts for a Warrant to Lauderdale in accordance with the recommendations of this letter may be seen in Mint Papers, III, fos. 104 and 146; his draft Warrant was adopted by the Treasury Lords word-for-word (even to the spelling 'tite') on 17 November 1718 (see Letter 1297, note (2)).

1302 OBSERVATIONS UPON THE STATE OF THE COINS OF GOLD AND SILVER

20 OCTOBER 1718

From the holograph draft in the Mint Papers[1]
Reply to Letter 1299

Obs 1. Standard Gold before sixpence was taken from the Guinea was worth 3*li*. 19*s*. 9¼*d* per ounce at the Mint, & by taking six pence from the Guinea became worth 3*li* 17*s* 11*d* per ounce. And standard silver is there worth 5*s* 2*d* per ounce. But the demand for exportation hath raised both species above the price at the Mint, & thereby hath carried out all the forreign silver for many years, & began to carry out some of the forreign gold the last November, & therefore raised the price of gold for exportation above the Mint price before the sixpence was taken from the Guinea. But it never raised it to more then 4*li*. 0*s*. 6*d* per ounce, nor kept it long at that price. For in March last, forreign gold fell down below the old Mint price, & hath ever since continued below it, being at 3*li*. 19*s*. 6*d* & for the most part at 3*li*. 19*s*. 0*d* & under. [2] And therefore the price of forreign Gold for exportation was raised the last winter by some other cause then the taking sixpence from the Guinea. And the price of Gold for exportation to forreign markets having been ever since March below the old Mint-price, that price was certainly too high.

Obs. 2. The price of gold for exportation depends upon the price in forreign markets & answers[3] to the course of exchange. When the Exchange is lowest the price of forreign gold is highest & on the contrary. And thence the coinage of gold hath of late years been greater or less according as the course of Exchange hath been higher or lower. In the year 1713 the course of Exchange & the coinage grew high together. In the years 1714 & 1715 the Exchange was highest, it being (for instance) with Amsterdam from 36 to 37 shillings, & then the coinage was greatest. In the year 1716 the Exchange was only from 35 to 36 shillings with Amsterdam & proportionally with other places, & the coinage abated accordingly. In the year 1717 the Exchange with Amsterdam was only from 34 to 35.2, & the coinage abated to one half of what it had been two or three years before. And in this present year the Exchange has been only from 33.10 to 34.10 till within this fortnight; & this low course of Exchange together with the discouragement of the coinage of gold by the taking sixpence from the Guinea, hath carried out almost all the Gold imported & thereby hath had the same good effect for paying our debts abroad in gold & preserving our silver, wch the Bill proposed [during] the last Sessions of Parliament would have had if it had passed into an Act for stopping the coinage of gold.

8

whence those debts arose is difficult to understand without more skill in Trade then I can pretend to. But considering that a good part of the Gold imported in the years 1713, 1714, 1715 & 1716 was in French money & Ducats, I suspect that after the war with France was at an end, great quantities of gold were sent hither to pay for Stocks untill the interest of Stocks was lowered by Act of Parliament: & since that discouragement some forreigners have been drawing their moneys back with the interest & some their stocks & some gold hath this year been sent to the Mints in France.[4]

Obs. 3. The course of Exchange was as low in November last before the 6*d* was taken from the Guinea, as it was afterwards in February last; & both times was at the lowest, being (with Amsterdam) at 33.10. And therefore the lowness of the Exchange last winter arose, not from the taking sixpence from the Guinea, but from the debts wch we had abroad before the sixpence was taken off. Which debts, if the coinage of gold had not been discouraged by taking 6*d* from the Guinea, might have remained unpaid till they could have been paid in silver with more advantage to private persons.

Obs. 4. By the payment of our debts abroad in Gold the demand for exportation hath abated ever since February last, & the exchange hath risen gradually to 35 shillings & gold hath fallen down gradually to the Mint price & begun to come to the Mint again, so that within a fortnight so much gold hath come to ye Mint as will make above 75000 *li*. Whence I gather that whenever the Exchange with Amsterdam is above 35 shillings it will bring gold to the Mint & would have brought gold to the Mint in the years 1713, 1714, 1715, 1716 & part of 1717 although the sixpence had been taken off before, the exchange with Amsterdam in all those years being above 35 shillings & for the most part above 36. And therefore in all the gold then coined, which was above five millions, the nation would have saved sixpence per Guinea had the sixpence been taken off before.

Obs. 5. The demand for exportation hath ever since the sixpence was taken from the Guinea, raised the price of silver about three times more then the price of gold & sometimes four or five times more or above. And therefore the temptation to export gold moneys hath all this year been three times less then the temptation to export silver moneys. And if this temptation hath not this year sensibly diminished the quantity of our silver moneys it hath much less carried out our gold moneys: And all or almost all the gold wch hath been exported this year hath been in forreign bullion. And by consequence the nation hath lost little or nothing by the exportation, because the bullion being forreign, went out at the same price that it came in at. Forreigners or their Agents who here receive guineas in payments, will lose above 3*d* per Guinea by exporting them, besides the danger they run by breaking the Law.[5]

9

Obs. 6. Since the demand of silver for exportation hath all this year been three times greater then that of Gold, no gold would have been exported this year had it not been for the want of forreign silver; & the exportation thereof hath prevented the exportation of the same value of forreign silver as fast as it could have been procured for paying debts abroad, & in the mean time hath saved the interest of the debts paid off.

Obs. 7. And this expenditure[6] hath been a further advantage to the nation by raising the course of exchange from 33.10 to 35. For when the exchange is low the nation loseth by it so much as it is under the par. And if the debts which have been paid in gold had continued till they could have been paid in silver, they would have caused the exchange to continue low.

Obs. 8. And to restore the sixpence to the Guinea would be to lose these advantages, & to give more by above ninepence in the Guinea for all the gold which shall be coined hereafter then it is worth in forreign markets, & to revive the corrupt trade of exporting silver to buy gold abroad & importing gold to buy silver at home.

Octob. 20. 1718

NOTES

(1) II, fos. 124–5. This Memorandum is in effect a reply to Letter 1299. We have not traced the official version submitted by Newton. There is at least one clerical manuscript copy, printed in Rigaud, *Correspondence*, II, pp. 430–4, showing quite marked revisions of the holograph here printed, and dated two days later. Probably this copy was taken from the official version.

(2) The manuscript copy is more detailed: 'Forreign Gold fell down to £3. 19s. 6d and in April to £3. 19s. 0d & in May to £3. 18s. 6d, all which was below the old Mint price...'

(3) The manuscript copy reads: 'depends upon our Debts abroad and answers'.

(4) The manuscript copy reads: 'drawing their Moneys back with the Interest of their Stocks & some Gold this year hath been sent to the Mint in France, & some Merchants are newly broke'.

(5) The manuscript copy continues:
there hath been above £110,000 Imported in Gold to be coyn'd since Christmas, and the 6d per Guinea sav'd in all this Coynage will recompense abundantly the loss of 6d per Guinea in all the Guineas Exported by Forreigners, and therefore there is nothing in the Objection, that in making payment to Forreigners in Guineas we lose 6d per Guinea, for we get the 6d again in receiving back all the Guineas which they do not Export.

(6) The manuscript copy reads: 'exportation'.

1303 MONMORT TO KEILL

c. 31 OCTOBER 1718

Extract from the holograph original in the University Library, Cambridge[1]

Jay cru Monsieur suivre vostre intention et celle de Mr Taylor en faisant part à Mr. J Bernoulli de cet endroit de vostre lettre a Mr Taylor ou vous distes en parlant de Mr J. Bernoulli I cannot deny but that M. Bernoulli...if he would apply his skill to something of use I desire he would solve the problem M. Leibnitz attempted but erroneously mistook and could not solve to find the curve a Projectile describes in the air in the most simple supposition of gravity and the density of the medium being both uniform: but the resistance in duplicate proportion of the velocity.[2] Quoy que je neusse point alors le bonheur destre en liaison avec vous je vous proteste que je ne laurois pas fait si javois cru que cela put vous deplaire mais je nai point doute que lintention quon avoit p[rinci]palement en menvoyant cet extrait de lettre etoit quil allait jusqua M Bernoulli [.][3]

je ne puis vous dire parquel hazard il est arrivé que mandant des Nouvelles a M[ess]rs Herman et N[icol]as Bernoulli et voulant leur faire part du probleme dont vous souhaittiez que Mr J. Bernoulli voulust entreprendre la solution, jy en ai substitué un autre par meprise.[4] mon etourderie a eu un effect heureux puisqu[']elle nous a valu de la part de ces deux M[ess]rs une solution parfaitte de ce probleme qui est fort difficile et auquel ils nauroient peut jamais pensé. Si vous aviez la moindre curiosité de voir [leurs] solutions j'aurois l'honneur de vous les envoyer ils mont laisse la liberte de les faire voir a mes amis avec les constructions.

NOTES

(1) Res. 1893(*a*) (Lucasian Professorship, Box 1, Packet 2). The letter is postmarked—seemingly in London—5 November; its contents give the year. Keill had written to Monmort on 3 September.

(2) For the history of this problem see A. R. Hall, *Ballistics in the Seventeenth Century* (Cambridge, 1952), Chapters v and vi. Fairly obvious mechanical considerations suggested that when a moving body such as a projectile strikes a number of air particles in rapid succession, its loss of energy by communication to these particles will be proportional to its velocity and the number of particles struck in a given time, that is to the square of the velocity. Huygens had examined this hypothesis of resistance as early as 1669, but the first to determine (approximately) the trajectory thus produced was Newton, in the often-mentioned tenth proposition of Book ii of the *Principia*. Newton assumed that both the density of the air and the gravitational acceleration were constant, an assumption also made by Newton's immediate successors; however, he recognized that an additional 'tenacious' component of resistance, arising from the viscosity of the medium, ought in principle to be considered as well.

Leibniz attacked the ballistic problem in 'Schediasma de Resistentia Medii, & Motu Projectorum, gravium in medio resistente' (*Acta Eruditorum* for January 1689, pp. 38–49; Gerhardt, *Leibniz: Mathematische Schriften*, VI, pp. 135–43). Presumably this, like the 'Tentamen', was written before Leibniz had read the *Principia* but he may have known in general terms of Newton's investigation. This paper argued that the resistance of the air to motion had two factors, one proportional to the velocity directly and the other to its square, that it would in principle be possible to treat this double proportion mathematically in a single equation containing the co-ordinates of the projectile's position at any time, and that the differential and integral calculus 'cujus elementa quaedam in his actis dedimus' would permit the problem to be solved. In the next year Huygens (for the first and only time) published a general account of his own work on resisted motion in the *Addition* to his *Discours sur la Cause de la Pesanteur*, which was the occasion of a long correspondence with Leibniz, and so indirectly caused the latter to produce an 'Additio' to his 'Schediasma' of 1689 (*Acta Eruditorum* for April 1691, pp. 177–8; reprinted in Gerhardt, *loc. cit.*,pp. 143–4). For Newton's correct recognition of Leibniz's basic error in separating the horizontal and vertical components of resisted motion, and his communication of this error to Keill in 1714, see Edleston, *Correspondence*, pp. 187 and 309–10 and Whiteside, *Mathematical Papers*, VI, pp. 70–1, note(109). Compare also vol. VI, p. 114, note (5).

Since Newton alone, at this time, had defined this curve, in 1717 the ballistic problem might well appear as both well-known and highly intractable, and as one in which the English achievement stood higher than the continental.

(3) Monmort and Taylor exchanged many letters, especially between 1717 and 1719. A volume of Brook Taylor's correspondence in the Library of the Royal Society (MS. 82) contains some of Monmort's letters to him, but not Taylor's to Monmort. Extracts from Monmort's letters on a paper in this volume were printed at the end of the 'Additamentum' to Keill's *Epistola ad Bernoulli* (London, 1720), p. 28 (see below); others, together with letters from other correspondents, were printed in Taylor's *Contemplatio Philosophica* (compare vol. VI, Letter 1194, and Letter 1361 above). However, we have found neither Keill's letter to Taylor, from which the immediately preceding English sentence was quoted by Monmort, nor Taylor's letter to Monmort in which it was conveyed.

Monmort was also in regular correspondence with Bernoulli now that Leibniz was dead; he admired Bernoulli and in writing to him frequently disparaged Taylor. A number of excerpts from Monmort's letters to Bernoulli (and also to his son Nikolaus) were printed by the publisher of the *Acta Eruditorum*, Johann Burchard Mencke, in an open 'Epistola ad Virum Clarissimum Broock Taylor' (*Acta Eruditorum* for May 1721, pp. 195–228; reprinted in Bernoulli, *Opera Omnia*, II, pp. 483–512). It can hardly be doubted that Keill conceived of the ballistic problem as one to stretch Johann's abilities; whether he actually meant Taylor to transmit it to Bernoulli as a challenge must remain uncertain. Thanks to Taylor and Monmort the latter received it as such.

Bernoulli's successful answer came under the title 'Responsio ad nonneminis provocationem, [et] solutio quæstionis...ab eodem propositæ de invenienda linea curva quam describet projectile in medio resistente' (*Acta Eruditorum* for May 1719, pp. 216–26; reprinted in his *Opera Omnia*, II, pp. 393–402). Here Keill appears as 'Homo quidem, natione *Scotus*, qui ut apud suos impuris inclaruit moribus, ita apud Exteros jam passim notus odio plus quam Vatiniano...' ('A certain individual of Scottish race who has become no less notorious among his own people for his immorality than odious everywhere to foreigners, who loathe him more than [the Romans loathed] Publius Vatinius'); clearly the evil gossip reaching Leibniz earlier was not wasted. In another paragraph Bernoulli deliberately distorts the issue over the error he pointed out in the *Principia* which was subsequently corrected by Newton in the second edition. he writes

> tulit ille [Keill] ægerrime quosdam a me notatos errores in Phil. natur. Principiis Newtoni, quos in Actis Lips. anni 1713 correxi, falsis vera substitui, ac defectos aliquos supplevi, tanto cætera candore & modestia usus, ut Newtonus ipse, qui in fine præfationis hoc a Lectore suo petit, non tantum non offensus meis annotationibus, sed ut postea intellexi ab Amico communi, qui Newtono familiariter utitur, iisdem haud parum devinctus esse videretur; nec miror, quis enim veritatis amans in errore licet inscius perseverare mallet, quam ejus admoneri? cum præsertim eo jam ventum esset, ut altera Principiorum paulo post in lucem prodiret, quod proin elegans opus iisdem nævis dedearatum comparuisset, nisi amica nostra interpellatio tempestive intervenisset.

> ([Keill] has taken very ill certain errors remarked by me in Newton's *Principia*, which I have corrected in the *Acta* [*Eruditorum*] of Leipzig for 1713, substituting the truth for what was false and supplying what was missing. This I had done, moreover, in so straightforward and modest a way that Newton himself (who at the end of his preface begs this from his reader) not only seemed not offended by my comments but, as I afterwards learned from a common friend who is intimately acquainted with Newton, seemed not a little obliged to me for them. Nor is this to be wondered at, for what lover of truth would wish to remain in error (though unconscious of it) rather than be advised of his mistake? Especially as preparations were already in train for the publication of the second edition of the *Principia*, a little later, which elegant book would have appeared spoiled by these same blemishes, if our friendly intervention had not been made in time.)

Bernoulli is correct in saying that his intervention was timely and effective, and that Newton accepted it politely (he could do no other). However, this intervention was made privately in October 1712 (see Number 951*a*, vol. v), not publicly in 1713. Bernoulli purposely confuses his 'intervention' with his publication of the correction. His articles could not affect the second edition of the *Principia*, and were *never* regarded by Newton as welcome.

Later in the paper Bernoulli states the solution to Keill's problem, but not its demonstration; this he expressed in a letter to Monmort of 13 July 1719, N.S., later printed in the *Acta Eruditorum* for May 1721, pp. 228–30 (and reprinted in Bernoulli, *Opera Omnia*, II, pp. 513–16).

Keill's reply to Bernoulli's 'Responsio' of 1719 was printed as an 'Additamentum' to his *Epistola ad Bernoulli* (London, 1720; see Letter 1321, note (3)). It was also read at a meeting of the Royal Society on 26 May 1720, when the following report was made,

> A Letter of Dr. Keill dated at Oxford the 21 May 1720 was read wherein he makes a complaint against Mr. Bernoulli a Fellow of this Society for affronting him with scurrilous language in the Lipsic Acta of the year 1719. And desires that this honourable Society would take such course to shew their Dislike of such foul proceedings, as has been customary when any of their members have been so abused by another.

> The president ordered that the Consideration of this complaint be differed until Dr Halley Secretary comes to town, and that Enquiry be made into the Presidents [*precedents*] for better Information and direction of the Society. [Journal Book of the Royal Society.]

The Royal Society does not appear to have taken the matter any further at its meetings. Newton was not much interested in the squabble between Bernoulli and Keill, which was in any case terminated by the latter's death on 31 August 1721.

Keill's 'Additamentum' to his *Epistola ad Bernoulli* ended with a few brief extracts of letters from Monmort to Brook Taylor, in which he expresses his distaste for Bernoulli's treatment of Keill. For example, in a letter of 11 June 1719, N.S., Monmort commented on Bernoulli's criticism of Keill's morals (see above) in the following terms, 'Les disputes litteraires ne doivent point aller jusques la. J'en suis fache pour M. Bernoulli, qui n'a pas eté en droit de faire un pareil reproche a M. Keill quelque sujet qui puisse avoir de s'en plaindre. Je vais luy ecrire aujourduy meme pour le prier instanment de retrancher un Medisance si grave.' The subsequent history of the disputes between Keill, Taylor, Monmort and Johann Bernoulli has no place in this *Correspondence*, since Newton had little involvement in them. Much may be learnt of Bernoulli's reactions to the activities of the English mathematicians from his correspondence with Varignon (see the Bernoulli Edition).

(4) It is not clear what problem Monmort substituted when he told Jakob Hermann and Nikolaus [II] Bernoulli of Keill's challenge. Possibly it was simply the generality of the problem that was in question: Nikolaus [II] Bernoulli, in a paper 'Responsio ad Taylori Querelas', *Acta Eruditorum* for June 1720, pp. 279–85, assumed that the question posed was the general one of finding the path of the projectile when the resistance of the medium varied as any given power of the velocity (see Bernoulli, *Opera Omnia*, ii, p. 477), and it was to this general form of the problem that Johann Bernoulli produced a solution; Keill, however, seems to have intended to propose only the limited case where the resistance varies as the square of the velocity.

1304 VARIGNON TO NEWTON

6 NOVEMBER 1718

From the holograph original in King's College Library, Cambridge[1]
Reply to Letter 1298
A possible answer is Letter 1305

Equiti Nobilissimo,
Viroque Doctissimo D.D. Newton
Regiæ Societatis Anglicanæ Præsidi Dignissimo,
S.P.D.
Petrus Varignon

Abhinc tres circiter menses tuo Nomine reddita sunt mihi novæ editionis mirandi tui libri Opticarum Exemplaria quæ ad usque tria mihi pro tua Benignitate ac Munificentia donasti. At cum Anglicæ linguæ sim imperitus & expers, eorum unum tunc Amico Anglo Mathematico,[2] quandoque me invisenti commisi, ut nimirum post meum rure reditum, quod proxime petiturus eram, doceret me quid hac in editione repererit quod nonsit in prima Latina[3] quam summa cum voluptate legi eoque majore quod Novum systema tuum de coloribus sit exquisitissimis experientijs stabilitum. Verum ubi advenienti mihi rure Parisios, die scilicet undecima mensis hujus Novembris N. St. redditæ sunt Clarissimi D. Moivræi litteræ cum inclusis quas ad me die 29. Aug. V. St.[4] scribere dignatus es, adeo me puduit gratias nondum egisse tibi pro tanto isto munere, ut decreverim (licet non reviso illo amico Anglo) statim eas tibi

14

maximas referre, ac pares pro Epistola tua Benevolentiæ erga me tuæ plena: Ago igitur immortales animo in æternum memore.

Trium autem illorum Exemplarium novæ editionis tractatus tui de Coloribus, quæ audivi a te perhumaniter destinata mihi & Amicorum meorum duobus ad meum arbitrium eligendis; unum tuo nomine dedi Illustrissimo ac doctissimo Abbati Bignonio,[5] qui pro eo gratias etiam agit tibi quam plurimas; alterum tuo etiam nomine misi ad D. Joh. Bernoullum, ut generosum ipsi cor tuum exhiberem: quod tibi & illi gratum fore autumavi. Responsum ab eo nondum accepi, cum tunc ei nuntiaverim me rus statim profecturum esse ad usque divi Martini diem.[6] Tertium tandem illorum Exemplarium servavi tua manu mihi subscriptum, si minus ad legendum ob meam linguæ Anglicæ ignorantiam, saltem ut honorificum mihi tuæ erga me Benevolentiæ Pignus.

Novas anteactis addo gratias tibi quod recte judicaveris me innocentem omnino esse factorum contra te in controversia Leibnitiana: tantum abest ut eorum quantulumcunque particeps fuerim, quin potius de illa controversia semper silui in meis ad D.D. Leibnitium & Bernoullum litteris, mecum & intus sollummodo dolens ea tantos viros agitari, quos, si penes me fuisset, in pristinam concordiam reduxissem; atque id unum in animo habui dum a te misi ad D. Joh. Bernoullum novam tui libri de coloribus editionem.

Quod spectat Acta Regiæ Scientiarum Academiæ, non dona sunt a me, sed debita tibi sunt ab eadem Academia, ut & cæteris ejus membris: Inde nulla pro his a te mihi gratiarum actio debetur. Sat mihi si grata tibi sit cura qua, statim ac edita fuerint, conabor ut ad te cum Ephemeri[di]bus quam promptissime perveniant. Salutem verbis tui nuntiavi Dno. de Fontenelle, qui perhonorificis te resalutat. Vale diu vir Nobilissime ac doctissime, mihique Beneficiorum in me tuorum æternum memori Benignus favere perge.
Dabam Parisijs Die 17. *Novemb.* 1718. *St. N.*

Translation

Pierre Varignon offers a grand salute
to that most noble Knight and learned man
[Sir Isaac] Newton,
most worthy President of the English Royal Society

About three months ago copies of the new edition of your marvellous book of *Opticks* were delivered to me on your behalf of which you have given me as many as three out of your kindness and generosity. But since I am ignorant and deficient in the English language, I then entrusted one of them to an English friend,[2] a mathematician, who visits me from time to time, so that doubtless after my return from the country, to which I was about to go off very shortly, he would tell me what he has found in this edition

15

which is not in the first Latin [edition],[3] which I have read with the greatest delight, and all the more so because your new system of colours is firmly established by the most beautiful experiments. Indeed on my return to Paris from the country (which was on the eleventh day N.S. of this present November), when I was handed Mr De Moivre's letter enclosing the one which you were kind enough to write to me on 29 August O.S.[4], I was so ashamed that I had not yet thanked you for so great a gift, that I have decided to return you my very best thanks at once (even though I have not seen my English friend) and the like for your letter, full of your kindness towards me: I therefore offer you my undying gratitude.

Now, of the three copies of the new edition of your treatise on colours, which I have heard from you are generously intended for me and two of my friends to be chosen as I please, I have given one on your behalf to the most famous and learned Abbé Bignon,[5] who also thanks you most heartily for it; the second I have sent, also on your behalf, to Mr Johann Bernoulli, in order that I might reveal your generous nature to him: which I have supposed will be pleasing to you and to him. I have not yet received any reply from him, since at that time I had told him I would be going at once into the country until St Martin's day.[6] Finally, the third of the copies, which is inscribed to me in your own hand, I have kept, if not for reading, because I am ignorant of the English language, then at least as an honourable assurance of your kindness towards me.

I add fresh gratitude to the thanks just offered because you have correctly judged me entirely innocent of acting against you in the Leibnizian controversy: quite the contrary, I have taken so little part in it, that I have rather always kept silent about that controversy in my Letters to Mr Leibniz and Mr Bernoulli, only bewailing, to myself and privately, that such great men are harassed by it, whom, if I had any influence, I would have brought back to their former cordiality; and this was the sole object I had in mind when I sent to Johann Bernoulli from yourself the new edition of your book on colours.

As regards the *Mémoires* of the Royal Academy of Sciences, they are not gifts from me, but are your due from the Academy itself, as they are due to its other members. Hence no return of thanks for them from you to me is needed. It is enough for me if you are pleased by the care with which I try to see to it that they, with the *Connoissance des Temps*, reach you as quickly as possible after they are published. I have given your respects to Mr Fontenelle, who most respectfully returns the greeting. Fare well long, most noble and learned Sir, and may you, kindness itself, continue to favour me, [who am] constantly mindful of your kindnesses towards me.
Paris 17 November 1718, *N.S.*

NOTES

(1) Keynes MS. 142(C); microfilm 1011. 26. Varignon had not yet received Newton's Letter 1300.

(2) Presumably an English resident in Paris, perhaps (like Francis Vernon in Huygens' time) a member of the English diplomatic staff.

(3) *Optice: sive de Reflexionibus, Refractionibus, Inflexionibus & Coloribus Lucis Libri Tres* (London, 1706). This was a translation by Samuel Clarke of the first English edition of 1704. The most important changes between the various editions of *Opticks* are in the Queries (see

Alexandre Koyré, 'Les Queries de l'Optique', *Archives internationales d'histoire des sciences*, **13** (1960), 15–29). In the first Latin edition Newton printed Quæstiones 17–23, not appearing in the first English edition. In the second English edition of 1717 (here in question) Newton extended the number of Queries to 31, placing his last or third group of Queries immediately after those (1–16) that had already been printed in the first English edition and giving the additional queries numbers from 17 to 24, then continuing with the second group of Queries that had been first printed in Latin, now renumbered 25 to 31. The existing Queries were also to some extent revised. This arrangement was of course repeated in the second Latin edition, received later by Varignon.

(4) Letter 1298.

(5) See Letter 1004, note (2) (vol. VI, p. 5).

(6) 11 November, N.S.

1305 NEWTON TO VARIGNON
?End of 1718
From a holograph draft in the University Library, Cambridge[1]
Reply to Letter 1304

Sr

I am much obliged to you for your Letter in order principally to bring Mr John Bernoulli & me to a better understanding.† Mr Leibnitz has attributed to him the Letter of June 7th 1713 [N.S.] inserted into a flying Paper[2] dated 29 July [N.S.] following & if that Letter be his he cites himself by the title of Eminens quidam Mathematicus & took upon him to act as judge between Mr Leibnitz & the Committee of the R. Society & in that sentence denied (in Effect[]) that when I wrote the book of Quadratures, (wch I can assure you was above 40 yeares ago, except the Introduction & Conclusion) I did not [*sic*] understand the first Proposition of the Book

† And I return you my hearty thanks for your kind present of the Elogia of the Academiciens.[3] In that of Mr Leibnitz Mr Fontenelle has been very candit. There are some mistakes in matter of fact but not by designe. I reccon that Mr Fontenell was not sufficiently informed. He seems to follow the Marquess de l'Hospital in saying that the differential characters are the more convenient,[4] but the Marquess whom I reccon a very candid person when he published his book of infinitely littles[5] had seen no other characters of mine then those in the Book of Principles. In pag. 44, lin 22[6] he should have mentioned Dr Wallis before Mr Fatio if he had known what the Dr wrote in the Preface to the two first Volumes of his works. Pag. 48 lin 15 he should have mentioned that I was first accused of Plagiary in the Acta Eruditorum for January 1705 pag. 34 & 35 & by Mr Leibnitz himself in his Letter to Dr Sloan dated 29 Jan. 1711 [N.S.], had he been advised of it.[7] For this was in opposition to my saying in the Introduction to the book of Quadratures that I found

the method of fluxions gradually in the years 1665 & 1666. Pag. 49, l. 20[8] they have given him a wrong information from Germany. For in the Elogium of Mr Leibnitz printed in the Acta Eruditorum for July 1717 page 335 we are told that a few days before his death he wrote to Mr Wolfius that he would refute the English after his historical labours were over. Pag. 51 lin. 22[9] Mr Leibnitz became acquainted with Mr Hugens in 1672 came to London & returned to Paris in the beginning of 1673, kept a correspondence with Mr Oldenburg about Arithmetical questions till June following & hither to knew nothing of the higher Geometry. The Horologium Oscillatorium of Mr Hugens was published in April 1673 & he begun with this book. And it doth not appear that he knew any thing of the Differential Method before his second coming to London at wch time he met with Dr Barrows works & my Letter of 24 Octob 1676 & saw in the hands of Collins the Letters & Papers sent to him by me & Gregory. Pag. 53. lin. 6,[10] the Letter here referred unto was writ by [read to] Collins in 1672 but not seen by Mr Leibnitz till 1676 [[11] Pag. 53 l. 22 Mr Leibnitz is blamed here in point of candor for concealing what he knew of other mens inventions. He received Gregories series twice from England & by a copy of Gregories Letter dated 15 Feb. 1671 knew that Gregory had invented it before the date of that Letter & yet printed it as his own in 1682 without making any mention that he had received it from London & that it was first invented by Gregory. He might have said that he invented it himself before he he knew that Gregory had invented it, but he should not have concealed his having receiving [sic] it from London, & much less should he have concealed his knowledge that Gregory had invented it before him. And so when he printed the elements of the Differential Method he should not have concealed his knowledge of what he [had] seen in my Letters concerning such a method. For its very plane by his Letter of 21 June 1677 [N.S.] that he then understood that my method was of the same kind with his own. And much less should he have concealed his knowledg after I had in the second Lemma of the second Book of Principles set down the elements of my method & in the Scholium upon that Lemma explained the sentence in wch I had concealed that Method in my Letter of 24 Octob. 1676 where I said that I had explained it in a Tract written five years before. He knew therefore that I was before him & should not have concealed his knowledge. In his Letter of 7 March 1693 [N.S.] he acknowledged of his own that I had shewed by my Book of Principles that I had such a method when I wrote it & to deny this afterwards proceeded from the want of candor.] Dr. Barrows works & Gregories Letters of 5 Sept 1670, & 15 Sept 1671 & mine of 10 Decem 1672[,] 13 June 1676 & 24 Octob 1676 were all of them seen by him when he was the second time in England or within less the[n] four months before & it doth not appear that he understood the

18

Differential Method before he saw all these Papers. And the Committee of the R. Society have affirmed nothing more than that he saw them before it appears that he had the Method, & that in them the Method of fluxions is sufficiently described to any intelligent person. But if the friends of Mr Leibnitz beleive that he did not understand those Papers the Gentlemen of [*draft ends here*]

NOTES

(1) Add. 3968, fo. 381. The draft is not dated, and we have not found a Latin version of the same letter; probably it was never sent, at any rate there is no mention of any communication from Newton later than 13 October 1718 in Varignon's next extant letter, of 15 July 1719 (Letter 1323). The opening sentence indicates that it may have been intended as a reply to Letter 1304. Note that Newton drafted a fresh start after the first sentence, where he placed the mark †. *De quadratura curvarum* was in fact composed in 1691.

(2) The *Charta Volans*; see Number 1009, vol. vi. The authorship of the letter there quoted (that is, Letter 1004, vol. vi) was to become the chief bone of contention between Newton and Bernoulli. Varignon clearly transmitted to the latter Newton's assertion that Bernoulli was author of the quoted letter. Bernoulli replied in Letter 1320.

(3) B. le B. de Fontenelle, *Histoire du renouvellement de l'Académie Royale des Sciences en M. DC. XCIX. et Les Éloges Historiques de tous les Academiciens morts depuis ce Renouvellement*, 2 vols. (Paris, 1717); Leibniz's 'Éloge' is in vol. ii, pp. 274–333. Fontenelle's *Éloges* of the Academicians, appearing first individually in the *Histoire et Mémoires de l'Académie Royale des Sciences*, were extremely popular, and a complete volume of them was first published in 1708. Fontenelle devoted pp. 42–57 of his 'Éloge' of Leibniz to the calculus dispute, making it clear that Leibniz's was the first publication but that Newton's was an independent and probably a prior discovery. Newton had already commented favourably on this 'Éloge' in his letter to Varignon of 29 August 1718 (Letter 1298), after seeing it in the *Histoire*; now his reaction is more critical. In Add. 3968(26), fos. 372–80 are a number of detailed drafts of 'Historical Annotations on the Elogium of Mr. Leibnitz'. In the present letter he merely summarizes these. Note that as usual at this time 'candid' signifies fair or impartial.

(4) Fontenelle points out that the mathematics of Leibniz and Newton are the same only in basic ideas, adding (incorrectly): 'Ce que M. Neuton appelloit *Fluxions* M. Leibnitz l'appelloit *Differences*, & le caractère par lequel M. Leibnitz marquoit...' (etc.)...'celui de M. Neuton.' (p. 44).

(5) G. F. A. de L'Hospital, *Analyse des infiniments petits;* see vol. vi, p. 15, note (4).

(6) Fontenelle mentions that Fatio in his 1699 tract *Lineæ Brevissimi descensus Investigatio Geometrica Duplex* claimed Newton as the first inventor of the differential calculus. Wallis had made the same claim earlier, in the unpaginated preface to his *Opera Mathematica*, i (Oxford, 1695), following the account of Newton's new methods he had already given in *ibid.*, ii (1693), pp. 389–95.

(7) Fontenelle writes that Leibniz was accused of plagiary in the *Commercium Epistolicum*, but does not mention Leibniz's earlier charges against Newton. On all this see vol. v, pp. xxiii–xxiv, 96–8 and 115–16.

(8) Fontenelle stated that 'M. *Leibnitz* avoit commencé à travailler à un *Commercium Mathematicum*, qu'il devoit opposer à celui d'Angleterre'. Newton's claim here seems to be that although Leibniz was considering such a publication, he never in fact began it. Newton based his view on the comment in the *Acta Eruditorum* for July 1717 (the year after Leibniz's death)

19

that 'Commercio Epistolico Anglorum aliud quoddam suum idemque amplius opponere decreverat'. It was clear that Leibniz's purpose had been frustrated by death. Newton could not know that Leibniz had actually begun his *Historia et origo calculi differentialis* (see vol. VI, p. 173, note (2))—which, however, was to remain unfinished.

(9) Fontenelle mentions only Leibniz's visit to Paris in 1672, and implies that his work on the calculus stemmed from his meeting with Huygens there.

(10) Fontenelle mistakenly wrote that Leibniz might have seen something of Newton's in 1672—Newton takes him to mean the letter to Collins of 10 December 1672—but was not yet a sufficiently competent mathematician to appreciate the subtleties of it.

(11) The square brackets are Newton's, indicating deletion as usual. In fact he probably meant to refer to p. 52, line 22, of the 'Éloge'. On this page (following Johann Bernoulli's inexact reconstruction of 1691) Fontenelle has traced Leibniz's mathematical development from his reading of Huygens, Pascal and Grégoire de Saint Vincent to his study of Gregory, Barrow and others: 'enfin [*he writes*] il avoit penetré jusqu'à des sources plus éloignées & plus fecondes, & avoit soumis à l'Analise ce qui ne l'avoit jamais été. C'est son Calcul dont il parle.' Newton's argument is that Leibniz should have acknowledged his masters and precursors, but Fontenelle nowhere blames Leibniz for failure in this respect, and the 'Éloge' ends with praise of his candour and openness.

1306 NEWTON TO HUMPHREY WANLEY
24 NOVEMBER 1718
From a printed notice, bearing Newton's signature, in the British Museum[1]

Sir,

These are to give Notice, that on Monday the first the Thirtieth Day of December[2] 1718 (being the next after ST. ANDREW'S DAY) the Council and Officers of the ROYAL SOCIETY are to be Elected for the Year ensuing; at which Election Your Presence is expected, at Nine of the Clock in the Forenoon, at the House of the ROYAL SOCIETY, in Crane-Court, Fleet-Street.

Novr. 24. 1718

ISAAC NEWTON P.R.S.

To
Mr Wanley
  These

NOTES

(1) Harley MS. 3780, fo. 254. Humphrey Wanley (1672–1726), antiquarian, was elected Fellow on 20 November 1706. He did not attend the Annual General Meeting on 30 December 1718, nor was he elected to the Council.

(2) The printed notice read 30 November, St Andrew's Day. In 1718 this was a Sunday, so it has been altered (not by Newton) to read 1 December, but the clerk has omitted to delete the words 'the Thirtieth day'. St Andrew's Day was traditionally the date of the Annual General Meeting of the Royal Society.

1307 MONMORT TO BROOK TAYLOR
7 DECEMBER 1718
Extract from a copy in the Burndy Library[1]

J'ay été fort surpris de trouver ce qui suit dans vostre lettre.[2] 'As to the owning of any one as Inventors or Improvers of the method, besides Sir Isaac Newton, I knew of none. I saw nothing anywhere that seem'd to me an Improvement upon what Sir Isaac had publisht. I was sensible that several had applied the Method with good success and understood pretty much of it: but I always took Sir Isaac Newton, not only for the Inventor, but also for the greatest Master of it.' Je pense comme vous Mr sur le merite de Mr Neuton. J'en parle toujours comme d'un homme au dessus des autres, & qu'on ne peut trop admirer. Mais je ne puis m'empecher de combattre l'opinion ou vous estes que le public a recu de Mr Neuton, & non de M[ess]rs Leibnitz et Bernoulli les nouveaux calculs & l'Art de les faire servir a touttes les recherches qu'on peut faire en Geometrie. C'est une erreur de fait. Il vaut mieux que moy, qui n'ay la dessus aucune prevention, ni rien qui me porte a en avoir, qui fais profession d'estre votre amis, & qui le suis plus sans comparaison que des Geometres Allemands que je nai jamais vu; il vaut mieux, dis je, que je vous fasse remarquer la fausseté qu'un adversaire a qui vous donneriez avantage sur vous & qui vous reprocheroit avec apparence de verité que vostre zele pour la gloire de vostre Nation vous rend partial & vous fait oublier touttes les regles de l'equité. Je n'examinerai point icy les droits de M[ess]rs Neuton & Leibnitz a la 1ere Invention du calcul differentiel & integral. Je vous rapporterai quand vous voudrez le detail des reflexions qu'un long & serieux examen ma fourni, & j'espere que vous n'en serez pas mecontent. Je veux seulement vous faire remarquer qu'il est insoutenable de dire que M[ess]rs Leibnitz & Bernoulli freres ne sont pas les vrays & presque uniques promoteurs de ces calculs. Voicy mon raisonnement, jugez en. Ce sont eux & eux seuls qui nous ont appris les regles de differentier & d'integrer, la maniere de trouver par ces calculs les tangentes des courbes, leur points d'inflexions & de rebroussement, leurs plus grandes & leurs plus petittes ordonnées, les developpees, les caustiques par reflexion & par refraction, les quadratures des courbes, les centres de gravité, ceux d'oscillation, & de percussion, les problemes de la Methode inverse des tangents tels que celuycy par ex: qui donna tant d'admiration a Mr Huygens en 1693 *trouver la courbe dont la tangente est a la partie interceptée de l'axe en raison donnée.*[3] Ce sont eux qui les 1ers. ont exprimé des courbes Mechaniques par des equations, qui nous ont appris a separer les indeterminées dans les equations differentielles, a en abaisser les dimensions, & a les construire par les logarithmes, ou par des rectifications des courbes quand cela est possible; &

qui enfin par de belles & nombreuses applications de ces calculs aux prob-
[lemes] les plus difficiles de la Mechanique, tels que sont ceux de la chainette,
de la voile, l'elastique, de la plus viste descente, de la paracentrique, nous ont
mis & nos neveux dans la voye des plus profondes decouvertes. Ce sont la des
faits sans replique. Il suffit pour sen convaincre d'ouvrir les journaux de
Leipsic. Vous y verrez les preuves de ce que j'avance. Personne hors Mr. le M.
de l'Hopital qu'on peut joindre en partie a ces M[ess]rs. quoi qu'il ait été
disciple de Mr. Jean Bernoulli, na paru avec eux sur la scene jusqu'en 1700 ou
environ. Je compte pour rien ce que Mr. Carré[4] en France & Mr Moivre en
Angleterre, de meme Mr Craige donnerent dans ce temps ou peu auparavant;
tout cela n'etoit rien en comparaison de ce qu'on nous avoit donné dans les
Actes de Leipsic. Il est vray, Mr, que les P[rinci]pes Math: de Mr Neuton ont
paru en 1686 [*read* 1687]; ce scavant ouvrage peut donner lieu de croire que
Mr. Neuton scavoit deslors de ces calculs tout ce qu'en scait aujourdhuy Mr
Bernoulli même; je ne veux pas disconvenir, et c'est une question a part. Mais
il est sur au moins que ce livre n'apprend rien de ces calculs, si ce n'est le lemme
2^d page 250, 1ere edit.,[5] mais vous scavez qu'il ne contient que la 1ere et plus
simple regle de prendre les differences, ce que Mr Leibnitz avoit fait avec plus
d'etendue en 1684. Je dois adjouter que dans le 2e vol de Mr. Wallis[6] imprime
en 1693 on trouve plus au long les regles de ce calcul, mais quoy-que ce
morceaux soit propre a nous donner une grande idée de ce qu'en scavoit alors
Mr Neuton, il n'en apprend pas plus que l'on en trouvoit dans les journaux de
Leipsic. On trouve en 1697 une solution de Mr Newton du probleme de la plus
viste descente,[7] mais comme il ny a point d'analyse, & qu'on ne scait point la
routte qu'il a suivie, cela ne touche point a ma Proposition qui est, que depuis
1684, 1ere datte publique de la naissance du calcul differentiel & integral,
jusqu'en 1700 ou environ, ou je suppose qu'il avoit acquis presque toutte la
perfection quil a aujourdhuy, personne na contribué a le perfectioner que
M[ess]rs Leibnitz & Bernoulli, a moins qu'on ny veuille joindre pour quelque
part Mr le M. de l'Hopital a qui ils avoient de bonne heure revelé leur secrets.
Qui apparement en seroient encore pour tous les Geometres d'aujourdhuy s'ils
avoient voulu les tenir cachez a l'imitation de Mr. Newton, qui a mon avis a du
avoir la clef de ceux la ou de pareils des le temps qu'il a donné son fameux
ouvrage, Ph: nat: p[rinci]pia math. On ne peut rien de plus beau ni de
meilleur en son genre que le traité de Mr. Neuton de Quadratura Curvarum,
mais il est venu bien tard.[8] La datte de l'impression de cet ouvrage est
facheuse, non pour Mr Newton, qui a acquis tant de gloire que l'homme le
plus ambitieux n'en pourroit desirer davantage, mais pour quelques Anglois
qui semblent porter envie a ceux qui ont decouvert & publié les 1ers ces
nouvelles methodes qui ont portes si long la Geometrie.

NOTES

(1) Printed in Brewster, *Memoirs*, II, pp. 513–15. For Monmort's correspondence with Taylor see Letter 1303, note (3), p. 12. In the present letter Monmort presents the view—fair so far as the general history of mathematics is concerned—that Leibniz and the Bernoulli brothers have been the chief promoters of the methods of the calculus, although Newton is indisputably the first inventor. Monmort himself had prepared a long essay directed against some of Newton's philosophical ideas (see vol. VI, p. 441, note (8)); certainly he may not be classed amongst Newton's 'toadies'. Nevertheless, he strongly criticized Johann Bernoulli's treatment of Keill and of Taylor, and in other letters to Taylor urged him to publish a reply to Bernoulli's 'Epistola pro Eminente Mathematico' (see Letter 1196, vol. VI) where Bernoulli had commented adversely on Taylor's *Methodus Incrementorum* (see vol. VI, p. 362, note (2)). In particular, Monmort wrote to Taylor on 5 November 1718 N.S., 'L'Auteur de la lettre pro Em. Math. vous fait un reproche grief honteux et injust dans l'essentiel auquel il faudra que vous repondiez.' In a further letter of 21 April 1719 N.S. he suspects Bernoulli of being not only author of the 'Epistola pro Eminente Mathematico', but also the moving force behind Kruse's reply to Keill in the *Acta Eruditorum* for 1718 (see Letter 1321, note (2)); he writes 'Je suis comme vous forte porté a croire qu'il [*Bernoulli*] a grande et tres grande part aux deux Memoires. Ep. pro Em. Math. et Responsio Crusii. Peut etre meme Crusius est il un nom en l'air. Il est certain que M. Bernoulli est trop jaloux d'honneur...' His criticism of Bernoulli is thus of his character and method of handling disputes, not of his mathematics. But it was Monmort who, by acting as intermediary between Keill and Taylor on the one hand, and Johann Bernoulli, his son Nikolaus and Jakob Hermann on the other, encouraged the continuation of the squabbles between the two parties.

(2) The letter is now lost.

(3) In the *Acta Eruditorum* for May 1693, p. 235, Johann Bernoulli had proposed the problem, 'Quæritur, qualis sit curva *ABC*, quæ hanc habet proprietatem, ut, ducta, ubicunque tangente *BD* terminata ad axe *AE*, portio ejus abscissa *AD* sit ad tangentem *BD* in ratione constante *M* ad *N*.' This is the problem to which Monmort here refers. Many mathematicians, including Jakob Bernoulli, L'Hospital, the Abbé Catelan, Leibniz, and Huygens himself, attempted solutions, or commented upon the problem, in the *Acta Eruditorum*. Huygens, in a short paper entitled 'De problemate Bernoulliano in Actis Lipsiensibus hujus anni p. 235 proposito' wrote, 'Præterea observanda venit in hoc problemate inusitata ac singularis analysis via, quæ ad alia multa in hac Tangentium doctrina aditum aperit, ut egregie jam animadvertit Vir celeberrimus *calculi differentialis* inventor, sine quo vix esset, ut ad hasce geometriæ subtilitates admitteremur.' ('Further, the most celebrated inventor of the *differential calculus* comes to this problem by observing the uncommon and extraordinary method of analysis which gives access to many other [problems] in this doctrine of tangents, as he has already excellently pointed out; without whom it would hardly have been that we should be admtted to these subtleties of geometry.') (See also Huygens, *Oeuvres Complètes*, X, p. 513 *passim*.)

(4) Louis Carré (1663–1711), whose primary interest was in the science of music, was a modest exponent of the Leibnizian calculus. In 1700 he published an introductory treatise, *Méthode pour la mesure des surfaces, la dimension des solides, leurs centres de pesanteur, de percussion, d'oscillation, par l'application du calcul intégral* (Paris, 1700). Later he published a number of mathematical papers in the *Mémoirs* of the Royal Academy of Sciences.

(5) That is, the fluxions Lemma following Proposition 7 of Book II.

(6) This is the Latin version of a paper, or rather a partial transcript, prepared by Newton

and printed in Wallis, *Opera Mathematica*, II (1693), pp. 393–6; for an English version of the passage, see vol. III, pp. 222–9. It was partially repeated in *De Quadratura Curvarum* (1704).

(7) Newton published anonymously his answer to this problem (which had been proposed by Johann Bernoulli) in *Phil. Trans.* for 1697, **19**, no. 224, 384–9; see also vol. IV, pp. 220–9, where the paper, originally sent as a letter to the President of the Royal Society, is printed,

(8) Newton's *De Quadratura Curvarum* was first published in 1704, although Newton had drafted it as early as 1691; see vol. V, p. 115, note (1), and Whiteside, *Mathematical Papers*, III, p. 372, note (1).

1308 SIR CHARLES GODOLPHIN TO NEWTON

17 DECEMBER 1718

From the holograph original in King's College Library, Cambridge[1]

Sr,

I have such due regard for that great character which you have justly obtayned in the world that I will not determine to withdraw my self from that part of a trust wherein I happen to be joyned with you till I have acquainted you with the reasons induceing me to thinke of doeing so. Sr. My late neighbour Mr. Thomas Hall haveing by his last will (wherein he made you one of his Executors for 3 yeares in trust for his son Mr Francis Hall) [2] given to his said son all his Estate after debts and Legacys payd to be accompted for to him with the improvement thereof at the end of the said 3 yeares, dureing which term he directs his Executors to pay his said son an Annuity of 200£ a yeare 'with a Liberty never ye less for his Executors, with the consent and approbation of *3* friends named in the will, (whereof I am one) [3] or the survivors or survivour of them to advance and pay to and for his said son such other summs of money as they shall thinke fitt for his benefitt and advancement:' Hereupon a Question arises concerning the extent of this Clause and all the lawyers who yet appeare to have been consulted hereupon agree so farr in opinion that it gives a discretionary latitude for the Executors dureing the *3* yeares to pay safely with the consent and approbation of the friends any part of the Testators Estate that they shall thinke fitt for the benefitt and advancement of the said Francis Hall; The said Mr. Hall being on a treaty of marriage and takeing notice of the Custome in the City of London wherby he being the only child left by his father, a freeman of the said City, became entitled immediatly on his father's death to one third part of his cleare personall Estate, and proposes That, in consideration of his Legall right to such third part if insisted upon his friends approve and the Executors assign a like proportion of his father's Estate to such trustees as he shall name in a setlement on his intended marriage with some small addition for his Equippment, and on such payments to release his Annuity and also the augmentation thereof.

24

To which Proposeall the Executors, or at least some of them, object that the will of the Testator is to be their Rule, and that the said will makes no mention of this Custome of the City or to the sons part, but directs them formally to pay him 200£ a yeare dureing the 3 yeares, The answer to which by such of the Testator's friends as doe not object to Mr. Francis Hall's proposall, is that tho' the will does not mention the sons third part yet it is highly probable that the Testator was not ignorant of the Custome of the City in that particular any more than in what related to his wifes part wherein he made a speciall Provision for obviating her pretension, in stead whereof he gives a discretionary latitude in his Will for advanceing his sons allowance, and what if the Friends & Executors in the exercise of this discretionary power take their measure from this Custome of the City or to the proportion to be advanced pursuant to the Will, without oblieging the heir to goe to Law for his acknowledged Right, where would be the hurt? or the danger? Mr. Francis Hall agreeing on assigning the summ proposed to be vested in Trustees for his marriage setlement. To discharge the estate of his pretension to a 3d. part by the Custome of the City, and also of his Annuity and the augmentation thereof, and in consideration of the other 2000£ he desires to be advanced for his Equippment, he will ask no more of the Executors dureing the residue of the 3 yeares, but will leave the Estate untouched by him to answer the intention of his father's Will with regard to the Charity under that contingency to which the Testator has subjected it.

But the Executors are advised it seemes by the same Lawyer (who told them just before they would be safe in paying any money to Mr. Francis Hall for his benefit and advancemt. with the consent & approbation of the friends named in the will and signifyed under their hands and seales) that it will be safest for them to have the direction of a Court of Equity in this matter and that Mr. Francis Hall may exhibit his Will against the Executors and the Company of Goldsmiths to have the sum he proposes payd to him, and Answers being put in the Cause may be immediately determined, but he does not say in what time precisely a Corporation may be compell'd to putt in their answer according to the course of the Court and had quite forgot his opinion of a quarter of an howre before that limiting the remainder of a personall Estate is voyd in Law and consequently the allowance proposed for Mr. Francis Hall being out of his fathers personall Estate only the Goldsmiths' Company can have no concern in it But I can tell him further that by an express Clause in the Will the gift to the Goldsmiths Company was not intended to commence 'till after the death of Mr. Francis Hall without issue liveing, for the *then* in that Clause is emphatical and by another express Clause his Will is *that the said Company of Goldsmiths or any other person whatsoever shall not give his son any disturbance in Law*

25

or Equity for or concerning his Estate or any part thereof on any pretence whatsoever: And when he thus prohibits the Company from troubling his son in any manner voluntarily; did he intend to permitt his Executors to incite them to doe it whether they are willing or not? This method of a Bill in Chancery is what seems to me to supersede the plain direction of Mr Halls Will; and some of his Executors insisting on it gives the occasion for my troubling you with this Long Letter and annexing to it a Copy of what Mr. Francis Hall proposes to be advanced for a setlement on his intended marriage as to the quantum, and in a scheme how the assignement may be made safely & easily practicable as to the modus of what he desires, and this with more immediate satisfaction to him self, with least disadvantage to the Estate, with greater ease to the active Executors and more security to the passive, and yet preserving the interest of the Charity designed by the Will under that Contingency to which the Testator has subjected it. [4]

Sr. I beg your pardon for running into this length the trouble whereof I should not have given you but from my being perswaded that Mr. Thomas Hall intended by his Will that his Friends should be Chancellours with his Executors as to the allowance to be advanced, within their *3* yeares of administration to his son, and not that any of them should goe at the charge of his Estate before a master in Chancery to ask what's o'th Clock when the sun shines on the Dyall or to ask whether 14 be a *3d.* part of *42.* for so easy, (you know Sr.) is ye acco[un]t of Mr. Halls Estate in the present situation of it being comprised in *3* articles only, and the value of all of them known every day on Exchange Alley. And tho' our Testator has exempted his friends from the penalty of loseing their Legacys, which the Executors are lyable to if they refuse or neglect to execute his Will; yet I thinke they will deserve them as litle if they doe not endeavour in whatever may be incumbent on them by the Trust they have in it to see it duly performed: For my own part (and in what I am now goeing to say I speake only for myself) Haveing known the Testator 48 yeares before the date of his will, I doe believe he would as soon have bequeathed a Gibbett to me as any other Legacy of Trust in his will if he thought I would consent to involve his Estate unnecessarily in any Law or Chancery-suit haveing valued his good fortune to me alitle before his death in that he had never any of either sort dureing his whole Life: But the Executors are advised that a Decree of a Court of Equity will be the safest way for them, tho' they are told by the same person at the same time that they may be safe without it; which is as if a man should raise any doubt within himself whether he might walk safely through fleet street because many other Houses in that street were built at the same time and, perhaps, by the same builder, and with like materials and scantlings as those which fell down last week without any warning, and tho' *2* or *3* very good Bricklayers should assure him there was no danger for that they had surveyed all the Houses in that street and found none of them in any such tottering condition, But another Bricklayer wishing to humour his apprehensions, tells him he could propose a method how he might walk that way more securely by his makeing a Gallery for him through the street and arching it strongly with double Brickwork, and then if any of the Houses should happen to fall as he went by such an arched gallery would secure him from harm: A Bill in chancery would be like such a Gallery as to the charge and tediousness of the work, but is no more necessary for the security of Mr. Halls Executor's in this case than the building such an Arche would be to the City-walker in the other. Sr. I should be very glad you would not insist upon such a Bill as it seems to be against the exprest will of the Testator in *the words I have recited out of it*; or if you doe, you will pardon what I have now written concerning it Liberando animam meam[5] I am Sr.

Your most humble servt
C GODOLPHIN

17 *Decr.* 1718
Sr Isaack Newton.

NOTES

(1) Keynes MS. 148 C(2); microfilm 1011. 32. The writer was elder brother to the statesman, Sidney, Earl of Godolphin. He was born about 1650 and died on 10 July 1720. He was a barrister, a Member of Parliament, and a commissioner of the customs.

(2) For Thomas Hall and his son Francis see vol. VI, p. 445, note (2). According to a holograph note of Newton's (in the same bundle of papers as this letter), Thomas Hall died on 25 February 1718. His estate was worth £42000. The executors named in his will, besides Newton, were Hopton Haynes, 'the Revd Mr Martin Rector of St Mildred Poultry' and Richard Walker.

(3) Besides Charles Godolphin, they were his nephew Francis, second Earl of Godolphin (1678–1766) and Dr Henry Godolphin (1648–1733), Dean of St Paul's, brother to Charles and Sidney.

(4) The drafts remain with the letter and other papers relating to the Halls; in brief, Francis proposed that £16000 be transferred to him from his father's investments in the Land and Malt Taxes.

(5) 'To relieve my mind.' There are orders for the payment of £200 and similar sums to Francis Hall, signed by Newton and Haynes, in Keynes MS. 148(D), and Trinity College, Cambridge, MSS. R. 4. 48, fo. 15 and R. 16. 38A², A³. Francis Hall himself wrote to Newton on 2 September 1719 (see Letter 1326).

27

1309 LOWNDES TO NEWTON
20 DECEMBER 1718
From a clerical copy in the Public Record Office[1]

Sr.

The Lords Comm[issione]rs of his Ma[jesty']s Treasury desire you to attend them here on Monday morning next at 11 aClock with a State of such Observations & difficulties as may arise to you in the Coynage of Quarter Guineas from the greater space of time necessary in making that Coyn than in Coyning Guineas & halfe Guineas I am &c 20th Dber 1718.

WM LOWNDES

NOTE

(1) T/27, 22, p. 383; no reply has been found.

1310 ROBERT SMITH TO NEWTON
23 DECEMBER 1718
From the holograph original in the University Library, Cambridge[1]

Hon[our]ed Sir

About a month ago I wrote to you to return you thanks for your kind advice to publish Mr Cotes's Papers relating to his Logometria.[2] At the same time I sent you the Title of the Book which I design to publish. I understand by Dr Bentley (who presents his service to you) that you did not receive that letter, but that you have seen one which I wrote afterwards in answer to Dr Taylor; who acquainted me that the Royal Society was pleased to desire the publication of those papers.[3] At that time I had thoughts of publishing the Book by Subscriptions[4] & am very much obliged to return You my hearty thanks for the great incouragement which I hear you are pleased to give to that design. Upon farther consideration I think it better to print it at my own charge & to run the risk of the Sale. I judge the book will be about the size of your Analysis by Fluxions[5] or somewt bigger. Mr Cotes by measures of Ratios & Angles

gives the Fluent of any Fluxion of this Form $\dfrac{d\dot{z}z^{\theta n+\frac{\delta}{\lambda}n-1}}{e+fz^n}$, by a very beautifull &

general Method; which extends itself even to this

$$\frac{d\dot{z}z^{\theta n+\frac{\delta}{\lambda}n-1}}{e+fz^n+gz^{2n}+hz^{3n}+\&c}$$

without any limitation: where d, e, f, g, h, &c are constant quantitys, z the variable, n any Index, θ any whole number affirmative or negative, $\frac{\delta}{\lambda}$ any fraction whatever, & whose denominator $e+fz^n+gz^{2n}+hz^{3n}+$ &c goes on in that progression ad infinitum. I design to print Tables of the Fluents of all inferior sorts of rational Forms at least as far as this $\dfrac{d\dot{z}z^{\theta n+\frac{3}{4}n-1}}{e+fz^n+gz^{2n}+hz^{3n}}$ Besides irrational ones of several sorts, with Theorems to continue them ad infinitum. My Cousen Cotes in his letter to Mr Jones,[6] which he desired might be communicated to You only, did not express himself so fully as to the extent of his method as he might have done. This I perceive Dr Taylor has seen, who writes me word he beleives he can do all that Mr Cotes has done.[7] Since he says he can demonstrate that the general Form which I last mentioned depends upon the Measures of Ratios and Angles; & that he has got a different way somewhat more simple than Mr Cotes's to take off the limitation from your 6th Form. He also tells me he shall be very ready & glad to communicate any thing he knows to render the Book more compleat. I shall thank him for his offer, but I think it more proper to publish only what the author himself has done, rather than a mixture of his & other people's: especially since the Book seems to me pretty compleat as it is, when I have finished some computations which were left imperfect. I intend to be at London shortly & to bring the papers along with me. I shall be extreamly obliged if you will please to take the trouble of a general view of them;[8] & so much the more, the more freedom you will please to use in telling me anything you think amiss. I desire you will beleive I shall make no other use of any advice you will please to give me than what you intend I should. I am sensible how little pretence I have to ask so great a favour; but am extreamly desirous [that] a very great love & esteem which I have for your admirable works, may be accepted as some excuse for using this freedom. I am

Sir Your faithfull Humble Servant

ROB SMITH

Trinity College
Dec: 23. 1718

NOTES

(1) Add. 3983, no. 40. Robert Smith (1689–1768)—son of Roger Cotes' mathematical uncle, John Smith—came up to Trinity in 1708, and was elected Fellow in 1715. Cotes thought very highly of him (Edleston, *Correspondence*, pp. 227 and 228) and in publishing his cousin's remains Smith was only repaying early kindness received. He became Master of Trinity after Bentley's death in 1742. He had then already published his best-known work, *A Compleat System of Opticks* (Cambridge, 1738), to be followed by *Harmonics* (Cambridge, 1749).

(2) This earlier exchange of letters between Newton and Smith is now lost. Clearly Smith was already preparing the material for *Harmonia Mensurarum, sive Analysis & Synthesis per Rationum & Angulorum Mensuras Promotæ: Accedunt alia Opuscula Mathematica: per Rogerum*

Cotesium (Cambridge, 1722). The work was an extension of Cotes' earlier lengthy paper, 'Logometria' printed in *Phil. Trans.* **29**, no. 338 (1714), 5–47.

(3) Taylor's letter to Smith, dated 27 November 1718, is printed in Edleston, *Correspondence*, pp. 231–2; he read it to the Royal Society on the same day. Taylor wrote, *inter alia*: 'All Lovers of Mathematical Learning do heartily joyn with me in this, particularly the Royal Society is so sensible of the great usefulness of those Tables [of integrals], that they have been pleased to order me' to ask Smith to publish them. Other papers of Cotes', he thought, could wait. Perhaps Taylor had the *Philosophical Transactions* in mind for the Tables? On 11 December Smith replied, writing that the papers were to be printed in a book by subscription, and giving its proposed title.

(4) At the Royal Society meeting on 18 December it was reported that there were already a hundred subscriptions offered. However, a week before Taylor had written (Edleston, *Correspondence*, pp. 233–4) doubting the wisdom of publishing by subscription and proposing that either Cambridge University or perhaps the Royal Society should publish the book.

(5) That is, William Jones (ed.), *Analysis per Quantitatum Series, Fluxiones ac Differentias...* (London, 1711). This was a work of 101 pages; in fact *Harmonia Mensurarum*, with 249 quarto pages, proved considerably more bulky. Further the *Opera Miscellania* were added to swell the book.

(6) Letter 1209, vol. VI, where the integral given above is briefly discussed. Taylor was to propose the integration of this fluxional expression as a problem to Monmort, for submission to the continental mathematicians. Both Johann Bernoulli and Hermann published solutions (*Acta Eruditorum*: for June 1719, p. 256; for August 1719, p. 351).

(7) See Taylor to Smith, 11 December 1718, Edleston, *Correspondence*, p. 234: 'I believe I can do all that Mr Cotes has done in his Tables; for I can demonstrate that any curve may be squared by Measures of Ratios and Angles' if its fluxion may be written in the form already stated. 'You know very well that the irrational forms depend upon the rational ones. I have a very different way from Mr. Cotes's, and something more simple, of supplying the defect in Sir Is: Newton's 6th form...'

(8) It does not seem that Newton ever did examine the papers.

1311   W. NEWTON TO NEWTON
4 FEBRUARY 1719
From the original in the University Library, Cambridge[1]

Sr Isaac

Bagdal near Whitby[2]
Yorkshire
Feby the 4th: 1718/19

    I once more importune your Goodness for which I humbly beg pardon, that you will be pleas'd in respect to the strait Circumstances of our Family that we now labour under and haveing an Aged Father lately come out of trouble, now in great Indigency, be so kind as to extend your favour to us. I am perswaded it is in your power several wayes to doe for us, except my late misfortunes, wherein I've been already so troublesom to you may have any wayes obstructed your Good Inclinations. But kind Sir I hope you will not impute it to my unworthiness or any omission of my duty, but rather to fatal necessitie, which I hope hath, and will appologise for me concludes

Yr most Humble and
Oblig'd Servant
WM NEWTON

Some business is desird & ye Interest is great, pray Sr remember me

My father Isaack Newton[3] p'sents his humble service to you

To Sr Isaack Newton at
his house in St Martin Street in
Lesterfields
        London

NOTES

(1) Add. 3965(18), fo. 726. The writer had already twice applied to Newton for financial assistance; see Letters 1220 and 1236, vol. VI.

(2) In Letter 1220 William Newton had mentioned his imminent removal to Whitby.

(3) His relationship to Sir Isaac is not known. Possibly he was descended from Newton's great-uncle of that name, the first Isaac in the family (see Appendix II, Newton's Genealogy, p. 485).

1312 THOMPSON AND NEWTON TO THE TREASURY
11 FEBRUARY 1719
From the clerical copy in the Public Record Office[1]

To the Right Honorable the Lords Commissioners
of his Majesty's Treasury

May it please your Lordships

In Obedience to your Lordship's Order of Reference of the 24th of July Last past upon the Bill hereunto annexed[2] of Mr James Gerard chief Engraver of his Majesty's Seales, We humbly report to your Lordships that we have considered and examined the same, and do find by the perusall of the Warrants he has produced to us that his Royall Highness the Prince of Wales when Guardian of the Kingdom, had directed the making of the several seales in the said Bill mentioned, and by the Receipt of the Duke of Roxburghe[3] it appeares that they were delivered to his Grace for the use of Scotland, and as they have been weighed in the Mint that they are of the Weight expressed in the said Bill.

We further humbly certifie your Lordships that we have examined the Impressions of the said Seales, and upon comparing them with those of former Seales remaining with us, made by his predecessors, we find the work now performed by Mr Gerard to be good, and no way inferior to what has been done before, but as he has overcharged £10. in the privy Seale mentioned in his Bill to be made for North Britain, we are humbly of Opinion that that sum ought to be deducted from his Bill, as likewise the forty shillings he sets down for making draughts of several Seales, there being no precedent for such an allowance.[4]

All which is most humbly submitted to your Lordships great Wisdom

Mint Office the 11*th.*   WM THOMPSON[5]
february 1718.   IS. NEWTON[6]

NOTES

(1) T/1, 220, no. 22.

(2) This follows at no. 22*a*. It was sent to the Mint on 24 July 1718 (T/4, 9, p. 401) by Charles Stanhope for consideration by the Officers. Gerard (or Girard) had made the Great Seal for 'North Britain' weighing over 121 ounces, the Privy Seal, Quarter Seal and Signet also for Scotland, his bill amounting to £411. 1*s*. 5*d* to which he had added £2 as a fee for preparing drawings for two other seals. Girard had been appointed Chief Engraver of His Majestie's Signets, Seals, Stamps and Arms, on 22 June 1716, in succession to John Roos (*Cal. Treas. Books*, xxx (Part II), 1716, p. 296). Soon after Newton's death he was himself succeeded by John Rolles (or Rollos).

(3) John Ker (*c.* 1680–1741), fifth Earl and first Duke (1707) of Roxburgh, was Keeper of the Privy Seal of Scotland (1714) and also one of the Lords Justices appointed during the absences in Hanover of George I.

(4) The Lords allowed Girard the £10 extra for which he had asked.

(5) William Thompson (?1680–1744) had been appointed Warden of the Mint on 20 March 1718; he served eleven years, and was then made a commissioner of the Victualling Office. He was M.P. for Scarborough 1701–22 and 1730–44, sitting as a loyal Whig. (Romney Sedgwick, *The House of Commons, 1715–54*, II (London, 1970), p. 467.)

(6) The signature is not autograph.

1313 LOWNDES TO NEWTON

15 APRIL 1719

From the original in the Public Record Office[1]
For the answer see Letter 1315

Whitehall Treasury Chambers the 15*th. of April* 1719.

The Right Honourable the Lords Commissioners of his Majesties Treasury are pleased to referr the within written Memorial of John Applebee and Henry Hines unto Sir Isaac Newton Knight, who is to consider the Same and Report his Opinion thereupon to their Lordships.

W. LOWNDES

NOTE

(1) T/1, 220, no. 44, written upon the Memorial itself which came up at the Treasury on the 13th. Appleby and Hines (see Index, vol. VI), the partners supplying the Mint with copper bars, complain that it is now three months since the coinage of halfpence and farthings was halted, and that they are losing money in paying men, holding a stock of half a ton of copper, and installing 'New Rooles' [Rolls] for drawing down the bars, which cannot be used for any other purpose.

1314 NEWTON CHAPMAN TO NEWTON
16 APRIL 1719
From the original in the University Library, Cambridge[1]

Most Hond Sr

I have since yesterday felt avery sensible concern and grief for my not being in ye way, to wait your honours commands Tho I was at Church offering my thanks and praise, for my recovery from my late dangerous indisposition of body. nothing but ye fear of having innocently dissoblig'd ye only patron and benefactor I have in the world to whom I am oblig'd (under God) for ye preservation of my life and my future fortune, could have made me presume so far upon your unparellelld candour and goodness. The uneasiness I labour under since ye moment I was told this, compels me to express my sorrow. Tho the place of divine worship is ye only place I have frequented, since my going abroad, except once to mr Booths [?] whom I have dealings with upon my brothers bussiness. I beg you may pardon this presumption and be pleas'd to accept my submissive gratitude and duty, for the innexpressible honour, and favour, you have shewn me for which I shall at all times, exactly comply with, and obey your Honrs. commands as becomes.

Your most obligd and most obedt humble
Kinsman[2]
NEWTON CHAPMAN

Aprille ye 16 1719
For
Sr. Isaac Newton
att his House in St
Martins Street near
Leicr. Feilds
London

NOTES

(1) Add. 3965(18), fo. 703.

(2) The genealogy of Newton's improverished family connections has never been established. It is clear from the older but undated letter that we print next that an aunt of Newton's,

née therefore either Ayscough or Newton (more probably the former), had a daughter who married a husband named Chapman; their son was Newton Chapman, and it appears that his father died during his infancy. From another document—a Newton holograph in Bodleian Library New College MS. 361, II, fo. 39 which seems to be a memorandum—it appears that one Jone Chapman was this correspondent's paternal grandmother, and that she left him some estate upon trust in which, however, other legatees were involved. (See Appendix II, Newton's Genealogy.)

1314A NEWTON'S AUNT TO NEWTON
n.d.
From the original in the University Library, Cambridge[1]

Deare Cousen

having an opertunity to right by Mr Billers I once more desire of thee to doe something for my poore newton chapman[2] who will be very miserable if you doe not by reason of his motheres many debts shee has to pay and that small estate shee Is left with[.] what little Estate I have I and Isaac make even whith what I have coming In so that I can be no way helpfull to my daughter and her chilldren If I could doe anythink I would willingly do It for my poor Newton Chapman who will have something of an Estate when I and my daughter Chapman are ded In the meantime he Is disolate and that Is my great greif I hope god will make thee an Instrument for his good my daughter Chapman haf kept him to Skool according to thy dissire and his master sayes hee Is very capable of learning and the child Is of an extraordinary humor hee Is twelve yearsold and haf ben In his lattin testtament a year or above.[3] no more but my kind love to thee and my cousen Barton which Is all from thy loving

Aunt
[signature illegible][4]

These to Sr Isaac Newton
  Living over ag[ain]st St
    Jameses church
      Londone

NOTES

(1) Add. 3970(8), fo. 613; written presumably before 1709, while Newton still lived in Jermyn Street.

(2) See the previous letter. It follows that Newton Chapman must have been born before about 1697.

(3) Despite this, Newton Chapman did not matriculate at either Cambridge or Oxford.

(4) The signature seems to be in a different, and very illiterate hand; possibly it reads 'Ande ever loving Cose.'

1315 NEWTON TO THE TREASURY
21 APRIL 1719
From a holograph draft in the Mint Papers[1]
Reply to Letter 1313

To the Rt Honble the Lords Comm[issione]rs of his Ma[jes]ties Treasury

May it please your Lordps

In obedience to your Lordps Order of Reference of 15 April 1719, upon the Memorial of Mr Appleby & Mr Hines, I have considered the same, & humbly represent that fine copper will not run close into Barrs like Gold Silver & coarse copper, but requires to be either battered or rolled thin by a Mill; & the cheapest way is to roll it. But a horse Mill being too weak & too chargeable for this purpose I advised the said Importers to procure a water Mill. And to encourage them to do it I promised to give them no disturbance my self so long as they kept to their covenants in the copper which they brought to me.[2] But at the same time I told them that I could engage nothing for my superiors. Hereupon they took a water Mill neare Maidenhead bridge three or four miles above Windsor at the rent of 52£[3] per an for two years and an half from midsummer last, besides a fine of 20 Guineas & 65£ paid for Tools left there by Mr Ayres [*Eyres*], as I understand by the writings. They took also another little place neare it for a Warehouse & lodging room & for building a Refining furnace, at 12£ pr an; & the furnace cost them almost 40£ as they inform me; besides the charge of Beds & furniture for servants & of new Rolls & iron moulds & boxes for the copper & other utensils amounting to 70£, & besides an hundred load of wood & four load of Charcoal upon their hands for this service. And during the intermission of the coinage they pay 39s per week retaining wages to a Clerk & Refiner & two other servants.

When a stop was put to the coinage, I informed your Lordps that 25 Tons of copper money were then coined & delivered, besides what was ready to be delivered of wch I did not then know the quantity: but it was just delivered and amounted unto 26 [cwt][4] weight. And as much copper was that day brought into the Mint as made 26 [cwt] more: so that there have been coined 27 Tonns & 12 [cwt] weight. There was also at that time 25 [cwt] in Barrs brought down the river ready to be delivered, & 45 [cwt] at the Mill in pickle & 35 [cwt] rolled but not cleaned, & 44¾ [cwt] rolled hot but not cold; & in Scissel Brockage & Cakes 128½ [cwt]; besides thirty Tons contracted for under hand & seale, as appears to me by the writing.

By the Bills of parcells several of wch I have seen, the copper hitherto imported cost about 13½*d* per Lwt & some almost 14*d*. The Blanks being hardned round the edges by the cutter & not nealed after cutting are apt to crack on the edges in stamping if the copper be not sufficiently fine, & this cracking has been promoted by two parcells of copper bought of Mr Briggs: but these cracks are very small & may be prevented by making the copper a very little finer. The Rolls at first were rough & made the copper rough, wch occasioned much complaint in the Mint till within a few days before the coinage was stopt: but the Rolls are now smooth, & the last parcell or two of copper imported was smooth & well cleaned.

Considering therefore, the charges that the Importers are & have been at; the quantity of copper upon their hands, some of which is prepared; & the demand of copper money by the people: it seems to me that the 25 [cwt] of copper barrs brought down the river to be imported when the coinage was stopt, & the 45 [cwt] then in pickle & now pickled; be forthwith imported, provided the copper beare the assays prescribed; & that the 35 [cwt] rolled but not yet cleaned be cleaned & imported if your Lordps think fit: the Moneyers being commanded by your Lordps, if you please to pick out all the brockage & blanks from the money. And then the rest of the copper may be prepared & imported, provided the barrs be smooth & cleane & well sized & beare the assays, & the Blanks beare the Press without cracking.

And in the mean time if any other Proposalls be made to your Lordp for coyning the money better or of better copper by the assay, the same may be compared wth what Mr Appleby & Mr Hines will undertake.

All wch is most humbly submitted to your Lordps great wisdome

ISAAC NEWTON

Mint Office. Apr. 21.
    1719

NOTES

(1) II, fo. 448; there are drafts at fos. 446 and 450.

(2) As he had been afterwards empowered to do by Number 1262, vol. VI, Newton had contracted with Henry Hines and John Appleby Junior on 22 August 1717 (Mint Papers, II, fos. 319 and 320) for thirty tons of copper fillets at eighteen pence per pound, at the delivery rate of not more than four tons per month. The coinage of this copper began in January 1718 (Craig, *Newton*, p. 99) and continued for about a year until (as Newton reports) some 25 tons had been coined (Craig makes the quantity rather greater). It was then stopped by Treasury order, but proceeded again shortly no doubt as a result of Newton's recommendation.

(3) We have replaced Newton's symbol *li* throughout by the now more familiar £, while (as usual) retaining Lwt for 'poundweight'.

(4) We have replaced Newton's symbol ⊕ by [cwt] throughout.

1316 TAYLOR TO KEILL
26 APRIL 1719
Extracts from the holograph original in the University Library, Cambridge[1]

Dear Sir

You having first put me upon proposing to Mr Monmort the Argument against the Cartesian Philosophy, taken from Comets, I think myself obliged to communicate his answer to you such a one as it is.[2] He writes thus: 'Je sçai certainement Monsr: que nos meilleurs Astronomes, & les plus consommez dans les observations regardent comme des réveries les assertions de Mr Whiston & autres sur la nature & les mouvements des Comets, par ex: que quelques unes ont passé plus pres du Soleil que n'est Mercure, & ont méme traversé le corps du Soleil. qu'elles decrivent comme les autres planettes autour du Soleil des aires p[ro]portionelles aux temps de leur revolution &c. ainsi je croirois qu'il faut attendre qu'on en connoisse mieux la theorie pour les employer a etablir un Systeme, ou a le detruire.' The Cartesians dont like this argument one bit, it makes so strong against them; so that you see they endeavor to avoid it by denying matter of fact. For this reason I wish you, (or some body else who is able & has leisure,) would take some pains to prove the Theory of Comets so clearly from their own observations, that they may have no power to deny it.

In answer to what you sent to Mr N. Bernoulli of Padua by me, he desired Mr Monmort to return me what follows, which I can hardly prevail upon myself to send you it is so disagreable.[3] I dont doubt but you will do your self justice, & I think you ought; but I beg I may not be concerned in it: for I will not meddle with any disputes when they come to this hight. Mr Bernoulli writes thus. 'J'accepte la promesse de M. Keil qui est de me donner 5 pistolles pour chaque mensonge dont je le pourrai convaincre. Si donc M. Keil tient sa parole je gagnerai au moins 20 pistolles car je soutiens qu'il na pas dit la veritè 1°. lorsqu'il a dit que depuis mon sejour a Londres J'avois publié le contraire de ce que M. Newton m'avoit demontré. 2°. lorsqu'il a dit qu'on a oublié par une faute d'Impression le mot *ut* dans le scholium qui est a la fin du traitté de quadraturis. 3°. lorsqu'il a dit que mon oncle (je passe sous silence ce qu'il dit de moy dans le meme endroit) n'entend pas le calcul differentiel. 4°. lorsqu'il a dit nouvellement dans sa lettre a M. Taylor qu'il peut me shew me lyes I have made for nothing. Je vous prie de luy faire notifier ces pretentions, & d'en demander sa reponse.'...

Since I have heard nothing from you in answer to my proposal of joyning with you against Bernoulli I have drawn up a paper, which I think soon to publish by it self.[4] Mr Innys did not give me your Astronomy as you told me you had order'd him. I am told you intend to publish it in English.[5] I should be very glad to see it finisht, & to know of any curious thing besides that you are preparing for the Publick.

I am

Dr Sir

Your most humble servant

BROOK TAYLOR

Bifrons
26 *April* 1719
Dr Keil
Savillian Professor of Astronomy
      at Oxford

NOTES

(1) Res. 1893(*a*), (Lucasian Professorship, Box I, Packet 3); Edleston, *Correspondence*, p. 235.

(2) Presumably Monmort replied privately to Taylor in a letter which is now lost. His arguments were, however, probably very similar to those he gave in his 'Dissertation' in *L'Europe Savante* for October 1718, **5** Part 2 (La Haye, 1718), pp. 209–94 (see also Letter 1366, note (10)). On pp. 273–7 of that paper Monmort claims that what is so far known from observations of the motion of comets in no way excludes the possibility that they move in independent vortices outside the solar system. He makes reference, however, to Derham rather than to Whiston.

(3) Keill's letter to Taylor, including his rather crude pecuniary challenge to Nikolaus [I] Bernoulli, Taylor's letter to Monmort and Bernoulli's answering letter to him are all now lost. Nikolaus had long supported his Uncle Johann in the disputes with the English mathematicians.

(4) Johann [I] Bernoulli in the 'Epistola pro Eminente Mathematico' had charged Taylor with plagiary in his *Methodus Incrementorum* (see Letter 1196, vol. VI). Taylor's answer was 'Apologia D. Brook Taylor, J.U.D. & R.S. Sec. contra V.C.J. Bernoullium, Math. Prof. Basileæ', *Phil. Trans.* for 1719, **30**, no. 360, pp. 955–63. In this Taylor claimed that he had employed only his own analysis 'inventis alienis sum minime usus ut meis', but admitted that he had been careless in the matter of acknowledging indebtedness to the authors from whom he had derived it. In dealing with the isoperimetrical problems he had used the analysis of Jakob Bernoulli 'viri a rebus mathematicis optime meriti, cui debitos nunc persolvo honores'. Nevertheless, Taylor went on to accuse Johann Bernoulli of egregious errors, and of falsely claiming too much for his own originality—such as the rules given in the *Analyse des infiniment petites* of the Marquis de l'Hospital.

Bernoulli himself did not reply to this paper, but both his son Nikolaus [II] and Johann Burchard Mencke, editor of the *Acta Eruditorum*, did so; see *Acta Eruditorum* for June 1720, pp. 279–85 and for May 1721, pp. 195–228. Both papers are reprinted in Bernoulli, *Opera Omnia*, II, pp. 474–512.

(5) *Introductio ad veram astronomiam, seu lectiones astronomiæ* (Oxford, 1718), republished in English in 1721. J. Innys was one of the stationers concerned in the edition.

1317 TAYLOR TO JONES
5 MAY 1719
Extracts from the holograph original in private possession[1]

Sir

Tho for reasons that you too well know, it was impossible for me to desire
that Sir Isaac Newton should see my paper against Bernoulli, before it was
publisht;[2] yet I am not displeased at what you have done: because I am sure
you would not have shew'd it him, if you had not been sure it could be no
inconvenience to me; and in what he says of it I find myself a gainer, for the
place that piece has in the Philosophical Transactions, will shew the World
that what I say has the approbation of the Royal Society, which is an advantage that I have no manner of reason to despise, tho' I never intended to
ask it, being much more desirous that my actions should subsist and be supported by their own reasonableness, than by the authority of any other persons
whatsoever. I am very sensible that Sir Isaac Newton is very glad that any
thing should be publisht which affects Bernoulli, & believe that is the only
reason which makes him willing this piece should appear in the Transactions.
But before I determine to let it appear, I have many other things to consider,
wherein I particularly desire your opinion and advice. 1. Whether I really
have sufficient provocation thus to attack Bernoulli, as my Passions make me
believe I have. 2. Whether my reputation would suffer or no, if I should let it
alone. 3. Whether I am not mistaken in any of the things I say. 4. Whether I
say too much, or enough. 5. Whether my expressions are clear, & my Language as it should be. 6. Lastly whether I have said any thing that is not
agreable to the character of one who is truly sensible of the duties of Honour
and Good manners. In these things I must claim your serious & free opinion;
for as I am fully determined, (as I say in the conclusion) never any more to
meddle in this wrangling way, it concerns me very much that what I say
should not be liable to blame. What reason is there that Sir Isaac Newton
should keep the paper? I had rather you had it, that I might have the benefit of
your opinion.

                ...   .

[Hermann][3] vindicates himself I think very handsomly against Dr Keil,
who had charged him with having made use of an invention of his without
naming him; & shows a disposition to treat every body with civility & Justice.
He gives greater testimonies to Sir Isaac Newton, than I have seen from any
of the Foreigners; though in one place he seems to wish that Leibnitz might

have a share in his discovery of universal gravitation. I would transcribe the passage, if I had not lent out the Book. By my correspondence with Mr Monmort, I find Hermann is not of Bernoulli's Party, but is rather under his displeasure.

Pray do what you can to make Mr Smith go on with Mr. Cotes's Works.[4] What says Sir Isaac Newton to Maclauren's paper about the description of Curves?[5] it seems to be very curious. That Gentleman certainly deserves to be encouraged.

> I am
> Dear Sir
> Your most obliged
> humble servant
> BROOK TAYLOR

Bifrons
5 *May*, 1719

NOTES

(1) Printed in Rigaud, *Correspondence*, II, pp. 278–80.

(2) Taylor's 'Apologia contra Bernoullium'; see Letter 1316, note (4). It was clearly Newton himself who ordered that this paper be so published.

(3) In the passage which we have omitted Taylor criticizes papers by Jakob Hermann published in the *Exercitationum Subsecivarum Francofurtensium* (Frankfurt-am-Oder), presumably in the volume for 1718. Taylor had written to Keill about the same publication at much greater length in Letter 1316, in a passage which again we have not printed. There he quotes a passage from the *Exercitationum* where Hermann answers Keill, who, in his 'Defense' (*Journal Literaire de la Haye* for 1716, p. 422; see vol. VI, p. 246, note (2)) had accused Hermann of borrowing from his own demonstration of the inverse problem of central forces (*Phil. Trans.* **26**, no. 318 (1708), 174–88); see also D. T. Whiteside, *Journal for the History of Astronomy*, **1** (1970), 134, note (32).

(4) See Letter 1310.

(5) Maclaurin's paper 'Nova Methoda Universalis Curvas Omnes Cujuscunque Ordinis Mechanicæ Describendi sola Datorum Angulorum et Rectarum ope' was read at the Royal Society on 8 January 1719 and printed in *Phil. Trans.* **30**, no. 359 (1719), p. 939.

1318 ARNOLD TO CHAMBERLAYNE
16 MAY 1719
Extracts from the holograph original in the British Museum[1]

On the other side you have what the M[ess]rs Leibnitz and Bernoulli writ me concerning what the Roial Society have thought fit to publish in their last Transactions[2] if You think proper You may communicate it to that Curious and Learned Body either with or without my name. I shal not be sorry to see it in Print if it may anyway serve to justifie my deceased Freind[3] for whose Memory I shal alwaies have a great Veneration. I wish the Roial Society and Dr Clarke[4] had acted more honorably with regard to him.

...   .

Exon May the 16th 1719

J. ARNOLD

[and on the reverse]

In the Philosophical Transactions Numb 359 pag: 923 &c I find *a Letter of M. l'Abbe Conti R.S. to the late M. Leibnitz concerning the dispute about the Invention of the Method of Fluxions or differential Method, with M. Leibnitz his Answer thereto*[5] I had the honour of correspondence with M. Leibnitz as You may well Remember. I have herewith sent You what he writ in several Letters concerning that affaire to which I have added what M. John Bernoulli of Basil hath sent me on the same subject.

[Excerpts of Leibniz's letters to Arnold, for the most part already printed in this Correspondence, *follow;*[6] *Arnold then continues...]* This is the last letter I had from M. Leibnitz, death putting an end to our correspondence the 15th November. Since M Bernoulli is declared the Author of the Probleme by M Leibnitz I shal take the Liberty of giving You what he was so kind as to send me from time to time on that subject.
[Excerpts of Bernoulli's letters to Arnold follow][7]

NOTES

(1) Birch MS. 4281, fos. 14, 12 and 13. For Arnold see vol. VI, p. 27, note (1). His close involvement in the affairs of the continental mathematicians is also indicated by several references to him in correspondence between Bernoulli and Varignon (see the Bernoulli Edition).

(2) *Phil. Trans.* for January and February 1719, **30**, no. 359, pp. 923–8, printing the correspondence we have given as Letters 1190 and 1202, vol. VI.

(3) Leibniz.

(4) Presumbably Arnold refers to Samuel Clarke, *A Collection of Papers which passed between . . . Mr. Leibnitz, and Dr. Clarke* (London, 1717); see vol. VI, p. 259, note (3).

(5) These are the letters already mentioned in note (2) above.

(6) The excerpts are taken from the following letters from Leibniz to Arnold: (i) 27 January 1716 (Letter 1179, vol. VI); (ii) 6 March 1716 (Letter 1193, vol. VI); (iii) 17 April 1716 (Letter 1204, vol. VI); (iv) 25 May 1716 (not printed; this letter merely repeats the contents of that immediately preceding it); (v) 24 July 1716 (an excerpt is printed in vol. VI, p. 349, note (1)); (vi) 29 September 1716 (not printed; this was Leibniz's last letter to Arnold. He wrote: 'on me dit que l'Abbé Conti s'est retiré a Oxford. Je ne scai s'il est reto[urné] a Londres').

(7) The excerpts are taken from the following letters of Johann Bernoulli to Arnold: (*a*) 4 August 1716 (Letter 1219, vol. VI); (*b*) 28 March 1717 (Letter 1238, vol. VI); (*c*) 9 March

1719 (not printed; Bernoulli thanks Arnold for sending him the *Philosophical Transactions* for 1717 containing Taylor's solution to the orthogonals problem (see vol. VI, p. 362, note (2)), which he finds inadequate since it fails to produce a construction for Leibniz's problem, which, he explains, he has himself succeeded in doing).

1319 LOWNDES TO NEWTON
4 JUNE 1719
From the original in the Mint Papers[1]

Sr

The inclosed Petition of W. Wood Gent. having been laid before the Lords Comm[ission]ers of His Ma[jes]tys Treasury, praying that he may be permitted to Furnish such Copper as shall be wanting for the Copper Coinage at 17½d per pound,[2] Their Lordps Command me to transmit the same to you, and are pleased to direct you to take the same into your Consideration and Report to them Your opinion what may be fit to be done therein.[3]
I am

Sr

Your most humble servt

W. LOWNDES

Trea[su]ry Chambers
4th. June 1719
Sr Isaac Newton: abt Mr Wood

NOTES

(1) II, fo. 368; there is a copy in P.R.O. T/27, 22, p. 424.

(2) Wood's petition to the Treasury is at Mint Papers, II, fo. 367. He complains that an offer of copper at 18d the pound was accepted by the Mint, although he offered it for 17½d. The Treasury had agreed that, after the existing contract for the supply of 30 tons was completed, Wood should supply a further 30 tons, but no order had yet been made for him to do so.

(3) Newton's answer is not extant, but compare Letter 1325.

1320 J. BERNOULLI TO NEWTON
24 JUNE 1719
From the holograph original in Yale University Library[1]
For the answer see Letter 1329a

Viro Illustrissimo atque Incomparabili
Isa[a]co Newtono
S.P.D.
Johannes Bernoulli

Opticam Tuam Anglia Te mihi dono datam nuper accepi missu Celeb. Varignonii, a quo etiam exemplar Lat. sicut intelligo, accepturus sum:[2] Pro utroque hoc egregio munere non minus quam pro aliis jam sæpius mihi acceptis tanquam totidem Tuæ erga me benevolentiæ signis nunc demum debitas persolvo gratias, quas quod fateor dudum persolvere debuissem. Noli quæso officii hujus neglecti causam imputare animo ingrato et beneficiorum immemori a quo semper quam maxime abhorrui; noli etiam credere, me ideo minus ingentia Tua merita coluisse. Quin potius, si quid fidei verba mea merentur, id tantum ex silentio meo colligas velim, quod Te divini ingenii Virum cui parem non habet ætas nostra ego præ summa veneratione compellare non audebam; Certe ne nunc quidem auderem, nisi nuper quod animum addidit, intellexissem, juxta stupendas ingenii dotes etiam comitatis et affabilitatis virtutem usque adeo esse Tibi connatam, ut ab inferioris conditionis hominibus, qualem me lubens profiteor, litteras accipere plane non detrectes. Cæterum quanti æstimaverim Tuam amicitiam, qua uti percepi ex Litteris Virorum Clariss. Monmortii[3] et Moivrei[4] me ante hac dignatus es, eosdam hos Viros antestor, ac præsertim quidem Moivreum, qui ea de re luculentissimum testimonium coram perhibere poterit. Sed nescio qui factum, ut post accensam facem feralis illius belli, quod maximo scientiæ Mathematicæ probro ante aliquot annos exortum inter quosdam utriusque Nationis Britannicæ et Germanicæ Geometras, ego nec Britannus nec Germanus sed Helvetius qui a partium studio alienissimus sum, et quidvis potius facerem, quam aliorum litibus me sponte immiscere, gratia tamen Tua, ut fama fert, exciderim.[5] Quod si ita esset, quamvis contrarium sperem, non possem non credere, hocce infortunium fuisse mihi conflatum a supplantatione quorundam sycophantarum, qui ex rabida quadam aviditate sibi suisque popularibus ædificandi monumenta ex ruderibus destructæ aliorum existimationis et famæ nos omnes non-Anglos insontes cum sontibus ni statim per omnia applaudere velimus, acerbissimis contumeliis proscindunt. Itaque non dubito, quin Tibi Vir maxime de me quoque multa falsa et conficta fuerint narrata, quæ gratiam, qua apud Te flagravi si non delere saltem imminuere potuerunt. Sed non est ut multis me excusem, provoco ad scripta mea quæ extant; docebunt quam singulari cum laude de Te Tuisque inventis quavis data occasione locutus fuerim; ecquis aliter posset, qui magnitudinem meritorum Tuorum considerat: quam mirabundus autem etiamnum illa deprædicem atque extollam quovis loco et tempore, privatim æque ac publice in Litteris, in sermonibus, in orationibus, in prælectionibus, illos loqui sinam qui me legunt, qui me audiunt. Sane si quid sapio, gratior erit Posteritati commemoratio meritissimæ Tuæ laudis a nobis instituta, utpote ex sincero animo et calamo profecta, quam nonnullorum ex Vestratibus[6] immodicus ardor (non dico Te laudandi, nam satis satis laudari

43

For both of these admirable gifts, no less than for others already so frequently received by me as just so many signs of your goodwill towards me, I now at last give due thanks, which I admit I ought to have done a little while ago. Do not, I beg you, attribute the reason for this neglect of duty to an ungrateful spirit forgetful of your kindnesses, from which [attitude] I have always shrunk back as much as possible; do not even believe on that account that I am any the less respectful of your enormous merit. Indeed I would prefer, if my words deserve trust, that you would only infer from my silence that I did not dare to address you, as a man of divine genius of whom our age has no equal, because of the highest respect [I have for you]; indeed I would not dare even now, except that recently I learned ([a report] which increased my courage) that you were also born with the virtue of friendliness and affability, as well as with extraordinary gifts of genius, so that you would not refuse entirely to accept letters from men of lesser condition, of which sort I freely acknowledge myself to be. For the rest, how greatly I valued that friendship of which you formerly considered me worthy (as I perceived from the letters of Monmort[3] and De Moivre[4]) I call those same men to bear witness, especially De Moivre, who will be able personally to produce most ample testimony on that score. But I do not know how it happened that, after the outbreak of that deadly feud which arose several years ago between the geometers of the British and German nations (to the great shame of mathematical science) I, who am neither British nor German but Swiss and am so very far from any inclination to take sides that I would rather do anything than involve myself of my own free will in the disputes of others, nevertheless have fallen from your favour, as rumour would have it.[5] If this has happened, contrary to my hopes, I cannot but believe such a misfortune to have been brought upon me by the machination of some toadies, who, from a raging desire to build monuments to themselves and their countrymen from the broken rubble of the destroyed fame and reputation of others, censure all of us who are not English (innocent and guilty alike) with the most bitter abuse unless we choose to praise [them] at once for everything. Therefore I do not doubt, great Sir, that you have also been told many lies and inventions about me, which may have at least weakened, if not destroyed, the good opinion which I formerly enjoyed with you. But it is not for me to excuse myself with many words; I appeal to my writings which are extant: they will show how I have spoken with particular praise of you and your discoveries whenever occasion arose: what else could anyone do, who considers the greatness of your merits? That I have praised and extolled them as wonderful at every time, in every place, in my letters, speeches, lectures and conversations, let those affirm who have heard me or read me. To be sure, if I have any sense of discernment, the memory of the well-deserved praise you have received from us arising from a sincere heart and pen will be more acceptable to posterity, than the immoderate passion of some of your countrymen[6]—I do not say immoderate in praising you, for you cannot be praised enough, but in claiming everything for you, even that which you yourself do not wish for, and in leaving nothing to foreigners.[7] Beyond all doubt they are mistaken who have reported me to you as the author of certain of those flying sheets[8] in which perhaps you were not treated with sufficient respect; but I earnestly entreat you, famous Sir, and implore you in the name manner of his writing of the gifts of *Opticks*, prompted by his desire for concord. Like De Moivre (who was independently in correspondence with Varignon, acting also as postman between Newton and the Abbé), Varignon was most eager to reconcile Newton and Bernoulli, urging the latter (in successive letters of May and June 1719) to seize the opportunity of the presentation of *Opticks* to write a conciliatory letter to Newton, and to use De Moivre's means to have it presented in the most propitious manner. Bernoulli delayed in following Varignon's advice (and did not write to De Moivre himself) but when his letter reached Varignon the Abbé was pleased with it. In September 1719 Varigon explained to Bernoulli how he had sent the present letter to De Moivre for Newton with his own Letter 1323 via a M. Battier, together with a letter of introduction for the bearer and a covering letter to De Moivre himself.

(3) Extracts only of Bernoulli's letters to Monmort survive (see Letters 1237 and 1280, vol. VI), so we do not know in what terms Monmort wrote to Bernoulli.

(4) The extant correspondence between De Moivre and Johann Bernoulli is published in Karl Wollenschläger, *De Moivre*. In the early letters (1708–13) compliments are exchanged between Newton and Bernoulli via De Moivre; later there are indications of a decrease in friendliness between the two men (see vol. VI, p. 74, note (1)).

(5) 'Rumour' presumably reached Bernoulli largely through the letters of Monmort and John Arnold. Certainly it was Monmort who told Bernoulli that Newton was behind Keill's activities (see Bernoulli's letter to Varignon of 1 August 1720, N.S., Bernoulli Edition, where Bernoulli quotes from Monmort's letters to himself).

(6) Bernoulli doubtless has Keill in mind: he was smarting under Keill's private challenge conveyed to him by Monmort (see Letter 1303, note (2)).

(7) A copy made by Newton of the passage 'Fallunt haud dubie...Te mei integritas'—most of the rest of the Letter—is also in Yale University Library. This portion is partially quoted in Brewster, *Memoirs*, II, p. 55, but he wrongly dates the extract 3 November 1719.

(8) Bernoulli means the letter printed in the *Charta Volans* (see Letter 1009, vol. VI). It was Leibniz himself who had let out the secret of Bernoulli's authorship of the letter, in an anonymous letter to the *Nouvelles Litteraires* (see Letter 1172, vol. VI). Later he mentioned the matter again in a letter to the Baroness von Kilmansegge (see vol. VI, p. 330, note (12)).

(9) Bernoulli refers to the 'Observations' written by Newton on a letter from Leibniz to Conti and first printed in the Appendix to Raphson's *History of Fluxions*. In these Newton mentions the letter in the *Nouvelles Litteraires* (see Number 1211, notes (6) and (9) (vol. VI, p. 350)). Bernoulli saw the 'Observations' only after their reprinting in Des Maizeaux's *Recueil*; see Letter 1357.

(10) Leibniz was not of course deceived; he had himself received permission from Bernoulli to publish the extract anonymously. This is downright falsehood on Bernoulli's part; Leibniz now being dead, only Bernoulli knew the truth of the matter, although, of course, he ran the risk of documentary evidence being found amongst Leibniz's papers.

(11) Bernoulli was elected a Fellow of the Royal Society on 1 December 1712, and the news of his election was conveyed to him by De Moivre shortly afterwards (see vol. VI, p. 65, note (4)). In Letter 1332 Bernoulli voices the fear that he has been expelled from the Society.

(12) Newton correctly refers to this letter as of 5 July 1719, N.S. Brewster added considerable confusion by giving a variety of dates for the letter, as follows: *Memoirs*, II: p. 54, 3 November 1719; p. 72, 10 July 1719; p. 295, 5 July 1714; and p. 502, 5 July 1719. All these dates are clearly intended to be new style.

1321 KEILL TO NEWTON
24 JUNE 1719
From the holograph original in the University Library, Cambridge[1]

June 24th 1719

Honoured Sr

I sent some time agoe to Dr Halley[2] my Answer to Mr Bernoulli and the
Lipsick Rogues[3] and desired him to give it you, that you might doe with it as
you please, since that I have been at Oxford, and looking there over some of
Mr Bernoulli's treatises I find he has all along been mistaken about 2d Differences particularly in the solution of the Isoperimetrical Problem as he calls
it. where by the paper[4] here inclosed you will find he has made the same mistake, as he did about your proposition in your Principia,[5] I desire this paper
I have here inclosed may be added to the rest, I beleive Johnson[6] at the
Hague has played the Rogue with me, for I have never heard from him since
last year I have a mind to write to Mr Gravesend[7] at Leyden, and see if I can
learn the matter from him, but I desire to have your advice about it.[8] I am

	Sr
Direct for me	with all respect
at Dr James Keilth[9]	your most faithfull
in Northampton.	and Obedient servant
	JOHN KEILL

For
The Honoured
Sr Isaack Newton
at his house in
St Martins Street
near Leicester Square
    Westminster

NOTES

(1) Add. 3985, no. 12. In Keill's somewhat careless writing, '4' and '9' are hardly dis-
tinguishable. The numerals on the postmark, however, are clearly '24' and as Keill was
writing in London (Oxford is 'there' in the letter) his letter was postmarked on the day of
writing. The year must, however, be read as '1719' not '1714': Keill did not yet suspect that
Bernoulli was a 'rogue' in 1714, nor had the isoperimetric problem become an issue in the
continuing debate.

(2) Halley was one of the two Secretaries of the Royal Society, John Machin (much junior)
being the other, appointed six months before. At a meeting of the Society on 28 May 1719
Halley had 'read part of a letter from Dr. Keil to Mr. Bernoulli in Vindication of himself from
some aspersions cast upon him in a Letter of Crusius printed in the Lipsic Acts of the year'

48

(Journal Book of the Royal Society, XI, p. 355). Johann H. Kruse (or Crusius) was author-amanuensis on behalf of his teacher, Johann Bernoulli—'si sit quisquam a te diversus, est certe intimus & familiaris tuus' as Keill wrote—of a 'Responsio ad Cl. Viri Johannis Keil,... Defensionem pro...Newtono,' published in the *Acta Eruditorum* for October 1718, pp. 454–66, which was in turn directed against 'Defense' (*Journal Literaire de la Haye* for 1716; see vol. VI, p. 246, note (2)).

(3) It seems likely that Keill means his answer to Kruse mentioned in note (2) above, which was later published separately as *Joannis Keill...Epistola ad Virum Clarissimum Joannem Bernoulli* (London, 1720). Drafts of this pamphlet, in both English and Latin, are in U.L.C. Res. 1893(a) (Lucasian Professorship, Box I, Packet 5). Compare Keill's 'Lettre à Bernoulli' in the *Journal Literaire de la Haye* for 1719, pp. 261–87 the publication of which had been long delayed by Newton (see vol. VI, p. 386, note (4)).

The particular offence of the editors of the *Acta Eruditorum*—the 'Lipsick Rogues'—was not only that of publishing Kruse's severe attack, but of indexing his abusive criticism of Keill and Newton in great detail in their volume. This is brought out in Keill's full title: 'Epistola...In qua Dominum Newtonum & seipsum defendit contra criminationes a Crusio quodam objectas, & in *Actis Lipsiensibus* publicatas; ubi etiam quæritur de nova calumniandi methodo ab Authoribus *Actorum Lipsiensium* inventa & usurpata, qua in Indicibus suis probra & convitia in alios fundunt.' ('A Letter...in which he defends Mr. Newton and himself against charges brought forward by a certain Kruse and published in the *Acta* [*Eruditorum*] of Leipzig, and in which also a complaint is made against the new fashion of slandering devised and adopted by the editors of the Leipzig *Acts*, whereby they pour infamy and abuse upon others in their indexes.') Keill deals with this offence on pp. 6–9 of the *Epistola ad Bernoulli*.

(4) There is now no attached paper with this letter; it was presumably returned to Keill. It must have been a draft of the passages on pp. 17–18 of Keill's *Epistola ad Bernoulli* (see note (3) above) where Keill criticizes his solution of the isoperimetrical problem (as given in the *Mémoires de l'Académie Royale des Sciences* for 1706, pp. 235–45) and accuses him of mis-stating the rule for determining second differences. (*Translating:* 'Thus you have twice in a single line erred with respect to second differences, and pronounced a false rule for determining them. I showed elsewhere [*note: Journal Literaire* for 1714, p. 348] that Leibniz fell into a similar mistake concerning differences, whence I came to the most confident conclusion that neither he nor you have properly understood the nature of the higher differences. Because of that imaginary and supposed error of Newton's concerning differences in his *Principia*, you argued that he was not the inventor of the calculus; are not the genuine and obvious mistakes of Leibniz and yourself with regard to the same subject, by the same logic, proofs that neither of you was the inventor of that same calculus?')

(5) Prop. 10, Book II; see Letter 951a, vol. V.

(6) Johnson was editor of the *Journal Literaire de la Haye*; but it was Newton, not Johnson, who had delayed publication of Keill's 'Lettre à Bernoulli'; see note (2) above.

(7) W. J. 'sGravesande, who had earlier promised Newton help in his disputes with Leibniz; see vol. VI, p. 145, note (1).

(8) No further letters between John Keill and Newton are extant.

(9) James Keill (1673–July 1719), John's younger brother, took up anatomy and medicine receiving the M.D. from Aberdeen (1699) and Cambridge (1705). He was elected F.R.S. in 1712. Besides practising medicine with success in Northampton, James published an *Anatomy of the Humane Body Abridged* (London, 1698) and important works on iatromechanics. He had been seriously ill since 1716.

1322 W. STOCKWOOD TO NEWTON

15 JULY 1719

From the original in the University Library, Cambridge[1]

July 15. 1719

Sr

You have now a Turn in King Street Schole,[2] The Child you shall please to put in ought to be nine years of age, reside in this Parish of St James's & read well in ye Bible

I am

Your Dutifull Servt
W. STOCKWOOD

Sr Isaac Newton

NOTES

(1) Add. 3965, fo. 721. We have not positively identified the writer, but it is possible that he was William Stockwood (*c.* 1690–1784), B.A. (Cantab.) 1712, ordained priest 1715. He became Rector of East Langdon Kent, and later Prebendary of Worcester and Westminster.

(2) This school (in Kingly Street, as it is now called, north of Piccadilly, not King Street, St James') for the instruction of sixteen 'poore Boys Natives & Inhabitants of the said Parish of St. James...to Read, Write, Cast accounts, and such other parts of mathematicks as may the better qualify [them] to be put out Apprentices' (the phrase suggests Newton's pen) had been founded *c.* 1687 by Thomas Tenison, later Archbishop of Canterbury, in association with what was to be known as the 'Golden Square Tabernacle'. Newton was one of the Trustees of the Tabernacle, and therefore also of its school, and had presumably continued to support the school financially after he left St James' parish (see F. H. W. Sheppard, *Survey of London*, XXXI (London, 1963), pp. 180–3). The Tabernacle and school had been rebuilt in 1702; by 1713 there were 36 scholars. In 1871 it was amalgamated with Tenison's other school in St Martin's n-the-Fields, and removed to Kennington (*ibid.*, p. 185).

1323 VARIGNON TO NEWTON

15 JULY 1719

From the holograph original in King's College Library, Cambridge[1]
Reply to Letter 1300; for the answer see Letter 1329

Equiti Nobilissimo,
Viroque Doctissimo D. D. Newton,
Regiæ Societatis Anglicanæ Præsidi Dignissimo,
S.P.D.
Petrus Varignonius

Vir Ilustrissime,

Fascem quinque Exemplarium mirandæ tuæ Optices, Latino sermone denuo vulgatæ, quem ad me Miseras die 13. Octob 1718 s.v. ut cognovi ex tuis Litteris in eo contentis, non accepi nisi die 6 s.n. Junij proxime elapsi. Statim, ut jusseras, horum Exemplarium a te dedi unum Regiæ Scientiarum Academiæ, alterum Dno. Fontenelle, tertium Dno. Montmort: Horum uterque gratias aget tibi; ac præterea D. Fontenelle Singulares tibi referet nomine Academiæ cui Donum hoc tuum gratissimum fuit.

Eorundem Exemplarium quartum iterum a te mihi Dno. Johanni Bernoulli eadem mente qua illi Anglice Scriptum anno proxime elapso miseram:[2] nimirum pacis & concordiæ inter te & illum conciliandæ causa; quod gratum tibi fore putavi æque ac illi, qui statim ac primum istud a te donum ipsi nuntiavi, rescripsit se nil gratius audire potuisse, rogavitque ut memorem ejus animum tibi significarem; sed Litteræ in quibus a me tibi gratias agebam, jam erant dimissæ. En tandem ille de secundo tuo dono monitus, qui pro utroque gratias nunc agit tibi in epistola his meis Litteris adjuncta,[3] quæ utinam valeat eum in tuam pristinam Benevolentiam reducere, de qua se invitam excidisse testificatur. Hanc autem illius epistolam paulo tardius ad te mitto ob varia & urgentia negotia quibus impeditus adeo fui ut per ea mihi non licitum fuerit citius ad te scribere.

Quod spectat quintum superiorum Exemplarium una cum quarto mihi a te pro tua erga me Benevolentia Donatum, raptim saltem legendum censui antequam pro utroque tibi gratias agerem. Sed cum in ejusdem Lib. 3. plures novas Quæstiones[4] invenerim adjectas veteribus, harumque non nullas pluribus experimentis auctas; ne quid novitatum istarum me fugere, Quæstiones has veteras attente denuo legi omnes cum novis, quas æque ac istas reperi in immensum augere numeram Effectuum seu Motuum exquisitissimis utrobique relatis experimentis manifestissimorum, quorum tamen causæ penitus ignorantur. Porro, his perlectis, non vidi (nec antea videram) qua ratione quis possit existimare Gravitatem a te inter essentiales corporum affectiones haberi, cum in quæst. 23 edit. 1. ut nunc in quæst. 31. edit. 2. Lib. 3. dixeris: *quam ego Attractionem appello, fieri sane potest ut ea efficiatur impulsu, vel alio aliquo modo nobis ignoto, &c.* Cumque jam ab annis 32. in edit. 1. aurei tui Tractatus de Princ. Math. ac dein in ejusdem edit 2. dixeris def. 8. Lib. 1. *Voces autem Attractionis, Impulsus, vel Propensionis cujuscunque in centrum, indifferenter & pro se mutuo promiscue usurpo.* Sed quo feror! cum his in Litteris non agatur nisi de gratiis quas tibi habeo agoque Maximas & plurimas pro tua erga me Benevolentia quam mihi continuo præstas & significas tuis pretiosis ac frequentibus donis, quorum omnium beneficiorum semper ero gratus ac Memor. Bene diuque vale Fautor benefice, mihique tui observantissimo favere perge.
Dabam Parisijs
die 26. Julij 1719. St. nov.

Translation

Pierre Varignon offers a grand salute
to that most noble Knight and learned man,
[Sir Isaac] Newton,
most worthy President of the English Royal Society

Most illustrious Sir,

I did not receive the packet of five copies of your wonderful *Opticks*, newly made public in Latin, which you sent me on 13 October 1718 O.S. (as I understood from your letter enclosed within it) until 6 June, N.S., last. At once, as you directed, I gave one of these copies from you to the Royal Academy of Sciences, a second to Mr Fontenelle and a third to Mr Monmort: Both of these return thanks to you and furthermore Mr Fontenelle thanks you particularly on behalf of the Academy, to which your gift was most welcome.

The fourth copy sent by you to me, I [have sent] to Johann Bernoulli, in the same spirit in which I sent [the version] written in English about a year ago:[2] namely for the sake of bringing about peace and concord between you and him; which I thought would be as welcome to you as to him. As soon as I had announced to him this gift from you to him, he wrote back that no news could be more welcome to him, and asked that I should let you know that he had not forgotten you. But the letter in which I thanked you had already been despatched. However, you may see how he thanks you for both gifts, after being advised of your second one, in a letter enclosed in this one of mine;[3] would that he may be restored into your former good opinion, which (upon his own evidence) he was reluctant to lose. However, I send this letter of his a little late in the day on account of a variety of urgent business by which I have been prevented from writing to you more promptly.

As for the fifth of the above copies of which, together with the fourth, your kindness made me a present, I decided to peruse it at any rate hastily before returning you thanks for both. But when I met its third Book with many new Queries[4] added to the old ones, and found that several of these were enlarged by many experiments, lest any feature of their novelty should escape me I read those old Queries again attentively with the new ones. These latter, like the former, I found enormously to increase the number of phenomena or motions [made] very manifest by very elegant experiments reported in both places, of the causes of which we are nevertheless entirely ignorant. Further, when I have read these [Queries] through I cannot see (nor could I see before) how anyone can suppose that you consider gravity to be among the essential attributes of bodies, since in Query 23, 1st edition, and now in Query 31, second edition, of Book III, you have written: 'What I call Attraction may be perform'd by impulse, or by some other means unknown to me, etc.' And thirty-two years ago in the first edition of your golden treatise on the mathematical principles, as now in the second edition of the same you have said, in Definition 8 of Book I, 'I use the words attraction, impulse or propensity of any sort towards a centre, indifferently, and promiscuously for one another.' But how

I am carried away! For nothing is in question in this letter except the very great and numerous feelings of gratitude which I have and offer to you on account of your kindness towards me which you continually show and indicate to me by your frequent precious gifts, of which great favours I will always be grateful and mindful. Good and lasting health be with you, kind patron, and may you continue to favour your most devoted [servant]

Paris, 26 July 1719, *N.S.*

NOTES

(1) Keynes MSS. 142(D); microfilm 1011. 26. This Letter, transmitting Letter 1320, was sent via De Moivre; see Letter 1320, note (2), p. 46.

(2) See Letter 1304.

(3) Letter 1320.

(4) Queries 17–24; see Letter 1304, note (3), p. 16. Varignon was familiar with Queries 25 to 31 already, as Quæstiones 17–23 of the first Latin edition. These new queries (Query 21 especially) revive Newton's ætherial hypotheses including the explanation of gravity as the pressure of an ætherial medium.

1324 JAMES STIRLING TO NEWTON
6 AUGUST 1719
From the holograph original in King's College Library, Cambridge[1]

Sir

I had the honour of your letter about five weeks after the date, as your generosity is infinitely above my merite, so I reackon my self ever bound to serve you to the outmost: and indeed a present from a person of such worth is more valued by me than tentimes the value from another. I humbly ask pardon for not returning my gratefull acknowledgments before now.[2] I wrote to Mr Desaguliers to make my excuse, while in the mean time I intended to send a supplement to the papers[3] I sent, but now I'm willing they be printed as they are, being at present taken up with my own affair here, wherewith I won't presume to trouble you, having sent Mr Desaguliers a full account thereof.[4]

I beg leave to let you know, that Mr Nicholas Bernoulli, proposed to me, to enquire into the curve which defines the resistances of a Pendulum. when the resistance is proportionall to the velocity.[5] I enquired into some of the most easy cases, and found that the Pendulum in the lowest point had no velocity, and consequently could perform but one half oscillation and then rest.[6] Bernoulli[7] had found that before, as also one Count Ricato,[8] which I understood, after I communicated to Bernoulli,[9] what occurred to me. Then he asked me, how in that hypothesis of resistance a pendulum could be said to

53

oscillat, since it only fell to the lowest point of the Cycloid and then rested. so I conjecture that his uncle[10] sets him on to see what he can pick out of your writings that may any ways be cavill'd against; for he has also been very busy in enquiring into some other parts of the Principles.

I humble beg pardon for this trouble, and prays God to prolong your daies, wishing that an opportunity shou'd offerr that I could demonstrate my gratefullness for the obligations you have been pleased to honour me w[i]t[h].

I am with the greatest respect

Sir

Venice 17 August   Your most humble
1719 N.St.   and most obedient
                                                 Servant
                                        JAMES STIRLING

P.S.

Mr Nicholus Bernoulli, as he hath been accused by Dr Keill of an ill will towards you, wrote you a letter some time ago to clear himself.[11] But having no return, desired me to assure you that what was printed in the Acta. Paris[ienses] relating to your 10. Prop. Lib 2. was writ before he had been in England, sent to his friend as his private opinion of the matter, and afterwards published without so much as his knowledge.[12] He is willing to make a full vindication of yourself as to that affair, whenever you'll please to desire it; he has laid the whole matter open to me, and if things are as he informs me, Dr Keill has been somewhat harsh in his Case. For my part I can witness that I never hear him mention your name without respect and honour. When he shewed me the Acta Eruditorum,[13] where his uncle has lately wrote against Dr Keill He shewed me that the Theorems there about quadratures are all corollarys from your Quadratures, and whereas Mr John Bernoulli, had said there that it did not appear by your construction of the Curve Prop. 4. Lib. 2. that the said Construction could be reduced to Logarithms, he presently shewed me Coroll. 2. of the said Proposition, where you shew how it is reduced to Logarithms,[14] and he said he wondered at his Uncles oversight. I find more modesty in him as to your affairs than could be expected from a young man, Nephew to one who is now become head of Mr Leibnitz's party. And among the many conferences I've had with him, I declare never to have heard a disrespectfull word from him of any of our Country but Dr Keill.

NOTES

(1) Keynes MS. 104; microfilm 931. 14. Printed in Brewster, *Memoirs*, II, pp. 516–17 and in Charles Tweedie, *James Stirling* (Oxford, 1922), pp. 11–12.

James Stirling (1692–1770) matriculated at Oxford in 1711, and entered Balliol College. He took a leading part in the political disturbances at the College, and his connections with the

Jacobites lost him his scholarship there when he refused to take oaths of allegiance. However, according to Tweedie, he was neither expelled from the University nor forced to flee the country. He was at this time invited to Italy to take up a professorship, probably at Padua, but refused the offer. In 1717 his *Lineæ Tertii Ordinis Newtonianæ* was published at Oxford, intended as a supplement to Newton's *Enumeratio Linearum Tertii Ordinis*. His movements at this period are uncertain, but from the present letter we know he was in Venice in 1719. According to a draft written by Newton many years later (see Brewster, *Memoirs*, II, p. 307, footnote) it was Conti who was responsible for 'sending Mr. Stirling to Italy, a person then unknown to me, to be ready to defend me there if I would have contributed to his maintenance', but Tweedie notes that it was actually Nicolas Tron, a Venetian ambassador at the English court (see Letter 1288, vol. VI) who invited Stirling to Italy.

Little is known of Stirling's stay in Italy (he came to be known as 'The Venetian'); he was back in Great Britain by 1724 and in London by 1725. There his reputation in English mathematical circles grew rapidly. He became a Fellow of the Royal Society in 1726. He returned to Scotland in 1735.

(2) We have not traced Newton's letter to Stirling. Tweedie (*op. cit.*, p. 12) assumes that Stirling had received financial aid from Newton, and that this is the 'generosity' referred to; the only corroborative evidence for this that we have found is the passage by Newton quoted in note (1) above. It seems just as likely that Stirling here is expressing gratitude merely for a letter from Newton, or for a copy of *Opticks*.

(3) Presumably his paper 'Methodus Differentialis Newtoniana illustrata,' which was read at the Royal Society on 18 June 1719, and printed in *Phil. Trans.* for November and December 1719, **30**, no. 362, 1050–71; it later became part of Stirling's *Methodus Differentialis, sive Tractatus de Summatione et Interpelatione Serierum Infinitarum* (London, 1730). No supplement was printed.

(4) J. T. Desaguliers (see Letter 1463) was at this time curator of experiments to the Royal Society. He had lectured on experimental philosophy at Oxford in 1712, so possibly Stirling had made his acquaintance there, and now used him as intermediary in his communications with the Royal Society.

(5) The extant correspondence between the nephew of both Jakob [I] and Johann [I] Bernoulli, Nikolaus [I] Bernoulli, and Stirling is printed in Tweedie, *op. cit.*, and includes a letter from Bernoulli to Stirling dated 29 April 1719, N.S., where Bernoulli poses the problem mentioned here.

(6) The results of Stirling's inquiries are lost. See Whiteside, *Mathematical Papers*, IV, pp. 441–2, note (4).

(7) Presumably Stirling here still means Nikolaus Bernoulli.

(8) Jacopo Francesco Riccati (1676–1754), was born in Venice, studied under the Jesuits, and obtained a Doctorate from Padua University in 1696. Despite several offers of teaching posts, he preferred to pursue his mathematical studies privately, having the financial means to do so. Two of his sons, Giordano and Vincenzo, were also highly respected mathematicians. Riccati is best known for his work on differential equations, in particular the equation which now bears his name (see M. Cantor, *Vorlesungen über Geschichte der Mathematik*, III (Leipzig, 1898), pp. 455–8).

Riccati discussed his work on the problem of a pendulum oscillating in a medium which resists in proportion to velocity, in a letter to Poleni, dated 18 March 1717, N.S. (See *Opere del conte Jacopo Riccati nobile Trevigiano*, III (Lucca, 1764), pp. 477–8.) He gave no details of his work, but wrote that he had studied the problem for mediums of both high and low viscosity, and suggested that Poleni pass the problem on to Johann Bernoulli. The details of Riccati's method were not printed until 1722, in vol. II of the Supplement to the *Giornale de' Letterati d'Italia* (see Riccati, *op. cit.*, pp. 376–95) in a paper entitled 'Sopra le leggi delle resistenze, con le quali i mezzi fluidi ritardano il moto de' corpi solidi'. There he refers briefly to work by Newton (in *Principia*, Book II, Prop. 26) and others, but rightly points out that they have produced no complete solutions to the problem. He describes Bernoulli's solution, but claims that his own is more elegant. He particularly emphasizes the critically damped and over-damped cases, since, he says, these occur very frequently in nature.

(9) Nikolaus [I] Bernoulli.

(10) Johann [I] Bernoulli, who wrote a paper, then unpublished, 'De oscillationibus penduli in medio quod resistit in ratione simplici velocitatis', (see Bernoulli, *Opera Omnia*, IV, pp. 374–7). The solution Bernoulli gives there, which covers lightly damped, critically damped and over-damped penduli, is certainly the one discussed by Riccati in his 1722 paper (see note (8)).

(11) The only extant letter from Nikolaus [I] Bernoulli to Newton, dated 20 May 1717 (Letter 1240, vol. VI) merely expresses Bernoulli's thanks for a gift of the second edition of the *Principia*. The letter referred to here is probably of a later date, written as a result of Nikolaus' argument with Keill referred to by Taylor in his letter to Keill of 26 April 1719 (Letter 1316, note (3)).

(12) For the correction to Book II, Prop. 10 see Letter 951*a*, vol. V. It is not known whether Nikolaus' 'Addition' to his uncle's paper was really published without his knowledge.

(13) The particular paper referred to is Johann Bernoulli's 'Responsio ad nonneminis provocationem', *Acta Eruditorum* for May 1719, pp. 216–26 (see Letter 1303, note (3)).

(14) See Whiteside, *Mathematical Papers*, VI, pp. 71–4, note (113).

1325 NEWTON TO THE TREASURY
12 AUGUST 1719
From the holograph draft in Trinity College Library, Cambridge[1]

To the Rt Honble the Lords Commissioners
of his Ma[jesty']s Treasury

May it please your Lordps

In answer to the Memorial of Mr Nicholson & Mr Briggs I humbly represent that as often as they have applied themselves to me I have told them that without a Warrant from the King I could not receive copper in blanks nor coin money with round edges for the people, & that without your Lordps Warrant I could not deliver Cutters to them. I have told them also that Mr Appleby & Mr Hines having all things ready for a triall, were to coin their five Tonns first; & that when I was ready for the Memorialists, I would give them notice, but it would take up some time first to prepare the Mint for a trial. There is no difficulty in rounding the edges of the Blanks, but the method we are bound to keep secret; & I have not delivered another Cutter to Mr Appleby & Mr Hines since the 3d of June last as is pretended tho I have a Warrant to justify me; nor are they Contractors but at discretion: nor have there been any delays but for want of Warrants which the Memorialists should have procured. But if the Memorialists instead of applying for Warrants have given themselves a great deal of trouble in solliciting me to act without any, if before they had Cutters to direct them in the measure of their work, if in opposition to the notice that I gave them that Appleby & Hines being first ready should have the first trial, they have built furnaces & prepared a sufficient quantity of copper ready for the Cutters & been at sufficient charges in preparing the same: they have done this contrary to my advice & without staying for sufficient authority, or knowing how to size their work, & can blame no body but themselves if they should lose their charges from acting in this manner.

In the last coinage of copper moneys the copper was worth about $7d\frac{1}{2}$ or $8d$ per Lwt,[2] & a pound weight was cut into $21\frac{1}{2}d$. The copper is now worth between $13\frac{1}{2}d$ & $14d$ per Lwt[2] & the coinage is difficulter & more chargeable & better performed, & yet the allowance for copper & coinage together is no greater then before, a pound weight being cut only into $23d$, whereof three half pence are reserved to the government. All the deficiency in the goodness of the copper hitherto complained of doth not amount to the forti[e]th part of the whole value of the copper wch is no more than the Remedy allowed in weight, & therefore would be within the Remedy in fineness if such a remedy could be stated by the Assays of copper. And for the form of the moneys it is before every bodies eyes, & cannot be judged of by a few fau[l]ty pieces culled out. For the nature of coinage is such that among the moneys newly coined there will be some pieces faulty in form. It always was & always will be so in the coinage of gold & silver, & the coinage of fine copper is more difficult. But it is the duty of the Moneyers to pick out the faulty pieces, & I have caused half a Tonn of such faulty copper moneys to be melted down again. If any faulty pieces escape the Moneyers we change them when they are brought back to us & coin them anew. And this has been the standing practise of the Mint time [out] of mind without any disparagement of the coinage of gold & silver. There will be accidents, but things are upon such a foot, that I can get nothing but discredit by coining the moneys ill.

Because I am accountable for the gold & silver moneys in weight & fineness, that coinage is performed by people under my direction: and his Ma[jes]ties Warrant has made me accountable also for the value of the copper moneys. And I humbly pray your Lordps that I may go on in the method wch is suitable to this Warrant, & in wch I have hitherto acted, which is of allowing me to try any mans copper, & to receive or reject it accordingly as I find it finer or coarser & manufactured better or worse, & to imploy those men first who serve me best.

To coin money of fine copper is a manufacture never before set up in England, & I have met with difficulties in setting it up for want of time to try experiments & settle things in the best manner before I began. For removing the main difficulties I perswaded Mr Appleby & Mr Hines to rent a water Mill, a horse Mill being too weak. And instead of nealing & cleaning the Barrs the Moneyers are now trying if they can learn to clean the blanks after nealing. And if the Refiner can but learn to keep the fumes of the sea-coal from his copper, the difficulties will be over. In the meantime, as I have brought the sizing of the gold & silver moneys to a much greater degree of exactness than ever was known before, & thereby saved some thousands of pounds to the government, & am not blameable for any accidental errors within the remedy of those moneys, & have found out a new assay for stopping the importation of copper not sufficiently refined or dammaged by the Sea-coal: so I hope that I shall not be blamed for want of better assays, or for such errors in the assays of copper as are only accidental & in value do not exceed the Remedies allowed in weight, especially while I am endeavouring to set up this coinage in the best manner.[3]

In the meantime it may be considered, that when this coinage was set on foot, it was recconed better to neale & clean the copper in the barrs then in the blanks; that the melting down of the blanks for an assay alters the goodness of the copper & thereby makes the assay less certain; that the Government pays no more for these moneys than she did for those in the reign of King William, & yet the copper & workmanship is much better, & within these seven years the price of copper is risen 3d per Lwt;[2] that no defect has hitherto been discovered in the value of the moneys wch exceeds the Remedy; that the persons hitherto complaining have wanted to be imployed; that they have not been able to discover any fraud in the management nor to help me to better assays; that their Proposals have hitherto tended either to lay aside the Assays or to weaken them or to enlarge the allowance for copper & coinage; & that the competition at present is between the Memorialists & the Moneyers about the form of the moneys & the brokage.

Your Lordps are impowered by his Ma[jesty']s Warrant to stop this coinage when you please, & when you please to stop it for any time I humbly pray that I may be authorized by your Lordps Warrant to stop it.

All wch is most humbly submitted to your Lordps great wisdome

Is. NEWTON

Mint Office
Aug. 12. 1719.

NOTES

(1) R. 4. 48, no. 10, signed and dated. This is so fair a copy, it appears that Newton intended to submit it as the letter, which we have not found in the Treasury papers. Since this paper is not endorsed, however, it cannot have been to the Treasury. An antecedent, much inter-lineated and altered draft is in Mint Papers, II, fo. 455. This, like the other two drafts next to be mentioned, is undated.

It is likely that, after preparing these lengthy explanations, Newton decided to be much more brief. One shorter version is in the Jewish National and University Library, Jerusalem, Yahuda Collection, Newton MS. 7(4). A still more brief letter is in private possession, and it may be of interest to reproduce it here for comparison (ignoring corrections etc.):

In answer to the Memorial of Mr Nicholson & Mr Briggs I humbly represent that I told them that without a Warrant from the King I could not receive copper in blanks, nor coin money with round edges for the people; that Mr Appleby & Mr Hines having all things ready for a triall were to coyne their five Tunns in the first place & that when I was ready for the Memorialists I would give them notice, but it would take up some time first to pre-pare for a tryall. There is no difficulty in rounding the edges of the blanks & I have not delivered another cutter to Mr Appleby & Mr Hines, & they are not contractors with your Lordps, but upon tryal & good behaviour. And if the Memorialists have built furnaces & prepared a sufficient quantity of copper ready for the cutters, & been at sufficient charges in preparing the same they have done it contrary to my advice & without staying for sufficient authority & can blame nobody but themselves if they should lose their charges.

Robert Nicholson and Thomas Briggs had evidently protested at the unique privilege granted to Appleby and Hines of supplying the Mint with copper. In June 1719 Newton met Nicholson at the Treasury, but as their (undated) memorial states, he continued to refuse metal which they were willing to supply, not in the form of fillets merely but as blanks with rounded edges, declaring that he had no Warrant authorizing such purchases (see P.R.O., T/1, 223, nos. 9 and 9a).

(2) Pound weight.

(3) The long draft in the Mint Papers shows some uninteresting variation from this para-graph. Newton at first intended to stop here, and began to write 'All wch is...' but went on with two further paragraphs, of which the former vanishes from the Trinity College draft, while the last is the same in both. In the deleted paragraph Newton emphasized his willing-ness to consider all technical improvements offered to him.

1326 FRANCIS HALL TO NEWTON
2 SEPTEMBER 1719
From the holograph original in King's College Library, Cambridge[1]

Septr. 2. 1719

Hond. Sr.

Mr Haynes[2] was with me yesterday as by appointment who told me he left the interest money with you which you put up with the Orders & Tallies,[3] it was part in paper & part in Gold & he believes that you will find it put up

59

together with them which if you do I beg You'l pay to the bearer hereof William Booth my servant & this shall be your discharge for so doing. I wish you all happiness, I think of going out of town to morrow, & begging pardon for this & other troubles do always remain, Hond. Sr, Yr

Most Obedient & Affect[ionat]e friend & humble Servant

F. HALL

To
Sr Isaac Newton Knt.
    humbly Present

NOTES

(1) Keynes MS. 148(C), 1; microfilm 1011. 32. For the writer see Letter 1308, and vol. VI, p. 445, note (2).

(2) See vol. VI, p. 177, note (1).

(3) Newton was an executor of the late Thomas Hall's will, and in consequence was involved in a curious legal wrangle to which this letter possibly relates, and which is more fully described in another of Newton's holograph drafts, also among the Keynes manuscripts. John Smith and his wife Elizabeth before their marriage in 1703 constituted a joint trust, one of the trustees being Newton's friend Thomas Hall. In 1711 the Smiths separated, the wife receiving a separate maintenance; but after her husband's death and that of Thomas Hall in February 1718 she claimed for herself a part of the trust fund left in the hands of Thomas Hall. Newton's draft noted that Hall's receipt for the sum from Smith appeared subsequently to have been cancelled, but added

> We have heard that the Tallies of the Annuities of [£]1100 per annum are in the hands of Mr [Francis] Hall the son of the late Mr [Thomas] Hall. But we [*Thomas Hall's executors*] are strangers to the Articles of separation, & remember not that any demand hath been made upon us for transferring the Annuities or Bank stock to the surviving Trustees of the Marriage Settlement. But we are ready to answer such demands as shall be made upon us so soon as they shall appear to be just & to submitt to ye Orders of this Honble Court.

These would seem to be the Tallies (with accrued interest) mentioned in this letter.

1327 PINCKNEY TO NEWTON
5 SEPTEMBER 1719
From the original in the Bodleian Library, Oxford[1]

5°: *Sept*: 1719

Sr Isaac

Justice Tracy[2] will be in Towne at his Chambers on Wensday next and towards the latter end next Week designes for Dorsett-Shire Therefore on Wensday Morn. or before I will take ye. freedome to Leave for or Deliver to you Mr. Tates Letter out of Leicestershire you gave me the other day that if

you please you may discourse with ye Judge on Wensday upon the contents
therein I p[re]sume to subscribe my Selfe

Hond. sr.

Yr. most Obet. humble Servt

CALVERLEY PINCKNEY

For
Sr. Isaac Newton att his
House In St. Martins Street on
  South side of
    Lester Feilds

NOTES

(1) Bodleian New College MS. 361, II, fo. 54; the writer was solicitor to the Mint, responsible
for the prosecution of currency offenders (see Letter 1296).

(2) Possibly Robert Tracy (1655–1735), of Gloucestershire, a Justice of the Common Pleas
since 1702.

1328 ALBERT HENRI DE SALLENGRE TO NEWTON
11 SEPTEMBER 1719
From the original in King's College Library, Cambridge[1]

Illustrissimo
Celeberrimo Viro
Isaco Newton
S.P.D.
A: H: de Sallengre

Iam pridem a Domino Des Maizeaux certior sum factus quod singulari tua
humanitate et beneficio inter eos sum collocatus, quorum nonnulli in proximo
Regiæ Societatis Conventu[2] Sociorum numero adscribentur, sed diuturno
morbo conflictatus non prius potui quam maximas tibi gratias agere pro in-
exspectato tanto favore, cujus memoria, dum vita suppetet, nunquam mihi
excidet. Licet autem probe conscius sim me illo honore minime esse dignum,
tamen nova illa dignitas maximum mihi erit incitamentum ad ulterius
provomenda studia mea, ut Regiæ Societati, si non decus addere possim,
saltem Dedecori non sim. Deum Oro ter Opt. Max. ut te bono Reipublicæ
Litterariæ cujus præcipuum decus et columen es, quam diutissime incolumem
servet. Vale Celeberrime Vir et me perpetuum nominis tui Cultorem puta.
Dabam Hagæ Comitum a.d. XXII. Septembris N.St. 1719

61

Translation

A. H. de Sallengre sends his greetings
to the most illustrious and celebrated [Sir] Isaac Newton

Some time ago I was informed by Mr Des Maizeaux that, by your extraordinary kindness and favour, I have been placed amongst those of whom several will be appointed to the number of Fellows at the next meeting of the Royal Society;[2] but, troubled with a long-lasting illness, I have not been able before to thank you for such an unexpected favour, the memory of which, while life remains, will never fail me. However, although I am fully aware that I am not in the least worthy of that honour, nevertheless that new position will be the greatest incitement to me for pushing forward further with my studies, so that at least I am not a disgrace to the Royal Society, even if I cannot add to its glory. I pray God, thrice greatest and most high, that he may keep you safe for a very long time, for the benefit of the learned world whose chief ornament and apex you are. Farewell, most celebrated Sir, and think me forever the worshipper of your name. The Hague, 22 September 1719, N.S.

NOTES

(1) Keynes MS. 93(C); microfilm 931.2. The writer, Albert Henri de Sallengre, a native of Holland, studied law at Leyden and was appointed counsellor to the Princess of Nassau-Orange and later Commissar of Finance of the States General. Besides these official posts, he was much occupied in editorial work; in particular he was one of the editors of the *Journal Literaire de la Haye*. He died in 1723. This letter was enclosed, with others addressed to Sloane and Halley, in a letter to Des Maizeaux; see British Museum, Birch MS. 4287, fo. 290 and Sloane MS. 4045, fo. 244.

(2) De Sallengre's name was put forward for possible election to the Society by Sir Hans Sloane at a meeting on 25 June 1719. His election took place after the summer recess, on 5 November.

1329 NEWTON TO VARIGNON
29 SEPTEMBER 1719
From the holograph draft in King's College Library, Cambridge[1]
Reply to Letter 1323

Literas tuas 26 Julij [N.S.] datas accepi, et gratias reddo tibi quammaximas quod exemplaria duo Optices ad D. Johannem Bernoulli meo nomine misisti, et eo pacto nos reconciliare conatus fueris: quod et fecisti ut ex literis ejus intelligo.[2] Nam D. Leibnitius Epistolis aliquot quas vidi, directis verbis affirmaverat D. Bernoullium authorem esse Epistolæ 7 Junij 1713 [N.S.][3] ad ipsum datæ et mox in Germania impressæ et per orbem literarium sparsæ.[4] Sed cum ex literis D. Bernoullij jam acceptis, intelligam *ipsum non fuisse authorem,* amicitiam ejus lubenter amplector et colo: et eo nomine literas inclusas ad ipsum scripsi, [5] quas oro ut ubi occasionem nactus fueris ad ipsum mittas. Oro etiam ut Academiæ vestræ gratias meas reddas ob munera historiæ suæ annuatim ad me missa, te curante: Sed et gratiæ meæ tibi ipsi debentur ob (La Connoissance des Temps) aliquoties ad me missum.

Inspiciendo literas quas olim a Dno Cotes, interea dum Editionem secundam Libri mei Principiorum curaret, accepi; animadverto quod schedæ primæ triginta septem (id est, usque ad paginam 296 inclusive) impressæ essent ante 30 Junij 1710 st. vet. id est antequam hæ lites incœperunt. [6] Et in eo tempore Prelum quievit per annum. Interea D. Cotes, in Epistola 21 Sept 1710 [7] ad me data, deduxit ex quantite aquæ dato tempore per datum foramen in fundo vasis factum effluente, juxta experimentum D. Mariotti (pag. 245 Traite du Movement des Eaux) sæpe et accurate repetitum, quod velocitas aquæ in ipso foramine ea sit quam corpus, cadendo a dimidia altitudine aquæ in vase stagnantis, acquirere potest. Et in alia epistola 5 Octob. 1710 [8] scripsit, se quoque experimentum idem bis cepisse eodem successu. Ego vero experimentum idem ceperam ante editionem primam hujus Libri. Et ex his omnibus Conclusio prædicta redditur certissima. Sed aqua post exitum acceleratur uti constat per alia experimenta a me sub initio anni 1711, [9] ac deinceps ab alijs, capta. Tandem mense Octobri anni 1712 ubi Liber totus usque ad pag 456 [10] inclusive impressa esset, D. Nicolaus Bernoulli me monuit quod error aliquis admissus fuisset in resolutione Prop. X Lib. II Edit 1, et Resolutionem subinde examinavi et correxi & correctam ei ostendi, & imprimi curavi non subdole sed eo cognoscente. [11] Cætera in lucem prodierunt uti fuerant ante has lites impressa Hujus Libri editio tertia forte lucem videbit, et in Exemplari quod hunc in finem correxi, [12] in fine Prop. XVII Lib. I addidi hæc verba *Nam si corpus revolvatur in Sectione Conica sic inventa, demonstratum est in Prop. XI, XII et XIII quod vis centripeta erit reciproce ut quadratum distantiæ corporis a centro.* [13] Cætera quæ correxi ad D. Bernoullium nil spectant. Si ansa aliqua adhuc maneat contendendi, rem totam judicio vel Orbis literarij vel Academiæ vestræ lubentissime permitto.

Præter verba quæ citasti, [14] extant alia in Scholio ad Prop. LXIX Lib. 1 Princip, [15] quibus clarissime patet me gravitatem nec corporibus essentialem nec sine impulsu Medij, olim statuisse. Quæ in Quæstionibus sub finem Optices in Editione secunda, circa causam Refractionis Reflexionis et Gravitatis addidi leviter attingebam in Editione prima Prop. XII Part. III Lib, II[.] Et inde etiam constat me ab initio nec vacuum absolutum statuisse nec Medium ætherium elasticum per cœlos universos diffusum negasse, cujus impetu corpora a motibus rectilineis detorqueri possint. [16]

Antequam literis tuis responderem cupiebam colloqui cum Dno. Keill qui

III Partie
II Discours
pag 421
Edit. ult.

longe aberat in agro Northamptoniensi; sed is jam in Urbem hanc redijt &
spero quod a litibus in posterum abstinebit, licet admodum irritatus. (17)

[In Libro tertio Principiorum non opus est ut vires centripetæ quibus corpora cælestia in orbibus suis retinentur, nominentur gravitas. Vocari possunt
vires cælorum; et probari potest quod vis cælestis qua Luna retinetur in Orbe,
si descendatur in Terram, æqualis evadat gravitati nostræ et æqualem corporum descensum efficiat, & quod similes vires in omnium Planetarum super-
ficiebus, descensum similem efficere possint. An vires quæ talem descensum
efficiant nominandæ sint gravitates, disputet qui volet. (18)]

Quæ sub finem libri D. Raphsoni ante triennium edita sunt, iterum impressa sunt idque in Hollandia. An lucem videbunt nondum scio. (19)

Cum D. Bernoullio lites ipse nondum habui, neque habebo; sed e contra
conabor, non tantum ut lites inter illum et amicos meos cessent, sed etiam ut
ejus amicitia fruar. Vale.

Translation

I have received your letter dated 26 July, [N.S.], and I return you the greatest
possible thanks because you sent two copies of the *Opticks* to Mr Johann Bernoulli on my
behalf, and in that way have made an attempt to reconcile us; and you have done so, as
I understand from his letter. (2) For Mr Leibniz had in express terms asserted in several
letters which I have seen that Mr Bernoulli is author of the letter to him dated 7 June
1713, [N.S.], (3) soon afterwards printed in Germany and circulated through the learned
world. (4) But since from the letter now received from M. Bernoulli, I understand that
he is not the author of it, I readily welcome and court his friendship, on which account I
have written the enclosed letter to him, (5) which I ask you to send him when an opportunity arises. I also ask you to give my thanks to your Academy for the present of its
Histoire sent to me each year by your care; but I owe to you yourself thanks for sending
me the *Connoissance des Temps* from time to time.

Looking over a letter which I formerly received from Mr Cotes while he was seeing
to the second edition of my book of Principles, I notice that the first thirty-seven sheets
(that is, up to p. 296 inclusive) were printed before 30 June 1710, O.S.; that is, before
these feuds began. (6) And at that time the press stopped for a year. In the interval
Mr Cotes, in a letter to me dated 21 September 1710, (7) deduced from the quantity of
water flowing in a given time through a given aperture made in the base of a vessel
(according to an experiment often and carefully repeated by Mr Mariotte (page 245
[of his] *Traité du Mouvement des Eaux*)) that the velocity of the water in that aperture is
that which a body can acquire by falling from half the height of the water standing in
the vessel. And in another letter of 5 October 1710 (8) he wrote that he had done the
same experiment twice with the same result. I too had in fact made this same experiment before the first edition of this book. And the aforesaid conclusion is most certainly
given by all these [experiments]. But the water is accelerated after leaving, as it appears
from other experiments made by me about the beginning of 1711, (9) and afterwards by

others. At last, by October 1712 when the whole book had been printed up to and including p. 456,[10] Mr Nikolaus Bernoulli advised me that some error had been passed
over in the proof of Proposition 10, Book II, first edition, and I examined the proof at
once and corrected it, and showed the correction to him, and saw to the printing of it,
not clandestinely but with his knowledge.[11] The rest was laid before the world as it
had been printed before these disputes. Perhaps a third edition of this book will appear,
and in the copy which I have corrected with this in mind,[12] I have added at the end of
Proposition 17, Book I, these words, 'For if the body be revolved in a conic section thus
discovered, it is demonstrated in Propositions 11, 12 and 13, that the centripetal force
will be reciprocally as the square of the distance of the body from the centre.'[13] The
remaining corrections I have made in no way pertain to Mr Bernoulli. If any excuse for
dispute still remains I most willingly submit the whole thing to the judgement of the
learned world or to that of your Academy.

Besides the words you have cited,[14] there are others in the Scholium to Proposition
69, Book I, *Principia*,[15] from which it is clear that I have not in the past claimed that
gravity is either essential to bodies or independent of the action of a medium. What I
have added in the Queries at the end of the *Opticks* in the second edition concerning the
cause of refraction, reflection and gravity, I touched on more lightly in the first edition,
Proposition 12, Part III, Book II. And there too it appears that I have neither insisted
upon an absolute vacuum nor denied an ætherial elastic medium diffused through the
entire heavens, by the action of which bodies can be deflected from their rectilinear
motion.[16]

I wished before I answered your letter to talk with Mr Keill, who has been away in
Northamptonshire for some time; but he has now returned to this city, and I hope he
will refrain from disputes in future, even though greatly provoked.[17]

[In the third Book of the *Principia*, it is not necessary that the centripetal forces by
which heavenly bodies are retained in their orbits should be called gravity. They could
be called forces of the heavens; and it can be proved that the heavenly force by which
the Moon is kept in orbit, if it penetrated to the Earth, would turn out to be equal to our
gravity, and would cause the same descent of bodies, and that similar forces at the surfaces of all the planets could cause similar descents. But that the forces which cause such
a descent should be called gravitational, let him dispute who will.[18]]

The things added at the end of Raphson's book three years ago are printed again in
Holland. But I do not know whether they have yet been published.[19]

I have hitherto entered into no dispute with Mr Bernoulli himself, nor shall I; but on
the contrary I will exert myself, not so much so that the dispute between my friends and
him should stop, but rather that I should have the advantage of his friendship. Farewell.

NOTES

(1) Keynes MS. 142(P); microfilm 1011.26. There are rougher Latin drafts in U.L.C.,
Add. 3968(42), fo. 596 and 615, and another Latin draft is printed in Rigaud, *Correspondence*,
II, pp. 436–7 from manuscripts in private possession. These manuscripts also include partial
English drafts, probably for the same letter; see note (6) below. We print here what seems to

us to be the fairest draft, but there is no means of telling which draft is closest to the version finally sent. We note major variants below.

In a letter to Bernoulli of 14 November 1719, N.S. (to be printed in the Bernoulli Edition) Varignon gives the date attached to the present letter and quotes from it extensively. He explains that it was accompanied by Letter 1329a, and also by letters from De Moivre to Bernoulli and himself, in the latter of which De Moivre reported Newton's desire to settle the quarrel with Bernoulli and his opposition to Keill's wish to publish in the *Philosopihcal Transactions* his 'Lettre à Bernoulli' (see vol. VI, p. 386, note (4)). Varignon again quoted from this letter in writing to Bernoulli on 17 July 1720, N.S.

Edleston (*Correspondence*, p. 187), having access only to the draft printed in Rigaud, *op. cit.*, assumed the letter to be addressed to Monmort.

(2) Letter 1320.

(3) Letter 1004 as printed in the *Charta Volans* (Number 1009, vol. VI). By now Newton had not only seen the letter to the *Nouvelles Litteraires* (Letter 1172, vol. VI), but also, in the proofs of Des Maizeaux's *Recueil*, Leibniz's letters to the Baroness von Kilmansegge and to Count Bothmar (see Letter 1203 and Letter 1295, note (3), vol. VI, p. 459).

(4) The two drafts in U.L.C. here differ considerably from the draft we print, continuing at length about the use of fluxions in the *Principia*; thus Add. 3948(42), fo. 596, continues (after '...literarium sparsæ'):

Et author hujus Epistolæ me plagiarij insimulat, quasi olim de calculo fluxionum et fluentium ne quidem somniassem cum in Epistolis in Commercio Epistolico editis nullæ occurrant literæ punctæ, uti nec in Principijs Naturæ Mathematicis ubi frequens erat occasio calculo fluxionum utendi: cum tamen in Lemmate secundo libri secundi Principiorum et Introductione ad librum de Quadraturis Elementa methodi fluxionum demonst[r]antur et methodus ipsa doceatur & exemplis illustretur nullis literis pun[c]tatis adhibitis. Affirmat etiam quod prima vice hæ literæ punctatæ comparuerunt in tertio volumine Operum Wallisij, multis annis postquam Calculus differentialis jam ubique locorum invaluisset: cum tamen comparuerint in volumine secundo operum ejus, desumptæ utique ex literis meis anno 1692 mense [ ] ad ipsum scriptis id est antequam calculus differentialis ubique celebrari cœpit.

(And the author of this letter accuses me of plagiary, just as if I did not in the past so much as dream about the calculus of fluxions and fluents, since no pricked letters occcur in the letters published in the *Commercium Epistolicum*, nor in the *Principia Mathematica Naturæ* where there was frequent occasion for using the calculus of fluxions: although in Lemma 2 of the second book of the *Principia* and in the Introduction to the book *De Quadratura Curvarum* the elements of the method of fluxions were demonstrated and the method itself taught and illustrated by examples, without the use of pricked letters. He even states that these pricked letters appeared for the first time in the third volume [1699] of Wallis's *Opera*, many years after the differential calculus had already proved itself everywhere: yet although they appeared in the second volume of his *Opera* [1693], they were assuredly extracted from my letter written to him in 1692 in the month of [   ] that is before the differential calculus began to become known everywhere.)

For Newton's letter to Wallis, see Letter X. 398. 2, note (5), p. 396.

The sense of this passage (which is in square brackets in the draft, indicating intended deletion) is repeated, but with considerably different wording in an alternative draft (Add. 3968(42), fo. 615). The following additional passage of self-justification is also found on that leaf:

Et quod Regulam circa gradus ulteriores differentiarum falsam dedi, et recta methodus differentiandi differentialia mihi non innotuit longo tempore postquam alijs fuisset familiaris cum tamen Regula circa gradus ulteriores in Propositione prima libri de Quadraturis a me data, et verissima sit et simplicissima et generalissima & omnium prima in lucem prodijt, (videlicet in secundo volumine operum Wallisij ineunte anno 1693) et Propositio illa in Epistola mea anno 1676 Octob 24 ad Leibnitium missa ijsdem syllabis expressa habeatur. Quod ea omnia quæ in Commercio Epistolico citantur ad methodum fluxionum nil spectant ob defectum literarium punctatarum cum tamen in Lemmate secundo libri secundi principiorum elementa hujus methodi verbis expressis proponam...

(And [he also asserts] that I gave a false Rule for the higher degrees of differences and [that] the true method of differentiating differentials was unknown to me until a long time after it was familiar to others and yet the Rule for higher degrees given by me in the first proposition of the book of Quadratures is both very true, very simple and very general, and was the first to be published (namely in the second volume of Wallis' *Opera* at the beginning of 1693) and that proposition may be found expressed in the very same words in my letter to Leibniz of 24 October 1676. That all those pieces quoted in the *Commercium Epistolicum* are irrelevant to the method of fluxions because of the absence [from them] of pricked letters and yet in Lemma 2, Book 2 of the *Principia* I propound the elements of this method in express terms...)

(5) Letter 1324a. Can Newton really have believed in Bernoulli's innocence? It was obviously in the interests of reconciliation that he thought better of such draft passages as those in note (4) above.

(6) See Letter 793, vol. v. In what follows Newton does not confine himself to replying to Bernoulli's Letter 1320, but answers a number of points raised in Bernoulli's 'Epistola pro Eminente Mathematico' (see Letter 1196, vol. vi) concerning corrections made in the second edition of the *Principia*. The 'Epistola' had come to Newton's notice in May 1717 (see Letter 1239, vol. vi), but Newton had hitherto kept silent, publicly at least, about his reactions to the paper. Possibly he had by now also seen Kruse's 'Responsio ad...Johannis Keil defensionem', *Acta Eruditorum* for October 1718, pp. 454–66, which covered much of the same ground (see Letter 1321, note (2)).

(7) Letter 805, vol. v. Newton's marginal note indicates that he had now seen the *Oeuvres de Mr. Mariotte* (Leyden, 1717), as well as the 1700 edition of the *Traité du Mouvement des Eaux*, which Cotes had used in 1710. In the 'Epistola pro Eminente Mathematico', *Acta Eruditorum* for July 1716, p. 310, Bernoulli had signalized Hermann's note (in his *Phoronomia sive de Viribus et Motibus Solidorum et Fluidorum Libri Duo* (Amsterdam, 1716), p. 394) that Newton, in the first edition of the *Principia*, Book ii, Prop. 37, wrote that 'aquam ea cum velocitate erumpere ex vasis, qua motu suo in altum converso ad dimidiam altitudinem aquae supra foramen evehi possit' ('water flows from a vessel with the velocity by which, changed into its own motion upwards, it could be raised to half the height of the water above the aperture'). Bernoulli there claimed that he himself, over four years ago (1712) had explained why in fact the velocity of efflux is greater than this theoretical value. Newton, in the present letter, shows from his correspondence with Cotes how he did experimental work, and discovered the 'vena contracta' *before* Bernoulli published his own results.

(8) Letter 809, vol. v.

(9) Newton described the experiment in a letter to Cotes of 24 March 1711 (Letter 826, vol. v).

(10) See Letters 956 and 958, vol. v. The printing of Proposition 36, Book ii, was resumed (after the year's delay mentioned by Newton) in June 1711 (see vol. v, p. 166).

(11) See Letter 951*a*, vol. v, for the error in Book II, Prop. 10, and Letter 961 for the re-printing of the Proposition. Newton may have been quite open (in private) about the correction, but he failed to make any public acknowledgement to Bernoulli in the second edition of the *Principia*.

An English draft in Newton's hand, in private possession, relates to the rest of this paragraph, and was possibly intended for the same letter. It reads as follows:

By a parcel of Letters wch passed between me & Mr Cotes while he was printing the second Edition of the Principles, I find that the first 28 sheets of that Edition ending with page 224, were printed off before April 15 1710 st.vet. And this I note that you may know that the alterations made in those sheets had no relation to the disputes about the method of fluxions but preceded them all. In order to a third Edition I have added to the XVIIth Proposition Lib. 1 the following words. *Nam si corpus in his casibus revolvatur in Conica Sectione sic inventa, demonstratum est in Prop. XI, XII, XIII quod vis centripeta erit reciproce ut quadratum distantiæ corporis a centro virium S, ideoque Linea quæsita recte exhibetur. Q.E.F.* But I can strike them out if you think it better to omit them.

When Mr N. Bernoulli was in London (wch was in Autumn 1712) & told me that there was a fault in the resolution of Prob. III, Lib. II, & that he took it [to] be a mistak[e] in second differences; I answered that I would examine it & toold [*sic*] him the next time I saw him (wch was within a day or two) that I had examined it & there was a mistake, but it lay in drawing the tangent of the Arch *lC* from the wrong end of the arch [*see the figure from the first edition printed in* Koyré and Cohen, *Principia*, I, p. 381]. For the tangents of both arches *lC* & *CG* should have been drawn the same way with the motion of the body because they represent the moments of the motion. I gave him also the scheme set right & the calculation suited thereunto, & added that I would cause that sheet to be reprinted. The Book of Principles was writ in about 17 or 18 months, whereof about two months were taken up with journeys, & the MS was sent to ye R.S. in spring 1686; & the shortness of the time in which I wrote it, makes me not ashamed of having committing [*read* committed] some faults. These things I men[tion]

The offer to strike out the quoted Latin sentence is curious. Did Newton besides seeking to defer to Varignon's wishes mean to insure that Bernoulli made no further complaint that Newton had benefitted from his advice without acknowledgement?

The concluding passage, indicating the period taken for composing the *Principia*, has been often quoted; see D. T. Whiteside, *Journal for the History of Astronomy*, **1** (1970), 131, note (4); Cohen, *Introduction*, p. 68; and Rigaud, *Essay*, p. 92.

(12) It is interesting that Newton tells us he had already corrected copies of the second edition in preparation for the production of a third; it has never been clear exactly when Newton began work on the third edition (see Cohen, *Introduction*, p. 258), but Newton's statement here indicates that he had made some progress at least by the end of 1719; see also the extract in note (11) above.

(13) See Koyré and Cohen, *Principia*, I, p. 132. Newton clearly had in mind earlier arguments with Bernoulli over the solution of the inverse problem of central forces (see Letter 1023, note (3) and Letter 1165, note (2), (vol. VI, pp. 38–9 and 246)). However it is difficult to see why Newton felt that the addition of this sentence to Proposition 17 in any way clarified the matter.

(14) In Letter 1323, *ad finem*, Varignon had commented upon Newton's discussion of gravity in Query 31 of the second edition of the *Opticks*.

(15) See Koyré and Cohen, *Principia*, I, p. 298.

(16) Proposition 12 (p. 78, first edition) stated that 'every ray of light in its passage through any refracting surface is put into a certain transient constitution or state, which in the progress of the ray returns at equal intervals, and disposes the ray at every return to be easily transmitted through the next refracting surface, and between the returns to be easily reflected by it'. He used this Proposition to explain a number of interference phenomena; but in his discussion of the possible explanation of these 'fits of easy reflexion and transmission' he makes no mention of gravity. He puts forward the possibility that they may be due to vibrations in the refracting or reflecting medium. In Queries 17–24 of the second edition he discusses the possible nature of this 'Ætherial Medium', and how refraction may be caused by it. In Query 21 he suggests it as a possible explanation of gravity.

(17) Newton's expressed wish that Keill should refrain from further dispute seems in this case to be consistent with his genuine feelings; he had discouraged Keill from publishing his 'Lettre à Bernoulli' (see vol. VI, p. 386, note (4) and Letter 1269) and apparently took no interest in his later *Epistola ad Bernoulli*, an answer to Kruse's paper (see Letter 1321, note (3)).

(18) The whole of this paragraph is struck out in the draft, and so presumably was omitted from the letter.

(19) Newton refers to the Appendix to Raphson's *History of Fluxions*; the letters between Conti, Leibniz and Newton contained there were reprinted in Des Maizeaux's *Recueil*, of which Newton had already received proofs, but which was not to be published until 1720. Newton himself was partly responsible for the delay; see Letter 1330, note (2), p. 73.

1329*a* NEWTON TO J. BERNOULLI

29 SEPTEMBER 1719

From the holograph draft in the University Library, Cambridge[1]
Reply to Letter 1320; for the answer see Letter 1332

Vir Celeberrime

Cum primum Literas tuas ad me mediante Dno Abbate Varignone missas acceperam, et ex ijs intellexeram te non esse authorem Epistolæ cujusdam ad D. Leibnitium 7 Junij 1713 [N.S.] datæ,[2] in animum statim induxi me non tantum lites mathematicas nuper commotas negligere velle (id enim prius feceram) sed etiam amicitiam tuam colere et ob ingentia tua in rem mathematicam merita magni æstimare. Famam apud exteras gentes nunquam captavi, sed nomen tamen probitatis, quod auctor illius Epistolæ, quasi autoritate magni alicujus Judicis, convellere conatus est, salvum esse cupio. Studijs Mathematicis jam senex minime delector, neque opinionibus per orbem propagandis operam unquam dedi sed caveo potius ne earum gratia disputationibus involvari.[3] Nam lites semper odi. Humanum fuit et gratias meretur quod libros Optices ad te missos benigne accipere dignatus fueris [:] Et hoc etiam nomine me tibi devinxisti, et humanitatem tuam amicitia mutua rependere conabor. Quæ sub finem libri D. Raphson ante triennium impressa fuerunt, iterum impressa sunt idque in Hollandia una cum nonnullis D.

69

Leibnitij Epistolis in quibus affirmat te authorem esse Epistolæ prædictæ.[4] Si ista lucem tandem videant (quod minime cupio)[5] spero quod tibi minime nocebunt cum non sis Auctor ille. In editione secunda Libri mei Principiorum postulabat D. Cotes ut Corol. 1 Prop. XIII Lib. 1 demonstratione munirem, et ea occasione Corollarium illud verbis nonnullis auxi:[6] SED hoc factum est antequam hæ lites cœperunt. Nam schedæ primæ viginti octo illius editionis, (id est usque ad pag. 224 inclusive) impressæ fuerint ante 13 Apr. 1710,[7] et schedæ primæ triginta septima (id est usque ad pag 296 inclusive) impressæ fuerunt ante 30 Junij 1710,[8] et prelum subinde quievit usque ad mensem Junium anni proxime sequentis, ut ex Literis Dni Cotes eo tempore ad me missis et adhuc asservatis intelligo.[9] Scheda igitur septima in qua Corollarium illud extat, impressa fuit anno 1709.[10] Et hoc annoto ut intelligas me animo candido Corollarium illud auxisse et hactenus nullas tecum lites agitasse. Litibus autem componendis quas cum amicis meas habuisti, quantum in meis est operam dabo. Vale.

Translation

Most famous Sir,

When first I received your letter sent to me through the Abbé Varignon, and understood from it that you were not the author of a certain letter to Mr Leibniz dated 7 June 1713 [N.S.],[2] at once I made up my mind not only to disregard the recently stirred-up mathematical feuds (for I had done that before) but also to court your friendship and to esteem [you] greatly on account of your enormous merit in mathematical matters. I have never sought fame among foreign nations but nevertheless I desire that my reputation for honesty be preserved, which the author of that letter (writing as though with the authority of some great judge) has endeavoured to wrest from me. Now that I am old I take very little pleasure in mathematical studies, nor have I ever taken the trouble of spreading opinions through the world,[3] but rather I take care not to allow them to involve me in wrangles. For I have always hated disputes. It was kind, and merits thanks, that you deigned to receive courteously the books on optics sent to you: and for this reason too you have placed me under an obligation to you, and I will try to repay your kindness with mutual friendship. What was printed three years ago at the end of Mr Raphson's book has been printed again in Holland, this time with several of Mr Leibniz's letters in which he affirms that you are the author of the letter referred to above.[4] If these should ever be published (which I do not in the least wish)[5] I hope they will do you no harm at all, since you are not that author. In the second edition of my book of *Principles* Mr Cotes requested that I should provide Corol. 1, Prop. 13, Book I, with a demonstration, and on that occasion I enlarged that corollary with a few words.[6] BUT this was done before these disputes began. For the first 28 sheets of that edition (that is up to p. 224 inclusive) were printed before 13 April 1710,[7] and the first 37 sheets (that is up to p. 296 inclusive) were printed before 30 June 1710,[8] and after that the press was at a stop until June of the next following year, as I gather from the letters of Mr Cotes, sent to me at that time and still preserved. [9] Therefore the seventh sheet on which that Corollary lies, was printed in 1709. [10] And this I mention in order that you should know that I enlarged that Corollary with a clear conscience, and to this day have stirred up no quarrels with you. Moreover, I shall make it my business to settle the arguments which you have with my friends, as far as in me lies. Farewell.

NOTES

(1) Add. 3968, fo. 614. There is an earlier, rougher draft at fo. 615, which contains long cancelled passages not in the fairer version. Brewster gives a partial translation in *Memoirs*, II, p. 72. Both drafts are undated, but the letter was enclosed with Newton's letter to Varignon (Letter 1329).

(2) That is, the letter printed in the *Charta Volans* (see Number 1009, vol. VI). Bernoulli *was*, of course, its author.

(3) Had Newton forgotten (or chosen to ignore) his anonymous involvement in the *Commercium Epistolicum*, in many of Keill's outpourings, and in the printing of the Appendix to Raphson's *History of Fluxions*?

(4) Newton refers to Des Maizeaux's *Recueil* (see vol. VI, p. 457 note (1)) which included Leibniz's letters to the Baroness von Kilmansegge and to Count Bothmar, where he reveals Bernoulli's authorship of the letter quoted in the *Charta Volans*. The *Recueil* was published in 1720. For the printing of the Appendix to Raphson's *History of Fluxions*, which took place in 1718, not 1716 as suggested here, see vol. VI, p. 254, note (2).

(5) It is hard to say whether Newton meant 'minime cupio' or 'nollem' to stand; the sense is the same.

(6) The corollary concerns the inverse problem of central forces; it is curious that Newton should mention this addition to the *Principia* to Bernoulli, but not to Varignon—although he did apparently mention to the latter another, later amendment concerning the same problem (see Letter 1329, note (13)). The extant correspondence between Cotes and Newton does not indicate that the addition mentioned here came from a suggestion of Cotes; Newton himself proposes it (see vol. V, p. 6, note (5)). The intention of the addition was to provide a proof of the inverse problem of central forces in the special case of an inverse square law of force (see also D. T. Whiteside, *Journal for the History of Astronomy*, 1 (1970), 134, note (33)).

(7) See vol. V, p. 27, note (2).

(8) See vol. V, p. 55, note (2).

(9) Newton sent Cotes the next part of the copy in September 1710 (see Letter 805, vol. V) but because of difficulties over Mariotte's experiment and the theory of efflux of fluids, the copy was not ready for the printer until June 1711 (see Letter 854, vol. V). Newton does not mention here why he is interested in the dating of *this* part of the printing, although he had done so in his letter to Varignon (see Letter 1329, notes (7), (8) and (9)).

(10) This is surely a *non sequitur*; all Newton has proved is that the seventh sheet was printed off *before* April, 1710. This, however, is quite sufficient to show that the addition to Proposition 13 was made before Bernoulli's commentaries on the matter were published.

1329*b* NEWTON TO FONTENELLE
?AUTUMN 1719
From the holograph draft in the University Library, Cambridge[1]

Vir Dignissime Illustrissime

Quod Academia vestra munusculum meum Libri Optices[2] benigne accipere dignatus est, gratissimum fuit ex literis tuis intelligere, et gaudium auxit plurimum quod Editio prima ipsis antea innotuisset & quod jam scire non dedignarentur quid additum sit in hac nova editione. Philosophiam experimentalem hic prosequor tanquam dignam quæ Philosophia vocetur, & philosophiam hypothet[ic]am tracto non ut scientiam sed per modum questionum.[3] Et quæ in hac Editione addidi sunt posterioris generis. Hac occasione gratias tibi persolvere debeo quam maximas ob honores quos in me contulisti in elogio D. Leibnitij.[4] Et[5] quo minus comemorationem tam honorificam a te merui, eo magis tibi devinxisti Domine

> Servum tuum humillimum et
> maxime obedientum
> I. N.

Translation

Most noble and illustrious Sir,

It is most pleasing to find from your letter that your Academy is good enough to accept my small gift of the book of *Opticks*,[2] and [my] pleasure has been increased because the first edition had been known to them before, and now they have not been above learning what has been added in this new edition. Here I cultivate the experimental philosophy as that which is worthy to be called philosophy, and I consider hypothetical philosophy not as knowledge but by means of queries. And the matters I have added to this edition are of the latter type.[3] On this occasion I should give you due thanks for the honours which you have conferred upon me in the 'Éloge' of Mr Leibniz.[4] And[5] the less deserving I am of that honourable notice you gave me, the greater the obligation that you, Sir, have laid upon

> Your humble and
> most obedient servant
> I[SAAC] N[EWTON]

NOTES

(1) Add. 3968, fo. 615 (bottom), on the same page as a draft for Letter 1329, and clearly composed after it. Possibly the letter, if sent, was enclosed with Letter 1329 to Varignon.

(2) We learn from Letter 1323 that Newton had presented the Royal Academy of Sciences with a copy of the second edition of Newton's *Optice* (London, 1719).

(3) Newton's comments seem to form a reply to the closing paragraph of Varignon's Letter 1323; perhaps Fontenelle had written to him in the same vein.

In U.L.C., Add. 3970, fos. 234–301, are many drafts of the new queries for the 1717 English edition. These drafts include a number of extremely interesting Queries which were never in fact printed in the *Opticks*; these relate to Newton's speculations about the porosity of transparent substances, cohesion, and electrical attraction, and also about the existence of a universal 'sensible potent elastic spirit' to explain the phenomena not only of light and electricity but also vegetation and generation. They also include drafts of the new Queries which were printed (see Letter 1304, note (3)), and in particular we find a number of versions of the passage in Query 31 where Newton speaks of experimental philosophy and the exclusion of hypotheses from it (see *Opticks*, Dover edition (New York, 1952), p. 404).

(4) See Letters 1298 and 1305.

(5) On the same folio Newton has drafted an alternative ending for the letter which includes a mention of the calculus dispute: 'Fundamentum posuisti terminandi lites circa methodos novas Analyticas & spero quod cito terminabuntur. Quo minus autem hæc a te merui, eo magis tibi devinxisti.' ('I have proposed a basis for finishing disputes about the new analytical methods, and hope that these will quickly be finished. The less deserving I am, the more am I obliged to you.')

1330 NEWTON TO [DES MAIZEAUX]
NOVEMBER 1719
From a holograph draft in the University Library, Cambridge[1]

Sr

It requiring some time to write Letters receive answers & do what else you & I were discoursing this morning before ye publishing of the Letters of Mr Leibnitz: I beg the favour to signify to your Bookseller that if he pleases to deferr publishing them till Lady [Day] next I will give him twelve Guineas as a recompense for the loss of his time.[2]

I am

Your humble Servant
Is. NEWTON.

NOTES

(1) Add. 3968(38), fo. 570.

(2) The context of this letter is supplied by the extensive correspondence between Des Maizeaux and Henri du Sauzet, the publisher of the *Recueil* (see British Museum, Birch MSS. 4287 and 4288). From this we learn that, some time before October 1719, Des Maizeaux decided to include in the *Recueil* the exchange of letters between Leibniz, Chamberlayne, and Newton (Letters 1062, 1072 and 1101, vol. VI). Then, presumably as a result of discussion with Newton, he changed his mind, but Du Sauzet wrote to him on 17 October 1719, N.S., 'Vous m'avertissez trop tard de ne point imprimer les Lettres de Mrs Leibniz et Newton à Mr Cham-

berlaine; je les ai mises à la fin de la dispute sur le Calcul differentiel, comme vous l'aviez marqué positivement, lorsque vous m'ecrivites de les ôter de l'endroit où vous les aviez placées...' (Birch MSS. 4288, fo. 33). He must have begged Du Sauzet yet again not to print them for in a letter of 1 November 1719, N.S., Du Sauzet stresses adamantly that he will not oblige in this way, and adds, 'Je suis faché que Mr. Newton ne soit pas content, et je croyais que vous agissiez de concert avec lui' (Birch MSS. 4288, fo. 34). Des Maizeaux must then have visited Newton, and received his letter suggesting that Du Sauzet might be induced by financial means to delay publication, for in his next letter to Des Maizeaux, dated 1 December 1719 N.S., Du Sauzet writes, 'Je reponds sans perdre tems à la lettre que vous m'avez fait l'honneur de m'écrire le 24 Novembre, dans laquelle vous me parlez de la proposition que Monsieur le Chevalier Newton vous a chargé de me faire de sa part; savoir qu'il me donneroit douze Guinées, si je voulois retarder la publication de mon Recueil jusqu'à la Nôtre Dame de Mars prochaine, ayant quelques pages à ajoûter aux feuilles qui le regardent. Je vous prie, Monsieur, d'assurer de mon respect Monsieur le Chevalier Newton, et de lui faire savoir que j'accepte sa proposition, esperant qu'il consentira que je publie le livre lorsque le terme qu'il demande sera expiré...' (Birch MSS. 4288, fo. 41). Later (see *ibid.*, fo. 42) the sum was raised to fifteen guineas.

It is not clear what Newton intended to add to this part of the *Recueil*; the letters were published without any additional commentary. Possibly Newton's real reason for delaying publication was his continued doubt over Johann Bernoulli's true rôle in the controversy since receiving, in June, Bernoulli's Letter 1320 disclaiming responsibility for the letter printed anonymously in the *Charta Volans* (Number 1009, vol. VI). On the other hand the letters to be printed in the *Recueil* clearly indicated Bernoulli as that anonymous writer. Des Maizeaux later assured Varignon that he had had no hint of Bernoulli's 'innocence' until the *Recueil* was almost finished (see Letter 1359). When the *Recueil* was finally published, a small addition to the Preface and an *erratum* to Newton's 'Observations' were inserted, presumably at Newton's behest, to indicate renewed doubts over the authorship of the letter.

1331 CATHERINE CONDUITT TO NEWTON
16 NOVEMBER 1719
From the original in King's College Library, Cambridge[1]

Cranbury, Nov. 16. 1719[2]

[I must] not omit thanking you for your letter tho' I have no [paper] fit to write upon. I am very much asshamed of the [trouble the] servant made whom I directed to carry no other message [to you than] an inquiry after your health which I bid him send [me an ac]count of. I am extreamly concerned for the loss of my nephew,[3] he was a very hopeful youth and just provided for. I hope soon to have the pleasure of paying my duty to you Mr. Conduitt begs you will accept of his. I am Sr

Yours most dutyfull neice
and most obliged servant
C CONDUITT.

NOTES

(1) Keynes MS. 125; microfilm 1011.18. The left-hand edge of the paper is torn, with the loss of some words.

(2) Catherine Barton had married John Conduitt (1688–1737) on 26 August 1717 (see vol. VI, p. xxi, and Appendix II, Newton's Genealogy). The early life of Conduitt is not well known, nor the manner of his acquaintance with Newton and his niece, who was nine years older than himself. He was a Hampshire gentleman, with a seat at Cranbury Park, south-west of Winchester, whence this letter was written. He had been educated at Trinity College, Cambridge, and this fact may conceivably have furnished his first link with Newton. He is said to have travelled on the continent after leaving Cambridge (in the middle of the War of the Spanish Succession?) and then to have served in the British army in Portugal (1711–12). Nothing more is known of him before his marriage, but Mrs Conduitt's one surviving letter to Swift renders it impossible that her husband was one of that witty, Tory circle. As is well known, Conduitt succeeded Newton as Master of the Mint, like him wrote on financial questions, and collected memoirs of his great connection.

(3) There is no evidence as to the identity of this young relative; possibly even Newton Chapman (see Letter 1314) might count as a 'nephew'.

1332   J. BERNOULLI TO NEWTON
10 DECEMBER 1719

From the holograph original in Yale University Library[1]
Reply to Letter 1329a; for the answer see Letter 1333

Prænobili ac toto Orbe
Celeberrimo Viro
D. Isaaco Newtono
S.P.D.
Joh. Bernoulli

Litteræ Tuæ insigni voluptate me affecerunt Vir Illustrissime; ex iis intellexi Te neglectis litibus mathematicis eadem me prosequi benevolentia et amicitia, qua me olim dignatus fueras: Facis certe prout decet Virum candidum et generosum, qui non facile patitur sibi eos denigrari, quos amore suo dignos judicat. Qualis sit epistola illa, de qua dicis quod sit 7 Junii 1713 [N.S.] data ad D. Leibnitium, mihi non constat: Non memini ad illum eo die me scripsisse, non tamen omnino negaverim, quandoquidem non omnium epistolarum a me scriptarum apographa retinui. Quodsi fortassis inter innumeras quas Ipsi exaraveram una reperiretur, quæ dictum diem et annum præ se ferrent, pro certo asseverare ausim, nihil in ea contineri quod probitatis nomen Tuum ullo modo convellat, neque me unquam ipsi veniam dedisse, ut quasdam ex Epistolis meis in publicum ederet, et talem imprimis quæ Tibi etsi contra spem

et voluntatem meam, non arrideret; Quocirca denuo Te rogo Vir Illustr.
velis Tibi persuasum habere, mentem mihi nunquam fuisse aliter de Te loqui
quam de Viro summo, nedum existimationem Tuam vel probitatem sugillare.
Absit ut dicam, Te famam apud exteras Gentes captasse, spero tamen, Te non
respuere elogia a nobis ultro oblata, utpote sincera et Te digna, atque adeo
magis acceptanda quam quæ ex immoderato partium studio offeruntur. Quod
Tibi jam seni (cui incolumitatem per novum quem propediem auspicabimur
annum et per multos secuturos ex animo apprecor) non liceat studiis mathematicis incumbere, acerbe dolebit Orbis eruditus, quem hucusque ditasti tot
stupendis inventis. Ego quidem nondum senex, ad senium tamen vergo,[2]
aliisque distringor negotiis, ut nec mihi amplius fas sit rei mathematicæ tam
sedulam operam dare, uti solebam. Quod memoras Vir Amplissime de Libro
Raphsoni,[3] eum scil. iterum impressum esse cum nonnullis Leibnitii epistolis,
in quibus affirmet, me Auctorem esse prædictae epistolæ, (quæ quid contineat
probitati Tuæ injurium hariolari non possum) hoc certe liti sopiendæ non
conducit; ipsum vero Librum Raphsoni fortasse nunquam videbo, quia
ejusmodi libri ex Hollandia huc raro deferuntur. Hoc interim considerari a
Vestratibus vellem, si per testimonia certandum esset, melius id fieri addu-
cendo alias epistolas quam a Leibnitio scriptas, quippe qui in propria causa
non haberi potest pro idoneo teste: Sunt mihi epistolæ Virorum quorundam
doctorum ex Nationibus nullam in hac lite nationali partem habentibus, quas
si publici juris facerem, nescio an ille ex Vestratibus, qui tanto cum fervore ad
injurias usque mecum expostulant, magnam inde gloriandi causam acquire-
rent; Habeo inter alia documenta authentica apographum a D. Monmortio
nuper defuncto Mathematico, ut nosti, dum viveret perdocto atque nulli Parti
addicto, utpote Gallo, habeo, inquam, apographum ab eo mihi transmissum
alicujus epistolæ, quam ipse ad Cl. Taylorum scripserat 18 Decemb. 1718
[N.S.],[4] et quae vel sola magnam litis partem dirimeret, sed non ex voto
Taylori cæterorumque ejus sequacium.[5] Ab istis autem evulgandis libenter
abstinebo, modo Vestri desinant, quod pacis causa optarem, nostram lacessere
patientiam. Lubens credo quod ais de aucto Corollario 1, Prop. XIII Lib. 1.
Operis Tui incomparabilis Princip. Phil. hoc nempe factum esse antequam
hæ lites cœperunt, neque dubitavi unquam, Tibi esse demonstrationem
propositionis inversæ quam nude asserueras in prima Operis Editione, aliquid
dicebam tantum contra formam illius asserti, atque optabam, ut quis analysin
daret, qua inversæ veritatem inveniret a priori, ac non supposita directa jam
cognita. Hoc vero, quod Te non invito dixerim, a me primo præstitum esse
puto, quantum saltem hactenus mihi constat.[6] Unum superest, quod pace Tua
monendum habeo: Retulit mihi nuperrime Amicus quispiam ex Anglia redux,
me esse ejectum ex numero Sodallum Illustr. Vestræ Societ. Reg. id quod

collegerit ex eo quod nomen meum non repererit in Catalogo Londini viso Sociorum (in ampla scheda annuatim imprimi solita) pro anno 1718.[7] Et quominus dubitarem, monstravit mihi Librum aliquem Anglicum impressum an. 1718, cui titulus *Magnæ Britanniæ Notitia*, ubi in parte postrema pag. 144 videre est catalogum Membrorum exterorum Societatis Regiæ[8] atque in illo nomen Agnati mei, sed meo nomine quod miror prorsus exulante: Liceat ergo ex Te quærere, utrum ex decreto Illustr. Societatis fuerim expunctus, et quid peccaverim vel quonam delicto ejus indignationem in me concitaverim, an vero Secretarius (qui ni fallor tum temporis fuit Taylorus) propria auctoritate me proscripserit. Quid? ideone locum in illustri hoc Corpore mihi non ambienti tam honorifice obtulissetis, ut postea tanto turpius ex eo me ejiceretis? Hoc equidem ob insignem vestram æquitatem suspicari vix possum; Quare enixe Te rogo Vir Nobilissime, ut quid ea de re sit me quantocius facias certiorem. Vale ac mihi studiosissimo Tui porro fave. Dabam Basileæ, a.d. XXI. Decemb. MDCCXIX [N.S.].

Translation

Johann Bernoulli presents a grand salute
to the very noble [Sir] Isaac Newton,
most celebrated throughout the world

Your letter gave me great pleasure, most illustrious Sir; I have learnt from it that, setting aside mathematical disputes, you honour me with the same kindness and friendship of which you once deemed me worthy. Indeed, you act in a way befitting an open and generous man, who does not readily allow those to be denigrated whom he judges worthy of his love. I am not certain of what kind that letter addressed to Mr Leibniz is of which you speak, which is dated 7 June 1713, [N.S.]. I do not remember having written to him myself that day, yet I would not deny it altogether, since I have not kept copies of all the letters I have written. But if perhaps, amongst the innumerable letters which I had written to him, one were found which bore the very day and year mentioned, I would have dared to assert with complete confidence that nothing was contained in it which would in any way weaken your reputation for honesty, nor did I ever give him leave to make public certain of my letters, particularly such as would not please you, albeit against my hope and wish. Therefore I ask you again, famous Sir, that you should convince yourself that it was never my intention to speak of you as other than a very great man, and that much less would I insult your reputation or good name. Let me not seem to say that you have grasped after fame among foreign peoples, nevertheless I hope you do not refuse eulogies offered by us beyond [the sea], since they are sincere and worthy of you and thus even more acceptable than those which arise from an immoderate partiality. The world of learning, which you have hitherto enriched by so many amazing discoveries, will sharply regret that you are not able to undertake mathematical studies, now that you are an old man (for whose well-being through the

New Year, which we will celebrate soon, and through many years to follow, I pray with all my heart). Even though I myself am not yet old I verge on old age[2] and am distracted by other business, so that it is not proper for me any longer to give such earnest thought to mathematical matters as I was wont to do. What you remark upon, most famous Sir, about Raphson's book,[3] namely that it has been reprinted with some of Leibniz's letters in which he affirms me to be author of the letter mentioned above (with regard to which I cannot foresee what it may contain that is damaging to your good name), certainly does not conduce to putting an end to the dispute. In truth, perhaps I shall never see Raphson's book, because books of this kind are rarely brought here from Holland. Meanwhile I wish this to be considered by your countrymen: if it were to be made certain by means of evidence, it were better done by alleging other letters than those written by Leibniz, for surely [he] who [speaks] in his own cause cannot be taken a proper witness. There are letters to me from certain learned men of nations having no part in this national strife; if I laid these before the public I know not whether those of your countrymen who quarrel so warmly with me as to become insulting would gain from them great reason for boasting. I have amongst other authentic documents a copy [of a letter] from Mr Monmort, a mathematician not long dead who was, as you know, learned and attached to neither party while he lived, since he was French; I have, I say, a copy (sent from him to me) of a certain letter which he had written to Mr Taylor on 18 December 1718 [N.S.],[4] and which alone would certainly dispel a large part of the dispute; but not to the taste of Taylor and the rest of his following.[5] However, I will willingly refrain from making these things public, if only your party will cease to try our patience, which I desire for the sake of peace. Gladly I believe what you say about the addition to Corollary 1, Proposition 13, Book I of your incomparable work, the *Principia*, that this was certainly done before these disputes began, nor have I any doubts that the demonstration of the inverse proposition, which you have merely stated in the first edition of the work, was yours; I only said something against the form of that assertion, and wished that someone would give an analysis that led *a priori* to the truth of the inverse [proposition] and without supposing the direct [proposition] to be already known. This indeed, which I would not have said to your displeasure, I think was first put forward by me, at least so far as I know at present.[6] One thing remains for me to advise you of, by your leave: very recently some friend or other returning from England reported to me that I have been expelled from the number of the Fellows of your illustrious Royal Society. He inferred this from the fact that my name was not to be found in a list of Fellows seen in London, for the year 1718, (which list is usually printed annually on a large sheet).[7] And lest I should doubt him, he showed me a certain English book, printed in 1718, with the title *Magnæ Britanniæ Notitia*, where in the last part of page 144 is to be seen the list of Foreign Members of the Royal Society,[8] and in it the name of my nephew, but my own name utterly banished, which astonishes me. May I inquire of you, then, whether I have been expelled by decree of that Illustrious Society, and how I have sinned or by what crime I have brought upon myself their indignation, or rather indeed if the Secretary (who, if I am not mistaken, was at that time Taylor) has outlawed me by his own authority. What? Did you so flatteringly procure for me a fitting place in that illustrious Society, without my soliciting it, in order afterwards to expel me in so disgraceful a manner? This indeed I can hardly suspect on account of your extreme fairness. Wherefore I ask you eagerly, most noble Sir, that you will as soon as possible tell me what the situation is. Farewell, and favour me further who am most devoted to you. Basel, 21 December 1719, [N.S.]

NOTES

(1) Newton MS. fo. 66; printed in Brewster, *Memoirs*, ii, pp. 504–6. The letter was received by Varignon on 19 December and by him forwarded on 22 December to De Moivre to deliver to Newton (see Varignon's letter to Bernoulli of 11 January 1720, N.S., in the Bernoulli Edition).

(2) Bernoulli was 52.

(3) Bernoulli had, quite naturally, misunderstood Newton's letter since not a reprint of Raphson's *History of Fluxions* but the issue of Des Maizeaux's *Recueil* was in question (see Letter 1320). Probably Newton could not have prevented the publication of this book, which Des Maizeaux did not compile in a spirit of partial hostility to Leibniz.

(4) See Letter 1307, full of compliments to Bernoulli and recognition of his vast contributions to the calculus.

(5) Taylor had been no friend of Bernoulli since the latter had criticized his *Methodus Incrementorum* and accused him of plagiary; see Letter 1217, note (2) (vol. vi, p. 362) and Letter 1316, note (3), p. 38.

(6) Bernoulli refers to his 'Extrait de la Reponse à M. Herman, datée de Basle le 7. octobre 1710 [N.S.]', *Mémoires de l'Académie Royale des Sciences* for 1710 (Paris, 1713), pp. 521–33. There Bernoulli had given a solution to the inverse problem of central forces with the definite implication that he was the first to do so. See Letter 1023, notes (2) and (3) (vol. vi, pp. 38–9), and D. T. Whiteside, *Journal for the History of Astronomy*, **1** (1970), 134, note (32).

(7) Lists of Fellows were published annually by the Society and included a list of Foreign Members. Johann Bernoulli's name is omitted from the lists for 1717, 1718 and 1719. It is reinstated in the list for 1720, and Newton reassured him in his reply that he had not been dismissed.

(8) Bernoulli refers to the twenty-fifth edition of *Magnæ Britanniæ Notitia, or, The Present State of Great-Britain* (London, 1718). The early editions of this work were by Edward Chamberlayne, but John Chamberlayne had taken over the task of editing it on the death of his father. The book comprises a number of separately paginated sections, the second of which, 'A General List or Catalogue of all the Offices and Officers Employed in the several Branches of His Majesty's Government...I. In *South-Britain*, or *England*', gives on p. 142–5 a list of Fellows of the Royal Society, omitting Johann Bernoulli's name from the list of Foreign Members.

1333 NEWTON TO J. BERNOULLI
EARLY 1720
From a holograph draft in the University Library, Cambridge[1]
Reply to Letter 1332

Sr

I do not take the controversy about the differential method to be national, nor is every thing written or printed in England written or printed by my order or advice, or with my consent. Mr Ralphson wrote & printed before the Commercium Epistolicum came abroad & his book was in the press before I knew of it, & I stopt its coming abroad for three or four years. It was published three years ago & is not reprinted.[2] The two Letters wherein Mr Leibnitz represents that you were the author of the letter dated 7 June 1713 [N.S.], were printed above a yeare ago by a friend[3] of Mr Leibnitz who has been collecting his remains & I knew nothing of the matter till I saw the Letters in print. I do not admitt Mr Leibnitz a witness against you, but have told my friends that I admit the author of that Letter, Mr Leibnitz, & you, to be three witnesses against Mr Leibnitz. You are not dismissed the R. Society.[4] The reduction of Problemes to Quadratures being the first degree of the inverse method of fluxions I wrote in the Introduction to that Book[5] that I found the method of fluxions gradually in the years 1665 & 1666. This has been contradicted by Mr Leibnitz & his friends, & particularly by the author of the aforesaid Letter of 7 June 1713. I never meant to affirm that [*draft ends*]

NOTES

(1) Add. 3968, fo. 617. We infer from the content of the draft that it is a reply to Letter 1332. Also compare the wording here with Newton's summary of the letter in Letter 1334, II.

It is virtually certain that the letter was never sent; it is not mentioned in the extant correspondence between Bernoulli and Varignon. However, De Moivre wrote an eye-witness account of Newton's reactions to Letter 1332 in a letter addressed to some third person, intended for Varignon's eye, and from which Varignon quoted in his own letter to Bernoulli of 8 March 1720, N.S. (see Bernoulli Edition): 'J'ai remis entre les mains de M. Newton la lettre de M. Bernoulli pour lui, il me fist la grace de la lire devant moy, & je remarquay avec plaisir qu'elle produisit tout le bon effet que M. Bernoulli & M. Varignon en pouvoient attendre: ces Messieurs peuvent compter que M. Newton est entierment satisfait, & qu'il ne lui reste rien sur le coeur.' Later, however, Newton must have re-read Bernoulli's letter, for Varignon, writing to Bernoulli on 6 July 1720, N.S., quoted from a new report from De Moivre, this time less favourable (see Letter 1334). De Moivre wrote to Varignon in confidence, but Varignon deemed it necessary to pass his words on to Bernoulli: 'Quoy que M. Newton voye avec plaisir que les choses s'acheminent à la paix, je vous diray en confidence qu'à la seconde lecture qu'il fist de la derniere lettre de M. Bernoulli, il trouva quelques petits traits qui lui firent de la peine.' The points to which Newton apparently took exception were, first, Bernoulli's implication that Monmort's letter (Letter 1307) would, if made public, influence opinion against Newton, and, second, that Newton, if he wished, could prevent Keill from publishing papers against Bernoulli. (As Varignon pointed out, the latter statement was not specifically set out in Bernoulli's letter; nonetheless De Moivre described Newton as having been considerably angered by it.) Varignon urged Bernoulli to write yet again to Newton to clarify his position.

Bernoulli's own attitude at this time is clearly expressed in his letters to Varignon; he felt that Newton should make the next move, and expressed irritation against Keill and Taylor for their continued mathematical disputes with him, and against De Moivre who had for some time owed Bernoulli a letter, and in any case seemed still to be aiding Keill. Newton, he believed, must have some control over these men. 'Mr. Newton souffre que je sois traité si indignement, et ne veut pas l'en empecher, comme il le pouroit et le devroit faire, si la reconciliation etoit sincere de son coté; je dis qu'il pouroit empecher ces mesdisances contre moy, car je me souviens que M. de Montmort m'a souvent ecrit, que tout ce que Keil, Taylor et quelques autres ecrivoient contre nous, cela faisoit sous les auspices et par l'Authorité de Mr. Newton.' (Bernoulli to Varignon, 29 June 1720, N.S.)

In the course of 1720, no fewer than fifteen long letters passed between Varignon and Bernoulli, in nearly every one of which were lengthy discussions of the calculus dispute.

(2) *Read*: 'and is not now being reprinted.' Bernoulli had become understandably confused, and Newton prevaricates. Newton had mentioned, in Letter 1329a, the reprinting in Des Maizeaux's *Recueil* of the letters in the Appendix to Raphson's *History of Fluxions*; Bernoulli took him to mean that the *History of Fluxions* was itself being reprinted. But Bernoulli was, in a sense, close to the truth; Newton had arranged a reprinting of Raphson's *History of Fluxions* after the death of Leibniz, and had himself added the Appendix of correspondence now under discussion (see vol. vi, p. 254, note (2)). For confirmation that Raphson's book was in being by 1711 see vol. v, p. 95.

(3) Pierre Des Maizeaux. Newton did, indeed, know nothing of Des Maizeaux's *Recueil* until he was sent the proof sheets in the summer of 1718 (see Letter 1295, vol. vi); but although he was given an opportunity to comment upon these printed sheets before they were published, he took little real advantage of it, although he did take steps to delay the publication (see Letter 1330, note (2)).

(4) See Letter 1332, note (7), p. 79.

(5) *De Quadratura Curvarum*. It seems that Newton is here yet again about to plunge into another lengthy claim of priority in the invention of the calculus.

1334 NEWTON TO ?DE MOIVRE
?EARLY 1720
From holograph drafts in the University Library, Cambridge [1]

I [2]

Sr

Understanding by yours of 28 August 1719 [3] that there is a good understanding between the Gentlemen of the Academy of Sciences & me, my desire to continue it is the occasion of this.

Mr John Bernoulli in a Letter to the late Mr Monmort dated 8 April 1717 [N.S.][4] excused himself for communicating a Probleme to Mr Leibnitz, representing that he had no design to challenge the English & was surprised that Mr Leibnitz should name him as the author of the Probleme & do this without his leave. And desired Mr Monmort to disabuse me & let me know how much he desired to live in amity with me, & then added Il seroit pourtant a souhaitter qu'il voulût bien prendre la peine d'inspirer a son ami Mr Kiel sentiments de douceur & de equité enverse les etrangers, pour laisser chacun en possession de ce que luy appartient de droit, et a juste titre. Car de vouloir nous exclure de tout pretention ce seroit une injustice criant. And Mr Monmort soon after the receipt of this Letter endeavoured to make us friends[5] but w[i]thout success because Mr Leibnitz had fathered upon Mr Bernoulli a scandalous Libel dated 7 June 1713 [N.S.];[6] & though I doubted whether Mr Bernoulli were the author because the author cited Mr Bernoulli as a person different from himself, & Mr Leibnitz when he began to father the Letter upon Mr Bernoulli left out the citation: yet I thought it reasonable that Mr Bernoulli should disown the Libel as publickly as Mr Leibnitz had fathered it upon him. And soon after I was shewed a Letter written by Mr Monmort to Dr Taylor concerning these Disputes with Dr Taylor's Answer:[7] but I took no copy thereof.

I have since received a Letter from Mr Bernoulli dated 3 Non. Julij 1719 [N.S.][8] in wch wth many compliments he endeavours to be reconciled, & for that end represents that he wrote no such Letters as that wch I complained of & I returned a friendly answer, but have since received another Letter from him dated 21 Decem. 1719 n.st[9] of a different humour from the former, expostulating wth me upon suspicion that the English are reprinting Raphsons book & making Mr Leibnitz a witness against him & that he has been dismissed the R.S. & telling me that he has Letters written by persons not concerned in this national controversy, & particularly an authentic copy of a Letter written by Monsr Monmort to Dr Taylor 18 Decem. 1718 [N.S.],[10] wch if he should print would decide a great part of the controversy contrary to the mind of Taylor and his followers, & [sic]

And[11] yet Ralpsons book was written & in the Press before I knew of it & I stopt the publishing of it four years together & could stop it no longer without paying for the edition & it is not reprinting. I never thought of making Mr. Leibnitz a witness ag[ains]t Mr Bernoulli but on the contrary look upon the Author of the aforesaid Libel in citing Mr Bernoulli as a person different from himself, Mr Leibnitz in printing this Libel with the citation and afterwards omitting the citation when he fathered this Libel upon Mr Bernoully, & Mr Bernoulli in denying that he was the author of the Libel to be three good witnesses against Mr Leibniz. Mr Bernoulli has not been dismissed the R.S. I do not take the controversy to be national: for Mr Leibnitz spent his life in corresponding wth learned men of all nations & particularly wth Mr Bernoulli. I never imployed Dr Taylor to write in my defence nor consented to it, nor think myself concerned in the dispute between him & Mr Bernoulli. The Letters printed at the end of Rapson Book were reprinted in Holland before I knew of it & many new ones were added which have been communicated to him by the correspondents of Mr Leibnitz & the person imployed to publish them is not an English man. And I have prevailed wth Dr Keill during the two last years to suspend publishing what he has written against Mr Bernoulli, tho I cannot in justice hinder him perpetually from defending hims[elf] from the usage he has met with from persons imployed by Mr Bernoulli.[12]

Mr Leibnitz from the beginning avoided medling wth Mr Keill & used his utmost endeavour to engage me in person, & Mr Bernoulli does the like. When the Court of Hanover came to London his[13] friends endeavoured to reconcile us in order to bring him over to London, but they could not get me to yeild.[14] Then he tried to get the original Letters out of the hands of the R.S. that he might print them entire in a new Commercium Epist: but I represented that I was so far from printing the Commercium Epist. my self that I did not so much as produce the Letters in my custody: for proof of this I produced two old Letters in my custody the one from Mr Leibnitz to me dated [blank][15] 1693 & the other from Dr Wallis to me dated [blank][16] 1695 wch I did not produce least I should seem to make my self a witness in my own cause. And there was the same reason why Mr Leibnitz should not be allowed to write a Commercium Epist. himself. And when these Letters were examined before the R.S. by those who knew the hands they were laid up in the Archives of the R.S. And the R.S. allowed only that if Mr Leibnitz had any old Letters wch he had received from England, & would send the originalls to any friend in London to [be] produced before the R.S. & examined by them who knew the hands & attested copies taken of them: the Originals might then be sent back to Mr Leibnitz & the Letters bee printed either in the Phil. Tr. or in Germany as Mr L. pleased. But no Letters were ever sent, but on the contrary Mr Leibnitz in his Letter[17] to Mr Abby Conti complaind that those Letters were either lost or involved in a heap of papers wch would require too [much] time & pains to search out: & [draft ends]

II[18]

Sr

Mr Bernoulli in a Letter to me of Decem 21 1719 [N.S.],[19] seems to write upon suppositions that Mr Raphson's book was printed with my consent &

has been reprinted by my friends; that the controversy about the differential method is national, & those of other nations then England & Germany are unconcerned some of whose letters he can publish unless the English desist to provoke him, that I can stop the publishing of letters against him in England; that the two Letters of Mr Leibnitz in wch Mr Bernoulli is said to be the author of the letter of 7 June 1713 [N.S.] are printed here by my friends; that I admit Mr Leibnitz to be a witness in his own cause against him; & that he has been dismissed the R. Society. You know how all this matter stands & I beg the favour that you would signify it to Mr l'Abby Varignon. [*del*: It is now above 40 years since I left of all correspondence by Letters about Mathematicks & Philosophy, & much more am I averse from disputes about those things]

Mr Raphson's book is not reprinted nor concerns Mr Bernoulli. Mr Leibnitz kept a general correspondence & has friends in England & France & other countries as well as in Germany. Some of those in England have been collecting his remains in honour of his memory & the two letters above mentioned are in this collection; & I have no hand in what they do. The Letter of 7 June 1713 [N.S.] was translated into French & printed in Holland & there ascribed to Mr Bernoulli four years ago;[20] that[21] it is reflected upon in the Elogium of Mr Leibnitz,[22] & therefore it was sent into France as well as into England; The author of this Letter, as it was at first printed in Germany, cited Mr Bernoulli as a person different from himself,[23] but in the french translation this citation is omitted & this omission of what was at first printed in order to father ye letter upon Mr Bernoulli, made me question the testimony of Mr Leibnitz in my answer[24] to one of his Letters three years before I received Mr Bernoulli's declaration[25] that he was not the author thereof. And after I received that Declaration I acquiesced therein without thinking of admitting Mr Leibnitz a witness against Mr Bernoulli & I gave Mr Bernoulli notice of what Mr Leibnitz had written not to question Mr Bernoullis Declaration, but that he might not be surprized hereafter at what the friends of Mr Leibnitz were publishing: And Mr J. Bernoulli has not been dismissed the R.S. but is still a Fellow thereof.[26] I believe that you know all this, & beg the favour that if you are satisfyed, you would signify your satisfaction to Abby Varignon.[27] For Mr Bernoulli will beleive you two[28] sooner than me.

I beg the favour also that you would send to Abby Varignon the inclosed copy of the letters of 7 June [N.S.] & 28 July 1673 [N.S.][29] printed in Germany & dispersed that autum in great numbers by the friends of Mr Leibnitz & desire him to communicate it to Mr Bernoulli that he may know what Letter it is that Mr Leibnitz has endeavoured to father upon him.

As for Mathematicks:[30] the Ancients had two methods, Synthesis & Analysis or Composition & Resolution. They invented things by their Analysis but admitted nothing into Geometry without a synthetical Demonstration. & when they had demonstrated any thing synthetically they made use of it as a Lemma for Demonstrating any thing else.

NOTES

(1) These two drafts are undated, but both comment upon Bernoulli's letter to Newton of 10 December 1719 (Letter 1332) so we place them immediately after the draft of his reply to it. Further drafts, similar in content, are at U.L.C., Add. 3968(8), fo. 617. It is not clear whether all the drafts were intended for a letter or letters to the same person, but they all seem to be addressed to someone with knowledge of the calculus dispute who nevertheless remained unaware of much of Newton's private and unacknowledged involvement. Possibly Fontenelle, secretary of the Royal Academy of Sciences, who had written at some length on the calculus dispute in his 'Éloge' of Leibniz, was the intended recipient; more probably it was Abraham de Moivre, who was at this time acting as intermediary between Varignon and Newton (see also Letters 1320, note (2), and Letter 1374).

(2) U.L.C. Add. 3968(8), fo. 607.

(3) We have not found a letter bearing this date addressed to Newton.

(4) See Letter 1237, vol. VI.

(5) Monmort forwarded the letter to Taylor for Newton, see vol. VI, p. 384, note (1).

(6) The 'scandalous Libel' is the extract of Bernoulli's letter printed in the *Charta Volans* (Number 1009, vol. VI); see Letter 1320, note (8), p. 47.

(7) These two letters are now lost.

(8) Letter 1320.

(9) Letter 1332.

(10) Letter 1307.

(11) Newton now repeats the arguments in his draft reply to Bernoulli; see Letter 1333. Th 'him' and 'the person...not an English man' are of course Des Maizeaux.

(12) See Letter 1329, note (17), p. 69.

(13) The remainder of the paragraph refers to Leibniz's actions, not Bernoulli's.

(14) George I and his retinue had arrived in England on 18 September 1714 from Hanover (see vol. VI, p. 260, note (5)). It was in fact earlier than this (contrary to Newton's implications here) that Chamberlayne had tried to reconcile Leibniz and Newton, and encouraged Leibniz to request letters from the Royal Society and to publish his own *Commercium*; see vol. VI, pp. 173–4, notes (1) and (2). Possibly the confusion arose in Newton's mind because Leibniz's request, dated 14 August 1714, was not read to the Royal Society until a meeting of 11 November 1714, *after* the arrival of the Hanoverian Court. The Society's view was that Leibniz should make good his charges against Keill, or 'ask pardon of the Society for suspecting their Judgement and Integrity in the Commercium Epistolicum already published by their Order & Approbation'. When Chamberlayne mentioned that Leibniz intended soon to visit England, 'further consideration of this Affair was referred to some other opportunity' (Journal Book of the Royal Society, Copy, IX, p. 25). Leibniz never paid his visit. Presumably with a view to the composition of the *Recensio*, in the following spring Newton ordered the original letters for the *Commercium Epistolicum* to be re-examined, and at a meeting of 5 May 1715 produced the two letters he mentions here to be added to the Royal Society's collection (see vol. VI, p. 242, note (2)).

(15) 7 March 1693; see Letter 407, vol. III.

(16) 10 April 1695; see Letter 498, vol. IV.

(17) Letter 1197, vol. VI. Newton paraphrases Leibniz's French.

(18) Add. 3968(8), fo. 616.

(19) Letter 1332. In what follows Newton covers much the same ground as in his letter to Bernoulli, Letter 1333.

(20) In the *Nouvelles Litteraires* for 28 December 1715, N.S.; see Letter 1172, vol. vi.

(21) Presumably Newton intended to delete 'that'.

(22) Fontenelle, in his 'Éloge' of Leibniz (see Letter 1305) merely touches upon the *Charta Volans*, and does not mention the letter which it contains. It is therefore not clear what Newton intends here.

(23) See vol. vi, p. 21, note (8). The citation in the original *Charta Volans* (Number 1009, vol. vi) was in square brackets, but in the French translation in the *Journal Literaire* (see Number 1018, vol. vi) the brackets were omitted, implying that the phrase was part of the original letter.

(24) Letter 1187. Newton thought the citation was purposely omitted from the second French version in the *Nouvelles Litteraires*, in order to convince the reader that the letter *could* be by Bernoulli.

(25) Letter 1320.

(26) See Letter 1332, note (7), p. 79.

(27) So that Varignon could send this additional testimony to Bernoulli.

(28) The reading is doubtful.

(29) *Read*: 29 July 1713, N.S. (that is, the *Charta Volans*).

(30) The paragraph which follows seems to be the reply to a previous inquiry made by the addressee.

1335 THE ORDNANCE TO THE MINT

23 FEBRUARY 1720

From the original in the Mint Papers[1]

Office of Ordnance
23d Febry 17$\frac{19}{20}$

Gentlemen,

Having directed our Officer to make a Plan of the late Master Smiths House & Shops in the Mint,[2] as also a Design for improving the same for the reception of Clerks, that they may the better attend their business, you have underwritten an abstract of his Report—

In improving this place, I humbly observe that ye Buildings adjoining to the premises on both sides ye street & belonging to the Mint, are all Timber Buildings & the Party Walls the same, so that in case of Fire, it may be of ill consequence On the North side of the Street, the Party Wall between the Smiths Shops & the House inhabited by the Widdow Lucuts is of Timber— A Stack of Chimneys irregularly built & ye Foundation not so low as the intended Cellar by Four Feet

This Stack of Chimneys makes a Break in the premises, & by its smallness is a great detriment, wch by Building a Brick party wall & new building the Chimneys may be prevented & no damage done, but rather an Improvement to the said House.

On the South side of the Street & the West end of the premises are the Milling Houses, which are built of Timber & will require a party Wall of Brick—There are some of the Windows in the upper Room above Stairs belonging to the Mint, which 'tis hoped may be well spared to be shutt up by the New party Wall.

Here is also on the back End of this Party Wall, a Break into the premises about 10 Feet square, which in ye first story is a Vault[,] belongs to these premises But above is a slight Timber Building & belongs to ye Mint, to preserve this upper Building when the Shops are pulled down will be a matter of some difficulty.

We therefore desire some of your Officers may soon meet wth some of ours, in order to settle these matters, which We hope may tend to ye mutual advantage of both offices.

> We are Gentlemen
>   Your very humble servants
>   CHA MILLS   T. WHITE
>   J RICHARDS   JOHN ARMSTRONG
>   T WHEATE

Principal Officers of his Majties Mint

NOTES

(1) III, fo. 434.
(2) For the background to this survey, and Newton's reaction to it, see Letter 1341.

1336 TILSON TO NEWTON
5 APRIL 1720
From the clerical copy in the Public Record Office[1]
For the answer see Letter 1337

Sir

The Crown of England by a late Convention with Sweden[2] having agreed to certain payments to be made in Imperial Dollars And some Dispute arising about the Value of the said Dollars; The Lords Commissioners of his Majesty's Treasury have commanded me to send you the inclosed Extract of the said

Convention [*missing*] as also an Extract of a Treaty with Sweden Ao. 1700[3] And desire you to consider the same and Report your Opinion to their Lordships with respect to the value of the said Dollars as well Intrinsically as by way of Exchange which you will please to do with all convenient speed This in the absence of ye Secretaries &c 5th April 172[0][4]

C. TILSON

NOTES

(1) T/27, 23, p. 69. For the writer, see vol. VI, p. 363, note (1).

(2) Britain had been involved in the Great Northern War (which only finally ended when Sweden made peace with Russia at Nystadt in August 1721), partly because of her maritime trade to the Baltic, partly because Hanover was a belligerent, and partly because the Swedish Crown gave support to the Jacobites. After diplomatic flurries extending over many years, Britain sent a fleet to the Baltic under Newton's friend Sir John Norris in the summer of 1715 which, in conjunction with the Dutch navy, protected their joint commerce. In subsequent years even more offensive orders were issued to the British admirals, and diplomats of either side were imprisoned by the other. The death of Charles XII of Sweden at the end of 1718 opened the way to peace, which was facilitated by the increasing fear of Russia felt by Denmark and Hanover. Hanover and Sweden concluded peace at Stockholm in November 1719: in return for major territorial concessions by Sweden, George I agreed to pay one million *reichthalers*. Further, on 21 January 1720 a treaty was concluded between Great Britain and Sweden, whereby the former was to mediate in peace negotiations between Sweden, Denmark and Prussia, to send a fleet to assist the Swedish navy in operations against Russia in the Baltic, and to pay a subsidy of 288000 *reichthalers* to Sweden. (See J. F. Chance, *British Diplomatic Instructions: Sweden, 1689–1727*; Royal Historical Society, Camden series, **32** (1922)).

(3) The extract is now with Newton's reply, Letter 1337. In a treaty of January 1700 Britain had agreed to support Sweden in the Great Northern War with a force of 6000 men, or the equivalent in ships and money. This had never been done, and circumstances had greatly altered since that time. Britain claimed that her embarking on war with France had abrogated this treaty with Sweden, and counter-claimed for the damages done by the Swedish privateers against British commerce.

(4) The date, left incomplete, is obvious from Newton's reply.

1337 NEWTON TO THE TREASURY
12 APRIL 1720
From the holograph original in the Public Record Office[1]
Reply to Letter 1336

To the Rt Honble the Lords Comm[ission]ers of
His Majties Treasury

May it please your Lordps

In obedience to your Lordps Order of Reference signified to me by Mr Tilsons letter of ye 5t Instant, that I should report the value of Imperial Dollars both intrinsecally & by way of Exchange with Sweden: I humbly represent that the specie Rix dollars are coined of several values by several Princes of the Empire from 4*s* 4*d* to 4*s* 8*d*. But in Books of Exchange the Rix dollar is valued at 48 schellings Lubs of Hamburgh, at 48 styvers of Antwerp, at 50 styvers of Amsterdam & at 4*s* 6*d* English. There is also a Common Dollar of the Empire in respect of wch the Gulde or Guilder is usually marked $\left(\frac{2}{3}\right)$ to signify that is two thirds of this Dollar. The Gulde is 24 Marien grosch the Common Dollar 36 Marien Grosh & the Rix dollar two Gulden, so that the common Dollar is three quarters of the Rix Dollar. The difficulty is to know whether by the Imperial Dollar be meant the Common Dollar of the Empire or the Rix Dollar.

If any payments have been made to Sweden since the Treaty of $\frac{6}{16}$ January 1700, the Precedent is to be followed as the best interpreter of the Treaty. If none; I am told that the Imperial Dollar is sometimes taken for the common Dollar but more usually for the Rix Dollar, & that the word Reichs or Rycks Thalere signifies imperial Dollar. But I am not skilled in the German language.

In the weekly Tables of Exchange with London, the number set over against Hamburgh signifies the number of Bank schelling and deniers to be paid or received at Hamburgh for one pound sterling at London: & $3\frac{5}{9}$ schellings are at a par with one pound sterling. How the exchange is between London & Stockholm I do not find in the Tables. But by the Treaty[2] the money is to be paid at London to the Order of the Crown of Sweden. He is to receive at London 288 000 Imperial Dollars, & if these be Rix Dollars they amount unto 64 800 pounds sterling.[3]

All wch is most humbly submitted to your Lordps great wisdome.

Is. NEWTON

Mint Office
12 *Apr* 1720.

NOTES

(1) T/1, 227, no. 31; printed in Shaw, pp. 196–7, who wrongly refers to the original as T/1, 217, no. 31. The letter is followed by the extract from the 1700 treaty mentioned by Tilson in Letter 1336.

(2) That is, the 1720 treaty; see Letter 1336, note (2), p. 88.

(3) Newton simply adopts the commercial value of 4*s*. 6*d* for the *reichthaler*.

1338 VARIGNON TO NEWTON

?MAY 1720

From the holograph original in King's College Library, Cambridge[1]

Equiti Nobilissimo
Viroque Doctissimo D. D. Isaaco Newton,
Regiæ Societatis Anglicanæ Præsidi Dignissimo
S.P.D.
Petrus Varignon.

Etsi quid novi scriberem non habeam, Vir Nobilissime ac Doctissime, tamen officioso huic Adolescenti[2] ex Illustri apud vos Mont-acutorum stirpe oriundo, hinc ad suos redeunti, quique summa te Observantia Veneratur, non potui Literas ad te non dare, quibus tuorum in me Beneficiorum animum perpetuo gratum ac in æter[n]um memorem tibi denuo testificarer. Inter illa maximi facio, & tanti sane quanti Cl. Bernoullus, redditam ipsi a te pristinam tuam Benevolentiam, qua Innocentem Malevolorum Obtrectationibus se excidisse conqueritur, quamque sibi charam[3] (ut ex postremis ejus ad me Literis[4] intelligo) sedulo colet: Quod ad me spectat hac in pace tantopere a me desiderata, dirimit ea Lites quas inter Vos ægerrime ferebam: quam ob rem de illa etiam tibi gratias ago, tranquillam vitam ac Diuturnam tibi peroptans. Vale.

Dabam Parisijs die [blank][5] *1720*

P.S. uno[6] e nostris bibliopolis, postulante jus prærogativum Eximij tui Libri Optices[7] Gallice redditi a D. Coste, typis nostris exarandi ad Exemplar Hollandicum; hujus examen mihi mandavit Illustrissimus Regiorum Sigillorum Custos,[8] cui approbatum[9] ac (ut par erat) Laudatum reddidi Die 28. Aprilis [N.S.] proxime elapsi.

Translation

Pierre Varignon presents a grand salute
to the most noble and learned Sir Isaac Newton
most worthy President of the English Royal Society

Although I do not have the news I would [wish to] write, most noble and learned Sir, nevertheless I could not fail to give to this courteous young man,[2] descended from the illustrious Montague family, your countryman, returning home from here, a letter to you by which my perpetual gratitude for your kindnesses to me and, further, my eternal remembrance of you, may be testified afresh. Among these I rate highest, and so surely does the famous Bernoulli,[3] that you have received him back into your former good graces (from which, he complains, he, though innocent, was removed by the disparagements of ill-wishers) and which, as being dear to himself (which I learn from his last letter[4] to me) he will zealously cherish. So far as I am concerned in this peace that I so much desired, it has destroyed those feuds between you which I greatly deplored; for which reason I give you thanks also on that account, desiring for you a long and peaceful life. Farewell.

Paris [blank][5] 1720

P.S. One of our booksellers,[6] seeking the prerogative right of publishing from our press your great book of *Opticks*[7] as translated into French by Mr Coste by following the Dutch edition, the examination of this [Amsterdam edition] was entrusted to me by the illustrious Keeper of the King's Seals.[8] I reported my approbation of it and, (as was but just) my praise also to him, on 28 April [N.S.] last.[9]

NOTES

(1) Keynes MS. 142(E).

(2) We have not identified the young traveller.

(3) Read: *caram.*

(4) Since Bernoulli's last letter to Varignon (of 11 May 1720, N.S.) contained little on the calculus controversy, but was more concerned with rumours of Newton's death and his own 'expulsion' from the Royal Society, it is possible that Varignon was reporting a general impression of Bernoulli's sentiments (see Letter 1333, note (1)). Bernoulli had often complained about Taylor and Keill as detractors from his good name, and clearly believed Newton to be the moving force behind their proceedings.

(5) Varignon has omitted the month; the postscript indicates that the letter was written shortly after the end of April.

(6) François Montalant, bookseller on the Quai des Augustins, Paris.

(7) There were two French editions of *Opticks*: (i) *Traité d'Optique sur les Reflexions, Refractions, Inflexions et les Couleurs, de la Lumière. Par M. le Chev. Newton. Traduit de l'Anglois Par M. Coste. Sur la seconde Edition, augmentée par l'Auteur, À Amsterdam, Chez Pierre Humbert, 1720.* Two volumes, 12ᵐᵒ. (ii) *Traité d'Optique...Par Monsieur le Chevalier Newton. Traduit par M. Coste, sur la seconde Edition Angloise, augmentée par l'Auteur. Seconde Edition Françoise, beaucoup plus correcte que la premiere. À Paris, chez Montalant, Quay des Augustins, 1722.* One volume, 4°.

The translator, Pierre Coste (see Letter 1350, note (6), and Letter 1365, note (1)) explains in his 'Preface du Traducteur' how he had invoked the aid of J. T. Desaguliers to ensure the greatest accuracy in presenting Newton's book to French readers. Varignon himself took charge of the second, Paris edition, and De Moivre was brought in as a stylistic adviser to Coste in revising the text. The preparation of this Paris edition figures largely in subsequent correspondence.

(8) The government secretary in charge of book privileges; censorship of the French press had originally been under the control of the Faculty of Theology at Paris University, but, in the field of scientific and technical literature, was increasingly becoming the prerogative of the Académie Royale des Sciences. Already the Académie could have books by its members printed

without the approval of the royal board of censors. Hence it would be natural for Varignon to be asked to examine the *Traité d'Optique* on behalf of the censors. (See Roger Hahn, *The Anatomy of a Scientific Institution: The Paris Academy of Sciences, 1666–1803* (London, 1971), p. 60.) The present *garde des sceaux* was Le Sieur Joseph Jean Baptiste Fleuriau d'Armenonville (1661–1728).

(9) Varignon's 'Approbation', which marks an important stage in the naturalization of the Newtonian philosophy in France, was printed in the Paris edition of the *Traité d'Optique*:

J'ai lû par l'Ordre de Monseigneur le Garde des Sceaux, le *Traité d'Optique sur les Couleurs*, &c. de Mr. le Chevalier Newton, traduit d'Anglois en François. Il m'a paru que ce Traité, par la nouveauté des choses qu'il decouvre, par les surprennantes expériences dont ces nouveautés y sont appuyées, & par la profonde capacité que son illustre & sçavant Auteur y fait paroître, comme depuis long-temps par tout ce qu'on a vû jusqu'ici de lui, méritoit fort d'être traduit en nôtre langue en faveur de ceux qui l'entendent, sans entendre l'Angloise, ni la Latine en laquelle il avoit été traduit. Ainsi je suis persuadé que l'Impression de cette Traduction Françoise fera d'autant plus de plaisir, qu'elle repandra davantage les connoissances merveilleuses dont ce Traité est rempli. Fait à Paris le 28 Avril 1720. Varignon.

1339 DES MAIZEAUX TO NEWTON
4 JUNE 1720
From the holograph original in the University Library, Cambridge[1]

London June 4, 1720.

Sir

The Impression of my Collection of Pieces of Mr. Leibniz &c,[2] is at last finish'd. I have received by the post the Preface enclos'd;[3] which I desire you would take the trouble to read. Before I sent it into Holland I communicated to Mr. Chamberlayne that part of it which relates to what pass'd in ye Royal Society;[4] and in a Letter that I have by me, he expresses his approbation of it.

As my design in giving an Account of ye dispute concerning ye Invention of Fluxions, was only to set right some Matters of fact mistaken or unknown beyond sea; and thereby endeavour to do you justice: so I hope you will find that I have said nothing but what is perfectly agreable to Truth. And I shall think those hours very well employed that I have spent in drawing up this Account, if it hath ye good fortune not to be dislik'd by you; and you are pleased to look upon it as a proof of ye perfect esteem and respect with which I am

Sir
Your most humble
and most obedient Servant
P. DES MAIZEAUX

As soon as I have receiv'd a compleat Copy of that Collection, I shall have ye honour to sent it to you.

Sir Isaac Newton

NOTES

(1) Add. 3968(36), fo. 507.

(2) That is, Des Maizeaux's *Recueil* (Amsterdam, 1720). For delays in its publication see Letter 1330, note (2), p. 73, and Letter 1281, vol. VI.

(3) The preface occupied pp. i–lxxxi of volume I of the *Recueil*, and included a lengthy account of the calculus dispute (pp. xi–lxvi). The proof copy mentioned here is not now with the letter; for Newton's reactions see Letter 1344.

The account began with a discussion of the early papers and correspondence of Leibniz and Newton in the seventeenth century, and then proceeded to describe in detail all the events eading up to the more recent quarrels, quoting at length from documentary evidence (including, unnecessarily, large excerpts from correspondence printed, *in extenso*, in volume II of the *Recueil*). The account is, on the whole, accurate and impartial, with Keill emerging as the major offender; but Des Maizeaux was, of course, unaware of how extensively Newton was involved in much of the controversy. Thus it is Keill who is described as the chief figure on the *Commercium Epistolicum* committee, and whilst Des Maizeaux seems aware that Newton, after Leibniz's death, spread privately, amongst his friends, printed copies of the Leibniz–Conti–Newton correspondence, he does not mention that its more public appearance in the Appendix to Raphson's *History of Fluxions* was Newton's doing. He attributes to Bernoulli the anonymous letter printed in the *Charta Volans* (see Letter 1344, note (12)).

(4) Des Maizeaux discusses the interchanges between Leibniz, Chamberlayne and Newton in the spring of 1714 and the reactions of the Royal Society, in the Preface of his *Recueil* pp. l–lvi, quoting at length from the relevant correspondence. The letters themselves are printed *in extenso* in *Recueil*, II, pp. 116–24. (See also Letters 1062, 1072 and 1101, vol. VI.)

1340 THE MINT TO THE TREASURY

5 JULY 1720

From the original in the Public Record Office[1]

To the Rt Honble the Lords Comm[ission]ers of his
Ma[jesty']s Trea[su]ry.

May it please your Lordps

In obedience to your Lordps Order of Reference of 5t May last We have Considered the Petition of the Widdow of John Roos Esq late Engraver of His Majts. Seals setting forth that his Bill for the silver & engraving work of publick Seales amounting to £788 15s 8½d was paid off by order of the Comm[issione]rs of his Majts. Treasury dated 7th Aug. 1717 excepting two Seals

93

for Ireland valued at £236. 18*s*. 10½*d* upon supposition that those two were to be paid for out of the Treasury of that kingdom And we are humbly of opinion that the said Graver be paid out of the Exchequer here for the Seals in question. For they were ordered & made here, & have formerly been paid for here as appears to us by an Order of the Treasury dated the 6 Sept. 1711, & some seals made since for Ireland have already been paid for here, & Gravers may hereafter prove unwilling to make seals here for Ireland if they must be sent thither for their money.

All which is most humbly submitted to your Lordps great wisdome

	WM: THOMPSON
Mint Office	IS. NEWTON
5 *July* 1720	M BLADEN

NOTES

(1) T/1, 228, no. 19, fo. 92; the letter is in Newton's hand. There is a copy at Mint/1, 7, p. 105. This is the answer to a petition for payment of £236. 18*s*. 10½*d* by Mrs Roos, the charge for making a new Great Seal for Ireland, transmitted to the Mint for an opinion by the Treasury on 5 May 1720 (see T/1, 228, no. 19*a*).

1341 NEWTON TO THE LORDS JUSTICES

c. JULY 1720

From the holograph draft in the Mint Papers[1]

To their Excellencies the Lords Justices
the Memorial of Sr Is Newton

Most humbly sheweth

In ye Indenture made between his Ma[jes]ty & the Master & Worker of his Mint there is a clause in these words. 'And his said Ma[jes]ty doth grant & confirm by these presents that the Officers of the Mint shall at all times have hold & peac[e]ably enjoy all places houses & grounds as well builded as unbuilded within the sd Mint wch heretofore have been called reputed or taken for the Mint without the medling let or disturbance by the chief Governour Constable Lieutenant or any other Officer or Minister of the Tower.['] This grant is of above 160 years standing as I find by copies of old Indentures. And no mention being made therein of any Officer of the Ordnance, it seems to have been made before that Office was erected.

About the year 1577 a Smith shop in the Mint was put into repair by the Office of Ordnance & may have been since frequently repaired by them. And the same smith has usually (if not always, till of late) been Smith of the Mint, & had a fourge at the end of the Mill-rooms for making Dyes & Puncheons for the Mint.

When the coinage was set on foot by the Mill & Press (wch was in ye year 1665,) there was an Order of Council for removing all strangers out of the Mint; but this Smith was not removed: whether because he was Smith to both Offices or for any other reason I do not know.

Mr Slingsby[2] about 40 years ago endeavoured to remove this Smith of the Ordnance out of the Mint & for that end a Committee of Council came to the Tower to view whether another place in the Tower might not be found for him to work in & the further end of the Mint was put into the hands of the Office of Ordnance & a new gate built for bounding the Mint at that end. And there is a tradition in the Mint that the Office of Ordnance was thereupon to have quitted the Smiths shop. But Mr Slingsby soon after falling into trouble & in the beginning of the reign of K. James the Mint being turned into a garrison, the shop continued in the hands of the Office of Ordnance, & about 21 years ago they rebuilt it [and defended their doing so by a Letter sent to the Treasury a copy of which is hereunto annexed. And on the 23d of Febr. last they sent us a Letter a copy of wch is also hereunto annexed,[3] to wch no answer has been returned in writing for want of a Board. But I told them that we had no authority to treat with them as desired or to do anything contrary to the above-mentioned Indenture. And upon my proposing to refer it to the Kings Counsel at Law, I was answered that they would submit to no determination but that of the King himself & directed [me] to acquaint your Excellencies with this matter before the Kings return.][4] and are now building more houses in the Mint.

Wherefore your Memorialist most humbly prays your Excellencies, that the Surveyor of the Ordnance may desist from building untill the case be examined & the bounds of the Mint be so settled as may prevent these disputes for the future, & render the coinage safe for encouraging Merchants to import their gold & silver, & leave sufficient room for building more Mills & Furnaces when ever it shall be necessary & also for building a house for our Porter, & that the Gates of the Mint may still remain in his custody for the safety of the coinage.

NOTES

(1) III, fo. 426r. There is an antecedent draft at fo. 425 in which Newton also complains that 'the Officers of the Ordnance can have no right to bring in [to the Mint] Carts with bricks & Timber for building' and, in reply to the accusation 'that the Officers of the Mint let their houses [there] to forreigners [*i.e. outsiders*] & why may not the Clerks of ye Ordnance as well as strangers live in the Mint' rejoins, somewhat weakly, 'disorders ought to be remedied & not drawn into a president [*read* precedent] & made incurable. The Mint is under the government of a board, & at present we have no board, but this disorder is not to be drawn into president.'

There are related drafts at fos. 426v and 427r; in the former Newton seems to argue, con-

95

trary to the draft printed above, that the smith was originally the Mint's Smith. This former draft, however, may well be connected rather with the similar dispute with the Ordnance of 1699 (see Mint Papers, III, fos. 429–30).

We have not traced this Memorial among the Treasury papers or the State Papers Domestic. It must have been written during the absence, in Hanover, of George I; that is (presumably) between 15 June and 10 November 1720. We place the draft arbitrarily here.

(2) Henry Slingsby (*c.* 1621–*c.* 1690), Master of the Mint 1662–80, an original F.R.S.

(3) See Letter 1335, and compare Letter 654, vol. IV.

(4) The brackets are Newton's and he has deleted the words between them.

1342 NEWTON TO FAUQUIER
27 JULY 1720
From the holograph original in the Library of the Royal Society[1]

Mint Office. 27th July. 1720

Sr

I desire you to subscribe for me & in my name the several Annuities you have in your hands belonging to me amounting in the whole to six hundred & fifty pounds per an for which this shall be your warrant.

ISAAC NEWTON

To Dr John Francis Fauquier

NOTE

(1) MS. MM 17.39. First printed in C. R. Weld, *History of the Royal Society*, I (London, 1848), p. 440; in More, p. 652; in De Villamil, pp. 19–20.

This instruction to Fauquier must be read in conjunction with a power of attorney Newton had granted to Fauquier on 19 April 1720 (Christ Church College, Oxford, Evelyn MS. 3, II, no. 97). By this Newton had empowered Fauquier 'to sell assign and transfer Three Thousand pounds Capital Stock being part of the Capitall and principall Stock I have in the Governoure and Company of Merchants of Great Britain trading to the South Seas and other parts of America...and to receive and give receipts for the consideration Moneys payable for the same ...'; this document appears to have been properly executed, stamped, and witnessed.

On 13 April, according to De Villamil (p. 23), the South Sea Company had offered for sale £2000000 of Capital Stock at a premium of 300 per cent; and on the following day declared a dividend of 10 per cent in stock, i.e. one free share given with every ten bought. The price was then £325, and it continued to rise rapidly. On the nineteenth, Newton decided to take advantage of the rising market by selling a portion of his holding, and one must presume this was done. There is no evidence to show how he employed the proceeds of this sale of stock, presumably about £10000.

De Villamil (pp. 23–6) interpreted the present document of July as an instruction to Fauquier to transfer into Capital Stock the long-term South Sea annuities he held, bringing him in £650 p.a.; the capital value of such an investment (at the fixed rate of 5 per cent) would have been £13000. Newton, after some hesitation, seems to have decided to take advantage of the Company's conversion terms, in order to exchange his annuities for Capital Stock. At this time the Company offered £770 in South Sea Stock and £575 cash for each £100 annuity, hence Newton might have received by the conversion £5005 stock and £3737. 10s. 0d cash. Effectively, Newton bought his stock at about 185, when the market price was at least 900; and reduced the extent of his investment. He could have sold out after the conversion for a large profit, but did not do so.

Whether or not De Villamil's interpretation of this particular document is correct, his conclusion that Newton was now left with £5000 Capital Stock, which he continued to hold to the time of his death, is certainly mistaken. The power of attorney to Fauquier shows that he continued to hold some Capital Stock after 19 April. Moreover, another document still (also unknown to De Villamil), Letter 1397, shows that up to August 1722 (after the Bubble had burst) Newton was still holding £21 696. 6s. 4d worth of South Sea Stock on which he was receiving a dividend of three per cent.

The estate inventory (De Villamil, p. 55) shows Newton possessed of £5000 Capital Stock valued (at 104) at £5200, as well as £5000 in 5 per cent annuities.

It is thus clear that far from Newton's holding of South Sea Stock remaining constant from 1720 to the time of his death, he must have disposed of over £16 000 of the stock between August 1722 and the time of his death. De Villamil's account of Newton's finances must therefore be treated with extreme reserve. Even the assumption that the present terse instruction refers to South Sea annuities, or to such annuities only, may well be questioned.

One further point may be made, as a caution to any one attempting to reconstruct Newtons' income, or the magnitude of his estate. The management of the Mint finances at times left large credit balances in the Master's hands, carried forward from year to year, which in accord with the custom of the age he was free to invest (at his own risk) in order to derive the interest or profit for his own advantage. In the years before 1720, for example, Newton's Mint accounts (P.R.O., Mint/19, 4) show sizeable credit balances, which, however, by 31 December 1720 had been reduced to a mere £10. 18s. 6½d, because the sums imprest to the Master in that year were less than his expenditure. This sharp return to near-par may explain why Newton sold South Sea Stock in 1720, to put money back into the Mint. In the years after 1720 his credit balance swelled monstrously; it was already by 31 December 1722 at £9286. 5s. 4d and the increase continued because imprests to the Master largely exceeded his disbursements. In the last account made while Newton was alive (on 28 May 1726), to 31 December 1725, the undischarged balance in the Master's hands was £18974. 13s. 11d. Presumably Newton did not hide this sum in his mattress but invested it securely; it is impossible to say what fraction of his holding of, say, South Sea Stock, at any time was 'really' his own.

The final Newton account (of 13 May 1727), to 31 December 1726, shows him owing the Crown over £22000, almost twice his total disbursements as Master for that year. The account bears an annotation by Conduitt: ''Mem[orandu]m. The said balance of £22277. 14. 4½d was by order of the Lords of the Treasury paid to his successor John Conduitt & charged upon him in his first Account. J.C.' The payment was presumably made, or to be made, out of Newton's estate.

1343 ROBERT SMITH TO NEWTON
12 AUGUST 1720
From the holograph original in the University Library, Cambridge[1]

Cambridge August. 12th. 1720.

Hon[our]ed Sr

Your kind enquirys after the progress of printing Mr Cotes's works[2] make me beleive that an account of it may not be unacceptable to you.[3] I have therefore sent You a copy of what is printed off. The New Tables of Fluents, which I calculated my Self and am going to print, are to make the 4th & last part of ye Harmonia Mensurarum.[4] The old Tables are printed in the 2d part,[5] just as Mr Cotes left 'em, because they are a finish'd piece in the Way which he first happen'd upon. For this reason I thought it proper the World should have 'em, notwithstanding they are all contain'd in the New Tables in a dress somewhat different & founded upon a more General Theorem or two, which the Author invented a little before his death, but had not time to build upon 'em. I was very glad to understand from Dr Bentley that a publication after the manner I have described was most agreeable to your sentiments. The Miscellaneous Works are to be bound up with the Harmony of Measures & I design to add 3 other small pieces[6] to 'em De Descensu Gravium, De Motu Pendulorum & De Motu Projectilium as soon as I can procure the best copy of 'em, which I understand is in Lord Harold's hands.[7] This is all I think proper to publish at this time. He had begun an Analytical investigation of the curves of ye first order & their principal propertys, but the piece is so imperfect that I shall reserve the publication of it & some other small papers of less moment to an other time.[8] I am with much respect your very obedient humble Servant.

Rob Smith

NOTES

(1) Add. 3983, no. 41, fo. 1r; compare Letter 1209, vol. VI, and Letter 1310.

(2) *Harmonia Mensurarum sive Analysis & Synthesis per Rationum & Angulorum Mensuras Promotæ: Accedunt alia Opuscula Mathematica per Rogerum Cotesium* (Cambridge, 1722), edited by Robert Smith.

(3) This does not necessarily imply that Newton wrote to Smith; from Smith's words below it seems likely that the message was conveyed by Richard Bentley.

(4) *Ibid.*, pp. 113–249, about a third of the book.

(5) *Ibid.*, pp. 43–76.

(6) These three short tracts occupy pp. 73–91 of the separately paginated *Opuscula Mathematica* appended to the *Harmonia Mensurarum.*

(7) Henry Grey (1671–1740) succeeded his father as Earl of Kent in 1702, was created Earl Harold and Marquess of Kent in 1706 and Duke of Kent in 1710. Two of these papers printed in the *Opuscula* were mentioned by Cotes in a letter to his uncle, John Smith, in 1708 (Edleston, *Correspondence*, p. 199) but the connection with 'Lord Harold' is obscure. The Duke's son, also Henry (*c.* 1697–1717), who perhaps bore that courtesy title after 1710, was admitted to Trinity College, Cambridge, in December 1713 and may therefore have been acquainted with Cotes.

(8) These were never published as such.

1344 DES MAIZEAUX TO CONTI
11 SEPTEMBER 1720
From the holograph in the British Museum[1]
For the answer see Letter 1360

a Monsr. L'Abbe Conti

Sept. 11. [1720][2]

Monsieur

Je me flate que vous aurez reçu le *Recueil de diverses Pieces de Messieurs Leibniz, Newton, &c*, que j'ai eu l'honneur de vous envoyer. Le public vous est redevable de la plus grande partie de ces Pieces, puisque vous m'avez fait la grace de me les fournir[3]...[4]

Vous me ferez beaucoup de plaisir, Monsieur, de m'aprendre ce que vous pensez de ma Preface.[5] Vous y trouverez l'histoire du démêlé de Mr. Leibniz avec Mr. Newton sur l'Invention des Fluxions, ou du Calcul differentiel. Je me suis borné à la simple qualité de Rapporteur; n'ayant aucun interêt à prendre parti dans cette dispute. Mr. Newton m'a paru assez content de ce Morceau de la Preface:[6] je me flate que les Amis de Mr. Leibniz n'y trouveront rien qui puisse leur déplaire. Mais s'il y a quelque chose qu'ils desaprouvent, & qu'ils veuillent bien prendre la peine de m'en informer; je profiterai avec plaisir de leurs éclaircissemens. Vous m'obligerez aussi beaucoup, Monsieur, de me communiquer vos remarques, sur tout à l'egard de ce que je puis avoir omis, ou n'avoir pas assez étendu. Car vous et[es] parfaitement au fait de toute cette affaire...[4]

Les Pieces qui regardent Mr. Newton n'ont point été imprimées sans son Aveu. Il m'a communiqué une Copie de la Lettre de Mr. Leibniz à Madame de Kielmansegg,[7] que j'ai comparée avec celle que vous aviez eu la bonté de me donner...[4]

A l'egard de la petite Lettre du 7 de Juin 1713 [N.S.],[8] inserée dans cette Lettre & atribuée à Mr. Bernoulli; vous n'ignorez pas, Monsieur, qu'elle avoit deja été publiée sous le nom de Mr. Bernoulli dans les *Nouvelles literaires* du 28 1715 [N.S.]:[9] & il y a lieu de croire que Mr. Leibniz lui-meme l'avoit envoyée au Journaliste. Mr. Leibniz l'atribue aussi à Mr. Bernoulli dans la Lettre qu'il vous écrivit le 9e d'Avril 1716 [N.S.],[10] pour répondre a celle de Mr. Newton, que vous lui aviez envoyée. *On connoit assez*, dit il (1) parlant de la feuille volante publiée en Latin, *le nom & le lieu de la Lettre y inserée d'un excellent Mathematicien que j'avois prié de dire son sentiment sur le* Commercium. Ainsi nous n'eumes pas le moindre soupçon, ni vous ni moi, que cette Lettre ne fut effectivement de Mr Bernoulli. Mais lorsque j'etois sur le point d'envoyer ma Preface en Hollande, Mr Newton me montra une Lettre[11] que Mr. Bernoulli lui avoit ecrite, ou il desavouoit cette petite piece. J'ai crûs que Mr. Bernoulli seroit bien aise que le Public en fut informé; & j'ai ajouté ce que vous trouverez page xlviii de cette Preface.[12]

La seconde edn. de l'Alg[ebre] de Mr Newton est fort avancée. &c.[13]

Nous aurons bientot les *Tables astronomiques* de Mr. Halley,[14] corrigées & augmentées. Elles fournissent une tres belle demonstration des Principes de Mr. Newton. On y verra que son systeme de la Gravitation (ou de l'Attraction) ne donne pas seulement la raison du mouvement des planetes; mais meme de celui des Cometes, & du progres des equinoxes, qu'on regardoit auparavant comme inexplicables. Car vous savez Monsieur, que ce ne sont pas ici des Hypotheses, ou des Romans philosophiques; mais des Principes qui répondent parfaitement aux phenomenes. Et après cela, je vous avoue que j'ai peine a comprendre que des Mathematiciens & des Astronomes, d'ailleurs très habiles, au lieu d'examiner si l'experience s'accorde avec ces Principes; aiment mieux se jetter dans des Speculations metaphysiques ou s'aheurter à la recherche de la Cause, ou de la nature de cette Attraction, qui sera peut etre toujours inconnue; lorsqu'il ne s'agit que de l'Effet, qui est visible & palpable. Des Cartes lui-meme abandonneroit aujourdhui ses ingenieuses Hypotheses, pour s'attacher à des Principes demonstres par l'experience. Il voudroit du moins connoitre à fonds un Systeme si simple, si general, & qui a de si grandes apparences de Verité. Mais ce n'est pas une petite affaire que surmonter la force des Prejuger & de l'habitude.

¹ *Recueil de diverses Pieces* &c Tom. 11. p. 51 [*Des Maizeaux's footnote*].

NOTES

(1) Birch MS. 4284, fos. 222–3; Des Maizeaux's own copy or draft.

(2) The year is not given. Printing of Des Maizeaux's *Recueil* was finished by June, 1720 (see Letter 1339), so we assume this letter was written in the same year.

(3) Conti had supplied much of the manuscript material for the *Recueil*. See Letter 1281, vol. VI, and associated correspondence between Conti and Des Maizeaux in the British Museum, Birch MS. 4282.

(4) The dots are in the copy; it is not clear whether or not they are intended to indicate some omission.

(5) See Letter 1339, note (3), p. 93.

(6) Compare Conti's later letter, Letter 1366.

(7) It seems unlikely that Newton would have sent Des Maizeaux a copy of this letter (see Letter 1203, vol. VI). He claimed to know nothing about the *Recueil* until he received proof sheets (including Leibniz's letter to the Baroness von Kilmansegge) from Des Maizeaux. In Letter 1359, however, Des Maizeaux does report that Conti had sent a copy of this letter to Newton before furnishing himself with one.

(8) That is, the excerpt of Bernoulli's letter to Leibniz (Letter 1004, vol. VI) which had previously been printed anonymously in the *Charta Volans* (Number 1009, vol. VI).

(9) See Letter 1172, vol. VI.

(10) See Letter 1197, vol. VI. Leibniz did not, in fact, reveal there the author of the letter, but only implied, as Des Maizeaux here goes on to say, that his identity would be obvious to everyone.

(11) See Letter 1330; Newton subsequently tried to delay the publishing of the *Recueil*.

(12) On p. xlviii of the preface, after giving his reasons for thinking Johann Bernoulli to be the author of the letter of 7 June 1713 [N.S.] (Letter 1004, vol. VI) printed in the *Charta Volans* (Letter 1009, vol. VI), Des Maizeaux abruptly concludes with the words 'Cependant je viens d'apprendre que M. Bernoulli la desavoue'; presumably this is the addition of which he speaks. It was later to cause Bernoulli considerable annoyance.

(13) Presumably *Arithmetica Universalis; sive De Compositione et Resolutione Arithmetica Liber*, 2nd edition (London, 1722), enlarged and improved by Newton himself.

(14) While no astronomical tables by Halley were published in his lifetime, the preface by John Bevis to his posthumous edition of them (*Edmundi Halleii Astronomi dum Viveret Regii Tabulæ Astronomicæ* (London, 1752 but begun 1749)) makes it clear that he was reproducing an earlier printed version which Halley had set in 1717–19. Newton clearly possessed a copy of these printed tables (Letters 1449 and 1460), but its location (if it survives) and that of any other extant copy of this first printing are not now known. Apparently Halley decided, after Flamsteed's death on 31 December 1719 and his own appointment as Astronomer Royal on 9 February 1720, to postpone the publication of his tables until he had completed a long cycle of lunar observations; his tables were, in fact, largely based on Flamsteed's observations though the computations were Halley's own. Halley was not, of course, concerned with the edition of Flamsteed's *Historia Cælestis* in 1725.

1345 CHARLES DELAFAYE TO NEWTON

11 OCTOBER 1720

From the original in the Mint Papers[1]
For the answer see Letter 1346

Whitehall Octr. 11th 1720

Sr

Having read to ye Lords Justices ye enclosed Mem[oria]l of John Rotherham proposing a new Method of Coining; Their Exc[ellenc]ys are pleased to refer to you ye Examination of that Matter. I am &c

CH. DELAFAYE

To Sir Isaac Newton

NOTE

(1) I, fo. 495r, written above a copy of the Memorial which (like another letter, dated 19 August 1720, and apparently not addressed by Rotherham to Newton, at fo. 492) promises to disclose, for a reward, a new and highly profitable method of making coin which cannot possibly be counterfeited.

Charles Delafaye was an Undersecretary of State in the Northern Department. Possibly the same person had proceeded B.A. at Oxford in 1696.

Rotherham describes himself as a gentleman of Barnet in Hertfordshire, and elsewhere claims to have invented the idea of burning 'the damaged and corrupt tobaccos' whereby the revenue is increased £100 000 p.a., for this he was promised (he says) a pension of £2000 per year, which he has never received.

1346 NEWTON TO DELAFAYE
OCTOBER 1720
From the holograph draft in the Mint Papers[1]
Reply to Letter 1345

In obedience to an Order of Reference of the late Lords Justices dated 11th Octob. 1720 upon a Memorial of Mr John Rotherham proposing a new method of coyning the moneys of Gold & Silver so as to prevent the counterfeiting thereof & to make it more durable provided he may be assured of a Reward before he discovers his secret & praying that it may be examined: I humbly represent that he offers nothing to be examined & without examination I am in the dark & know not what report to make. I take him to be a trifler more fit to embroyle the coinage then to mend it.

NOTE

(1) I, fo. 494. In 1728 Rotherham again sought a recompense, claiming that Newton and the other Mint Officers would make no report on his invention before he had disclosed the whole secret, which he refused to do (*Cal. Treas. Papers, 1720–8*, p. 509).

1347 NEWTON TO JAMES CRAGGS
24 OCTOBER 1720
From the original in the Public Record Office[1]

Mint Office 24th Oct 1720

Sr

The Annext contains ye Names of 16 persons in London Accus'd before Mr Delafay for Keeping Engines or presses in their houses Usefull for makeing of Counterfeit Money[2]

As such a Number cannot be Apprehended at once without a suitable Force Nor ye Constables alone be trusted in the severall Divisions where these persons live I therefore desire you'll please to order the assistance of some of his Majties. Messengers and begg leave to subscribe myselfe

Sr.

Your most humble & most obedt. Servt:

ISAAC NEWTON

NOTES

(1) State Papers Domestic, SP/35, 23, no. 119. The signature is probably Newton's own, the rest is written by a clerk, as is the list of craftsmen enclosed, as follows: Keyling, engine-maker, Blackfriars; Long, engraver, Old Bailey; Buntin, Key bow maker, Foster Lane; Stanley, button-maker, Cripplegate; Cooth, engraver, Aldersgate Street; Trewboan (*sic*) button-maker, Aldersgate Street; Lane, bow key (*sic*) maker, Old Street; Boddington 'att ye presse in St. John's street'; Leversage, St John's Street; Auksford, button-maker, Peter's Street, Saffron Hill; Dowsett, Key bow maker, Shoe Lane; Clement, button-maker, New Street Square; Rolls, button-maker, same address; Grilliard, engraver, Katherine Street, Strand; Perquett, watchmaker, Beaufort Buildings, Strand; Wallis, gold chain maker, Kensington.

The appeal for help was signed by Newton (presumably in the absence of the Warden) as a mere formality, as is made plain by the covering note in which Pinckney, the Mint Solicitor, sent it to Delafaye (SP/35, 23, 120), in which he apologizes for 'Sr Isaac Newton's delay' in producing the necessary signature.

James Craggs (b. 1680), son of the Postmaster-General, had been appointed Secretary at War in 1717, and became Secretary of State on 16 March 1718. He died on 16 February 1721 after having been (like his father) scandalously implicated in the affairs of the South Sea Company.

(2) As early as 14 September Thomas Dearsley had laid information before Delafaye against these craftsmen (the names variously spelled), who kept engines and tools that could be used for coining 'being the like Instruments & Tools for which Ralph Mansfield was lately convicted of High Treason'.

The chief offender was the fly-press, conveniently if illegally employed for stamping out buttons and other small articles of metal.

For the outcome of this affair see Letter 1358.

1348 VARIGNON TO NEWTON

17 NOVEMBER 1720

From the holograph original in King's College Library, Cambridge[1]
For the answer see Letter 1353

Nobilissimo Doctissimoque viro
D.D. Newtono, Equiti aurato,
Regiæ Societatis Anglicanæ Præsidi Dignissimo
S.P.D.
Petrus Varignonius

Exoptatissimam mihi Effigiem[2] tui, qua me donare dignatus es, Vir Humanissime ac Munificentissime, Gaudenti gratissimoque animo nuper accepi. Tui spectandi percupidus capsam statim distraxi, evolutaque tela, in hujus Effigiei vultu & fronte & oculis quasi spirans mihi visum est tuum summum atque eminens Ingenium cum Oris dignitate conjunctum, etiam-numque videtur. Paucis post diebus venit ad me Cl. Taylorus[3] (quatuor abhinc vel quinque mensibus hic habitans) qui eam intuitus attente, suo usus conspicillo, tibi simillimam esse pro certo mihi affirmavit; quod admodum me delectavit ac delectat. Porro sculptam alteram tui Imaginem,[4] jam inde a Decem circiter annis habebam ex Dono amici Angli (Oxoniensis nomine *Arnold*)[5] qui cum me sæpius de te magnalia loquentem audisset, reversus Londinum, illinc eam ad me misit, pergratam mihi fore existimans: recte quidem; Sed cum Sculpta tui similitudinem ex vero non effingat æque ac picta, hanc nihilominus semper exoptavi, qua nunc mihi datur videre tandem Illustrissimum ac Doctissimum eum virum quem amplius triginta annos summa veneratione colebam ob ingentia ejus merita præsertim in Mathesim quam promovit & auxit immensum, cujusque legibus astrictam primus demonstravit esse Naturam. Quantas autem pro tanto Dono (quod antea pecuniæ summa quavis emissem si aliunde quam a perhonorifica mihi tuæ liberalitatis magnificentia obtinere potuissem) gratias agere tibi debeam, optime intelligo & intime sentio, sed tantas ut eas expedire verbis nequeam; nec etiam eas quas habeo tibi maximas pro eo quod me monuit Cl. Moivræus[6] te non dedignari mei quoque imaginem[7] quam nudiusquartus idcirco misi Do. Ayres (capellano D.D. Equitis Sutton,[8] Excellentissimi legati vestri apud nos) in longiore capsula volutatam, quam pridie mihi officiosissime promiserat se missurum fore Londinum ad Dum. Preverau (apud D.D. Craggs[9] Sanctioris consilij Anglicani commentariensem) ut eam tibi reddat, quam benigne accipias Rogo. Vale, mihique tuorum in me Beneficiorum æternum memori favere perge.

Dabam Parisijs die 28 Novemb. 1720. *N.S.*

P.S. Post Scriptam hanc Epistolam D. Nicole[10] ex Anglia recens me invisit ac monuit, dum apud te pranderet aut cœnaret, propinasse te toti generatim Academiæ Nostræ Parisiensi, speciatimque Cl. Fontenello, ac etiam mihi; pro quo honore novas habeo tibi gratias & ago maximas. Contemplatus etiam D[omin]us Nicole pictam Effigiem tui, de ea censuit penitus idem ac D. Taylorus, nimirum eam tibi persimilem esse; quod meum de ea obtenta gaudium auxit.

Translation

Pierre Varignon presents a grand salute
to the very noble and learned Sir Isaac Newton
most worthy President of the English Royal Society

I have recently received, with a most grateful and rejoicing heart, a portrait[2] of you, which I very much desired and which you were so good as to present to me, most kind and generous Sir. I tore open the parcel at once, wanting so much to look at you, and when the canvas had been unrolled your most high and lofty intellect together with the dignity of your aspect seemed to me as it were alive in the countenance and forehead and eyes of this likeness, and still so seems. A few days afterwards Mr Taylor[3] came to see me (he has been living here for the last four or five months) who assured me, having looked at it attentively and used his eyeglass, that it was certainly most like you, which greatly delighted and still delights me. Further, about ten years ago now, I had another engraved likeness of you,[4] the gift of an English friend (an Oxford man named Arnold)[5] who, since he had often heard me speaking of you as a great man, sent it to me from London after his return there, judging, and rightly so, that I would be most delighted with it. But since an engraving cannot represent your likeness so fully to the life as a painting does, I none the less always greatly longed for the latter, by which I am now at last enabled to see that most famous and learned man whom I have respected for more than 30 years with the greatest veneration on account of his extreme merit in mathematics, which he has furthered and added to enormously, and by whose laws he first demonstrated nature to be bound. Moreover, I do very well understand and sense within my being the very great gratitude I ought to feel for so great a gift for which I would formerly have paid any sum of money if I could have obtained it by other means than through the magnificent honour of your liberality to me; but it is too great for me to express in words; nor [can I find words for the very great gratitude] I feel towards you on account of what Mr De Moivre[2] told me, [namely] that you would not refuse a portrait of me also,[6] which I therefore sent three days ago by Mr Ayres (chaplain to His Excellency Sir [Robert] Sutton,[7] your ambassador here with us), rolled up in a rather long box, which the day before he very kindly promised me to send to London to Mr Preverau (at Dr Craggs',[8] English Secretary of State), so that he might deliver it to you; I ask you to accept it graciously. Farewell, and may you continue to think well of me, who am ever mindful of your kindnesses toward myself.

Paris, 28 November 1720, N.S.

P.S. After this letter was written, Mr Nicole,[9] recently come from England, visited me and told me that, when he took dinner or supper at your house, you drank to the whole of our Parisian Academy in general, and to Mr Fontenelle in particular, and also to myself; for which honour I am most grateful to you and give you my greatest thanks. Mr Nicole too, having examined your portrait, gave entirely the same judgement of it as Mr Taylor, namely that it is a good likeness of you; which increased my joy in having it.

NOTES

(1) Keynes MS. 142(F); microfilm 1011.26. Printed in Brewster, *Memoirs*, II, pp. 496–7.

(2) From a letter addressed to Bernoulli by Varignon, 19 January 1721, N.S. (see Bernoulli Edition), we learn how the portrait was obtained. In 1715 Varignon had asked Conti to procure one for him, but to no avail. Then he enlisted De Moivre's help to have a copy made of an already existing portrait, but Newton had given away the one he had in mind. Newton promised, however, to have one specially painted for Varignon, and also expressed a desire for Varignon's picture in return.

On receiving the portrait Varignon wrote, 'Dans ce portrait de M. Newton, il ne paroist pas avoir plus de 50 ans, tant il y a l'air frais & vigoureux: cependant M. Taylor, qui etoit encore ici lorsque ce portrait m'arriva...me dist qu'il etoit fort ressemblant.'

The painter of this portrait was Sir Godfrey Kneller, as William Stukeley relates:

in the year 1720 Sr. Isaacs picture was painted by Sr. Godfry Kneller to be sent to Abbé [Varig]non in France; who sent his picture to Sr. Isaac. both Sr. Isaac & Sr. Godfry desired me to be present at all the sittings. it was no little entertainment, to hear the discourse that passd between these two first of men in this way. thô it was Sr. Isaacs temper to say little, yet it was one of Sr. Godfrys arts to keep up a perpetual discourse, to preserve the lines and spirit of a face.

On the same occasion Newton refused to allow Kneller to paint a profile portrait of himself for Stukeley, as the latter afterwards wrote to Conduitt. (A. Hastings White, *Memoirs of Sir Isaac Newton's Life by William Stukeley M.D. F.R.S. 1752* (London, 1936), pp. 12–13; Brewster, *Memoirs*, II, p. 414, note (3). We owe these references to D. T. Whiteside.)

In Stukeley's manuscript (Royal Society MS. 142), whence the quotation above is transcribed, the Abbé's name is written 'Bignon', clearly as the result of misunderstanding or misrecollection—there can be no doubt that Varignon was meant.

Despite the article by David Eugene Smith, 'Portraits of Sir Isaac Newton' (in W. J. Greenstreet (ed.), *Isaac Newton, 1642–1727: A Memorial Volume edited for the Mathematical Association* (London, 1927)), the history of Newton's portraiture is far from clear. It seems that Kneller painted Newton on at least three occasions. There is the splendid informal portrait of 1689 (the property of the Portsmouth Settled Estate) reproduced as the frontispiece to vol. I; a portrait of 1702, in the National Portrait Gallery, London; and this of 1720, sent to Varignon and acquired in the eighteenth century by Lord Egremont, which has hung at Petworth House ever since. This last is signed and dated by Kneller on the back, hence there is no doubt as to its identity.

(3) Towards the end of 1720 Brook Taylor was invited to spend some time at La Source, Lord Bolingbroke's country seat near Orleans, where he stayed until the spring of 1721; see William Young's biographical introduction to Brook Taylor, *Contemplatio Philosophica*, pp. 31–32. Taylor found opportunities to visit Varignon in Paris, as we learn from Varignon's correspondence with Bernoulli.

(4) It is not possible to identify the engraving.

(5) For John Arnold see vol. VI, p. 27, note (1). Arnold was also a correspondent of Johann Bernoulli and Leibniz; see Letter 1318.

(6) We have not identified this portrait.

(7) Sir Robert Sutton, 1671–1724, ambassador in Holland, Constantinople and Paris; M.P. for Nottingham.

(8) For James Craggs see Letter 1347, note (1), p. 103.

(9) François Nicole (1683–1758), French mathematician and close friend of Monmort, under whom he studied.

1349 JAMES WILSON TO NEWTON
15 DECEMBER 1720
From the holograph original in King's College Library, Cambridge[1]

London Decemb. 15. 1720.

Sir,

I saw the other day, in the hands of a certain person, several Mathematical Papers, which, he told me, were transcribed from your Manuscripts.[2] They chiefly related to the Doctrine of Series and Fluxions, and seemed to be taken out of the Treatises you wrote on those Subjects in the years 1666 and 1671.[3] I was not permitted to peruse them thoroughly, but one of the papers I particularly took notice of, and it contained a deduction of your Binomial Theorem from a corallary in your Quadratures, with some improvement, as a Series for the Rectangle under any two Dignities of two Binomials.[4] These papers, I observed, had been very incorrectly copied, so that I endeavoured all I could, to dissuade the Possessour of them from getting them printed, of which nevertheless he seemed very fond. I therefore thought it behouved me to acquaint you with this Matter, and as I have not the honour to be known to you, I believed the less troublesome way to do it, would be that of a letter.

And now, Sr, permit me to say, that it is the earnest desire of every Body conversant in these Subjects, that you would be pleased to publish yourself what you have formerly written. For this would effectually prevent their being ever printed incorrectly, and unworthy of yourself. Nor ought you to deny that pleasure to your Well-wishers and Admirers, in reading your noble Inventions of this kind, whereof you have expressed yourself to have been so sensible, at the time you first made these Discoveries. Your Analysis per æquationes numero terminorum infinitas, Quadratures &c. give us so exquisite a delight, that when we read your Letters to Mr. Oldenbourgh,[5] and your Remarks[6] on Leibnitz's Reply to your Letter to the Abbot Conti, we glow, as it were, with a desire of seeing all that you wrote on these subjects in 1671 and the years preceeding. I have heard indeed, that you were prevented from publishing one Treatise by reason it had in it the Determination of the Radius of Curvity, which Huygens published afterwards in his Horologium.[7] But this objection can now be of no force, since the world has been lately informed by your Letter written to Mr Collins, in 1672 Decemb. 10.[8] that you had applied your Method ad resolvendum abstrusiora problematum genera de curvitatibus &c. long before the publication of Huygens's book. Nor indeed was this unknown to Apollonius in respect to the Conick Sections, as appears from his 5th Book.[9]

It was a very great satisfaction to your Country-Men and Friends, to observe, that by the happy Discovery of Mr Collins's papers,[10] you had an Oportunity of triumphing over such disingenuous persons, who laboured all they were able, to defraud you of the honour of some of your Inventions. But such is the force of Prejudice on some minds, that you cannot but observe, that there are in the world dishonest men, who contrary to their own Conscience and Knowledge, still raise a Clamour on this head. To shame therefore such obstinate people, to make your Right to these Inventions evident to all, even the least knowing in these matters, and to put an end for ever to all Disputes, the best Method would be to publish all that you have formerly written on this Subject. Whereby we should have an exact and adequate notion of Fluxions and their Uses, which cannot be had from what has been delivered by others. But as they produced these Things abroad first, those that are Learners have recourse to their Writings, and consequently mention them as the Authours thereof. This has happened through your own backwardness in giving to the world what you had discovered so long ago. However, there is still a way left to retrieve all; for as their pretended methods are grounded on the Notion of Indivisibles, so they have given a wrong Idea of what you alone had found out, and have erred egregiously, when they ever attempted to apply Second &c Differences, as they call 'em, to Mathematical figures.[11] This tho' they cannot but be now at length very sensible of, yet by their cavils they would dissemble their being conscious of their errors. But these would be so apparent to every Eye, if you would publish all your papers, that those who had a mind to be rightly instructed in these Matters, would have recourse alone to your immortal Writings; so that all succeeding Mathematicians would constantly mention with honour the Series, Fluxions, &c., of the great Newton, when the differentials and integrals of Leibnitz and Bernoulli shall be quite forgotten.

> I am your most obedient
> and most humble servt,
> A.B.

P.S.

I had just the liberty to transcribe the Theorem[12] I mentioned at the Beginning of my Letter; and it was this:—

$$\overline{P+PQ}|^{\lambda} \times \overline{M+MN}|^{\mu} = P^{\lambda} M^{\mu}$$

$$+ \dfrac{\dfrac{\lambda \times Q}{\mu \times N} A}{1}$$

$$+ \dfrac{\dfrac{\overline{\lambda-1} \times Q}{\overline{\mu-1} \times N} B + \overline{\lambda+\mu} \times NQA}{2}$$

$$+ \dfrac{\dfrac{\overline{\lambda-2} \times Q}{\overline{\mu-1} \times N} C + \overline{\lambda+\mu-1} \times NQB}{3}$$

$$+ \dfrac{\dfrac{\overline{\lambda-3} \times Q}{\overline{\mu-3} \times N} D + \overline{\lambda+\mu-2} \times NQC}{4}$$

&c

Amongst the papers I likewise observed there were some, which deduced even the first principles of Geometry from the Fluxions of Points &c.[13]

I have since met with another Person, who told me, he had likewise a Copy of your Manuscripts. But he would not let me see them, or inform me how he came by them. I imagine, when you sent any of your Friends your papers, the person they got to transcribe them, took a double copy, which is a frequent practice, in order to make profit by it. So that they are in different hands. To prevent these things being ever published incorrectly, the only way is to let them come abroad yourself. For to declare that such papers as shall be published without your knowledge or consent, are imperfect and faulty, will not be sufficient to deter some Bookseller or another, from adventuring on the printing them for the hopes of Gain. Nor need the publishing [of] them be any Trouble to you; for any of your Friends, would gladly undergo the Labour of seeing them correctly printed.

In the introduction to your Quadratures, you have given us an exact Idea of Fluxions, but it is too short, and does not instruct us how the superior Fluxions are represented by Lines.[14] The truth is, the trifling objections that are made by your Antagonists would never have been raised, if you had given us your papers, where your Fluxions are illustrated by various Examples. And is not this a Pity, since you have pleased to permit the publication of your

[work] illustrating the common Algebra. Nothing less than this can make all Foreigners and prejudiced persons acquiesce, and at length to acknowledge you to be the Inventor of a Method that is so admirably suited both to the investigating and demonstrating the most difficult Mathematical Truths.

NOTES

(1) Keynes MS. 143.1; printed in Brewster, *Memoirs*, II, pp. 440–3.

Wilson signed a subsequent letter to Newton with his own name, from which it is obvious that he was the writer of this. Presumably he was the James Wilson (*c.* 1690–1771) who entered Trinity College, Cambridge, in 1707; he graduated in medicine at Leyden in 1713 and proceeded M.B. (1718) and M.D. (1728) at Cambridge. He was a close friend of Pemberton, and published *A Course of Chemistry...formerly given by the late learned Doctor Henry Pemberton* (London, 1771); ten years previously Wilson had also published (in London) the *Mathematical Tracts of the late Benjamin Robins, Esq.*, the second volume of which not only reprints Robins' *Discourse concerning the nature and certainty of Sir Isaac Newton's Methods of Fluxions and of Prime and Ultimate Ratios* (London, 1735) but contains (in an Appendix, pp. 340–80) an account and defence of Newton's mathematical discoveries based on the manuscripts described in this letter, and others, which Wilson had examined.

(2) Several copies of Newton's early mathematical manuscripts passed into Wilson's own hands, taken from other copies (deriving from Newton or John Collins) in the possession of William Jones. These copie sare frequently mentioned in Whiteside, *Mathematical Papers*, notably II, pp. 293–4 and III, pp. 10–13. Newton steadfastly refused Wilson permission to publish these papers, nor did he succeed in doing so after Newton's death. The first of them appeared in Horsley's *Opera Omnia*.

(3) That is, the October 1666 tract on Fluxions (see Whiteside, *Mathematical Papers*, I, pp. 400–48) and the tract 'De Methodis Serierum et Fluxionum' (*ibid.*, III, pp. 32–328).

(4) See note (12) below.

(5) The *Epistola Prior* and *Epistola Posterior* of 1676; see Letters 165 and 188, vol. II.

(6) See Number 1211, vol. VI.

(7) See Whiteside, *Mathematical Papers*, III, p. 165, note (308). Huygens' *Horologium Oscillatorium* was published in 1673, but the results on radius of curvature he gives there he had discovered in 1659. It is not known to what extent Newton was aware of Huygens' work prior to its publication.

(8) See Letter 98, vol. I.

(9) *Apollonii Pergæi Conicorum Lib. V. VI. VII Paraphraste Abalphato Asphahanensi nunc primum editi* (Florence, 1661), edited by G. A. Borelli, V, props 51 and 52. See also Whiteside, *Mathematical Papers*, III, p. 152, note (273), and p. 174, note (335).

(10) Collins had died in 1683, and in 1708 his papers fell into the hands of William Jones. See also note (2) above.

(11) This is a reflection on Leibniz's 'Tentamen de motuum coelestium causis', *Acta Eruditorum* for February 1689, pp. 82–96, where Leibniz had made a mistake in the use of second differences which Newton had noticed (see vol. VI, p. 119, note (1)). Keill had made the error public, and used it to imply Leibniz's general inability to manipulate second differences, in his 'Answer' (*Journal Literature de la Haye* for July and August 1714, p. 352; see vol. VI, p. 90, note (1)), and Newton had repeated the accusation in the 'Recensio' (*Phil. Trans.* for January and February 1715, **30**, no. 342, 208–9; see vol. VI, p. 242, note (2)).

(12) The quotation is apparently taken from an addendum by William Jones to a partial transcript by him of Newton's early work on fluxions. The transcript (U.L.C., Add. 3960(1), fos. 1–55) is mainly of the October 1666 tract, but the addendum relates to the 1693 version of Newton's *De Quadratura Curvarum*.

The portion Wilson transcribes in his letter is from fo. 54, and is preceded, on fos. 52–53 by a brief explanation. Jones used as a starting-point Propositions 4 and 6 of the *De Quadratura Curvarum*. These deal with the differentiation and integration of functions of the form $z^\theta R^\lambda S^\mu$ where R, S are polynomials in z. Thus having constructed a series expressing the differential of zRS in Proposition 4, using this the series for $z^\theta R^\lambda S^\mu$ is constructed. However, the expansion Wilson gives in his letter is a special case related to these Propositions, using the same device for expressing the coefficient of each term by means of the terms preceding it. Hence (using modern notation) $A = P^\lambda M^\mu$, $B = (\lambda Q + \mu N)A$,

$C = \frac{1}{2}[\{(\lambda-1)Q + (\mu-1)N\}B + (\lambda+\mu)NQA]$ and so on. The general term is thus

$$\frac{1}{K}[\{(\lambda-K+1)Q + (\mu-K+1)N\}C_{K-1} + \{(\lambda+\mu-K+z)NQ\}C_{K-2}].$$

(13) Wilson probably refers here to the opening passage of the October 1666 tract (see Whiteside, *Mathematical Papers*, I, pp. 400–2), or the earlier one of May (*ibid.* pp. 319–9).

(14) The introduction to Newton's *De Quadratura Curvarum* makes no mention of second or higher fluxions

1350 SEBASTIEN TRUCHET TO NEWTON

?1721

From the holograph original in King's College Library, Cambridge[1]

Vir clarissime

Non sine singulari, et gratitudinis, et voluptatis sensu, tuo nomine accepi ab eruditissimo De Varignon viro academico et socio, Exemplar, illius optimi et exquisitissimi de optica libri,[2] quem quidem iterata vice in lucem a te prolatum, Dominus Coste, pro sua in patriam benignitate, in sermonem gallicum convertit.

Vix tuum illud opus de lumine et coloribus, meas ad aures pervenit,[3] cum animum subiit ardens et experrectum ea tentandi experimenta studium, quæ in tuo tractatu tam nitide congeris. Nec mora; Duce et quasi auspice tum tui tractatus exemplari anglici, cum maxime opera famosi inter nos academici domini Geofroy,[4] quo quidem tui idiomatis interprete uti fas fuit et utile, citata a te experimenta pleraque ita tentavi, presente et plaudente Eminentiss-imo cardinali de Polignac,[5] nec non frequentissima optimatum et in rebus phisicis versatissimorum hominum Corona, ut tuorum quasi sint germana;[6] quam quidem fortunam debeo meo domicilio, quod meridiem ita aspicit, ut a sole per septem aut octo æstate, hyeme tres aut quatuor horas recreari soleat et illuminari, adeo ut præsidio speculorum metallicorum directionem radiorum

solarium, pro ut opus est, permutem; cumque paries distet ab altero, longitudine 42 pedum galliæ; penes me est, intercipere, separare et frangere, etiam
vicibus repetitis, eundem luminis radium, vel antequam ad extremum cubiculi
pervenerit.

Meæ, (fatebor quod res est,) invideo Epistolæ, illud quod ad te facit iter,
quam nitidius mihi affulgeret ille dies quo tuo gratissimo conspectu, et
colloquio fas esset frui, Vir experientis[si]me, sed obstant fata, in ea versor
vitæ conditione et sorte, quam mihi vix sexdecim annos nato objecit ratio, in
qua tanta voluptate et solatio frui non datur.

Concedas velim ut huicce meæ epistolæ contexam singularem modum, [7] ad
explicanda quæ in solaribus radijs observamus phænomena aptissimum; eo
gratior esse debet, quod illum, tuo de optica libro, quasi obstetricante par-
turierim. Summam votorum attigero, si illa mea quamvis exigua industria, tibi
paululum arriserit tuamque acuat et accendat.

Si illa Epistola jamdiu scripta tantam moram fecerit, excusatum me habeas
vir clarissime, gravissimo et repetito morbo per plures menses tentatus fui, ac
pæne fractus.

Vale Vir clarissime, Det tibi Deus longam annorum seriem, ut ex officina
ingenij tam locupletis et fœcundi nova in phisicis arcana detegas.

Sum cum summa reverentia, et singulari studio, tibi adjectissimus.

F[RATER] SEBASTIANUS TRUCHET
R. C. hon. Reg. Scien. acad.

[*The enclosed sheet*] [8]

Prima figura quæ videtur demonstrare radiorum solarium in cubiculo obscuro
per foramen receptorum, diversam esse directionem, vim, et celeritatem.

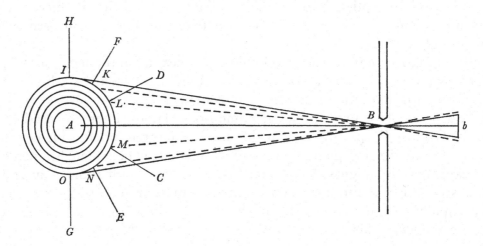

Sit *A* corpus solare. *Ab*. Radius directus et primarius, receptus in foramine *B*. cubiculi obscuri *b*.

AC, AD, AE, AF, AG, AH. alij radij directi quidem, sed qui præcipua sua directione non ferantur ad foramen *B*. prædicti cubiculi.

IB, KB, LB, MB, NB, OB. Radij qui oblique ferantur in cubiculo obscuro, per foramen *B*.

Constat radios circumferentiæ solis cum cæteris subire cubiculum obscurum quoniam solis imaginem 30′ circiter Minut. depingunt; ergo radij obliqui non ea vi et celeritate donantur, qua radius primarius *AB*. quippequi suo in itinere reperiant plures radios directos se fortiores a quibus debilitantur et retardantur, sicque possunt producere varios illos gradus refractionis, nec non illos diversos colores qui observantur in radiorum exitu a prismate.

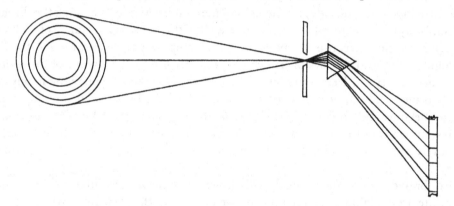

Præsens figura demonstrat, Radios solares pari directione aut celeritate donatos, in prismate quod trajiciunt a se mutuo separari et assimilari, et producere diversos gradus refractionis et colorum.

Certum est radios solares in angusto foramine collectos, aliter subire corpus solidum quale est Prisma, aliter liquidem aera, radiosque celeriores in eo transitu minus debilitari quam tardiores, hoc patet evidenter ex illa imagine et spectro quod in longum depingitur et formatur in radiorum exitu a Prismate.

Ergo radij majore vi donati sese suo in statu conservant diutius quam qui sunt debiliores.

Cæteras demonstrationes silentio prætermittam, ne fatigare videar Virum physicorum omnium nostræ Ætatis longe Principem.

Translation

Most famous Sir,

With an extraordinary sense of gratitude and delight I have received in your name from the most learned Varignon, Academician and Fellow [of the Royal Society], a copy of that best and most exquisite book about optics,[2] which indeed, after you had published it again a second time, Mr Coste has translated into the French tongue, as an act of kindness to his country.

Scarcely had that work of yours on light and colours come to my notice,(3) when a deep and ardent eagerness entered my mind to try those experiments which you bring together so brilliantly in your treatise. Without delay, taking as my guide and as it were protector an English copy of your treatise, and especially benefiting from the assistance of Mr Geoffroy,(4) an Academician famous amongst us, whom it was proper and useful to employ as interpreter of your tongue, I tried the majority of the experiments cited by you in the presence and with the approval of the most eminent Cardinal de Polignac,(5) and also that of a very numerous group of aristocrats and men deeply versed in physics, who are, as it were, blood-brothers of yours.(6) This good fortune I indeed owe to my dwelling, which faces south, in such a way that it is usually warmed and illuminated by the Sun for seven or eight hours [a day] in the summer and three or four in the winter; accordingly, by the help of metal mirrors I may change the direction of the Sun's rays as is required, and since one wall of the house is a distance of 42 French feet away from the other, it is in my power to intercept, separate and refract, the same ray of light, even several times over, before it reaches the furthest point of the room.

To confess the truth, I envy my letter the journey it makes to you; how much more glorious to me would appear that day on which I might be allowed to enjoy a most welcome meeting and conversation with you, the wisest of men; but fate prevents it, for I pass my days in such a condition and kind of life, which prudence thrust upon me when barely sixteen years old, that it is not given to me to enjoy so great a pleasure and reward.

Please forgive me for including in this letter of mine a particular method(7) most appropriate for explaining the phenomena we observe in the solar rays; which ought to be the more welcome because your book on optics served as midwife, so to speak, when I gave birth to the idea. I will have attained my highest wish, if my exceedingly paltry effort pleases you a little, and interests and provokes you.

If this letter, written a long time ago, has been so greatly delayed, you have my excuse, most famous Sir, in the most serious, repeated sickness I have suffered for many months, and which has nearly broken me.

Farewell most famous Sir, and may God grant you a long span of years in which to elucidate fresh secrets of nature from the workshop of that fertile and rich talent of yours.

I am, with the greatest reverence and outstanding devotion, your greatest disciple,

[Brother] Sebastien Truchet, Carmelite Brother
Honorary Member of the Académie Royale des Sciences

[*The enclosed sheet*](8)

The first figure, which is seen to show how the direction, strength and speed of the solar rays admitted through a hole into a darkened room, are various:

Let A be the solar body, Ab the direct and primary ray admitted through the hole B into the darkened room b.

114

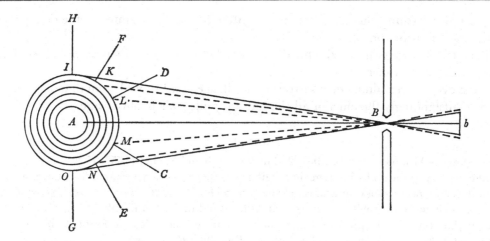

Let *AC*, *AD*, *AE*, *AF*, *AG*, *AH* be other direct rays, but ones which do not travel in their particular directions to the hole *B* of the said room.

Let *IB*, *KB*, *LB*, *MB*, *NB*, *OB* be rays which travel obliquely into the darkened room, through the hole *B*.

It is well known that the rays from the Sun's circumference pass into the darkened room with the rest, since they depict an image of the sun of about 30′ [angular diameter]; therefore the oblique rays are not endowed with that strength and speed which the primary ray *AB* [has]; for they certainly meet in their path many direct rays, stronger than themselves, by which they are weakened and retarded, and thus they can produce those various degrees of refraction, and also those varied colours which are observed in the rays issuing from the prism.

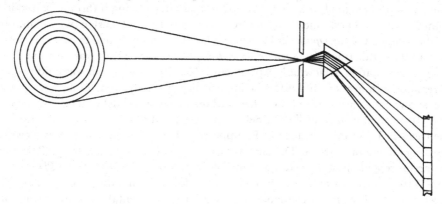

The figure given here shows that the solar rays, endowed with a like direction or speed, are separated from each other and assimilated, and produce various degrees of refraction and colour.

It is certain that the solar rays, gathered into a small hole, penetrate in one way into a solid body such as a prism, but in a different way into the fluid air, and that the swifter rays are less weakened in their passage than the slower; this is clearly obvious from the image and spectrum, which is given an elongated form as it is painted at the emergence of the rays from the prism.

Therefore the rays endowed with the greater force maintain their state longer than those which are weaker.

I pass over the remaining demonstrations in silence, lest I should seem to weary a man who is outstandingly the chief of all the natural philosophers of our age.

NOTES

(1) Keynes MS. 93; microfilm 931.2. The writer, Jean Truchet (1657–1729), a Carmelite Brother generally known by his name in religion as Father Sebastien, was an influential engineer and advisor on public works in France. He became an honorary member of the Paris Académie Royale des Sciences in 1699, and there is an 'Éloge' of him by Fontenelle (*Éloges des Académiciens*, II (La Haye, 1740), pp. 359–74). Some mechanical innovations of Father Sebastien are described in *Machines et Inventions Approuvées par l'Académie Royale des Sciences* (Paris, 1735–77).

(2) It is impossible to be certain whether Truchet refers to the Amsterdam 1720 issue of the French translation, or to the Paris 1722 edition (see Letter 1338, note (7)); certainly in Letter 1372 Newton asked Varignon to present a copy of the latter to Truchet. The date of the letter is of the less significance since it is known that the experiments described in it were made in 1719 (see note (6) below).

(3) Newton's early accounts of his optical experiments had not won credence in France, where they at once met criticism from Huygens, Pardies and Lucas. In the words of Newton's translator, Pierre Coste: 'quoique Monsieur le Chevalier Newton n'ait fondé la Theorie des Couleurs que sur des Experiences très-sensibles, l'art de les faire a été, pour ainsi dire, renfermé assez long-temps dans l'Angleterre; & il se trouva d'abord en France, en Allemagne & ailleurs, des Sçavans qui n'ayant pu séparer exactement les differentes especes de Rayons dont la Lumiere est composée, regardent toute cette Theorie comme une simple Hypothese.' Thus Mariotte, attempting to analyse a 'red' ray, found that after passing it through a second prism he obtained both violet and blue, as well as red. To quote Henry Guerlac (whose unpublished study of the reception of the *Opticks* in France was kindly communicated to us): 'Mariotte published [in 1681] a little tract, *De la nature des couleurs*, which henceforth was taken in France to be the decisive refutation of Newton's experiments.' No further attention was paid to Newton's optical researches in France until after the appearance of the *Optice* in 1706, a work which in turn was known for some years only to a few authors, of whom Malebranche was the most influential. (It seems obvious that Father Sebastien had failed to come across it.) Malebranche read the *Optice* in 1707, and a little later Philippe de la Hire attempted to repeat Newton's experiments, again without success. The chief reason for Newton's encouraging J. T. Desaguliers to do a careful repetition of them (an account of which was published in the *Phil. Trans.*, **29**, no. 348 (1716), 433–52) was to satisfy continental doubts. Similarly, the translator, Pierre Coste, refers in his Preface to Desaguliers' course of experimental lectures, where 'il a... demontré en particulier les Experiences sur la Lumiere & les Couleurs, qu'on trouvera dans cet Ouvrage.'

Probably Father Sebastien first saw one of the copies of the second English edition of the *Opticks* that Newton sent to France in 1718 (Letter 1298); the second Latin edition arrived only in June 1719 (Letter 1323).

(4) Étienne-François Geoffroy (1672–1731), a distinguished medical writer, had received a presentation copy of the first edition of the *Opticks* and certainly read English, for he

prepared a French version which was read to the Paris Academy of Sciences in 1706–7. Probably he is meant rather than his brother Claude-Joseph (1685–1752), who had been in the French party (which also included Pierre Rémond de Monmort and the Chevalier de Louville) which travelled to England in 1715 to observe the solar eclipse; all three were elected Fellows of the Royal Society and became familiar with the English scientific scene.

(5) Cardinal Melchior de Polignac (1661–1742), French philosopher and diplomat. In his *Éloge*, 1747, p. 66, Dortous de Mairan writes of Polignac (who was elected to the Académie in 1715): 'personne n'honora jamais nos assemblées d'une assiduité plus flatteuse, personne n'y fut jamais plus attentif...' He was 'Zèle Cartésien par choix, par habitude & même par principe de religion, le Newtonianisme, tel qu'il le concevoit, lui avoit toûjours paru dangereux par sa conformité avec les points fondementaux de la Physique d'Epicure.' Polignac was nevertheless probably the chief mover in the repetition of Newton's experiments; indeed, Montucla (*Histoire des Mathématiques*, II (Paris, 1758), p. 626) records that he bore the expense of them. According to Dortous de Mairan (*loc. cit.* p. 79) Polignac intended to write something about Newton's experiments on light in his pro-Cartesian poem *Anti-Lucrèce* (Paris, 1748), but this was never completed by its author. Montucla also reports that Newton wrote Polignac a letter of thanks, but we have not found such a letter. It may be that Montucla was misinformed about Newton's letter to Daguesseau (see p. 141), now lost.

(6) With respect to the confirmation of Newton's optical experiments, Coste wrote in his Preface to the 1720, Amsterdam *Traité d'Optique*: 'C'est ce que M. Desaguliers fit voir distinctement [*footnote*: En 1715] à Londres à M. Remond de Montmor, M. le Chevalier de Louville, & autres Membres de l'Académie Royale des Sciences; & qui a été demontré depuis quelque temps à Paris par le P. Sebastien, lequel en presence de plusieurs personnes très-intelligentes a verifié la plûpart des Experiences de ce *Traité des Couleurs*, avec une entiére exactitude.' A second footnote indicates that the spectators were M. le Cardinal de Polignac, Varignon, Fontenelle &c.

In the revised 'Preface du Traducteur' in the second, Paris, edition Coste adds further details assigning Sebastien's experiments to the year 1719, besides mentioning others earlier and later: '...& ce qui a été demontré à Paris en 1719 par le P. Sebastien, & depuis quelquetemps par M. Gauger, lesquels en présence de plusieurs personnes très-intelligentes, ont verifié la plûpart...' Again, footnotes indicate that the spectators with Father Sebastien were M. le Cardinal de Polignac, M. Varignon, M. Jaugeon, M. Jussieu &c; while with Nicholas Gauger were M. le Chancelier (Daguesseau), Messieurs ses Fils, M. le Nepveu, M. de Lagny, Le P. Reyneau, &c. Coste adds that Dortous de Mairan had successfully repeated the experiments at Beziers in 1716 and 1717. It seems clear that Gauger's show was put on for the benefit of the Chancellor, after he had become interested in Newton's work in 1721 and Father Sebastien had left Paris for Lorraine. Montucla's brief narrative seems to have confused the two demonstrations; he does not mention Father Sebastien at all and neither (it must be added) does Fontenelle mention the optical experiments in his 'Éloge' of Father Sebastien. If we may suppose Coste correct in omitting Fontenelle from his revised list of spectators, there seems little reason to distrust his account. See p. 215

(7) The explanation is given on the enclosed sheet, printed here immediately after the letter.

(8) Again, Father Sebastien implies (rather than explicitly states) that (i) some rays are more refracted by the prism than others; (ii) various colours are associated with varying degrees of refraction. Thus he accepts the barest elements of Newton's theory of colours, without (however) either implicitly or explicitly agreeing with Newton that the separated, coloured,

homogeneous rays existed previously, intermingled, in the heterogeneous white light. On the contrary, his hypothesis (which will hardly bear serious examination) seems to suppose that adventitious differences in the strength or velocity of the rays coming from the Sun are modified by refraction to produce the colour discrimination. Even if one supposes Father Sebastien to mean that weaker or slower rays are more markedly refracted than stronger or swift ones, it is hard to understand why swift (least refrangible, or red) rays should always come from the centre of the Sun, and the slowest (blue) from the periphery. These assumptions seem wholly arbitrary.

1351 C. STANHOPE TO NEWTON AND BLADEN

7 JANUARY 1721

From the original in the Mint Papers[1]
For the answer see Letter 1358

Whitehall Treasury Chambers 7th Janry 1720[/21]

The Right Honoble the Lords Comm[issione]rs of his Ma[jesty']s Treasury are pleased to referr this petition[2] to Sr: Isaac Newton Master and Worker and Martin Bladen Esqr. Comptroller of his Ma[jesty']s Mint who are to consider the same and report to their Lordps a true State of the Petitioners case together with their opinion what is fit to be done therein

C. STANHOPE

3d[3] *Janry* 1720[/21]
To the Master & Compt[roller]
of the Mint

NOTES

(1) I, fo. 463 v.

(2) Thomas Dearsly's petition is at *ibid.*, fo. 462. He claims to have informed Delafaye of a number of illegal presses (see Letter 1347) and demands payment. In particular he mentions one Mansell, who, he says, was prosecuted and whose presses were seized. The petition is endorsed by Delafaye, certifying that Dearsly brought information against a number of counterfeiters, whom he believes are now under prosecution. The endorsement is dated 10 December 1720.

On fo. 460 Newton has sketched the following comments:

'Several persons have been taken up with Presses sufficient to coin money & plead in excuse that they use them only in their trades of making metal-buttons bowes for watch keys, middles for Dial-plates of watches & pillars for watches

Quære 1. Is this a sufficient excuse to free them from being prosecuted for high treason.

Quære 2. May not the Presses be demolished by order of one or more justices of the Peace without a Prosecution

Quære 3. May not one or two Artificers be authorised to have Presses for doing this sort

of work for Watch-makers & Button-makers & others, without leaving every man at liberty
to have coining Presses who can pretend that he uses them in his Trade.'
Clearly Newton considered the possibility of licensing legal presses, but in his reply to the
Treasury he does not mention this idea (see Letter 1358).

(3) Presumably '7th' was intended.

1352 RICHARD HOWLETSON TO NEWTON
16 JANUARY 1721
From the copy in the Jewish National and University Library, Jerusalem[1]

Rec[eive]d aboard my good ship the Richard & Ann, a box marked glass in
3X wch I promiss to deliver to Christopher Montague Esqr danger of the sea
& Custome house excepted.[2] Rotterdam 27 Jan 1721, [N.S.]

RICHARD HOWLETSON

NOTES
(1) Yahuda Collection, Newton MS. 7(4); a copy in Newton's hand. We know nothing of
the writer.
(2) A possible meaning of this note is that the master of the vessel was advising Newton by
post of his intention to deliver a parcel of glass from Frans Greenwood (vol. VI, p. xxii, note),
intended as a present to Newton, to the named agent in London. '10 Dozen of Glasses of
divers sorts' are mentioned in the inventory of Newton's house (De Villamil, p. 53). Or of
course a looking-glass might be meant.

1353 NEWTON TO VARIGNON
19 JANUARY 1721
From a holograph draft in the University Library, Cambridge[1]
Reply to Letter 1348

Viro celeberrimo D. Abbati de Varignon, Regio
Matheseos Professori, et Academiæ Scientiarum
Socio
Isaacus Newton. S.P.D.

Vir celeberrime

Historiam Academiæ vestræ accepi una cum Ephemeride pro Anno hocce
1721, pro quibus gratias reddo tibi quam maximas. Sed et Pictura tui in manus
meus tandem pervenit, elegans sane, et vultus venustate pingentisque artificio
pulcherrima. Unde et tui similem esse concludo licet spectatorem nondum
nactus sum qui te de facie novit. Quod Picturam mei tam benigne acceperis
amicitiæ et humanitati tuæ debetur.

Collectio chartarum D. Leibnitij quam D. Desmaiseaux nuper edidit,[2]

119

cœpta fuit concilio D. Leibnitij ipsius, schedas[3] aliquot eo fine ad ipsum mittentis, ut ex epistola ejus ad D. Desmaizeaux 21 Aug. 1616 [*recte* 1716, N.S.] data & in secundo Collectionis Tomo, pag 355 impressa, colligere licet. Ineunte anno 1717 D. Abas de Comitibus (qui me mystice admodum tractavit) Epistolas[4] D. Leibnitij ad Dnam Kilmansegg & D. Bothmar aliasque nonnullas cum D. Desmaizeaux communicavit ut in lucem etiam ederentur; et alias[5] anno proximo ex Gallia eodem consilio ad ipsum transmisit, ut ex Epistola D. Desmaizeaux ad Abbatem illum 18 Aug. 1718 [N.S.] data[6] et in Collectionis illius Tomo secundo pag. 362 impressa, facile discas. Hæc omnia me inscio facta sunt donec schedæ quatuor vel quinque priores Tomi secundi impressæ essent & ex Hollandia missæ et mihi ostensæ,[7] in quibus erant Epistolæ duæ prædictæ D. Leibnitij mense Aprili anni 1716 scriptæ; quarum illa quæ ad D. Kilmansegg scripta fuit, tribuebat Epistolam 13 Junij 1713 [N.S.][8] datam Dno. J. Bernoullio, eandemque Gallice versam recitabat. Sed hæc nullius sunt momenti cum Epistola eadem ijsdem verbis Gallice versa, prius impressa fuisset in Belgio in *Novelles Litterairs* 28 Decem. 1715 [N.S.] pag 414 sub hoc Titulo

Lettre de M. Jean Bernoulli a Bale du 7 Junij 1713

In charta volante quæ anno 1713 alicubi in Germania impressa fuit & per orbem sparsa, Auctor hujus Epistolæ D. Bernoullium citabat tanquam hominem a se diversum his verbis [quemadmodum ab eminente quodam Mathematico dudum notatum est,] & D. Leibnitius citationem tunc probabat; jam vero in ejusdem Epistolæ versione Gallica citationem omisit ut Epistolam Bernoullio tribueret. Si Epistolæ originali & primo editæ fides adhibenda sit, D. Bernoullius non est auctor.

Post hæc omnia, cum Collectio prædicta prope impressa esset, D. J. Bernoullius Litteris[9] per manus tuas ad me transmissis, negavit se auctorem esse illius Epistolæ, & cum fama quod is Auctor esset a D. Leibnitio per orbem sparsa, non aliter dilui posset quam per testimonium D. Bernoulli quod is auctor non esset: ostendi Epistolam illam D. Keill ut illi suaderem D. Bernoullium non esse auctorem. Ille autem cum adduci non posset ut crederet, me inconsulto in lucem edidit quæ in ejus Appendice vel P.S.[10] hac de re contra Bernoullium leguntur.

Onus quod in te suscepisti (inter multa tua negotia) iterum edendi Versionem Gallicam[11] Optices meæ, summam in me benignitatem tuam arguit, pro qua gratias satis reddere non valeam. D. de Moivre emendationes suas Versionis illius Gallicæ ad te jam missurus est. Aliorum correctiones (si quas acceperis) minime cures, ne tibi vel molestiam aliquam vel moram creent. Si quæ aliæ occurrerent, hic examinabuntur & hinc mittentur. Vale

Dabam Londini 19 *Jan. st.v.* 172$\frac{0}{1}$[12]

120

Translation

Isaac Newton sends a grand salute
to the most celebrated Abbé Varignon,
Regius Professor of Mathematics
and Member of the Academy of Sciences

Most famous Sir,

I have received the *Histoire* of your Academy together with the *Connoissance des Temps* for this year, 1721, for which I return you the greatest possible thanks. But also the picture of you has come into my hands at last, an elegant [picture] indeed, and with the face [done] with very beautiful skill in its comeliness and colouring. Whence I conclude it to be a good likeness of you, although I have as yet found no-one to look at it who has known you by sight. It is a mark of your friendship and good nature that you accepted my picture so kindly.

The collection of papers of Mr Leibniz, which Mr Des Maizeaux has recently published[2] was begun according to a plan of Mr Leibniz himself, who sent several papers[3] to that end to Mr Des Maizeaux, as may be gathered from his letter to Mr Des Maizeaux dated 21 August 1716 [N.S.], and printed in the second volume of the collection, p. 355. At the beginning of 1717 the Abbé Conti (who has certainly behaved in a very strange way towards me) communicated to Mr Des Maizeaux letters[4] from Mr Leibniz to Madame Kilmansegge and Mr Bothmar, and to several others, in order that they, too, should be published. And he sent others[5] to him the next year from France for the same purpose, as you may easily learn from the letter from Mr Des Maizeaux to the Abbé, dated 18 August 1718 [N.S.][6] and printed in the second volume of that collection, p. 362. All these events took place without my knowledge until the first four or five sheets of the second volume had been printed and sent from Holland and shown to me,[7] in which were the two letters of Mr Leibniz mentioned above, written in April of the year 1716, of which the one which was written to Madame Kilmansegge attributed the letter dated 13 June 1713[8] to Mr J. Bernoulli, and quoted it in a French translation. But these things are of no importance, since the same letter, translated into the same words of French, was formerly printed in Holland in the *Nouvelles Litteraires* for 28 December 1715 [N.S.], p. 414, under this heading:

Lettre de M. Jean Bernoulli à Bale du 7 Junij 1713

In the flying paper which was printed somewhere in Germany in 1713 and spread through the world, the author of this letter cited Mr Bernoulli as a man different from himself in these words, 'as was noted a little while ago by a certain eminent mathematician,' and at that time Mr Leibniz approved the quotation whereas here in the French translation of the same letter he has omitted it, so that now he may attribute the letter to Bernoulli. If the first and original form of the published letter is to be trusted, Mr Bernoulli is not its author.

After all this, when the above-mentioned collection was almost printed, Mr Bernoulli denied in a letter[9] sent to me through you that he was author of that letter, and, when the rumour that he was author was spread abroad by Mr Leibniz, it could be weakened only by Mr Bernoulli's own testimony that he was not the author: I showed Mr Keill that letter in order to persuade him that Mr Bernoulli was not its author. However, since he could not be brought to believe [this], he published, without consulting me, the things against Bernoulli which may be read in his Appendix or postscript about this affair.[10]

The trouble which you have taken upon yourself (among your many commitments) in publishing a second time the French translation[11] of my *Opticks* demonstrates the greatness of your liberality to me, for which I am unable to return adequate thanks. Mr De Moivre is now about to send his corrections of that French translation to you. You should pay very little attention to the corrections of others (should you receive any), in case they cause you trouble or delay. If any others turn up, they will be looked into here and sent there. Farewell.

London, 19 *January* 172$\frac{0}{1}$, *O.S.*[12]

NOTES

(1) Add. 3968(42), fo. 604. The letter reiterates many of the points made in Letters 1334 and 1354. No further letters passed between Newton and Varignon directly until August (see Letter 1363). However contact was maintained through the intermediary of De Moivre, and Varignon in turn kept Bernoulli well informed of affairs in England. Below, we print one of Varignon's letters to Bernoulli during this periods (Letter 1357.)

(2) Des Maizeaux, *Recueil* (Amsterdam, 1720). De Moivre had written to Varignon concerning the *Recueil* in September 1720; Varignon reported this letter to Bernoulli in October (see Varignon's letter of 1 October 1720, N.S., Bernoulli Edition). Bernoulli's reaction, sent to Varignon in a letter of 17 October 1720, N.S. (*ibid.*), was strongly worded. De Moivre had written that Newton had tried to stop, or at least delay the publishing of the *Recueil*. Bernoulli expressed his incredulity in no uncertain terms. 'Mais me croyez vous', he wrote, 'assez simple pour donner dans le panneau, non! je voy bien qu'on veut jouer à mes depens une farce, dont Mr. Newton doit etre le principal Acteur derriere la scene, pour n'etre pas vû de moy, pendant que les autres Acteurs, qui paroissent instruits par lui, me jouent.' He insisted that unless Newton publicly denounced the *Recueil*, he would take it that he tacitly approved of the re-publication of letters concerning the calculus dispute which it contained—although Bernoulli did not at this stage know which letters these were. He asked Varignon to inform De Moivre of his disapproval, so that he in turn could bring it to Newton's notice. Presumably this was done, and is the reason why Newton discusses the matter in the present letter to Varignon.

Varignon quoted the section 'Collectio...Bernoullium leguntur' in Letter 1357.

(3) These concerned Leibniz's correspondence with Pierre Bayle and Samuel Clarke.

(4) See Letter 1203, vol. VI.

(5) Des Maizeaux does not specify in his letter which letters these were, but it is likely that Leibniz's correspondence with Nicolas Rémond is meant, some of which concerned the calculus dispute. See also Des Maizeaux, *Recueil* (Amsterdam, 1720), I, p. lxvii.

(6) *Read*: 21 August 1718. See Letter 1359, note (9), p. 136.

(7) Probably in August 1718; see Letter 1295, vol. VI.

(8) *Read*: 7 June 1713, N.S. Newton returns yet again to the subject of the authorship of the letter extract printed in the *Charta Volans* (Number 1009, vol. VI), in the *Journal Literaire* (Letter 1018, vol. VI) and in the *Nouvelles Litteraires* (Letter 1172, vol. VI).

(9) Letter 1320.

(10) Newton means the 'Additamentum' to Keill's *Epistola ad Bernoulli* (London, 1720) (see Letter 1303, note (2) *ad finem*). On p. 22 Keill wrote, 'Præterea quo facilius in gratiam tecum redeat *Newtonus*, negas te fuisse auctorem istius Epistolæ *Latine* Editæ & 29 *Julii* 1713 [N.S.] datæ...Nemo profecto melius noverit unde prodivit illa Epistola, quam *Leibnitius*, qui te ejus auctorem fuisse scripsit.' ('Besides, in order to ingratiate yourself with Newton, you deny that you are author of this same Latin letter dated 29 July 1713...No-one indeed knew better than Leibniz from whence the letter came, who wrote that you were its author.') Keill sent a number of copies of the *Epistola* to France, and De Moivre brought it to Varignon's notice in October 1720, who then passed the information on to Bernoulli. See Varignon's letters to Bernoulli of 1 October 1720, N.S., and 14 November 1720, N.S. (Bernoulli Edition).

Bernoulli treated as 'frivolous' Newton's suggestion that Keill was acting against Newton's wishes when he made public Newton's denial (see Bernoulli's letter to Varignon of 17 May 1721, N.S., Bernoulli Edition).

(11) See Letter 1338, note (7), p. 91.

(12) After the date Newton has added, and then deleted, the sentence 'D. Desmaizeaux exemplar Collectionis suæ ad te misit. Ab eo caveas:' ('Mr Des Maizeaux has sent you a copy of his *Recueil*. Be on your guard against it...'). Varignon did receive a copy of the *Recueil* (see Letter 1356) and puzzled over who had sent it. Possibly Newton thought that to reveal his knowledge of Des Maizeaux's actions would indicate too great an involvement in the affair of the *Recueil*.

1354 NEWTON TO VARIGNON

?1721

From a holograph draft in the University Library, Cambridge[1]

In autumn 1713 I received from Mr Chamberlain[2] (who then kept a correspondence with Mr Leibnitz) a flying paper[3] in Latin dated 29 July 1713 [N.S.], in wch it was pretended that Mr Leibnitz being then at Vienna, had not seen the Commercium Epistolicum nor had time to examin it himself, but had referred it to the judgment of a very famous Mathematician who was impartial & very able to judge of it & that he had received his judgement in a letter dated 7 June 1713 [N.S.]. And this Letter was inserted into the flying paper, & in the end of the Letter Mr John Bernoulli was cited by the author of the Letter as a Person different from himself in these words (quemadmodum ab Eminente quodam Mathematico dudum notatum est.) For these words referred to a Paper of Mr John Bernoulli published in the Acta Eruditorum of Feb. et Mart. 1713. This Latin Edition was dispersed in loose sheets, & a translation thereof into French was printed also in Holland in Mr Johnsons Journal literaire Novem & Decem 1713 pag 448, 449, 450, 451.

About two years after this, Mr Leibnitz [[4] began to declare that the said Letter of 7 June 1713 [N.S.] was writ by Mr J. Bernoulli & for that end to omit the said citation in the copies of that Letter wch he then sent to his friends. For in November or December 1715 he wrote a letter to M. Abbé Conti with a Postscript in wch were these words; *suivant ce que M. Bernoulli a tres bien jugé.*[5] And at the same time he][4] sent into Holland the aforesaid flying paper translated into French & ascribed the Letter of 7 June 1713 [N.S.] to Mr J. Bernoulli, & therein omitted the abovementioned citation & about four months after he did the like in a letter to Madam Kilmansegg dated 18 Apr: 1716 [N.S.]. And at the same [time] he wrote also to Count Bothmar that Mr J. Bernoulli was the author.[6] What he sent into Holland was printed in the Novelles Literairs Decem 28 1715 [N.S.] pag. 414,[7] but the Letters sent to Madam Kilmansegg & Count Bothmar were not printed till [*blank*] 1718.[8] Mr. Des Maizeaux received copies of them from Abbé Conti in Spring 1718 in order to print a collection of the remains of Mr Leibnitz & I knew nothing of the design of printing them till I saw them in print which was about [*blank*] 1718.[9] But the Collection was not yet complete. Mr De[s]maizeaux received some other Letters & Papers of Mr Leibnitz from Paris as you may understand by his Letter to M. L'Abbé Conti 21 Aug. 1718 printed in this collection, Tome II, pag. 362.[10] And about eleven months after this when the Collection was almost printed off, except the Preface, I received[11] Mr John Bernoulli's Letter dated July 5 1719 [N.S.] in wch he assured me that he wrote no such Letter to Mr Leibnitz as that dated 7 June 1713 [N.S.]. And in my Answer to him I acquiesced in that Declaration & have ever since told my friends that I am satisfied that Mr John Bernoulli was not the Author of that Letter. And tho some of them will not yet beleive me, yet I have given none of them authority to contradict me.[12]

NOTES

(1) Add. 3968, fo. 610. This is one of a group of drafts (see Add. 3968(34), fo. 482, and Add. 3968(27), fos. 408–9) which are evidently addressed to Varignon, and closely related to Letter 1353. Possibly they were composed after Newton knew that Varignon had been sent a copy of the *Recueil* by Des Maizeaux (Varignon received it on 26 December 1720; see Letter 1353, note (12), and Letter 1356) but before Newton had received Varignon's letter of 17 November 1720 (Letter 1348), which necessitated thanks for various gifts from Varignon. The draft could, however. be of a considerably later date. There is no evidence to show that a letter in these terms was ever sent.

We print here the most finished draft; the remaining drafts differ considerably from it.

(2) It is doubtful that Newton's recollection is correct; see vol. VI, p. 72, note (3).

(3) The *Charta Volans*; see Number 1009, vol. VI.

(4) The square brackets here are Newton's own, to indicate a possible deletion.

(5) Letter 1170, vol. VI.

(6) See Letter 1203, vol. VI.

(7) See Letter 1172, vol. VI.

(8) In Des Maizeaux's *Recueil*, which was printed in 1718, although not published until 1720. See Letter 1295, vol. VI and Letter 1330, note (2), p. 73.

(9) In another draft (Add. 3968, fo. 482) Newton specifies that it was 'the first four or five sheets of the second part of M. Desmaizeaus collection' which he saw; see also vol. VI, p. 457, note (1).

A further draft (on a slip attached to fo. 409) mentions Newton's involvement in the printing of Raphson's *History of Fluxions*. He writes 'When I heard that Mr Leibnitz was dead I caused what had passed between him & me to be printed at the end of Ralphsons book because copies thereof had been dispersed by Mr Leibnitz. And while this was doing Mr Desmaizeaux consulted with D. Abby Conti about reprinting the same in Holland together wth other Letters of Mr Leibnitz...' (See also vol. VI, p. 254, note (2).)

(10) On Add. 3968, fo. 482 Newton adds 'Mr Des-Maizeaux had a correspondence with Mr Leibnitz & was his friend & on that account published this collection of his Remains...'

(11) On Add. 3968 fo. 482, Newton inserts 'from you', thus indicating that the letter drafted here is intended for Varignon, who had forwarded Bernoulli's letter of 5 July 1719, N.S. (Letter 1320), to Newton.

(12) Newton presumably refers to Keill; see Letter 1353, note (10).

1355 WILSON TO NEWTON

21 JANUARY 1721

From the holograph original in King's College Library, Cambridge[1]

London 21st January 1720.21

Sir,

As some time ago I presumed to let you know that I had seen copies of several of your manuscripts;[2] so having since been permitted to transcribe some of them, I take the Liberty to send them to you, that you may compare them with the Originals, to see after what manner they have been copied. They contain 3 Problems, which I take to be the 2d, 3d and 4th of your Treatise wrote in 1671.[3] Here is also a paper in english containing 5 Problems, which I guess to be part of that which you have mentioned, in your Remarks[4] on Leibnitz's Letter to the Abbot Conti, as dated 13 Novemb 1665.[5] The other papers seemed to have not been so well copied as these, so I did not write them out. They contained, I observed, several Problems, as to find the Curvature and Areas of Curves, and to compare Curves together, &c. There was also a paper containing 6 Examples, shewing how to deduce the Areas of Curves from the Tables in your Quadratures, with Constructions and Synthetick Demonstrations. It concluded with saying that it was here judged proper to demonstrate by the means of Moments, as it had an Analogy to what the

Ancients have done on the like occasions. I have been likewise told by One that he had a copy of a Manuscript of yours entitled Geometria Analytica, which he highly prized, but this I never saw.[6]

When I had the honour of seeing you at Mr Innys's shop[7] you was pleased to object against publishing these Manuscripts; that you apprehended, it would occasion Disputse concerning their Antiquity. The followers of Leibnitz are, it is true, an obstinate sort of people, and no Proof, however clear, seems sufficient to make them lay aside their Prejudices. Yet on such an Occasion, I cannot think they should be more than ordinarily exasperated. For thereby you will not do more, than by what you have said, when you published your Quadratures, and in your remarks[4] on Leibnitz's reply to your Letter to the Abbot Conti. The publishing indeed of the Commercium Epistolicum raised their Fury; because that not only proved you to be the inventor of fluxions, but moreover made it appear that their Master was a Plagiary. However, notwithstanding this, the Defamatory Writings they spread abroad on that Occasion, were without a Name, as if they were ashamed of them; and the person who has been charged as the Author of them, has since thought fit to deny it.[8] But suppose this should raise ever so great a Clamour, I cannot see that you need be concerned the least about it. For in publishing these papers, you would not pretend to vindicate to your self the Right to these Inventions, from their Antiquity. For that you relie on the Arguments that are drawn from the papers contained in the Commercium Epistolicum, which the Leibnitians themselves do not pretend to say, are not of an older date than their Master's Letter of June 21. 1677 [N.S.].[9]

But then I think these papers ought to be published on many accounts. By that means young Mathematicians will be able readily to perceive the Force of the Arguments contained in the Commercium Epistolicum, and in its admirable Abridgment before they receive the least Prejudice from the Cavils of your Antagonists. These I think are now all reduced to this, that it does not appear from the Commercium that you were acquainted with the true Characteristics and Algorithm of Fluxions or Differences, before their Master. The Weakness of this Cavil would appear evident even to the most prejudiced, if you would publish all your papers. Again, your Book of Quadratures, which all intelligent persons must own is the perfectest peice that ever saw the light, seems not to be well understood by Foreigners (and perhaps not by some at home); for otherwise a certain confident Person[8] durst not lay claim to many things contained in it, under the Notion of his integral Calculus. But the publishing your Papers would enable all to see the Beauties of that noble Treatise, and this is now absolutely necessary, since there are Pretenders in the world to these Inventions. Lastly, as various Copies of your Manuscripts, more or less imperfect, are got abroad, nothing but causing them to be printed your-self, can prevent their coming out incorrect and Mangled, which ought not to be the Fate of such excellent Things.

I am, Sr, with the profoundest Respect,
Your most obedient and most humble servt
JAMES WILSON

P.S.

I humbly desire that, when you have perused these papers, you would be pleased to seal them up, and to leave them with your servants, that I may have them again upon calling for them some time or other.[10]

At page 48 of the Commercium Epistolicum, it is said by Mr Collins, that the Doctrine of Series &c was the subject of your Lectures at Cambridge, and that these Lectures were reserved there, which if so, they might afford con-vincing Proofs of your Right to these Inventions.[11]

In your remarks[4] on Leibnitz's reply to your Letter to the Abbot Conti, I think you seem too readily to acknowledge that Leibnitz might have found out by himself your Method of an arbitrary Series; for in a Scholium of the Principia you say that one of the Things which you concealed under a Cypher, in your Letter of Octob 24. 1676 was, Data Æquatione Quotcunque Fluentes quantitates involvente, Fluxiones invenire; et vice versa. Now might not Leibnitz by that means be helped to decypher what was besides concealed in that Letter?[12] Amongst which was that very Method of assuming a Series, which he did not publish till some years after you had helped him to a key in the Scholium above-mentioned.

I hope you will pardon this Freedom, for it is not my purpose to go on in troubling you thus with impertinent Letters.

SIR

I am your most obedient and
most humble servant
JAMES WILSON

NOTES

(1) Keynes MS. 143(B); microfilm 1011.27; printed in Brewster, *Memoirs*, II, pp. 443–6.
(2) See Letter 1349.
(3) The treatise 'De Methodis Serierum et Fluxionum', apparently never given a title by Newton himself, first published in English translation by John Colson as *The Method of Fluxions and Infinite Series; with its Application to the Geometry of Curve-lines* (London, 1736) and in Latin by Samuel Horsley in his *Opera Omnia* (London, 1779), I (see Whiteside, *Mathematical Papers*, III, pp. 32–3). William Jones had been allowed to transcribe it in 1710, and from him the extracts had passed to Wilson; his copy is now U.L.C. Add. 3960.4.
(4) See vol. VI, p. 346.

(5) For the fragment 'To find ye velocitys of body by ye lines they describe', dated 13 November 1665, see Whiteside, *Mathematical Papers*, I, pp. 382–9. It is more likely that Wilson had seen 'Problems of Curves' (Whiteside, *op. cit.*, II, pp. 177–82); his copy is now U.L.C. MS. Add. 3958.4, fos. 75–6.

(6) These manuscripts are respectively Problems 1 to 17 of Newton's October 1666 tract on fluxions (Whiteside, *op. cit.*, I, pp. 400–48; Wilson's copy is Add. 3960.1) and the Addendum (Whiteside, *op. cit.*, III, 328–52) to 'De Methodis Serierum...' (see note (3) above), the title *Geometria Analytica* having been conferred by William Jones upon the copy he made of it in 1710 (the same title was used by Horsley again in 1779). Pellet's copy, of course, came from Jones.

(7) The stationer; see Letter 1413, note (4), p. 269.

(8) Johann Bernoulli; Wilson refers to the excerpt of Bernoulli's letter of 7 June 1713, N.S., to Leibniz published in the *Charta Volans* (Number 1009, vol. VI).

(9) Letter 209 (vol. II, pp. 212–19), Leibniz's reply to the Epistola Posterior.

(10) This was not done, since Newton retained Wilson's papers. Wilson obtained other copies in later years and used them in his edition of the *Mathematical Tracts of the late Benjamin Robins Esq.*, II (London, 1761), Appendix. Newton evidently discouraged Wilson as far as possible.

(11) Wilson had misunderstood; Newton had never lectured on series as Lucasian Professor. His only lectures in pure mathematics were those on algebra, published already by Whiston under the title *Arithmetica Universalis* (Cambridge and London, 1707) (see Whiteside, *Mathematical Papers*, V).

(12) Again, Wilson misunderstands (though many followed him afterwards in this mistake). The Epistola Posterior contains no decipherable statement about fluxions; see vol. II, p. 129.

1356 VARIGNON TO DES MAIZEAUX
31 JANUARY 1721
From the holograph original in the British Museum[1]
For the answer see Letter 1359

Monsieur

Le 26. Decembre [N.S.] dernier, en rentrant chez nous, un de nos Portiers me rendit un paquet contenant le Curieux Recueil[2] que vous venez de donner au public, avec une liste des membres de la Societé Royale, sans pouvoir me dire de quelle part, mais seulement qu'il lui avoit été donné pour moy par deux Etrangers qui ne le lui avaient pas dit. Comme ce paquet ne contenoit point de lettre qui me l'apprist, & que ces deux Etrangers (apparemment Anglois) ne sont point revenues, je pensay d'abord que c'etoit à M. Newton ou à M. de Moivre que j'etois redevable de ce present, ne me croyant pas assez connu de vous pour vous soupçonner de me l'avoir fait: mais la difference qui me parut entre les ecritures de leurs lettres & celle de l'inscrition de ce paquet, me fist cesser de le penser. Quelques jours apres, ayant occasion d'écrire à M. de Moivre, je le priay de vouloir bien me dire ce qu'il en scavoit, ou ce qu'il en pouvoit decouvrir, & de vouloir bien aussi remercier de ma part celui qu'il scaura m'avoir honoré de ce present. Il m'a écrit[3] l'avoir fait, & que c'est à vous, Monsieur, que j'en suis redevable. Je vous en rend donc moy même tres humbles graces, & avec d'autant plus de Reconnoissance que je n'ai jamais merité de vous cette distinction.

Ce Recueil m'a paru fait de pieces tres curieuses que j'ay lues avec beaucoup de plaisir, excepté (il faut vous avoüer) celles ou il est parlé de M. Bernoulli: j'aurois surtout souhaité aussi bien que M. Newton, que celles ou cet Illustre Scavant fait mention de M. Bernoulli, n'y eussent point été.[4] Celui-ci s'en est plaind a moy, me priant de m'en plaindre aussi de sa part à M. de Moivre pour le faire scavoir à M. Newton;[5] ce que j'ay fait pour le bien de la paix, & seulement dans la crainte que M. Bernoulli ne prist l'edition que vous venez de faire de ces pieces de M. Newton, comme faite de son aveu, & en consequence comme une infraction faite à leur Reconciliation. Mais M. Newton vient de m'ecrire,[6] ce que j'ay toujours cru, qu'il n'a eu aucune part à cette edition, qu'elle a été faite sans sa participation, & même à son inscu: en voila assez pour appaisser M. Bernoulli, & pour me tranquiliser sur la rupture que j'apprehendois de leur reconciliation.

Pardon, Monsieur, de la franchise avec laquelle je vous avoüe ce qui m'a fait de la peine dans votre ouvrage en consequence de mon amour pour la paix que je souhaite autant pour mes Amis que pour moy même: cette aveu est une marque de ma sincerité, laquelle doit me rendre aussi croyable dans la pro-testation qui je vous fais que cet ouvrage m'a fait beaucoup de plaisir dans tout le reste, & même, à cela près, dans les pieces ou il se trouve. Je vous en rend donc tres humbles graces, & suis avec une parfaite Reconnoissance,

Monsieur

Votre tres humble & tres

obeissant serviteur

VARIGNON

A Paris le 11 *fevr.* 1721. [*N.S.*]
A Monsieur
Monsieur Des Maizeaux
De La Societé Royale
   A Londre

NOTES

(1) Birch MS. 4288, fos. 202–3.

(2) Des Maizeaux, *Recueil* (Amsterdam, 1720).

(3) In a letter dated 5 January 1721, from which Varignon quotes in his Letter 1357.

(4) Newton had remarked on the anonymous writer of the letter printed in the *Charta Volans* in his letter to Conti for Leibniz (Des Maizeaux, *Recueil*, II, p. 17; also printed as Letter 1187, vol. VI), and revealed his suspicion that Bernoulli was its author in his 'Observations' (*ibid.*, II,

pp. 79–80; also printed as Number 1211, vol. VI). In his Preface to the *Recueil* Des Maizeaux also had strongly implicated Bernoulli (see Letter 1344, note (12)).

(5) See Letter 1353, note (2), p. 122.

(6) Letter 1353.

1357 VARIGNON TO J. BERNOULLI
5 FEBRUARY 1721
From the original in the University Library, Basel[1]

Monsieur

Apres vous avoir dit que dans le Recueil de M. des Maizeaux je n'ay rien trouvé de M. Newton, qui n'eust été imprimé avant votre Reconciliation avec lui; & apres les extraits que je vous ay envoyés de ce qui peut avoir donné lieu au raport en partie faux, en partie infidelle qu'on vous a fait d'y etre traité de *pretendu Mathematicien*, de *Homo Novus*, & de *Chevalier Errant*: je vous croy presentement satisfait sur cet article.[2] Quant a l'opinion ou vous me marquiez être que ce Recueil de M. des Maizeaux n'avoit été fait que par l'instigation de M. Newton, ou du moins de concert avec lui; je croy aussi vous en avoir desabusé par l'extrait que je vous ay envoyé de la lettre où M. de Moivre me marquoit que M. Newton en avoit été au contraire fort mécontent jusqu'à avoir fait tout son possible pour faire supprimer ce qu'il y a de lui dans ce Recueil, en offrant pour cela à M. des Maizeaux de payer tous les frais de l'impression, & de le dedommaguer du Guain qu'il en pouvoit esperer, sans avoir jamais pu l'engager à cette suppression.[3] Ce mécontentement de M. Newton me vient encore d'être confirmé par M. de Moivre, & par M. Newton lui même, à l'occasion de la plainte[4] que j'avois faite de votre part sur ce que ce Recueil de M. des Maizeaux renouveloit la memoire de choses qui apres votre Reconciliation avec M. Newton, devoient être en enselevier dans un eternel oubli. La premiere de ces deux Lettres qui m'an[on]çoit l'autre que je recu quelques jours apres, est de M. de Moivre: voici des extraits de touttes les deux, lesquels ne vous laisseront (je croy) plus rien sur le coeur touchant cette affaire.

Voici ce que M. de Moivre m'en dit dans sa lettre qui est du 5. Janvier 1721. 'Je n'ay pas jugé à propos de faire voir à M. des Maizeaux votre premiere lettre' (ou etoit la plainte precedente) 'sans scavoir si vous y consentiriez: il vint il y a quelques jours expres chez moy pour me dire que M. Newton lui avoit renouvelé les Reproches qu'il lui avoit faits par le passé d'avoir publié ce Recueil, & qu'il avoit ajoute que M. Bernoulli, avec qui il vouloit vivre en bonne intelligence, ne manqueroit pas aussi bien que le Publique de regarder cette impression comme étant faite par son ordre, ou du moins avec son consentement; que la dessus, lui M. des Maizeaux, s'étoit offert de vous écrire, à vous Monsieur, pour vous instruire exactement du Fait, & pour disculper M. Newton du soupçon qu'il eust eu la moindre part à cette impression qui étoit presque achevée avant qu'il en eust aucune connoissance; mais que pour cela il étoit necessaire que je lui communiquasse la lettre que j'avois reçue de vous, & qui avoit donné lieu à ce renouvellement de plaintes; que M. Newton lui avoit repondu qu'il pouvoit me voir affin de consulter lui & moy sur ce qu'il seroit à propos de faire: qu'ainsi il venoit me trouver pour cela. Mais connoissant le Sr. de Maizeaux pour un homme fort affairé & plus propre à faire Naitre des difficultès qu'à les aplanir, je pris le pretexte que j'etois obligè de sortir en diligence: je me contentay de lui dire en termes generaux que nous nous reverrions à loisir & que nous reparlerions de cette affaire là. Le jour suivant j'allay trouver M. Newton à qui je raportay la conversation que j'avois eue avec M. des Maizeaux, & j'ajoutay que je croyois qu'il n'etoit pas necessaire que M. des Maizeaux entrast dans aucune confidence, que ce pendant je vous ecrirois pour sçavoir votre sentiment la dessus. Je pris occasion d'essayer s'il ne voudroit point lui même publier le Fait dans quelques uns des Journaux de Hollandes; mais je luy trouvay tant d'eloignement pour cela, que je n'y voulus pas insister: aussi m'etois-je bien doute que cela seroit inutile; mais j'etois resolu de ne rien negliger qui fust tendre à faire voir à M. Bernoulli que je souhaiterois de toutte mon Ame que la paix fust retablie entre M. Newton & lui. Je pris donc un autre tour, & lui dis que dans le Remercimens qu'il vous envoyroit pour votre Portrait, il pouvoit vous marquer lui même qu'il n'avoit pas eu la moindre part à la publication de ce Recueil de M. des Maizeaux, & que l'impression en étoit desja comme achevée avant qu'il en eust aucune connoissance; qu'à l'egard de M. Keill, il pouvoit vous asseurer qu'il ne lui avoit fait voir la lettre de M. Bernoulli que dans la vue de l'obliger, s'il étoit possible, de finir cette dispute; puisque lui, M. Newton se trouvoit satisfait de la declaration de M. Bernoulli, qu'il ne s'etoit point attendu que M. Keill eust cité cette lettre, & qu'il lui en avoit fait des Reproches tres amers. M. Newton approuva ce que je disois,[5] & me promit de vous écrire conformement à cela, ajoutant que ç'avoit été sa propre pensée, & me protesta que c'étoit la verité. Je souhaite que cet Expedient suffise pour contenter M. Bernoulli: si apres cela vos efforts & les miens pour le retablissment de la paix se trouvent inutiles, nous aurons la consolation de penser que les Parties interessées nous sçauront quelque gré de nos soins, & s'apperceveront que nous avons agi en Gens qui les Aiment & qui les Respectent.'

Voici presentement ce que M. Newton m'ecrit sur le meme sujet dans la lettre que M. de Moivre m'annonce ci-dessus, & que je recu quelques jours apres la sienne: elle est datée du 19. Jan. V. St. 1721.
'Collectio Chartarum...contra Bernoulli leguntur'[6]

[P.S.] J'oubliois de vous dire que dans la Lettre de M. de Moivre, d'ou est tiré le premier des deux extraits precedents, il me mande que ça été M. des Maizeaux lui même qui m'a fait present du Recueil des pieces qu'il vient de donner au public, & qu'il l'en a remercié pour moy, Je l'en remerciay aussi moy meme Mardy:[7] en lui avouant franchement que son Recueil curieux d'ailleurs, mavoit fort deplu dans ce quil y avoit par raport à vous, & que je m'en etois pleind de votre part à M. de Moivre pour le faire savoir à M. Newton: cet aveu n'est guere propre a faire bien recevoir mon remerciment; mais j'ay voulu faire voir à M. des Maizeaux que je suis sincere.

NOTES

(1) MS. L. I. *a*, 156*.

(2) Newton had used the phrases which Varignon quotes here in Letter 1187 and Number 1211, vol. VI (see also vol. VI, p. 350, note (11)); these were now printed in Des Maizeaux's *Recueil*. All three phrases were applied to the anonymous author of the letter of 7 June 1713, N.S., printed in the *Charta Volans* (Number 1009, vol. VI); Des Maizeaux both implied that Bernoulli was this author, and stated that Bernoulli had denied the implication (see Letter 1344, note (12)).

Bernoulli reported in his reply to Varignon (20 February 1721, N.S., Bernoulli Edition) that he had now himself seen a copy of the *Recueil*; he added that he had never disclaimed authorship of the letter of 7 June 1713, N.S., and wrote 'je ne me souviens pas, si je l'ai ecrite ou non.' He intended, he said, when he had time, to look through his old correspondence to find out the truth of the matter.

In a further letter to Varignon of 17 May 1721, N.S. (see Bernoulli Edition), Bernoulli, still unsatisfied by Varignon's attempts at explanation, raised yet again the question of the three offending phrases; his chief objection was to the appellation 'Chevalier errant' (see vol. VI, p. 344). The other phrases, he said, could be interpreted as intended in no way to be insulting; the third could only be taken as a direct provocation. He would only be satisfied if Newton made a public retraction. So Varignon, as he informed Bernoulli in a letter of 20 June 1721, N.S. (see Bernoulli Edition), again asked De Moivre to try to persuade Newton to write a satisfactory explanation for his use of the phrase; this explanation could then be published. Bernoulli in his reply to Varignon (10 July 1721, N.S., see Bernoulli Edition) supported this procedure. At first Varignon received no response from De Moivre (see Varignon's letter to Bernoulli, 9 August 1721, N.S., Bernoulli Edition); eventually, however, he received the reply he discusses in Letter 1374.

(3) De Moivre's letter is lost, but we have already much evidence of Newton's attempts to delay the publication of the *Recueil* although it is not clear that he intended to suppress it entirely (see Letter 1330, note (2)).

(4) See Letter 1356.

(5) Newton's letter to Varignon (Letter 1353) followed the lines De Moivre suggests here.

(6) Varignon quotes, with minor orthographical differences, the passages concerning Bernoulli from Newton's Letter 1353. For the English translation see p. 121.

(7) See Letter 1356.

1358 NEWTON TO THE TREASURY
1 MARCH 1721
From the holograph copy in the Mint Papers[1]
Reply to Letter 1351

To the Rt Honble the Lords Comm[issione]rs
of his Ma[jesty']s Trea[su]ry

May it please your Lordps

In obedience to your Lordps order of Reference of 7th Janry upon the Petition of Tho. Dearsly for a Reward for discovering several Presses sufficient for coining of money: I humbly represent that all the persons taken up on this Information, pleaded that they used those presses in their lawfull trades, & that in the opinion of the Attorney General this plea was a sufficient excuse, & the Presses could not be destroyed without a suspicion of coining. And thereupon the men have been set at liberty without a tryall, & their Presses delivered back to them, except one or two which belonged to a person who was suspected of coining & fled. And the men by their not being prosecuted are encouraged to go on. The Law appoints Rewards for prosecuting Coiners to conviction: but I know of no Precedent for rewarding the Petitioner, & feare consequence of making new Precedents

All wch is most humbly submitted to Your Lordps great wisdome

ISAAC NEWTON

Mint Office Mar. 1st
172$\frac{0}{1}$

NOTE

(1) I, fo. 464. There are drafts at fo. 452, and in U.L.C., Add. 3965(10), fo. 144v.

1359 DES MAIZEAUX TO VARIGNON
30 MARCH 1721
From a holograph copy in the British Museum[1]
Reply to Letter 1356

Mr. l'Abbé Varignon   *A Londres le* 30 *de Mars* 1721

Monsieur

Mr. de Moivre m'a remis la Lettre que vous m'avez fait l'honneur de m'écrire le 11 du Mois dernier. Je suis tres sensible à la bonté avec laquelle vous y parlez du Recueil que j'avois pris la liberté de vous envoyer. Desormais je puis me flater qu'il aura une heureuse destinée, puisque vous ne l'avez pas trouvé indigne de votre aprobation. Ma joie seroit parfaite, si ce qui regarde Mr Bernoulli ne vous avoit pas fait de la peine. Je vous suis infiniment obligé d'avoir bien voulu me l'aprendre. C'est une marque de cette bonté de coeur, & de cette droiture qui vous est si naturelle; & dont vos Amis ont ressenti les bons effets en tant d'occasions. si tous les gens de Lettres avoient des Amis communs aussi genereux & aussi sinceres que vous, Monsieur, leurs demêles finiroient bien tot, & ne produiroient pas cet eclat, qui [est] rarement avantageux. A l'egard de Mr. Bernoulli, je puis vous assurer que je ne croiois pas qu'il se trouvât la moindre chose dans ce Recueil [qui] put déplaire à cet illustre Mathematicien. Vous n'[. . .   po]int[2] Monsieur, si vous me permettre de faire ici, en peu de Mots, l'histoire de l'Impression des pieces qu'il contient.

Mr. le Dr. Clarke ayant publié en 1717 un Recueil[3] contenant quelques Ecrits de Mr. Leibniz sur les Principes de la Philosophie et de la Religion naturelle, avec ses Reponses; je conseillai au Sr. du Sauzet,[4] Libraire de la Haye, de les reimprimer: & il les annonça dans ses Nouvelles literaires du 27 Mars 1717[N.S.], p. 208. Je lui proposai aussi d'y joindre une Traduction Francaise des Recherches sur la Liberté de l'Homme par Mr. [Anthony] Collins,[5] & des Remarques de Mr. Clarke sur ces Recherches. Ces Remarques faisoient partie du Recueil publié par Mr. Clarke.

Dans ce tems-là Mr. le Chevalier Newton fit imprimer une Traduction Française de deux Ecrits sur l'Invention des Fluxions ou du Calcul differential, qu'il avoit opposez à deux Lettres de Mr. Leibniz adresseès à Mr. l'Abbé Conti.[6] Ces 4 petites pieces, qui avoient deja eté publiées ici, me parurent très propres à etre ajouteès aux autres. Je ne doutai point qu'elles ne fussent bien reçues dans les Pays etrangers, où l'on n'avoit que des ideès confuses de cette Dispute. Mr. l'Abbé Conti gouta fort ce dessein; & pour le rendre plus utile, il me donna quelques Lettres de Mr. Leibniz qui avoient du rapport avec celles dont j'ai parlé, & pouvoient leur servir d'eclaircissement; une entr'autres ecrite a Mad[am]e. de Kielmansegg,[7] où Mr. Leibniz faisoit l'histoire de son demêle avec Mr. Newton, & une autre à Mr. le Comte de Bothmer, qui rouloit sur la même matiere. Il avoit deja fourni des Copies de ces mêmes Lettres à Mr. Newton.

Mr l'Abbé Conti etant ensuite allé à Paris au commencement de l'année 1718, il m'envoya par Mr. L'Abbé Greco, plusieurs autres Pieces de Mr Leibniz pour augmenter mon Recueil; savoir, des Dissertations, & des Lettres ecrites à Mr. Rémond.[8] Je l'en ai remercié par la Lettre[9] qui est imprimée dans le II Tome pag. 362 & suiv. Aussi tot que les 4 premieres feuilles de ce second Tome furent imprimeès, le Libraire me les envoya le 14 de Juin 1718 [N.S.]. Je les communiquai à Mr. Newton,[10] a fin qu'il fut à tems de faire corriger les fautes qui auroient pû se glisser dans les deux Ecrits qui etoient de lui; & pour savoir en même tems ce qu'il pensoit sur l'assemblage des pieces contenues dans ces quatre feuilles; en général, il n'aprouva pas la maniere dont on les avoit placées: [11] il auroit souhaité que les deux Lettres de Mr. Leibniz qui le regardoient, & ses deux Réponses, eussent été mises ensemble, & que la suite n'en eut pas été interompue par les Lettres à Mad[am]e de Kielmansegg, & à Mr. le Comte de Bothmer. Il ne parut pas même fort content de l'impression de ces dernieres Lettres, parce qu'il y trouvoit certains faits dont il ne convenoit point, & qu'il ne jugeoit pourtant pas à propos de refuter. C'est ce qui m'a oblige d'en avertir les Lecteurs dans la Preface, pag. lxvi. [12]

Divers incidens ont fait trainer l'impression de ce Recueil, qui n'a été finie qu'au mois de May de l'annèe derniere. [4] Au reste, [13] la petite Lettre du 7 de Juin 1713 [N.S.], inserée dans la Lettre à Mad[am]e de Kielmansegg, pag. 36, & atribuée à Mr. Bernoulli; avoit deja été publiée sous le nom de Mr. Bernoulli dans les Nouvelles literaires du 28 Decembre 1715 [N.S.]; comme je l'ai remarqué à la marge: & il ne faut que lire ce qui la precede dans ces Nouvelles pour se convaincre que Mr. Leibniz l'avoit envoyée au Journaliste. Mr. Leibniz atribue encore cette Lettre à Mr. Bernoulli dans celle qu'il adressa à Mr. L'Abbé Conti le 9 d'Avil 1716 [N.S.], pag. 51 pour repondre à la Lettre de Mr. Newton. *On connoit asses*, dit il parlant de la feuille volante publieé en Latin, *le nom & le lieu de l'Auteur de la Lettre y insereé d'un excellent Mathematicien que j'avois prie de dire son sentiment sur le* Commercium. Ainsi nous n'eumes pas le moindre soupçon, Mr. l'Abbé Conti & moi, que cette Lettre ne fut effectivement de lui. Cependant, comme j'appris avant que d'envoyer ma Preface en Hollande, que Mr. Bernoulli la desavouoit, je crûs que l'equité demandoit que j'en avertisse le Public, comme j'ai fait pag. xlviii. Après cela j'ose esperer, Monsieur, que vous ne trouverez pas que j'aie manqué en rien de ce qui est dû au rare merite de Mr. Bernoulli. La liste des Membres de la Societé Royale etoit pour Mr l'Abbé Conti, qui me l'avoit demandèe: mais si vous souhaitez de la garder, je lui en enverrai une autre qu'on a imprimée depuis celle-lá. Oserois je vous prier, Monsieur, de demander à Mr. l'Abbé Bignon sil a reçu le Recueil de Mr. Leibniz &c. & un Recueil d'Ouvrages de Mr. Locke [14] que jai fait imprimer ici. Je crains que ce paquet ne se soit perdu. Faites moi la grace de continuer à m'honorer de votre Bienveillance & d'etre persuadé que je suis tres parfaitement

Monsieur

Votre tres humble & tres
obeissant serviteur
DES MAIZEAUX

NOTES

(1) Birch MS. 4284, fos. 217–18. A copy of the second and third paragraphs concerning the publication of the *Recueil* in Newton's hand and dated 'Le 9e de Xbre 1720' is in U.L.C., Add. 3968(27), fo. 399. The date is puzzling; perhaps Des Maizeaux drafted this history of the *Recueil* earlier, and showed it to Newton, who copied it.

(2) Two or three words have been obliterated by a large ink-blot.

(3) *A Collection of Papers which passed between the late learned Mr. Leibnitz and Dr. Clarke...* (London, 1717); see also vol. VI, p. 259, note (3). The letters and papers are printed in French in Des Maizeaux, *Recueil*, I, pp. 3–409.

(4) For the correspondence between Du Sauzet and Des Maizeaux, see Letter 1330, note (2), p. 73.

(5) First published, in English, in 1715, and also appearing in Clarke's *Collection of Papers*; see note (3) above.

(6) Newton had published these four items (Letters 1170, 1187, 1197 and 1211, vol. VI) as an appendix to Raphson's *History of Fluxions* in 1718 (see vol. VI, p. 254, note (2)), but his own letter to Conti, and his 'Observations', appeared in English. Printed copies of a French translation, never issued, exist in U.L.C. MS. Add. 4005.3.

(7) Letter 1203, vol. VI.

(8) These writings, and Leibniz's correspondence with Nicolas Rémond (see vol. VI, p. 216, note (5)) are printed in Des Maizeaux, *Recueil*, II, pp. 129–389.

(9) Dated 21 August 1718, N.S. (see Des Maizeaux, *Recueil*, II, p. 362).

(10) See vol. VI, p. 457, note (1).

(11) This confirms our conjecture that Newton did send some communication to Des Maizeaux about August 1718, possibly a version of Letter 1295a, vol. VI. Compare vol. VI, p. 457, note (1).

(12) Des Maizeaux wrote, 'Je ne saurois me dispenser de dire, que M. Newton ne convient pas de tous les faits qui sont raportez [dans les lettres de M. Leibniz]: mais il n'a pas voulu pousser après la Mort de M. Leibniz, une Dispute ou il n'étoit entré qu'à regret, pendant qu'il étoit en vie.'

(13) In the discussion of Bernoulli's letter to Leibniz, first printed in the *Charta Volans* (see Number 1009, vol. VI), Des Maizeaux repeats the same details as those given in his letter to Conti; see Letter 1344, pp. 99–100, using almost the same wording.

(14) John Locke, *A Collection of several Pieces never before printed* (London, 1720); published by Des Maizeaux.

1360 CONTI TO DES MAIZEAUX

21 APRIL 1721

Extract from the holograph original in the British Museum[1]
Reply to Letter 1344

Monsieur

je ne verray donc jamais de vos lettres? et je vous ècris inutilement. j'ay priè Mr. pultenay de vous chercher au Caffè, et je prie presentement Mr. Taylor de vous faire des reproches de ma part[.] vous avez oubliè dans la preface[2] une

l'idee de Mr. Kirmansegger; et le Roy même a qui on la proposa le soir, l'approuva, ayant dit tout cela a Mr. Newton, cinque ou six jours après il m'escrivit une lettre[3] pour envoyer a Mr. Leibnitz a Hanover. Mr. Newton, peut il dire que je l'ay prié de m'adressér cette lettre? cependant la necessité de l'envoyer a Hanover, et de l'accompagnér d'une des miennes[4] m'engagea dans la querelle. La lettre qui fût portee à Hanover par le Baron de Discau, resta plus d'un mois a Londres. Mad[a]me la Comtesse de Kirmansegger la fit traduire en François par Mr. Costa:[5] le Roy la lût, et l'aprouva fort, en disant que les raisons etoient très simples et tres claires, et qu'il etoit difficile de repondre a des faits. J'ay lu a Mons. Newton la lettre que j'escrivois a Mons. Leibnitz; c'est Mr. de Moivre que me l'avoit corrigé et j'en conserve encore la brouillon: Mr. de Moivre y-avoit ajouté quelquechose a l'egard de la maniere equivoque dont Mr. Leibnitz avoit proposé le probleme.[6] Mr. Leibnitz fut fort irrité de la lettre que je luy avois envoyé, comme il paroit par sa reponse,[7] et par des expressions assez fortes qu'il avoit avancé contre moy dans ses lettres a S.A.R. la Princesse de Galles.[8] Il ecrivit plusieurs lettres pour sa fustification que j'ay donné a Mons. Newton a proportion qu'elles m'ont tombé dans les mains;[9] Mr. Newton en fit une espece de reponse[10] qui fut imprimée avec la premiere lettre a la fin de l'Histoire des Fluxions;[11] les lettres que Mr. Leibnitz m'avoit adressé, furent aussy imprimées dans le même livre; et Mr. Leibnitz[12] en fit non seulement otér mon nom; mais encore ne me fit aucune part qu'on les imprimoit. Quand Mr. des Mesaus[13] luy proposa de les imprimer de nouveau en Hollande, il luy donna son approbation, et dit même qu'il luy fourniroit quelque autre petit papier.[14] J'ignore ce qui est arrivé d'après, parce que j'ay quitté l'Angleterre. On dit que Mr. Newton a changé de sentiment et qu'il se plaint de moy de l'avoir engagé dans la querelle avec Mr. Leibnitz:[15] je le prie tres humblement de reflechir a des faits qui sont incontestables; et par lesquels il paroit assez que je n'ay eu d'autre part a la question qu'autant qu'il voulut bien m'en faire. J'ay essuyé tous les reproches des Allemans, et de Mr. Leibnitz luymeme pour soutenir ses raisons. Je les ai aussy soutenu en France ou malgré tout ce qu'on à l'adresse de luy ecrire en Angleterre, on n'est pas trop dans ses interets comme il pense. J'ai pensé un jour me brouillér avec un grand Mathematicien,[16] chez une Dame, ou on parloit de cette dispute; il soutenoit que tous les argumens du *Commercium Epistolicum* n'etoient pas concluans; et que Mr. Newton n'y avoit aucune part, non plus qu'aux lettres qu'on avoit imprimées par son ordre. J'aurois bien d'autres choses a dire la-dessus: mais je suis las d'entendre parlér d'une matiere qui n'est pas agreable. On a voulu me commettre avec Mr. Newton, et je ne sçay pas pourquoy: je l'ay toujours honoré et respecté; et je luy ay toujours dit la verité sans aucun interest: mais si les plaintes continuent, je ne pourray pas me dispensèr de faire imprimér la simple histoire d'un fait, qui fera voir au public que je n'ay pas pretendu me meler dans cette querelle pour acquerir du nom, Mr. je suis,

Votre très humble,
et très obeissant serviteur,
CONTY.

A Paris ce 22 *May,* 1721 [*N.S.*].

NOTES

(1) Also printed in Brewster, *Memoirs*, II, pp. 432–3. From this letter we learn many details of Newton's dispute with Leibniz in 1715 and 1716.

(2) There is no record of this meeting in the Journal Book of the Royal Society. It must have taken place early in 1716; vol. VI, p. 288, note (1).

(3) Letter 1187, vol. VI. Compare Conti's letter to Des Maizeaux (Letter 1366), where Conti writes that Newton composed the letter ten days after the suggestion was made.

(4) Letter 1190, vol. VI.

(5) Presumably Pierre Coste (see Letter 1365, note (1)).

(6) Leibniz had posed to the English the problem of curves cutting each other orthogonally in a letter of 25 November 1715; see vol. VI, p. 253.

(7) Letter 1197, vol. VI.

(8) The correspondence between Leibniz and Caroline, Princess of Wales, is printed in Onno Klopp (ed.), *Die Werke von Leibniz, 1st Series*, XI. *Correspondenz von Leibniz mit Caroline* (Hanover, 1884; reprint Georg Olms, Hildesheim, 1973). Leibniz accused Conti of inconstancy in a letter to Caroline dated 12 May 1716, N.S. (Klopp, *op. cit.*, p. 100) where he wrote, 'Il [*Conti*] ne paroist pas avoir des principes fixes, et ressemble à un caméléon qui prend (dit-on) la couleur des choses qu'il touche. Quand il repassera en France on le fera retourner du vuide au plein.' In another, undated letter (Klopp, *op. cit.*, p. 188) he wrote, 'Les sermons de Mr. l'Abbé Conti pourroient donner matiere à un Roman philosophique. Quoyqu'il m'ait quitté pour Mr. Newton, je ne lasse pas de le plaindre.' Caroline, who frequently met Conti, defended his integrity, but was somewhat annoyed when he lost some papers she had lent him, and did not scruple to gossip to Leibniz about Conti's unreciprocated attachment to an 'ambassadrice'.

(9) Presumably Conti means the letters to Baroness von Kilmansegge and to Count Bothmar (see Letter 1203, vol. VI), copies of which Conti also sent Newton (see Letter 1344, note (7)).

(10) Newton's 'Observations (Number 1211, vol. VI), which Newton wrote *before* he saw the letters to the Baroness von Kilmansegge and Count Bothmar.

(11) That is, in the appendix to Raphson's *History of Fluxions*; see vol. VI, p. 254, note (2).

(12) Presumably Conti meant to write 'Mr. Newton' here.

(13) Pierre Des Maizeaux.

(14) Conti refers here, perhaps, to the French translations noted on p.136, note (6).

(15) Newton justifiably held Conti responsible for engaging him in the interchange of letters with Leibniz in 1715–16. Later Conti fell completely out of his favour over the affair of the publication in 1725 of the *Abrégé de Chronologie de M. Le Chevalier Newton* (see Letter 1436, note (2)).

(16) Possibly Conti means Brook Taylor himself, who, according to his biographer, William Young, was honoured with 'a particular, but virtuous, regard' by Marcilly de Villette, the wife of Bolingbroke, his host at La Source (see Letter 1348, note (3), and Taylor, *Contemplatio Philosophica*, p. 25). Conti was also acquainted with Bolingbroke, and was himself at La Source later in 1721 (see Bolingbroke's correspondence with Taylor in *Contemplatio Philosophica*).

1362 NEWTON TO LEVINUS VINCENT
c. MAY 1721
From a holograph draft in the University Library, Cambridge[1]

Viro Honorando D. Levino Vincenti Isaacus Newtonus S.P.D.

A Monseignr Levino Vincent

Clarissime D[omi]ne

Librum[2] a Te missum accepi, et Societati Regiæ tradidi tanquam donam illis in amicitiæ tuæ testimonium datum. Illi vero eundem amicissime acceperunt, & absenti eorum Secretario me rogarunt ut eorum nomine gratias tibi redderem. Id quod lubentissime facio. Vale Vir optime & Societatem nostram amare perge

Translation

Isaac Newton sends a grand salute to the honourable
Mr Levinus Vincent

To Mr Levino Vincent

Most famous Sir,

I have received the book you sent, and have passed it on to the Royal Society as a gift given to them in testimony to your friendship. They have indeed accepted it with the greatest friendliness, and in the absence of their secretary have asked me to thank you in their name. Which I most gladly do. Farewell, excellent Sir, and continue to think fondly of our Society.

NOTES

(1) Add. 3965(10), fo. 145v.

(2) The Royal Society possesses a copy of Vincent's *Elenchus Tabularum Pinacothecarum, atque Nonnullorum Cimeliorum, in Gazophylacio Levini Vincent* (Haarlem, 1719) inscribed as a gift from the author, 11 May 1721; there is also an appropriate entry in the Journal Book of the Royal Society for the same day. Thus we assume that Newton's letter to Vincent is of about this date. The book is a catalogue of the curiosities to be found in Vincent's collection.

1363 NEWTON TO VARIGNON

EARLY AUGUST 1721

From holograph drafts in the University Library and King's College Library, Cambridge[1]
For the answer see Letter 1369

Dignissimo Viro Dno Abbati Varignonio
Isaacus Newtonus S.P.D.

Gratiæ meæ multis nominibus tibi debentur, tum quod mediante R. Patre Reynau & ejus amico applicasti te ad D. Cancellarium Galliæ & prælongo Memoriali causam Bibliopolæ Montalan ei explicuisti, et ejus authoritate obtinuisti ut Bibliopola flecteretur; tum quod negotium cum Bibliopola ad finem perduxisti; tum etiam quod magnum illum laborem in te suscipere digneris conferendi librum cum correctionibus missis, in quo faciendo spero et precor ut proprio utaris judicio & libere corrigas quæcunque tibi in correctionibus missis corrigenda videbuntur. Nam D. Moivreus omnia judicio tuo submittit.[2] Pactum tuum cum Bibliopola ratum habeo, & vice duodecim librarum sterlingarum mittam viginti.[3] Et nummus solvetur quandocunque volueris et expensa compingendi libros amicis donandos ex nummis missis desumi possunt. Epistolam inclusam Dno Cancellario tradas,[4] precor, una cum gratijs meis quamplurimis pro favoribus quibus me decoravit. Gratias etiam meas reddas, precor, R. Patri Reynau, ut et ejus amico cujus mediatione negotium delatum fuit ad D. Cancellarium, et D. Abbati Daguesseau D. Cancellarij fratri quod ægre tulerit Bibliopolam non adeo bene mecum egisse.[5]

In Literis ad Abbatem de Comitibus,[6] non vocavi Bernoullium equitem erraticum sed Leibnitium reprehendi qui Problema Bernoullij ex Actis Erud. desumptum Mathematicis et alijs Anglis ad veritatem determinandam miserat & sic ut equites erraticas tractabat.[7] Contra Leibnitium enim scripsi et non contra Bernoullium. D. Des Maiseaux commercium cum Leibnitio olim habuit, & Literas inter Leibnitium & Clarkium in primo Collectionum suarum Tomo impressas accepit a Leibnitio ut easdem in lucem mitteret. Ille mihi ignotus erat antequam me adiret cum Literis inter Leibnitium et me in prima parte Tomi secundi ad usque paginam 88 impressis. & in secunda parte quæ priore major est edidita [sic] est Collectio chartarum Leibnitij quæ ad me nil spectant. Et ex his omnibus demonstratur eum amicum fuisse Leibnitij. Præfationem qua me defendit, moleste habui tum quod non esset Mathematicus, & controversiam non satis intelligeret tum quod amicus esset Leibnitij et meus tamen amicus videri cuperet tum denique quod me defendendo causam meam ab exe[r]citu Leibnitij explodendam præbuit.[8]

Translation

Isaac Newton sends a grand salute
to the most noble Abbé Varignon

I owe you my thanks on many accounts: because you have approached the Chancellor of France through the Reverend Father Reyneau and a friend of his and explained to him in a very long memorandum the situation respecting the bookseller Montalant, and by his authority have managed to prevail with the bookseller; and further because you brought the negotiations with the bookseller to a close; and again because you are good enough to take upon yourself the great labour of collating the book with the corrections sent [to you], in doing which I hope and pray that you will use your own judgement and freely make any alterations which seem necessary in the corrections sent to you. For Mr De Moivre submits everything to your judgement.[2] I have considered your contract with the bookseller, and instead of twelve pounds sterling am sending twenty.[3] And this sum may be paid [you] whenever you like, and the costs of binding the books to be given to friends can be taken out of the cash transferred. Please give the enclosed letter to the Chancellor,[4] together with my greatest possible thanks for the favours with which he has honoured me. Also please give my thanks to the Reverend Father Reyneau, as also to the friend by whose intervention the business was brought to the notice of the Chancellor and also [give my thanks] to the Abbé Daguesseau, brother of the Chancellor, because he was vexed that the bookseller did not treat me fairly.[5]

In [my] letter to the Abbé Conti,[6] I did not call Bernoulli a knight-errant, but rebuked Leibniz, who, taking a problem of Bernoulli's from the *Acta Eruditorum*, sent it to the mathematicians and others of England for the determination of truth, and so treated them as knights-errant.[7] For I have written against Leibniz and not against Bernoulli. Mr Des Maizeaux once had a correspondence with Leibniz, and received from Leibniz the letters [passed] between Leibniz and Clarke printed in the first volume of his *Recueil*, in order that he should publish them. He was unknown to me until he approached me with the letters between Leibniz and myself printed in the first part of the second volume up to page 88, and in the second part, which is larger than the first, was printed a collection of Leibniz's papers which have nothing to do with me. And by all these things he is shown to have been a friend of Leibniz. I took offence at the preface whereby he defended me, partly because he was no mathematician and did not sufficiently understand the controversy, partly because he was a friend of Leibniz and yet wished, nevertheless, to appear as my friend, and lastly because in defending me he gave an opportunity for the forces of Leibniz to explode my case.[8]

NOTES

(1) Add. 3968(27), fos. 410–11. See also Keynes MS. 142(T); microfilm 1011.26. The former is a much longer draft or rather series of draft passages, very much crossed out and overwritten. The latter, which we print here, seems to be almost a final draft, though still with much reworking. It is obvious that this must be the letter mentioned by Varignon at the

beginning of Letter 1369; the preceding letter from Varignon, however, to which the present one is a reply, is missing. For this period, apart from Letter 1348 (whose sole topic is the portrait of himself presented by Newton to Varignon) the only extant letter from Varignon is Letter 1338, where there is a brief allusion to Varignon's concern with a new Paris edition of the *Opticks* in French. Varignon evidently wrote again giving details of the negotiations with the bookseller, possible costs, and the benevolent intervention of powerful friends.

(2) Compare Letters 1365 and 1369. Varignon proved less eager than Newton had hoped to assume final editorial responsibility. The original translator, Pierre Coste, seems to have been overlooked by all concerned.

(3) On the draft in U.L.C., Add. 3968 at fo. 411r is a note in De Moivre's hand: 'To ye 12*li* already promised to ye Bookseller I shall add 8*li* more to be paid you when you think fit, Mr De Moivre will send you a List of about 15 Names, I desire that ye Charges of ye bindings may be taken out of ye 8*li*, as for ye rest of ye Copies in sheets I desire to have 6 for my self, and that you would dispose of ye rest of ye Copies as you think fit.' Beneath, Newton has drafted a translation of this passage into Latin.

(4) This letter to Daguesseau is now lost; for the reply to it see Letter 1370.

(5) So far the King's College draft follows closely the antecedent U.L.C. draft, but in the latter the paragraph following differs markedly. Newton began to write: 'Ex epistola tua ad D. Moivreum data intelligo quod D. Bernoullio displicet me ipsum vocasse [the Mathematician or pretended Mathematician]...' (compare Letter 1372 to Varignon; the square brackets are Newton's own). He then embarks upon a much longer discussion of Leibniz's and Bernoulli's conduct in the calculus dispute than appears in the King's College draft. On fos. 410v and 411r are repeated attempts at this same paragraph.

(6) See Letter 1372, notes (15)–(20), p. 166, where Newton discusses the issue at greater length.

(7) It is very difficult to decide what words Newton meant to stand here. At first he wrote of Leibniz's boasting of Bernoulli and other friends as knights-errant on his own behalf. Then, altering *jactabat* to *tractabat*, he turned the sense round to accuse Leibniz of treating the English mathematicians as knights-errant on Newton's behalf. Compare Letters 1357 and 1374.

(8) Compare Letter 1356. In this sentence also the exact wording is not easily made out. Newton crossed out *ab amicis* in the last line, writing instead (after *causam*) *meam ab exercitu Leibnitij*. At U.L.C. Add. 3968.42, fo. 603r are the opening lines of an English draft letter to Bernoulli himself, never sent, corresponding to the second paragraph of the present letter.

1364 'sGRAVESANDE TO NEWTON

7 AUGUST 1721

From the holograph original in King's College Library, Cambridge[1]

Monsieur

Le docteur Desaguliers vous aura sans doute fait voir une lettre que le Baron Fischer lui a ecrite il y a quelque tems, touchant la roüe d'Orfirous,[2] que l'inventeur assure estre un mouvement perpetuel.[3]

Monseigneur le Land-Grave a voulu que j'examinasse aussi la Machine. [4] Ce Prince qui aime les sciences et les beaux arts, et qui par les secours qu'il donne à ceux qui s'y attachent avec quelque succes, ne neglige aucune occasion de rendre utiles au public les inventions qu'on lui presente, souhaiteroit de voir cette Machine connue de tout le monde, et entre les mains de gens plus habile que l'inventeur afin qu'on en retire l'utilité qu'on doit naturellement attendre d'une invention aussi particuliere.

J'ai cru, Monsieur, que vous ne seriez pas faché d'avoir une relation un peu detaillée de ce qu'on observe dans un examen exterieur d'une Machine sur la quelle les sentiments sont si partagez, et qui a presque tous les Mathematiciens les plus habiles contre elle. Un tres grand nombre soutient l'impossibilité du mouvement perpetuel, d'ou est venu le peu d'attention qu'en a fait jusques ici à la roue d'Orfirous. Je conois combien je suis inferieur à ceux qui ont donné leur demonstrations sur l'impossibilité de ce mouvement; cependant, pour vous expliquer les sentiments avec lesquels j'ai examiné cette Machine, j'aurois l'honneur de vous dire, qu'il y a environ sept ans que je cru decouvrir le paralogisme de ces demonstrations en ce qu'elles ne peuvent pas estre applicable à toutes les machines possibles. [5] depuis je suis toujours resté tres persuadé qu'on peut demontrer que le mouvement perpetuel n'est pas contradictoire; et il m'a paru que Monsieur Leibniz avoit tord de regarder comme une axiome l'impossibilité du mouvement perpetuel; ce qui sert neanmoins de fondement a une partie de sa Philosophie.

Malgré cette persuasion j'estois fort eloigné de croire qu'Orfirous fut assez habille pour faire la decouverte du mouvement dont il s'agit ici, Je regardois cette invention comme ne devant estre decouverte qu'apres plusieurs autres, au cas qu'elle le fut jamais.

Depuis que j'ai vu la roue en question je suis dans un etonnement que je ne scaurois exprimer. l'auteur a du genie pour les Mechaniques mais n'est rien moins qu'habile Mathematicien, cependant cette invention a quelque chose de surprenant quand même ce seroit une fourberie.

Voici ce qui regarde la Machine même, dont l'auteur ne laisse voir que l'exterieur de peur qu'on ne lui vole son secret; c'est un tambour d'environ quatorse pouces depaisseurs sur douse piéds de diametre, il est tres leger etant fait de quelque planches assemblées par d'autres pieces de bois de maniere qu'on en veroit l'interieur de tous costez sans une toile cirée qui couvre tout le tambour. Ce tambour est traversé d'un axe d'environ six pouces de diametre, dont les extremitez sont fermées par des axes de fer sur les quels roule toute la

144

machine. J'ai examiné ces axes et je suis tres persuadé qu'il n'y a rien en dehors qui contribue au mouvement de la Machine. J'ai tourné le tambour tres lentement et il est reste en repos aussitot que j'en ai retire la main, je lui ai fait faire un tour ou deux de cette maniere. Ensuite je l'ai fait mouvoir tant soi peu plus vite, je lui ai fait faire de même un tour ou deux mais alors j'estois obligé de la retenir continuellement, car l'aiant lachée elle a pris en moins de deux tours sa plus grande celerité de maniere qu'elle a fait 25 à 26 tours dans une minute; c'est le mouvement quelle a conservé ci devant pendant deux mois dans une chambre cachetée dans la quelle il estoit impossible qu'il y eut aucune fraude. apres ce tems-la S.A.S.[6] fit ouvrir la chambre et arretter la Machine; comme ce n'est qu'un essai elle n'est pas assez forte pour que les materiaux ne s'usent par une longue agitation.

Monseigneur le Land-Grave a voulu estre present à l'examen que jai fait de la Machine. j'ai pris la liberté de demander a S.A.S., qui à vu l'interieur du tambour, si apres avoir eté agitée pendant un certain tems rien n'estoit changé, comme aussi s'il n'y avoit pas quelques pieces dans les quelles on pouroit soupsonner de la fraude; S.A.S. m'a assuré que non et que la Machine etoit tres simple.

Vous voiez, Monsieur, que je n'en ai pas assez vu par moi même pour assures que j'aie une demonstration, que dans cette machine, le principe du mouvement (qui est certainement dans le tambour) soit tel qu'il le faut pour rendre le mouvement perpetuel: mais je crois aussi qu'on ne scauroit me nier d'avoir des presomptions tres fortes en faveur de l'inventeur.

Monseigneur le Land-Grave a donné une recompense digne de sa generosité a Orfirous pour voir le secret de cette invention avec promesse de ne point se servir de ce secret ni de le decouvrir avant que l'auteur ait retiré d'autres recompenses pour la rendre publique.

Je sais tres bien, Monsieur, qu'il n'y a qu'en Angleterre ou les sciences fleurisent assee pour que l'auteur peut retire de sa decouverte une recompanse qui y soit proportionée. Il sagit simplement de la lui assurer suppose qu'il n'y ait pas de fourberie dans sa roue; il ne demande a toucher l'argent qu'apres que sa machine aura été examinée en dedans, examen qu'on ne scauroit raisonablement exiger avant que la recompense soit promise. Comme il s'agit d'une chose utile au public, et à l'avancement des sciences de decouvrir l'invention ou la fraude J'ai cru que cette relation ne vous seroit pas desagreable. Je suis avec respect

Monsieur

Vostre tres humble
et tres obeisant serviteur
G. J. 'sGRAVESANDE.

de Cassel ce 18 *d Aoust* 1721. N.S.

NOTES

(1) Keynes MS. 94(B); microfilm 931.3. Printed in the *Mercure Historique et Politique* for September 1721, p. 363 and in J. N. S. Allamand, *Œuvres Philosophiques et Mathematiques de Mr. G. J. 'sGravesande*, Part 1 (Amsterdam, 1774), pp. 302–3. An English translation (from *Annual Register*, **6** (1763), pp. 126–8) is reprinted in Henry Dircks, *Perpetuum Mobile* (London, 1861), pp. 39–42.

(2) Following J. C. Adelung, *Fortsetzung und Ergänzungen zu Christian Gottlieb Jöchers allgemeinem Gelehrten-Lexicon*, II (reprinted Georg Olm, Hildesheim, 1960), p. 1115, this was Johann Bernard Fischer von Ehrlach (d. 1723), Austrian architect, creator of the Schönbrunn Palace, Vienna. Following F. Hoefer (ed.), *Nouvelle Biographie Générale*, XVII (Paris, 1856), p. 756, with greater probability, this was his son Joseph-Emmanuel, who was, however, not ennobled until 1731. He also was an architect. The Fischer associated with Orffyræus also erected the first steam-engine in central Europe. An English version of a part of Fischer's letter to Desaguliers (who was to publish a paper critical of all perpetual motion claims in *Phil. Trans.* for September–December 1721, **31**, no. 359, pp. 234–9) is reprinted in Henry Dircks, *Perpetuum Mobile, Second Series* (London, 1870), pp. 110–12. Fischer concludes by asking Desaguliers to show his letter to Newton.

(3) The circumstances are explained in the life of 'sGravesande prefaced to Allamand's *Œuvres*, pp. xxiv–xxvi (see note (1) above). Johann Ernst Elias Orffyræus (1680–1745), apparently born in Germany with the name of Bessler (see C. G. Jöcher, *Gelehrten Lexicon*, III, p. 1093, and *Ergänzungsband*, V, pp. 1164–7; *Nouvelle Biographie Générale*, XXXVIII, p. 780), was one of the most notorious inventors of a perpetual motion machine. He had experience as a clockmaker, and a rich marriage gave him the means to put his ideas into practice. His first machine was completed by 1712, but he soon after destroyed it because it was said to be too small. Then he attracted the patronage of the Landgrave Charles of Hesse-Cassel (ruled 1670–1730) for whom, at his *Schloss* at Weissenstein, he constructed a new machine, which 'sGravesande proceeds to describe. Presumably it was concerning this device that Orffyræus wrote his *Grundlicher Bericht von dem...Perpetua ac per se Mobili* (Leipzig, 1715); this was followed by other publications on the same theme. The machine at Weissenstein was said to have been started, to have accelerated itself to a steady speed, and to have run for two months in a sealed room. This, or perhaps another machine, was destroyed by its inventor in a fit of pique at 'sGravesande's curiosity as to its construction (see, in general, the two volumes by Dircks listed in notes (1) and (2) above).

(4) 'sGravesande visited the Landgrave Charles by invitation in the summer of 1721.

(5) 'sGravesande's 'demonstration' that proofs of the impossibility of perpetual motion were false may be found in Allamand, *op. cit.* Part 1, pp. 305–12. He did not distinguish between a machine that moves itself for ever (so long as its parts last) and a perpetually moving machine that is capable of doing work; since Orffyræus' machine is described as accelerating after being given initial rotation and as raising weights it was presumably of the second type. 'sGravesande's definition reads: 'une Machine dont le principe du mouvement ne dépend d'aucun

agent étranger, & dont le mouvement ne s'arrêteroit jamais si les materiaux ne s'usoient pas'. He starts initially from the supposition that the 'force' of a moving body is proportional to mv, arguing that it is possible for a unit weight falling 4 ft to raise a weight of four units 1 ft; conversely, the 4-unit weight falling back 1 ft can raise the unit weight to 16 ft! Hence spare 'force' is available for other uses.

'sGravesande next departs from the supposition that the force is proportional to the *square* of the velocity, to which (he explains) he was converted soon after writing his letter to Newton. Now in (inelastic) collision if body A ($m = 10$, $v = 1$) meets body B ($m = 1$, $v = 10$) moving oppositely, both come to rest because mv is of magnitude 10 in both cases, although the *vires vivæ* differ by a factor of ten. Hence a small *vis viva* can nullify a large *vis viva*; it has not been proved, he argues, that the reverse is impossible.

Then, 'sGravesande maintains, our knowledge of the laws of nature is not perfect, therefore we are not entitled to assert that any proposition is impossible.

His first erroneous refutation arises from a simple failure to manipulate correctly dynamical concepts that were already established; his second arises from ignorance of the first law of thermodynamics; he could not know that the 90 units of *viris vivæ* lost in the inelastic collision were transformed into heat, and that a 'reversal' of the collision would require the addition to the resting system of energy equivalent to those 90 units of heat.

(6) 'Son Altesse Sérène': the Landgrave.

1365   PIERRE COSTE TO NEWTON
16 AUGUST 1721
From the holograph original in King's College Library, Cambridge[1]

Sr

I am at Richemond, where I am obliged to stay about a fortnight longer. I have just now received a letter from Mr. Varignon, by which I learn that the impression of your Opticks is began at Paris, but interrupted by an accident. The Lord Chancellour there[2] having seen the first sheet with a letter of yours will have the Book to be printed with a finer caracter, & has order'd the Bookseller to make use of a caracter perfectly new.[3] *Vous voyez par là*, saith Mr. Varignon, *que nous sommes encore en état de faire tout ce que M. Newton souhaitera par rapport à cette Edition. Ainsi c'est à luy que vous devez vous adresser pour l'employ des corrections que vous me dites avoir faites sur cet Ouvrage; & à moy de n'y rien introduire que ce qu'il m'ordonnera, & que ce qui me viendra directement de sa main pour cela. Ainsi si vous voulez que j'y introduise vos corrections, engagez le à me les envoyer lui-même, & à m'ordonner de m'en servir.*[4] I have writt to him already that you had promise to me to send him my reflexions upon some corrections of Mr. de Moivre, & my own corrections.[5] I know you are full of bounty & humanity. I think 'tis just, that promise of yours should be perform'd; & so I intreat you again to do it. I am used coarsely enough in not having had the perusal of the

corrections of Mr. de Moivre before they were printed, by the slanders of
Mr. Des Maizeaux, an effect of that denyal as I fortold it, & in not having the
perusal of one third of least of the corrections of M. de Moivre. But as I hate
quarells, & know that tho' some of those corrections will perhaps be worse than
that which was already printed, the Book will be as good, since a nicety of
language is not required in such a Book. I am with a sincere & [profound]
respect, & the great[est] gratitude imaginable, Sir

> Your most obedient
> & humble Servant
> COSTE

Mr. Varignon seems resolved not so much as to examin the corrections of
Mr. de Moivre, but consider them all as directly approv'd by you: which if it
was true, I should have nothing to say but to intreat M. Varignon not to
recede in the least from them. Mr. de Moivre, whom I saw th'other day at his
lodging, told me he would not disapprove, I should take notice in my Preface
of the care he has taken of this new Edition, which I shall do with great deal of
pleasure. Gratitude is a vertue I love very much as well as true civility.

From Richemond at Mr. Fetherston upon the Green
the 16th of August 1721

I saw to days ago her Royal Highness the Princess of Wales, who did inquire to
me concerning your health, in a very obliging manner.

    To
Sir Isaac Newton at his
    House in Martin's Street
        London

NOTES

(1) Keynes MS. 94(A); microfilm 931.3. The paper has been slightly damaged by the breaking of the seal. Newton has annotated the letter: 'Recd Friday Aug. 18. Answered Saturday Aug. 19'. We have not found the answer.
Pierre Coste (1668–1747), the translator of Newton's *Opticks* into French, was a Huguenot educated at Geneva, first visited England at the invitation of John Locke in 1697 and spent much time here thereafter. He had already translated Locke's *Some Thoughts concerning Education* (London, 1693) into French as *De l'éducation des enfans* (Amsterdam, 1695). This was to be followed by many other translations from Coste's pen. He succeeded Locke as tutor to the heir of Sir Francis and Lady Masham at Oates (where he may well have met Newton) and later accompanied various young noblemen on the grand tour. (See James L. Axtell, *The Educational Writings of John Locke* (Cambridge 1968), pp. 88–97 and Letter 1338, note (7).)
(2) Henri François Daguesseau, see Letter 1370.

(3) For discussions about the printing of the book see Letter 1369. The bookseller was Montalant.

(4) Varignon reports to Newton in Letter 1369 this answer which he gave to Coste.

(5) In Letter 1372 Newton thanks Varignon for attending to the insertion of these corrections.

1366 CONTI TO DES MAIZEAUX
21 AUGUST 1721
Extracts from the holograph original in the British Museum[1]

A Paris ce 1er 7bre 1721 [N.S.]

Monsieur,

  ... ...

Comme je n'ay pas vû vôtre lettre; je ne vous puis rèpondre rien en particullier; je vous diray seulement qu'on a dit icy que Mr Neuton n'avoit pas estè trop content de vôtre livre;[2] et que même il avoit fait tous ses efforts pour en supprimer l'edition;[3] j'ay disputè longtemps là dessus, disant toujours, que si Mr. Neuton n'avoit point estè content de vôtre ouvrage, il ne vous auroit pas donné deux pieces originales de sa façon,[4] et qu'on n'avoit pas imprimè ailleurs.[5] les Géometres d'icy ont peur de Mr Bernouille, et n'osent pas se declarer nettement contre Mr. Leibniz; les mêmes idèes subsistent toujours. je riois de gayeté de coeur de la paix qu'on vouloit faire entre Mr Néuton, et Mr. Bernouille, ce n'estoit qu'une paix forcèe de costé et d'autre: mais a la fin on m'a avouè sincerement que M. Bernouille n'en vouloit point entendre parler; il a prevenu les Anglois, qui a leur tour, dit-on, prépare un ecrit,[6] qui contiendra l'examen de ses œuvres, et de son procedé: qu'ils se battent a la bonheur, je seray spectateur indifferent.

je suis bien aise qu'on ait fait voir a Mr. Neuton l'endroit des actes de Leipsic ou on parle de moy.[7] j'ay toujours agi de bonne foy, et me suis exposé aux railleries de toute une cour pour soûtenir la verité, qui me paroissoit estre du coté de Mr. Néuton. on m'a neantmoins soupçonné, comme on me l'a dit après en france; et des esprits malins n'ont pas manqué de me brouiller avec Mr. Néuton autant qu'ils ont pû; ils ont voulu luy faire croire que je me suis mêlé de son affaire avec Mr Leibniz pour me donner un nom, et le bonhomme, dit-on, l'a crû, lorsqu'il sçavoit luy même que je ne luy avois jamais parlé de m'addresser la lettre qu'il m'a addressée. Mr. de Kirmanseguer [*sic*] ayant dit a Mr. Néuton que pour finir sa querelle avec M. Leibniz il faloit qu'il se donnât la peine de luy écrire Mr. Néuton, dix jours après m'envoya la lettre que vous avez imprimé. il y a d'autres particularitez dans les faits que j'ay exposé a Mr. Taylor[8]...si vous faites jamais une seconde edition de ces lettres il faut adjouter cette particularité dans la préface:[9]...je veux que la posterité sçache comme les choses se sont passées; et que Mr Néuton a un grand tort de me soupçonner lorsque j'estois son meilleur Amy...j'ay pensé me broüiller avec Mr. de Montmor pour la phylosophie de Mr. Néuton. vous avez vû la lettre qu'il a écrit a Mr Taylor, et la réponse de Mr Taylor; elles sont imprimées dans l'Europe Scavante;[10] et il faudroit les ajouter aux autres lettres dans la nouvelle edition.[11] la lettre de Mr. de Montmort est le dernier effort du Cartesianisme. l'Abbé Terrasson[12] et Mr Saurin[13] y ont travaillé et je crois même le Pere Regnaut;[14] il y a bien apparence qu'ils ne répondront pas a Mr. Taylor[15] a qui Mr Néuton est bien obligé...

NOTES

(1) Birch MS. 4284, fos. 224–6.

(2) Des Maizeaux, *Recueil*. Possibly De Moivre was Conti's informant.

(3) See Letter 1330, note (2), p. 73.

(4) See Letter 1344, note (7), p. 101. Newton only sent Des Maizeaux copies of the letters of Leibniz to Baroness von Kilmansegge and to Bothmar after he knew of their intended publication.

(5) Presumably Conti refers to the appendix to Raphson's *History of Fluxions*.

(6) Possibly Keill's *Epistola ad Bernoulli* (London, 1720; see Letter 1321, note (3)),which Conti may not yet have seen.

(7) Conti refers to a brief mention in Mencke's paper, 'Epistola ad...Broock Taylor', *Acta Eruditorum* for May 1721, p. 219 (see Letter 1303, note (3)). Mencke writes 'Provoco...ad ipsum qui vestratibus impensius favet Abbatem Contium'. ('I appeal to the Abbé Conti himself, who more strongly favours your side'.)

(8) See Letter 1361.

(9) No addition was made to the Preface of the second edition.

(10) See *L'Europe Savante* for October 1718, **5**, part 2 (La Haye, 1718), 209–94. Monmort's letter, mentioned also in Letter 1281, vol. vi, was entitled 'Dissertation de M. de Montmor [*sic*], sur les Principes de Physique de M. Descartes, comparez à ceux des Philosophes Anglois'. It is followed on pp. 294–304 by the anonymous 'Reflexions de M.... sur la Dissertation de M. de Montmor'. A second reply appears in the volume for May 1719, **9** (Part i), 83–134, 'Reponse a la Dissertation de M. de Montmor,...Traduite de l'Anglois de M. Taylor'. Possibly both replies came from Taylor.

(11) No such additions were made to the second edition of the *Recueil*.

(12) No member of the Terrasson family is obviously appropriate here, but as Jean (1670–1750), who had once been a brother of the Oratory, was later an *associé* of the Académie Royale des Sciences perhaps he (though a literary man and philosopher) was meant.

(13) Joseph Saurin (1659–1737), theologian and mathematician. He contributed several papers on mathematics and on the Cartesian system to the *Mémoires de l'Académie Royale des Sciences*.

(14) Noel Regnault (1683–1762), author of *Entretiens physiques d'Ariste et d'Eudoxe, ou physique nouvelle en dialogues*...(Paris, 1729), and other Cartesian treatises.

(15) No answer to Taylor appears in *L'Europe Savante*, which discontinued publication in 1720.

1367 TAYLOR TO KEILL

26 AUGUST 1721

Extracts from the holograph original in the University Library, Cambridge[1]

Bifrons 26 *August*
1721

Dear Sir

The enclosed is just come to me from Abbé Conti,[2] who desires me to convey it to you. He tells me, that he disputes continually with the French in favor of Sir Isaac Newton and the English Mathematicians; but that he can by no means make them sensible of the true nature of Sir Isaac's method, they not yet rightly understanding what he means by first and last ratios of nascent and evanescent quantities. I should be glad to see your form of expressing the Radius of Curvature which M. Conti desires of you, the Dutch Journals not coming to my hands.[3] I shall trouble you with no more at present, not knowing how unwelcome this little may be to you from me upon account of what Bernoulli has publisht out of my letters to Monmort in hopes to provoke your resentments against me.[4]

...   .

I had almost forgot to tell you, that his [*Bernoulli's*] accusing me of charging you with partiality is a mere calumny.

I am, Dear Sir,

Your most faithful and
most humble servant
Br[ook]Taylor

Dr Keil
Savilian Professor of Astronomy
in the University of Oxford

NOTES

(1) Res. 1893 (*a*) (Lucasian Professorship), Box 1, Packet 3; partially printed in Edleston, *Correspondence*, p. 236. Keill died on 31 August.

(2) Presumably not Letter 1361, in which Conti recounts to Taylor his part in the calculus dispute, which could hardly have 'just come' and makes no allusion to Keill. The subsequent letter from Conti, sent to Keill, is perhaps now lost.

(3) Possibly the meaning of this compressed utterance is as follows: Conti had come across one or other of the anti-Keill articles in the *Journal Literaire de la Haye* or *Nouvelles Litteraires* in which the determination of radii of curvature (involving second differentials) was brought in as a criterion of mathematical proficiency; Conti remembered that Keill had told him of his own success in handling such questions, and so Conti now wants a paper from Keill on this

matter which (one supposes) he meant to employ as evidence of Keill's mathematical excellence. Taylor, too, wished to see what Keill had accomplished.

(4) Taylor refers to Burchard Mencke's 'Epistola ad Taylor', see Letter 1303, note (3), p. 12. He goes on to apologize for the possible offence this may have caused Keill.

1368 ELIZABETH JOHNSON TO NEWTON
2 SEPTEMBER 1721
From the original in the University Library, Cambridge[1]

Hono[ure]d Sir

With humble submission i beg leave to trouble You once more being in ye greatest trouble Imagenable. my Son, who has layn Sick these two Years, have Reduced me to ye lowest Extremity Yesterday he Departed, haveing tasted of Yr Honours Charity [I] beg once more to Consider my Deploarable Condition

And yr: Distress'd, but
much Oblidgd Servt
Shall ever Pray
ELIZTH: JOHNSON

Sept: ye: 2d:
1721
    To
  The Honble
Sir Isaac Newton
    Kt:

NOTE

(1) Add. 3965, fo. 551. The letter (including the signature) is written (with defects of orthography as printed) in an ornate clerical hand. Newton used the paper for a short draft relating to the position of the comet of 1680 on 26 November and an alternative to the paragraph (*Principia*, Book III, Prop. 41; Koyré and Cohen, *Principia*, II, p. 732) beginning 'Congruunt igitur hæ observationes...' (not used) that cannot have been written before the date of this letter.

1369 VARIGNON TO NEWTON
7 SEPTEMBER 1721
From the holograph original in King's College, Cambridge[1]
Reply to Letter 1363; continued in Letter 1371; for the answer see Letter 1372

Nobilissimo Doctissimoque
Viro D.D. Isaaco Newton.
Equiti Aurato nec non Regiæ Societatis Anglicanæ
Præsidi Dignissimo
S.P.D.
Petrus Varignon.

Vir nobilissime ac Doctissime.

Die 19. Augusti [N.S.] proxime elapsi binas a te literas[2] accepi, quarum alteras Illustrissimo D. Cancellario[3] inscriptas, postridie cum plurimis & maximis a te gratijs ipsi reddidi: Has affabiliter acceptas illico legit ea frontis hilaritate ex qua conjicere fas erat gratissimas illi esse; qua de re, ijsdem perlectis, certiorem me fecit dicens se maxime lætari tibi aliquatenus significasse quanti te faciat. Dein quæsivit a me quo in statu esset tuæ Optices[4] impressio, & quales caracteres sint suffecti in locum primorum quos jam, utpote nimium attritos, respuerat. Tum ei novum obtuli, idcirco allatum, folij primi specimen novis caracteribus impressum qui nec visi sunt ei satis nitidi: Quamobrem, ijsdem pariter rejectis, accivit quarto Montalanum cui altiori voce jussit ut in horum locum alios denuo substitueret omnium perfectissimos. Tandem hic Bibliopola, cum diutius tergiversari non posset, suffecit eos quibus impressum vides hoc quod adest, primum operis tui folium quodque absolutum non accepi nisi hesterno die post specimen ejus approbatum ab Illustrissimo D. Cancellario. His in omnibus Bibliopolæ tergiversationibus consumptus est mensis totus Augusti cum parte presentis ad hunc usque diem. Hoc mitto tibi folium ut de presenti libri tui editione judices, quam spero fore tibi placituram. Folia autem cætera in fasciculos collecta, ut lenta suppeditabit Typographia, identidem mittam ad te per Cursorem Excellentissimi vestri apud nos Legati. Quod spectat pecuniam a te huic rei destinatam, nil urget: post absolutum opus dicam quam in summam sumptus abeant.

Ecce præterea tibi debitum Exemplar Actorum Nostrorum anni 1719, quæ data sunt nobis in ultimo cœtu Academico diei 6. mensis hujus, a quo die usque ad Martinalia vacabit Academia: has inducias tuæ Optices editioni promovendæ totas libens impenderem, Si cunctanti Bibliopolæ gradum maturare possem; sed strigosorum Equorum more, calcaribus sæpius etiam ac validius admotis, vix concitari potest.

Vides autem primum hujus operis tui folium, quod in Actorum Nostrorum exemplari conclusum est, expectare frontem[5] (Gallice *Vignette*) in ære incisam, quæ singalum [*read* singularum] quoque partium librorum singulorum frons erit & ornamentum, cujus solvendo pretio plus quam sufficiet excessus viginti tuarum librarum sterlingarum supra pecuniæ summam Bibliopolæ promissam, & eam quæ insumenda erit compingendis exemplaribus a te donandis. Porro cum Nemo te sit aptior ad excogitandam hujus frontis ideam Operi tuo convenientem, nos maxime juvabis si te duce adumbratam a perito delineatore Anglo, ad nos hic sculpendam mittere cures.[6]

Gratias etiam quam maximas egi a te D. Abbati d'Aguesseau, quibus perofficiose respondit. Tuo quoque Nomine tuisque verbis salutavi P. Reyneau[7] & Amicum ejus, qui ambo summa cum veneratione te resalutant. Cordatus hic

Amicus, Auctoritate Illustrissimi D. Cancellarij fretus, identidem instat
Montalano ad promovendam diligentius libri tui editionem cui huc usque
Bibliopola hic indormire visus est.

Post actas alijs tuo nomine gratias, jam meo tibi ago maximas & plurimas
pro alteris tuis humanissimis ad me literis, quibus grata tibi mea erga te studia
scribis esse: sane nil Jucundius atque Optatius audire poteram, cum nihil mihi
sit antiquius quam aliquid agere quod tibi gratum sit.

Diebus 21a. Augusti, & 5a. mensis hujus, literas[8] accepi a D. Coste petente
ut in corrigendis Optices tuæ folijs utar etiam Notis quas ad id paravit, quasque
dicebat se ideo missurum ad me, si concesserim. Has ad binas literas idem
Responsum dedi, scilicet inserere quicquam huic operi tuo mihi non licere, nisi
prius a te sit approbatum: qua de causa suas tibi offerat Notas: & si jusseris,
me ijsdem usurum fore perinde a Moivreanis, nimirum ut utræque mihi videbuntur eidem operi tuo convenire. Sed ex Novis tuis ad me literis[9] repertis in
fasciculo quem die 12a. [N.S.] mensis hujus accepi ab eodem D. Coste,
intelligo te postulationi ejus indulsisse: quamobrem locutionum emendationi-
bus eodem in fasciculo conclusis utar etiam prout ad tuos hoc in opere sensus
exprimendos apta mihi videbuntur. Vale.

Dabam Parisijs die 18, *Septemb.* 1721, *n.s.*

P.S. Memineris, quæso, promissarum a te mihi literarum[10] de D. Bernoullo,
cujus querelas non repeto ne molestus tibi sim; quod absit.

Translation

Pierre Varignon sends a grand salute
to the most noble and learned
Sir Isaac Newton, most worthy President
of the English Royal Society

Most noble and learned Sir,

On the 19th [N.S.] of last August I received two letters[2] from you, the second of
which, addressed to the most illustrious Chancellor,[3] I delivered to him the next day,
with your repeated and most profound thanks to him: receiving the letter in a friendly
manner, he read it there and then with so cheerful an expression that it was easy to
guess that it was very welcome to him. When he had perused it he told me that it was
so, saying he was greatly delighted that he had succeeded in conveying to you a little of
his high esteem for you. Then he asked me how far the impression of your *Opticks*[4] had
proceeded, and what sort of type-fount had been substituted for the first which he had
rejected as being excessively worn. I then presented to him a new specimen of the first
sheet which I had brought [with me] for that purpose, printed in a new fount which did

not seem handsome enough to him; whereupon, having rejected this [type-fount] also, he fetched Montalant for the fourth time, whom he commanded in a rather imperious tone to substitute for it yet another, absolutely unblemished. At last this bookseller, since he could make excuses no longer, supplied [the type] in which you see printed here before you this first sheet of your work, which I did not receive finished until yesterday, the day after a specimen of it [was] approved by our illustrious Chancellor. The whole of the month of August, and part of the present [month] until today, have been wasted in all these evasions of the bookseller. I send to you this sheet so that you may form an opinion of this present edition of your book, which I hope will please you. However, I shall from time to time send the remaining sheets to you by the messenger of His Excellency your Ambassador here, in gatherings, as our sluggish printer supplies them. There is no urgency concerning the money that you have set aside for this purpose; when the work is finished I will tell you how much the expenses amount to.

Moreover, here is the copy of our *Mémoires* for the year 1719 to which you are entitled; these [are the *Mémoires*] which I presented at the last meeting of the Academy on the sixth of this month, from which day the Academy will recess until Martinmas [*11 November*]: this whole break I would willingly devote to pressing on with the edition of your *Opticks*, if I could hasten the slow pace of the bookseller; but, like broken-down nags, he can hardly be hurried even when the spurs are applied often and fiercely.

You see, however, that the first sheet of this work of yours, which has been enclosed in the copy of our *Mémoires*, is lacking an engraved headpiece[5] (in French a *vignette*) which will also be the headpiece and adornment of each section of each book; the cost of this will be more than met by the excess of your £20 sterling above the sum of money promised to the bookseller and what will be spent on binding the copies which you are to give away as presents. Further, as no one is better able than yourself to think of an idea for a headpiece suitable for your work, you will help us greatly if you will take steps to send us a sketch [prepared] under your direction by a skilful English draughtsman, to be engraved here.[6]

I have given your best thanks to the Abbé Daguesseau to which he has responded most cordially. Also I have greeted Father Reyneau[7] and his friend in your name and in your own words, who both return your greeting with the greatest respect. This wise friend, armed with the authority of our illustrious Chancellor, continually urges Montalant to push on more diligently with the publication of your book, since hitherto the bookseller has seemed to him to be drowsing over it.

After giving these [people] thanks on your behalf, I now on my own account offer you the most profound and reiterated gratitude for your other very kind letter to me, in which you write that my zeal towards you is pleasing to you: indeed I could hear nothing more agreeable and pleasant, since nothing is dearer to me than to do what you like.

I received letters from Mr. Coste on the 21st August and the 5th of this month,[8] asking me to employ the annotated pages which he had made ready with that end in view when [proof-]correcting your *Opticks*; these he said he would send to me accordingly, if I agreed. To these two letters I have given the same reply, namely that I am not permitted to add anything to this work of yours unless it is first approved by you; for which reason he may offer you his annotations; and if you so order me I shall make use of them just as I do of De Moivre's; that is, to the extent that either of them seems to fit in with this same work of yours. But from your recent letter[9] to me discovered in the package which I received on the twelfth [N.S.] of this month from that same Mr Coste, I understand that you have been indulgent to his request: for which reason I shall also make use of the emendations of style contained in that package in so far as they seem to me fitted to express your meaning in this work. Farewell.

Paris, 18 *September* 1721, *N.S.*

P.S. I beg you to remember the letter[10] you promised [to write] me about Mr Bernoulli, whose complaints I do not repeat lest I be troublesome to you, which Heaven forfend!

NOTES

(1) Keynes MS. 142(9); microfilm 1011.26.

(2) Letter 1363; the enclosed letter from Newton to Daguesseau is now lost.

(3) Henri François Daguesseau; see Letter 1370.

(4) The second French edition; see Letter 1338, note (7), p. 91.

(5) A sketch for this vignette is printed in vol. I, Plate III (facing p. 107) of this *Correspondence*, wrongly ascribed to a much earlier date. It is also discussed, and compared with the final engraving, in J. A. Lohne, 'Experimentum Crucis', *Notes and Records of the Royal Society of London*, **23** (1968), 193–6. The printed vignette is a mirror image of the sketch; see Plate I.

(6) Newton in fact sent a rough sketch, but preferred Varignon to provide a draughtsman. Varignon employed the services of John Arlaud; see Letter 1400.

(7) Charles-René Reyneau (1656–1728); see vol. VI, p. 189, note (1). Reyneau was a great friend of Chancellor Daguesseau, and hence, too, probably of his brother.

(8) That is, on 10 August and 25 August O.S. Coste also wrote direct to Newton; see Letter 1365.

(9) Presumably Newton had given this letter to Coste to forward to Varignon with his (Coste's) corrections. It is now lost.

(10) This seems to indicate that the letter Newton drafted to Varignon in January 1721 (Letter 1353) was never sent, or at least did not arrive. Possibly, too, Letter 1363 was sent without the final comments Newton had drafted concerning Bernoulli.

1370 HENRI FRANÇOIS DAGUESSEAU TO NEWTON
17 SEPTEMBER 1721
From the original in King's College Library, Cambridge[1]

Henricus franciscus Daguesseau
Isaaco Newton Equiti aurato

Eximias illas tuas, Vir clarissime, de Lumine Colorumque, tum natura, tum varietate lucubrationes novis accessionibus a te locupletatas, gallica lingua donari, et, apud nos, prælo iterum subjici, non sine summa cum animi voluptate acceperam. Non diu igitur apud me laborandum fuit, ut segnem, nec satis ad vota nostra properantem Typographi manum acrius urgerem.[2] Sed pace tua dixerim, hanc qualemcunque operam, non tibi, Vir clarissime, sed mihi præstare visus sum; Imo, ut verius dicam, non mihi sed toti galliæ, quæ suas tibi voces, suumque sermonem libens et gratulabundæ commodat, Nec galliæ tantum sed omnibus omnium ubique gentium Philosophis Mathematicisque viris qui te, ut sui ordinis facile principem, jamdudum suspiciunt veneranturque. Non committam igitur, ut diutius ab eis desideretur nova illa tam præclari operis editio, et hoc saltem in communem letteratorum utilitatem conferam, ut typis quam emendatissimis nitidissimisque in lucem prodeat. Sint qui arcana scientiæ et ingenii tui scrutentur subtilius, aut fœlicius recludant. Mihi satis sit, nobilissima tua inventa, sin minus intelligentia, at certe admiratione prosequi. Vale, Vir clarissime, et ut scriptis tuis æternitatem spondere ausim, sic tibi ætatem quam longissimam, litterariæ Republicæ semper profuturam, toto animo adprecor.

Dabam in fraxineto nostro,[3] ubi primum post multiplices negotiorum curas, quæ hoc meum erga te officium diutius, quam par erat, distulerunt, respirare cœpi, IIII*d*. Kal. Octobr. ann. 1721 [N.S.]

Translation

Henri François Daguesseau to Sir Isaac Newton

I had learned (not without the greatest joy to my spirit) that those distinguished investigations of yours, most famous Sir, concerning light and both the nature and variety of colours were to appear in French with the addition of your latest embellishments, and that the book was to be printed here this second time. It was no matter of long debate for me, therefore, to urge on pretty sharply the printer's sluggish hand, so much less speedy than we desired.[2] But, by your leave, most famous Sir, I should have said that I saw fit to act in this particular way not for your sake but my own, or rather to speak more truly not for my own sake but for that of all France, who freely and with applause offers her tongue and her speech to you; nor for the sake of France alone, but on behalf of all philosophers and mathematicians of every nation everywhere, who already look up to you and venerate you as obvious master of them all. Therefore, I cannot allow this new edition of such an outstanding work to be any longer desired by them [in vain], and this at least I may accomplish for the common benefit of the world of letters that it may be published in the most correct and elegant typography. There are those who have examined the profundities of your learning and intelligence more subtly [than I], or who have expounded them with greater success; it will be enough for me that I have followed your most noble discoveries with admiration certainly, if with less comprehension.

Farewell, most famous Sir, and just as I have been bold enough to promise eternal fame to your writings so I pray with all my heart that your life may be far extended, to the continual benefit of the Republic of Letters.

Written from my ash-grove,[3] where I first began to breathe again after the multiple cares of business affairs, which have postponed this service of mine owing to you for longer than was proper, the fourth Kalends of October 1721. [N.S.]

NOTES

(1) Keynes MS. 94, fo. 77; microfilm 931.2. Henri François Daguesseau (or d'Aguesseau, 1668–1751) was at this time one of the most powerful men in France. His father was *intendant* of Bordeaux; he himself was appointed procurator-general of the Parlement de Paris (1700). He was distinguished by his brilliant legal rhetoric, by his administrative skills in the famine years after 1709, and by his opposition to the registration by the Parlement of the papal Bull *Unigenitus*, which largely destroyed the privileges of the Gallican Church. Unyielding to Louis XIV, he was rewarded for his devotion to the Duke of Orleans (regent on behalf of Louis XV) by appointment as Chancellor of France (February 1717). Within less than a year, however, he had been deprived of his office because of his opposition to the financial schemes of John Law; his re-instatement as Chancellor (June 1720) occurred after Law's collapse. There followed a new battle over *Unigenitus*, with Daguesseau this time favouring its registration, which was accomplished in December 1720. He was to be dismissed a second time in February 1722, and again re-established by Cardinal Fleury in 1727. He was a patron of literature and himself a copious writer.

(2) See Letter 1369.

(3) This is a pun, Daguesseau's estate in Brie, south-east of Paris, was called Fresnes, *frêne* being an ash-tree.

1371 VARIGNON TO NEWTON

21 SEPTEMBER 1721

From the holograph original in King's College Library, Cambridge[1]
Continuation of Letter 1369

Nobilissimo Doctissimoque Viro
D.D. Isaaco Newton,
Equiti aurato, nec non Regiæ Societatis Anglicanæ
Præsidi Dignissimo
S.P.D.
Petrus Varignon.

Vir nobilissime ac Doctissime,

Obsignatas has ad te Literas[2] heri accepi ab Illustrissimo D.D. Cancellario ut eas ad te mitterem, quas nimirum, lectis tuis ad illum,[3] dixerat mihi se ad te rescripturum fore. Iisdem D. Cancellarij Literis adjungo tria presentis editionis Optices tuæ folia *B, C, D*, subsequentia primum *A* quod jam tibi misi cum Literis ad te[4] & ad D. Moivreum in fasce convolutorum Academiæ Nostræ commentariorum an. 1719. quem a me acceptam die 13. Septemb [N.S.] D. Laurenzy, ab Epistolis Excellentissimo D.D. Legato vestro apus [*read* apud] nos, promisit mihi se missurum ad te ut primum sese offerret occasio, quam jam ab isto tempore oblatam esse autumo. Porro ejusdem tuæ Optices editionis folia cætera, in fasciculos collecta, prout suppeditabit Typographia, identidem eadem D. Laurenzy opera mittam tibi. Vale.

Dabam Parisijs die 2. Octob. 1721, n.s.

Translation

Pierre Varignon presents a grand salute
to the most noble and learned Sir Isaac Newton
most worthy President of the English Royal Society

Most noble and learned Sir,

Yesterday I received a sealed letter[2] to be sent on to you from our most illustrious Chancellor. He had indeed told me (at the time when he read your letter to him) that he was going to write to you.[3] I add to this letter of the Chancellor the three sheets *B*, *C* and *D* of the present edition of your *Opticks*, following the first sheet, *A*, which I have already sent with letters to you[4] and to Mr De Moivre in a bundle of rolled up *Mémoires* of our Académie for the year 1719, which [bundle] Mr Laurenzy, having received them from me on 13th September [N.S.], promised me in a letter to His Excellency your

Ambassador here with us that he would send to you at the first opportunity, which I think will have presented itself already in the lapse of time since then. Further, I shall send you from time to time by Mr Laurenzy the remaining sheets of this same edition of your *Opticks*, collected into gatherings, just as the printer supplies them. Farewell.

Paris, 2 October 1721, *N.S.*

NOTES

(1) Keynes MS. 142(H); microfilm 1011.26.
(2) Letter 1370.
(3) Newton's letter to Daguesseau is now lost, but for Daguesseau's reactions to it see Letter 1369.
(4) Letter 1369.

1372 NEWTON TO VARIGNON
26 SEPTEMBER 1721
From a holograph draft in the University Library, Cambridge[1]
Reply to Letter 1369; for the answer see Letter 1381

Viro celeberrimo Dno Abbati Varignon Regio Mathesis Professori
& Academiæ scientiarum Socio apud Parisienses
Is. Newtonus S.P.D.

Clarissime Dne,

Accepi Historiam et Commentaria ex Archivis Academiæ Scientiarum pro Anno 1719, pro quibus gratias tibi reddo quammaximas. Accepi etiam schedam primam Libri de coloribus[2] elegantem sane & specie nobilem. Et ne Dnus Montalanus[3] expensa moleste habeat dabo illi libras viginti sterlingas, et expensa compingendi libros insuper solvam. Gratias tibi reddo quamplurimas quod insinuasti libros plures amicis donandos esse, scilicet Cardinali Polignac,[4] et filio Cancellarij,[5] et Bibliothecæ Academiæ. Vellem et alios donandos esse filio et nepoti D. Joannis Bernoullij,[6] et alios Abbati de Comitibus & P. Sebastian,[7] & D. Remond.[8] Sed et gratias tibi maximas reddo quod onus in te suscipere digneris conferendi Correctiones Dni Coste & Dni Moyvre inter se, et quod optimum videbitur eligendi; ut et emendandi quæcunque alia occurrerint.[9] Metuebam utique ne correctiones Dni Coste, inter plurima tua negotia molestiam nimiam tibi crearent. Sed cum hocce onus in te suscipere non dedigneris, eo magis me tibi obligasti. Schema tuum libris singulis præfigendum probo, sed nondum a Pictore delineatum est. Pictorem mox adibo.[10]

In sententia Mathematici Judicis quam D. Leibnitius D. Joanni Bernoullio

TRAITÉ
D'OPTIQUE,
SUR
LA LUMIERE
ET LES COULEURS.

Plate I. (*Upper*) The vignette from *Traité d'Optique* (Paris, 1722). The motto reads 'Nec variat lux fracta colorem'. See Letter 1391.

(*Lower*) The Tower of London, *c.* 1730. From a wash drawing attributed to Samuel Buck. On the far right is the Wharf (with ships), next the Gate Tower, in the centre the Beauchamp Tower with, to the left, the buildings of the Mint extending to the Devlin Tower.
Courtesy of the Museum of London.

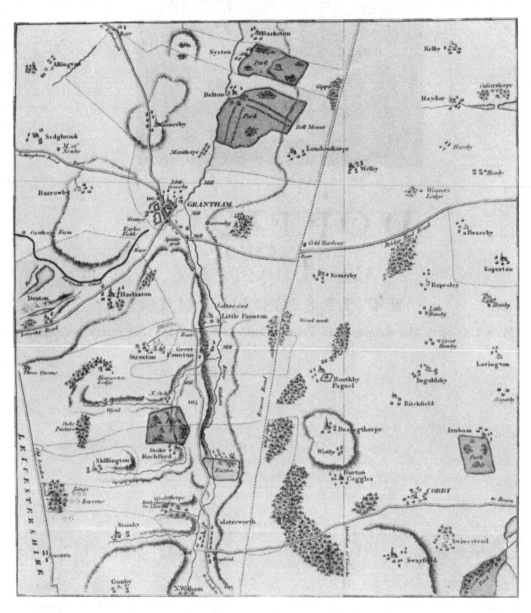

Plate II. Newton's countryside near Grantham, Lincolnshire. Woolsthorpe is at the lower left of centre. After Turnor; see Appendix II, Newton's Genealogy.

ascripsit, publice accusor plagij.[11] Et epistola quam D. Bernoullius ad me misit[12] & qua se talem sententiam scripsisse negavit, videbatur ad me missa ut remedium contra injuriam illam publicam: et eo nomine licentiam mihi datam esse putabam diluendi injuriam illam auctoritate D. Bernoullij, præsertim cum is me non prohibuerit.[13] Attamen Epistolam illam non nisi privatim communicavi, & Keilio nullam dedi licentiam aliquid evulgandi ex eadem, et multo minus scribendi contra Bernoullium ob ea quæ in Epistola illa mihi amice scripserat. Et hac de causa Keilium quasi liti studentem vehementer objurgavi:[14] sed ille jam mortuus est.

Conqueritur D. Bernoullius quod ipsum vocavi *hominem novum*, & *Mathematicum fictum*, & *Equitem erraticum*.[15] Sed contra Bernoullium nondum cœpi scribere. Hæc omnia dixi scribendo contra Leibnitium, & ejus argumenta repellendo. 1 Dixerat utique D. Leibnitius *Keilium esse hominem novum et rerum anteactarum parum peritum cognitorem*;[16] id est hominem qui floruit post tempora Commercij quod Leibnitius habuit cum Oldenburgio: et idem objeci Leibnitio Bernoullium judicem constituenti, cui utique commercium illud antiquum annis plus triginta post mortem Oldenburgij ignotum fuerat. 2 Cum D. Leibnitius sententiam Judicis mathematici Bernoullio ascriberet, vocavi judicem illum *Mathematicum* vel *fictum Mathematicum*,[17] id est Mathematicum qui vere author esset sententiæ illius, vel fingebatur esse author. Nam cum Bernoullius ab Authore sententiæ illius citabatur tanquam ab Authore aversus, dubitabam utrum ille author esset, necne. Et Bernoullius ipse literis ad me datis affirmavit se non fuisse authorem. 3 D. Leibnitius in Epistola sua prima ad Abbatem de Comitibus[18] Quæstionem de primo methodi differentialis inventore deseruit & ad disputationes novas confugit de gravitate universali & qualitatibus occultis & miraculis & vacuo et atomis, et spatio & tempore & perfectione mundi; Et sub finem Epistolæ Problema Bernoullij ex Actis Eruditorum desumptum proposuit mathematicis Anglis: Et initio proximæ suæ ad Abbatem Epistolæ[19] contulit hanc novam controversiam cum *duello*, scribens se nolle in arenam descendere contra milites meos emissarios, sed cum ipse apparerem, se lubenter mihi satisfactionem daturum. Et ad hæc omnia alludens non contra Bernoullium sed contra Leibnitium scripsi in Observationibus meis in hanc ejus Epistolam, ubi dixi quod[20] *Epistolæ et chartæ antiquæ* [ex mente Leibnitii scilicet] *jam abjiciendæ sunt, et Quæstio* [de primo methodi inventore] *deducenda est ad rixam circa Philosophiam et circa res alias: et magnus ille Mathematicus quem D. Leibnitius Judicem sine nomine constituit, jam velum detrahere debet* [secundum Leibnitium scilicet] *et a partibus Leibnitii stare in hac rixa, & chartam provocatoriam ad Mathematicos in Anglia per Leibnitium mittere quasi Duellum, vel potius bellum inter milites meos emissarios* [uti loquitur] *et exercitum discipulorum in quibus se felicem jactat; methodus esset magis idonea ad Quæstionem de primo inventore dirimendam quam examinatio veterum et authenticorum scriptorum, & scientiæ Mathematicæ imposterum factis nobilibus equitum erraticorum vice argumentorum ac Demonstrationum implendæ essent.* Hoc totum contra Leibnitium scripsi, et non contra Bernoullium. Leibnitius Bernoullium constituit judicem, Leibnitius eundem ex judice constituit advocatum. Leibnitius Commercium Epistolicum fugit quasi a Judice suo condemnatum. Leibnitius vice Quæstionis de primo Inventore disputationes novas de Quæstionibus Philosophicis proposuit, et Problema tanquam a Bernoullio misit a Mathematicis Anglis solvendum. Leibnitius fuit Eques ille erraticus qui vice argumentorum ex veteribus et authenticis scriptis desumendorum, introduxit alias disputationes, quas ipse contulit cum duello. Ad hoc duellum ille me provocavit Methodi infinitesimalis gratia. Hæc methodus erat virgo illa pulchra pro qua Eques noster pugnabat. Quæstionem de primo methodi hujus inventore per victoriam in hoc duello dirimere sperabat, & Virginem lucrari non examinatis veteribus & authenticis scriptis in Commercio Epistolico editis, per quæ Quæstio illa dirimi debuisset. Problemata Mathematica proponi possunt exercitij gratia, sed non ad dirimendas lites alterius generis: Et solus Leibnitius eadem in hunc finem proposuit.

Hæc tibi scripsi non ut in lucem edantur, sed ut scias me nondum cum Bernoullio lites habuisse. Contra illum nondum scripsi, neque in animo habeo ut scribam: nam lites semper fugi.

D[ominu]s Moivreus mihi dixit D. Bernoullium picturam meam optare:[21] sed ille nondum agnovit publice me methodum fluxionum et momentorum habuisse anno 1672 uti conceditur in Elogio[22] D. Leibnitij in Historia Academiæ vestræ edito. Ille nondum agnovit me in Propositione prima Libri de Quadraturis, anno 1693 a Wallisio edita, & anno 1686 in Lem. 2 Lib. 2 Princip. synthetice demonstrata, Regulam veram differendi differentialia dedisse & Regulam illam anno 1672 habuisse, per quam utique Curvaturas Curvarum tunc determinabam. Ille nondum agnovit me anno 1669 quando scripsi Analysin per series, methodum habuisse quadrandi Curvilineas accurate si fieri possit, quemadmodum in Epistola mea 24 Octob. 1676 ad Oldenburgium data et in Propositione quinta Libri de Quadraturis exponitur; et Tabulas Curvilinearum quæ cum Conicis sectionibus comparari possunt per ea tempora a me compositas fuisse. Si ea concesserit quæ lites prorsus amovebunt,[23] picturam meam haud facile negabo.[24] Vale

Dabam Londini
26 Sept. 1721. St. vet.

Translation

Isaac Newton sends a grand salute
to the most famous Abbé Varignon, Regius Professor of Mathematics
and member of the Académie des Sciences of Paris

Most famous Sir,

I have received the *Histoire* and *Mémoires* [extracted] from the records of the Academy of Sciences covering the year 1719, for which I give you the greatest possible thanks. I have also received the first sheet of the book of colours, [2] [which is] indeed elegant and of a noble appearance. And lest Mr Montalant[3] should suffer from the cost of it, I will give him £20 sterling, and over and above [that] pay the expenses of binding the books. I give you very many thanks for hinting that many books must be given to friends such as the Cardinal Polignac,[4] the Chancellor's son,[5] and the library of the Academy. And I wish others to be given, too, to the son and the nephew of Mr Johann Bernoulli,[6] and others to the Abbé Conti, to Father Sebastien,[7] and to Mr Rémond.[8] But I give you the greatest thanks also because you have been so good as to take upon yourself the task of comparing together the corrections of Mr Coste and of Mr De Moivre, and choosing what shall seem the best; and of amending anything else that may occur.[9] I was quite fearful that Mr Coste's corrections would make too much trouble for you, busy as you are with many other things. But since you were so kind as to take on this task, you put me the more heavily in your debt. I approve your [idea of] prefixing each book with an illustration, but it is not yet drawn by the artist. I will soon spur the artist on.[10]

In the opinion of the mathematical judge, which Mr Leibniz attributed to Mr Johann Bernoulli, I am publicly accused of plagiary.[11] And the letter which Mr Bernoulli has sent to me,[12] where he has denied that he wrote any such words, seemed sent to me so that I may publish a defence against that injury: and for that reason I thought that liberty was given to me to undo that injury by the authority of Mr Bernoulli, especially as he had not forbidden me [to do so].[13] However, I have only communicated that letter privately, and I did not give Keill permission to divulge anything from it, and much less to write anything against Bernoulli on account of what he had written out of friendship in that letter to me. And for that reason I have reproved Keill strongly for pursuing the quarrel;[14] but he is now dead.

Mr Bernoulli complains because I have called him a 'novice', a 'pretended mathematician', and a 'knight-errant'.[15] But I have not yet begun to write against Bernoulli. I have said all these things writing against Leibniz, and in refuting his arguments: 1. Beyond question Mr Leibniz said that 'Keill was a novice and a poor witness to what had happened in earlier years',[16] that is, a man who reached maturity after the time of the correspondence which Leibniz had with Oldenburg: and I raised the same objection to Leibniz's appointing Bernoulli as judge, to whom that early correspondence had surely been unknown for more than thirty years after the death of Oldenburg. 2. Since Mr Leibniz attributed the opinion of the mathematical judge to Bernoulli, I have called that judge a 'Mathematician', or 'pretended Mathematician',[17] that is, a mathematician who was really author of that judgement or was feigned to be its author. For since Bernoulli was cited by the author of that judgement as different from that author, I was uncertain whether or not he was himself that author. And Bernoulli himself has stated in a letter written to me that he was not the author. 3. Mr Leibniz, in his first letter to the Abbé Conti,[18] disregarded the question of the first inventor of the differential method, and took refuge in new disputes about universal gravity, occult qualities, miracles, the vacuum, atoms, space, time, and the perfection of the world; and at the end of the letter posed to the English mathematicians a problem of Bernoulli's chosen from the *Acta Eruditorum*. And at the beginning of his next letter[19] to the Abbé he compared this new controversy with a *duel*, writing that he did not wish to enter the ring against my soldier-guards but if I would appear myself he would willingly give me satisfaction. And, alluding to all this in my 'Observations' on this letter of his, I wrote not against Bernoulli but against Leibniz when I said[20]

> The ancient Letters and Papers must now be laid aside, [that is to say, in Leibniz's opinion] and the Question [of the first inventor of the method] must be run off into a Squabble about Philosophy and other matters: And the great Mathematician, whom Mr. Leibniz put forward as a judge without a name, must now pull off his Mask [that is, according to Leibniz], and become a Party-Man in this Squabble, and send a Challenge by Mr. Leibniz to the Mathematicians in England, as if a Duel, or perhaps a Battel between my soldier-guards [as he calls them] and his Army of Disciples, to whom he happily gives himself up; were a fitter way for deciding the Question of the first inventor than an Appeal to ancient and authentick writings, and Mathematicks must henceforward be filled with Achievements in knight-errantry, instead of Reasons and Demonstrations.

I wrote all this against Leibniz, and not against Bernoulli. Leibniz appointed Bernoulli a judge; Leibniz then converted the same person from a judge into an advocate. Leibniz has avoided the *Commercium Epistolicum* as though it had been condemned by his judge. Leibniz has put forward instead of the question of the first inventor new disputes concerning philosophical questions, and has sent to the English mathematicians a problem for solution as though from Bernoulli. Leibniz was that knight-errant who, instead of drawing his arguments from ancient and authentic writings, introduced other disputes, which he himself compared with a duel. And for the sake of the infinitesimal method he challenged me to that duel. This method was that beautiful maiden on whose behalf our knight did battle. He hoped to settle the question of the first inventor of this method by victory in this duel, and to gain the maiden without examining the old and authentic writings printed in the *Commercium Epistolicum*, by which the question ought to have been settled. Mathematical problems can be put forward as exercises, but not for resolving quarrels of another kind; and only Leibniz has put them forward for this purpose.

I have written this to you not in order that it might be published, but so that you should know that I have not yet had any disputes with Bernoulli. I have not yet written against him, nor do I have it in mind to do so: for I have always avoided quarrels.

Mr De Moivre has told me that Mr Bernoulli desires a picture of me:[21] but he still has not publicly acknowledged that I had the method of fluxions and moments in the year 1672, as was admitted in the 'Éloge' of Mr Leibniz published in the *Histoire* of your Academy.[22] He still has not acknowledged that I gave a true rule for differentiating differentials in the first proposition of the book on quadratures, published by Wallis in 1693, and demonstrated synthetically in 1686 in Lemma 2, Book 2 of the *Principia*, by means of which I did, beyond question, at that time determine the curvatures of curves. He still has not acknowledged that in 1669, when I wrote my *Analysis per Quantitum Series*, I had a method for accurately squaring curved lines, if it can be done, just as I explained it in my Letter to Oldenburg of 24 October 1676 and in the fifth proposition of the book on quadratures; and that the Tables of curved lines, which can be compared with conic sections, were composed by me at that time. If he will submit on those points[23] so as to entirely remove the cause of disputes, I will not readily refuse [him] my picture.[24] Farewell.

London, 26 September 1721, *O.S.*

NOTES

(1) Add. 3968(42), fo. 599; printed in Brewster, *Memoirs*, II, pp. 497–501. From Varignon's report of it to Bernoulli in Letter 1374, we know that a letter very much following this draft was sent.

There is an antecedent Latin draft at fo. 601, and a fragment at fo. 602v. There is an English draft corresponding approximately to the Latin at fo. 597; the arguments in this are not in exactly the same form or in the same order, but it is a closely analogous text. Less similar and more fragmentary English drafts are at fo. 411v.

(2) The second edition of the *Traité d'Optique* (Paris, 1722). See Letter 1338, note (7), p. 91.

(3) The bookseller and publisher; see Letter 1338, note (6), p. 91.

(4) See Letter 1350, note (5), p. 117.

(5) Daguesseau's son.

(6) Presumably Nikolaus [II] Bernoulli and Nikolaus [I] Bernoulli. Varignon reported Newton's intentions to Johann Bernoulli in Letter 1374. Bernoulli later pointed out that he had five sons and could not be certain which one Varignon intended, although he supposed it was the eldest (see Bernoulli to Varignon, 22 November 1721, N.S., Bernoulli Edition).

(7) Sebastien Truchet; see Letter 1350, note (1), p. 116.

(8) Nicolas Rémond, brother of Pierre Rémond de Monmort who had recently died; see vol. VI, p. 216, note (5).

(9) Compare Letter 1369, p. 154.

(10) The artist was Jacques-Antoine Arlaud; see Letter 1400, note (1), p. 213. Newton's attention to the sketch for the vignette (see Letter 1369, note (5)) is shown by a note of two mottoes for it, at U.L.C., Add. 3968(42), fo. 602, amongst the drafts for the present letter: 'Nec variat lux fracta colorem' and 'Dispescit sed non variat lux fracta colores.' ('Refracted light does not vary its colour' and 'Light splits but the refracted light does not vary its colours.')

(11) Newton harks back to the calculus controversy and Bernoulli's authorship of the anonymous letter printed in the *Charta Volans* (see Number 1009, vol. VI, and compare Letter 1353); he had promised that he would write to Varignon a detailed account of the dispute

(see Letter 1374, the P.S.) and Varignon had reminded him to do so (see Letter 1369, p. 168, top).

(12) Letter 1332.

(13) The English version on fo. 597 reads: 'I looked upon Mr Bernoullis Letter to me as intended for a Remedy But if he intended it only as a private Letter he should have told me so & then I had kept it private.'

(14) At fo. 597 Newton uses the phrase 'I chid him for it as a breach of friendship.' For Newton's reactions to Keill's continued attempts to champion him, see above Letter 1353, p. 120, and Letter 1329, note (17), p. 69. Keill had died on 31 August.

(15) Newton had used these three phrases in his earlier writings now reprinted in Des Maizeaux's *Recueil*; see notes (16), (17) and (20), below. 'Knight-errant' is Newton's own English expression. Varignon had taken pains to bring these expressions to Bernoulli's notice in Letter 1357.

(16) Leibniz used these words to describe Keill in Letter 884, vol. v; Newton used them to describe Bernoulli (not Leibniz) in his printed 'Observations' (vol. vi, p. 343).

(17) Newton used this phrase in Letter 1187, vol. vi, p. 285; he explained his meaning in Number 1211, vol. vi, p. 342. Compare also Letter 1374, note (7), p. 170.

(18) Letter 1170, vol. vi.

(19) Letter 1197, vol. vi. Leibniz wrote 'C'est sans doute pour l'amour de la verité que vous [*Conti*] vous etes chargé d'une espece de cartel de la part de M. *Newton*. Je n'ay point voulu entrer en lice avec des enfans perdus...'

(20) Newton here translates into Latin, with some minor changes, the English he used in Number 1211, vol. vi, p. 344. In our translation we have tried to retain the style of the original English passage whilst following Newton's revised, Latin version. In his English draft on fo. 597 (see note (1) above) Newton followed exactly the wording given in Number 1211. The square brackets in the passage are Newton's.

(21) Presumably this information had been passed on via Varignon and De Moivre; see Letter 1374, note (3), p. 170.

(22) Fontenelle's 'Éloge' of Leibniz; see Letter 1298, note (4), p. 4.

(23) Compare the excerpts from Newton's drafts, which we quote in Letter 1329, note (4), p. 66. Newton had discussed all these points in detail in his 'Observations'; see Number 1211, vol. vi.

(24) In a letter to Varignon of 20 February 1721, N.S. (see Bernoulli Edition), Bernoulli had expressed his desire for a portrait of Newton.

1373 KATHERINE RASTALL TO NEWTON
7 OCTOBER 1721
From the original in the Bodleian Library, Oxford[1]

I give you a thousand thanks for ye many civilities yt I received from You I have been ill & can never get well since, I had ye happiness to bounteously hear from you & how ye Lord designs to deall with me, He alone knows. The Lords will be done. I magnifie Gods Holy name to continue so good a freind in

ye land of ye liveing amongst us I hope God will spare you with a longer
continuence amongst us is ye hearty Prayer of her, who is

Your obedient servant
CA: RASTALL

Basingthorpe[2] *October ye* 7th
        1721

If you please to send to me ye Bearer will Faithfully convey it.

To
The Honorable Sr Isaack Newton
At ye signe of ye two lamps near
Red Lyon Square.[3] In London
      Humbly Present

NOTES

(1) Bodleian New College MS. 361, fo. 66. For a previous letter from the writer, who was evidently often a recipient of Newton's alms, see Letter 1110, vol. VI, and Appendix II, Newton's Genealogy.

(2) Now Bassingthorpe (in Kesteven, Lincolnshire, near Corby).

(3) Red Lion Square is a good way north of St Martin's Street; perhaps the 'Bearer' of the letter to Newton lived there, and was familiar with Newton's true address.

1374 VARIGNON TO J. BERNOULLI
10/11 OCTOBER 1721
Extracts from the holograph original in the University Library, Basel[1]

A Paris le 21. *Octob*. 1721 [*N.S.*]

Monsieur,

Le 9. Aoust [N.S.] dernier je repondis à votre lettre du 10. Juillet, & entre-autres choses je vous dis que je n'avois point encore reçu de Reponse d'Angle-terre par raport à la satisfaction que vous demandiez à M. Newton sur les Griefs que vous m'aviez chargé de faire scavoir a M. de Moivre pour les lui representer; ce que j'avois fait le 24 Juillet [N.S.] ainsi que je vous l'avais promis, & en vos propres termes pour lui en exprimer toutte la force.[2] Le 19. Aoust M. de Moivre me repondit à cela:

'J'ay lu de point en point votre Lettre à M. Newton,[3] j'ay eté bien aise qu'il sçust comme de la propre bouche de M. Bernoulli lui même, quels étoient ses sujets de plainte, & sur quoi il insistoit pour obtenir reparation de

167

l'injure qu'il dit avoir reçue. Apres cette lettre démarche jay supplié M. Newton de vouloir répondre lui même sur ce qui le regarde, & de m'oter de dessus
les épaules un fardeau qui me devient trop pesant. Il me l'a promis:[4] je
souhaite de toutte mon Ame que M. Bernoulli ve[u]ille bien se contenter de ce
que vous verrez dans la lettre que M. Newton vous écrira. Mais si apres les
peines que vous et moy nous nous sommes données pour amener les choses au
point d'une Reconciliation parfaite, Monsieur Bernoulli persiste à exiger une
plus ample satisfaction, je souhaite qu'à l'avenir il veuille porter ses plaintes
directement à M. Newton lui même, il ne m'est pas possible de rien faire de
plus pour M. Bernoulli que ce que j'ay fait, sans me rendre suspect de par-
tialité pour lui.'

Je ne vous ay point fait scavoir cette Reponse de M. de Moivre, attendant
toujours celle qu'il m'y promettoit de M. Newton pour vous la faire scavoir
comme le capitale que vous demandez. M. de Moivre me fist encore esperer
cette Reponse de M. Newton par raport a vous, dans une Lettre du 16 Sep-
tembre, que je reçu le 7. [N.S.] de ce mois-ci, dans laquelle il me disoit que *la
Mort de M. Keil, arrivée depuis quinze jours, leve un grand obstacle à la paix.*

Enfin je reçu hier au soir cette Reponse de M. Newton: elle me fut rendue
par un Etranger, non Anglois, arrivant de Londre ou il me dist avoir passé
quatre mois frequemment avec M. Newton, dont il me dist tout le bien
imaginable du coté de sa droiture, de sa douceur, & de son bon coeur. Voici
les propres termes de cette lettre de M. Newton par raport a la satisfaction
que vous lui demandez.[5]

'In sententia Mathematici...mortuus est.'[6]

'1° Dixerat...

'2°[7] Cum D. Leibnitius sententiam Judicis Mathematici Bernoullio ascriberet, vocavi illum Mathematicum vel fictum Mathematicum non ut ipsi
derogarem, vel vel peritiam ejus minuerem; sed quia dubitabam utrum esset
Mathematicus iste. Nam cum Auctor sententiæ illius Bernoullium citaverat
tanquam a seipso diversum, dubitabam utrum Bernoullius auctor esse posset.
Et Bernoullius tandem literis ad me datis affirmavit se non fuisse auctorem.

'3° D. Leibnitius...nam lites semper fugi.'

Voila, Monsieur un ample Ecclaircissement de M. Newton, lui même, sur
tous vos Griefs, excepté sur celui qui regarde M. des Maizeaux, sur lequel je
vous en ay desja envoyé un pareil de M. Newton, il y a environ un an,[8] Vous
voyez par celui-ci qu'il n'a jamais rien écrit contre vous, mais seulement contre
M. Leibnitz, que c'est M. Leibniz seul qu'il [a] traité de *Chevalier errant*, que
c'est le Mathematicien imaginaire, qui vous citoit, qu'il a traite de *soy disant
Mathematicien*; qu'il ne vous a traité de *Homo Novus* que dans les faits dont vous
n'etiez pas instruit, & cela sur ce que M. Leibnitz vouloit vous faire le Juge de

la contestation qui rouloit toutte sur ces faits; qu'en faisant voir votre lettre à M. Keill, il n'a pas cru que celui ci en publiast rien, & qu'il l[']a fort reprimandé d'en avoir parlé dans sa Reponse, la traitant d'homme qui aimoit trop les contestations; quenfin il a cru etre endroit pour sa propre justification de faire voir votre lettre pour montrer que vous y desavouez celle du 29. Juillet 1713 [N.S.] publiee contre lui, dans laquelle il a (dit il) eté traité de *Plagiaire*: Grief beaucoup plus grave que les votres, ne fust ce que par conjecture qu'il ait ainsi éte traité. Je souhaite de tout mon coeur pour votre repos, que vous soyez satisfait du desaveu que M. Neuton fait ici, & dans ce que je vous ay envoyé de lui par raport à M. des Maizeaux,[8] des griefs dont vous vous plaignez comme il paroist l'être du desaveu que vous lui avez fait dans votre lettre, de celui dont il se plaind: vous vivrez par-là Amis ensemble; & en paix, Dieu venant d'en lever le plus grand obstacle.[9] De sorte que M. Newton asseurant ici qu'il aime trop la paix pour écrire contre vous, votre repos ne depend plus que de vous même.

Si vous n'etes pas content de cette satisfaction de M. Newton, j'en seray tres faché, M. Newton ne paroissant pas disposé à la porter plus loin, ny (comme vous venez de voir) M. de M[oivre][10] à lui en demander davantage, ny moy non plus, qui n'avois que M. de Moivre pour porter vos plaintes, n'osant le faire par moy même, faute d'être à portée pour cel[a][10]

Au reste vous voyez, Monsieur, qu'à la fin de l'extrait que je viens de vous faire de la lettre de M. Newton, il me defend absolument que rien d'elle paroisse en public, apparement de peur que le recit qu'il y fait du procedé de M. Leibnitz à son egard, n'excite quelqu'un à renouveler leur contestation qui n'a desja fait que trop de bruit. Ainsi je vous prie, Monsieur de ne point rendre ceci public de peur que M. Newton n'en fist retomber le contre-coup sur moy qui ne veux de démêlé avec personne.

Je vous ay dit qu'on imprime ici L'Optique francoise de cet Auteur qui m'a prié d'y veiller. Il vous en destine trois Exemplaires: un pour vous, un pour M. Votre Fils, & le troisieme pour M. Votre Neveu le Mathematicien. M. de Moivre dans sa penultime lettre m'ecrivit que M. Neuton vous en destinoit un; & dans sa derniere il ajoute que M. Newton vous en destine deux de plus, un pour M. Votre fils & l'autre pour M. Votre Neveu. C'est ce que M. Newton me dit lui même dans sa lettre. *Vellem* (dit il, en parlant de ceux à qui il en destine) *alios* (libros) *donandos esse Filio & Nepoti D. Joannis Bernoullij.*[11]

NOTES

(1) MS. L.I.*a*, 159*; the letter bears the date 21 October 1721, [N.S.], at the head, and 22 October 1721 at the foot.

We do not print Bernoulli's lengthy reply, dated 22 November 1721, N.S. (see Bernoulli

Edition). There he repeated his claim that he had forgotten whether or not he was author of the letter of 7 June 1713, N.S. Bernoulli's anger was directed chiefly at Keill (who had asserted that Bernoulli, and not Leibniz had composed the *Charta Volans* (Number 1009, vol. VI) itself, not only the letter extract quoted in it). He also confessed himself still deeply wounded by the expression 'Chevalier Errant' (see Letter 1357, note (2)), and was dissatisfied because, despite Varignon's plan for publication, Newton asked that Letter 1372 should not be published. Despite further protests from Bernoulli, Newton remained adamant (see Letter 1385).

(2) See Letter 1357, note (2), p. 132.

(3) De Moivre refers here to a letter, now lost, from Varignon to himself, intended possibly as a reply to Newton's Letter 1353 of 19 January 1721, where Newton had discussed the calculus dispute, and mentioned that Bernoulli had written to him concerning it. See also Letter 1357, note (2), p. 132, and Letter 1363, note (5), p. 143.

(4) Newton did eventually reply on 26 September 1721 (Letter 1372).

(5) Varignon quotes from Newton's Letter 1372 *verbatim*, beginning at the second paragraph.

(6) The first two sentences of the next paragraph in Newton's draft, 'Conqueritur D Bernoullius...argumenta repellendo' are omitted. Either Newton himself left them out of his final letter to Varignon, or Varignon chose to omit them.

(7) The wording Varignon quotes is different from that in Newton's draft (see Letter 1372). Varignon's version reads, '2. Since Mr Leibniz attributed the opinion of the mathematical judge to Bernoulli, I have called that judge a 'Mathematician', or 'pretended Mathematician', not in order to dishonour him, or to belittle his skill, but because I doubted that he was himself that other mathematician. For since the author of that opinion cited Bernoulli as a writer different [from himself], I doubted that Bernoulli could be that other author. And Bernoulli has finally asserted in a letter to me that he is not that author.'

(8) Letter 1357.

(9) Varignon reiterates De Moivre's words; the 'obstacle' was Keill.

(10) The paper is torn.

(11) See Letter 1372, p. 160.

1375 THE TREASURY TO NEWTON
11 OCTOBER 1721
From the copy in the Public Record Office[1]
For the answer see Letter 1378

Whitehall Trea[su]ry Chambers. 11th *Oct* 1721

The Rt Honoble the Lords Comm[issione]rs of his Ma[jes]ties Treasury are pleased to Referr this Memoriall to Sir Isaac Newton Master & Worker of his Ma[jes]ties Mint who is to Report to their Lordps a State of the Mint in Scotland and the Summs due to the Offices there As also An Account of the Moneys that have been Imprested for the Service of the said Mint And Whether any or what part thereof hath been Accounted for

W LOWNDES

Mint in Scot[lan]d Ref to Sir Isaac Newton

[The enclosed Memorial] [2]

To the Right Honble the Lords Commissioners
of Treasury

Whereas Your Lordships were pleased to Imprest the sum of two thousand one hundred pounds for the use and Service of his Majesties Mint in North Britain ending at Christmas 1719, I humbly Move Your Lordships to direct that the sum of One thousand eight hundred therty seven pounds ten shillings may be Imprest to me upon accot. of the Salarys of the offices and expences of the said Mint for one year and three Quarters ending at Michaelmas 1721

Which is most humbly
Submitted to Your Lordships
LAUDERDALE

NOTES

(1) T/1, 235, fo. 179v.

(2) Lauderdale's original memorial, in a clerical hand but bearing his signature, is on the recto page, and was clearly forwarded to Newton with the Treasury reference.

1376 BURCHETT TO NEWTON
11 OCTOBER 1721
From a copy in the Public Record Office [1]
For the answer see Letter 1377

11th Octo[be]r 1721

Sir

The Bearer Mr Laurans having communicated to the Lords Commiss[ione]rs of the Adm[iral]ty the inclosed Paper, [2] being a Scheme towards finding out the Longitude by the Means of a Piece of Watch Work of New Construction which he proposes shall go with a regular Motion at Sea, and their Lordships being willing to encourage any Endeavours towards so useful a Discovery, they desire You will please to give Your self the trouble, at Your convenient leisure to Examine into this Scheme and to give them Your Opinion of it. I am

Sr Your most &c J BURCHETT

Honble Sr Isaac Newton

NOTES

(1) ADM/2, 455, p. 375. The Admiralty decided to refer Laurans' letter to Newton at a meeting on 10 October 1721; see P.R.O., ADM/3, 33 (unpaginated). For a similar reference see Letter 1176B, vol. VI, pp. 266–7.

(2) A letter from P. Laurans to the Treasury, dated 'hanover street Longacre october ye 6 1721', is at U.L.C., Add. 3972, fo. 17. The author (about whom we know no biographical details) clearly had enclosed a paper describing his invention, but this is no longer with the letter. It was read at a meeting of the Royal Society on 19 October 1721, but 'for want of due Explication was referred to the Author to be further Explained.' Certainly Laurans' covering letter gives little clue as to the exact nature of the invention, stating only that 'by the nature of its Construction it wille show the exact mesure of time upon sea notwithstanding all its various motions[:] in this scheeme is seen the Combination of two powers moving togethere by equality of periodes making att the same time a réduction of theire quantity of rubbing the Cheef obstacle to the régularity of a watch.' Laurans, who was possibly a Huguenot immigrant as his letter is freely sprinkled with acute accents, had apparently been working on his invention for several years.

1377 NEWTON TO BURCHETT

?OCTOBER 1721

From a holograph draft in the University Library, Cambridge[1]
Reply to Letter 1376

Sr

I received last week your Letter conteining a reference from the Board of Admiralty upon a proposal of Mr Laurans for finding the Longitude at sea by Watch work. The Longitude has been found at Land long ago by Astronomy & the method of late years much improved so as to rectify Geography thereby. It is not to be found at sea by any method by wch it cannot be found at land. And it is not yet found at land by Watch-work. The method of finding it at land must be improved before it be fit for sea. A good Watch may serve to keep a recconing at Sea for some days & to know the time of a celestial Observ[at]-ion: & for this end a good Jewel watch may suffice till a better sort of Watch can be found out. But when the Longitude at sea is once lost, it cannot be found again by any watch. By Astronomy it may be found at land without erring a quarter of a degree & by this method the longitude of the Harbours in ye world may be settled. And I told the Committee of Parliament that by the same method the longitude might be found at sea without erring above two or three degrees.[2] And that if the Method could be improved so as not to erre at sea above one degree, it would be useful; if so as not to erre above 40' it would be more usefull; if so as not to erre above half a degree, it was as much as could be expected. The Rewards in the Act of Parliament are adapted to these three cases, & therefore relate to the finding it by Astronomy. But Astronomy is not yet exact enough. It has been much improved of late & must be a little further improved before it be exact enough for the sea: & this improvement must be made at land, not by Watchmakers or teachers of Navigation or people that know not how to find the Longitude at land, but by the ablest Astronomers.

But the References have been all of another kind & should not have been made before a general meeting of the P. commissioners to agree about the meaning of the Act & the best manner of putting it in execution, & appoint a Committee (if it was thought fit) to see things done.[3]

NOTES

(1) Add. 3972, fo. 37. There are other rougher drafts at fo. 40. All the drafts indicate clearly Newton's growing irritation at the ridiculous, and often unspecific, methods of finding longitude at sea which were continually referred to him by the Board of Admiralty or privately. (See Letters 1124 and 1137, vol. VI; for the Government prize offered for a successful method of determining longitude at sea, see Number 1093a, vol. VI.)

Newton's impatience at being pestered by the Admiralty for his opinion about various inventions concerning the determination of longitude is further expressed in an undated draft in U.L.C. on an unpaginated scrap in Add. 3972(2).

If anything should be proposed more advantageous for bringing ships safe into harbour then that of light houses, & extend to the distance of 80 geographical miles from the shore & seem reasonable to be tried it might be adviseable for a sufficient number of the Commissioners to meet upon the same. But as for Clocks or other instruments for keeping time, I do not think it advisable for the Commissioners to meet upon Proposals of that kind untill the Instruments have been made & sufficiently tried both at land & sea & in various latitudes & witnesses are ready to attest the success. And as for Astronomical Theories & Observations, I do not think it advisable for the Comm[issione]rs to meet you upon them before the Theories have been examined by Astronomers & approved as sufficiently exact & the sea-Captains of Trinity house are satisfied that the Observations can be made at sea with sufficient exactness.   I.N.

Despite Newton's negative reply, which may not have been transmitted by the Admiralty to Laurans, Laurans persisted in recommending his invention. He directed a petition to the Treasury on 27 July 1722 (see P.R.O., T/1, 240, fos. 60–2) demanding a reply to a previous petition. On 14 October 1723 he wrote to Lowndes (P.R.O., T/1, 244, fo. 163) enclosing another petition, in which he claimed to have heard that certain Londoners, impressed by his 'invention', had sent money to the Treasury for him, which he would now like to have. The Treasury denied any obligation.

(2) Newton's evidence spoken in June 1714 is printed as Number 1093a, vol. VI. It is obvious that Newton distinguished sharply between *finding* the longitude and *keeping* the longitude. The former (he rightly argued) could only be done by astronomical means, since only the heavens provide a universal clock. A chronometer therefore could not *find* the longitude but only *keep* it, and Newton despaired of the possibility of making a chronometer accurate at sea for more than a few days. Since half a degree of longitude is equivalent to two minutes in time, his despair is not unreasonable in terms of (say) a West Indies voyage lasting several weeks. For those of less subtle intellects, however, his insistence that chronometers could not find the longitude inevitably confused the issue. See also Letter 1476.

(3) To judge from the number of drafts Newton made of this paragraph, he experienced considerable difficulty in composing it. In one version (U.L.C., Add. 3972, fo. 40) he was particularly scathing:

But the chairman of the Committee of the House of Commons [*set up to prepare the Acts of Parliament authorizing the rewards*] being a seaman represented that they did not want the longitude, & so far as I can observe the seamen are generally of that opinion. He represented further that the motion of the Moon is too swift to be made use of in finding the Longitude whereas if it were swifter it would be fitter for the purpose. And the first step for putting the Act in execution should have been to consider how Astronomy might be sufficiently improved before it be applied to sea affairs.

1378 NEWTON TO THE TREASURY
20 OCTOBER 1721
From the holograph original in the Public Record Office[1]
Reply to Letter 1375

To the Rt Honble the Lords Comm[ission]ers of his
Maj[es]ties Treasury

May it please your Lordps

In obedience to your Lordps order of Reference of 11 October upon the Memorial of the Rt Honble the Earle of Lauderdale General of his Ma[jesty']s Mint in Scotland I humbly represent that his Lordp by the last renewal of the coinage Act, hath been Treasurer of the said Mint ever since the first of March 1714, & hath received out of the Coynage Duty by an Order of the Treasury dated 6 Aug 1717 the summ of 5415*lb*. By his Ma[jesty']s Warrant dated 17 Novem. 1718 he was to pay the salaries & other charges of that Mint commencing at Christmas 1714. The salaries are annually as follows: vizt to the General 300*lb*, to the Master 200*lb*, to the Warden 150*lb* to the Counterwarden 60*lb*, to the Assayer 100*lb*, to the Engraver 50*lb* to the King's Clerk 40*lb*, to the Smith 30*lb*: in all 930*lb*. The charges of Coinage & keeping the coining tools in repair are paid by the pound weight of the moneys coind, & there has been no coinage. The charges of keeping the Offices in repair & the dwelling houses wind & water tyte may amount to about 80*lb* or 100*lb* per an. And this charge added to the salaries makes the annual charge of that Mint about 1020[*lb*].[2] At wch rate the charges of that Mint from Christmas 1716 to Michaelmas 1721 will amount unto about 6885*lb*, whereof 5145*lb*[3] is already imprest. I do not find that the moneys already imprest to his Lordp or any part thereof are yet accounted for but I hear that his Lordp is ready to make up the account.[4]

All which is most humbly submitted to your Lordps great wisdome.

Mint Office. Octob. 20.
            1721

                                                              IS. NEWTON

NOTES

(1) T/1, 235, fo. 177; related drafts may be found in Mint Papers, III, fo. 217.

(2) For a copy of Lauderdale's Warrant of 17 November 1718 see P.R.O. T/17, 5, p. 7—the terms being as stated by Newton. Compare Letters 1297 and 1301.

(3) Presumably an error for £5415, as above.

(4) Lauderdale did indeed proceed to this; see Letters 1423 and 1425. Subtracting £5415 from £6885 shows £1470 as still owing to Lauderdale for the period up to Michaelmas 1721; an Order for the payment of this sum, to cover the stated period, was made on 22 January 1723 (P.R.O., T/53, 30, pp. 148–9).

1379   DU QUET TO NEWTON
?OCTOBER 1721
From the original in the Library of the Royal Society[1]

Monsieur

Le bien qui resulte de ma decouverte qui perfectionne la navigation, me fait esperer que vous me pardonnerez la liberté que je prens de vous adresser le discours[2] que j'ai fait a lacademie royale des sciences pour en prouver l'utilité, et qu'aprez l'avoir leu vous voudrez bien avoir la bonté de le Communiquer a vostre sçavante societé royale; si cette advantageuse decouverte est protegee de vostre nation, que je crois plus attentive quaucune autre a recœüillir les fruits des sciences, je tacherai de meriter de plus en plus son estime et la vostre, Monsieur, par une autre decouverte encor tres interessante pour la navigation.[3] je suis tres sincerement

Monsieur

Vostre tres humble et
tres obeyssant serviteur
DU QUET

NOTES

(1) MS. Early Letters, Q–R, I, no. 6. This letter was sent to Newton as President of the Royal Society together with a paper (still with the letter) entitled 'Moyen tres assuré pour faire route aux vaisseaux de guerre et mesme au plus gros.' The only clue to the date of the letter is the record in the Journal Book of the Royal Society that the paper was read on 2 November 1721.

Du Quet, who describes himself simply as a 'French engineer', was a protégé of the Duc d'Orleans. He succeeded in completing, after a fashion, the artificial hands that Father Sebastien (Truchet) had begun to contrive for a Swedish officer (see Fontenelle, *Éloges des Académiciens*, II (La Haye, 1740), p. 367). A number of his inventions are listed in *Machines et inventions approuvées par l'Académie Royale des Sciences*, I–IV (Paris), and he also later sent a biographical paper to the Royal Society which enumerates further devices (see Letter 1487).

(2) Du Quet proposed paddlewheels powered by men for the propulsion of ships, and gave a report on a full-scale trial of the method. The Society judged his proposal too unspecific for detailed evaluation, and certainly the paper enclosed with his letter is extremely brief. The discourse addressed to the members of the Académie was considerably longer, and was approved by them in 1702. (See *Machines...approuvées par l'Académie Royale des Sciences*, 1 (Paris, 1735), pp. 173–86. The article is entitled 'Rames Tournantes inventées par M. du Quet en 1699. Approuvée en forme en 1702. Et Comparaison de l'effet de ces rames a celui des rames ordinaires'; it was followed by a short supplement.) Four oars 12 ft long were joined together to form a vertical paddle-wheel, and a crank arrangement on the shaft allowed a large number of men, using what was more or less a rowing motion, to operate it. In the full-scale trials two hundred men and, presumably, several paddlewheels, were used in all, but the performance was not much better than that of a normal galley using the same number of men. In the supplement Du Quet described a device intended to alter the angle of the blades as they entered the water.

(3) At the beginning of his paper Du Quet mentioned that he saw the two chief remaining problems in navigation to be, first, how to propel a becalmed ship, and, secondly, how to determine longitude at sea. The latter may be the 'other discovery' he had in mind; compare Letter 1487.

1380   WARRANT TO NEWTON
1 NOVEMBER 1721
From a copy in the Public Record Office[1]
For the answer see Letter 1383

George R:

Whereas the proprietors of our patent for fishing for Wrecks between ten and fifteen Degrees North Latitude, which was granted by us to Jacob Row Esqr.[2] have been putting our said patent in Execution and are returned with a considerable Treasure of Bullion Plate, Silver Coyn and other Goods taken up from a Wreck, in ye Sea at the Isle of Mayo, whereof one tenth part clear of all Deductions whatsoever is Reserved to Us,[3] Our Will and Pleasure is and wee do hereby direct authorize and Command You (together with John Sydenham Esqr who was appointed our Officer for inspecting the said fishing & securing our part reserved as aforesaid) to go to the house of Andrew Drummond Goldsmith at Charing Cross where (as Wee are informed) the said Plate is lodged, and in the presence of such of the Proprietors, or their Agents as shall be at the said house to cause the said Plate to be weigh'd and to receive and take from them one Tenth part clear of all Deductions whatsoever, and give to the said Proprietors or their Agents a proper Receipt for the same, and transmitt to the Commissioners of our Treasury under your hands a Certificate thereof, and you the Master & Worker of our Mint are to carry the Plate so received into our Mint, and to cause the same to be melted into clean

Ingotts and then after to be converted into the Current Coyns of this Kingdom & being so Coyned to pay the proceed thereof (Except as hereafter is Excepted) unto our Trusty and well beloved Casper Frederick Henning Esqr. for the use and service of our privy Purse without Account,[4] That is to say, Except so much out of the said proceed as shall be sufficient to defray the Charge of Melting as aforesaid, and so much more as shall be sufficient to satisfie and pay unto the said John Sydenham an Allowance after the rate of Five Pounds percent upon the Neat Proceed, Which allowance Wee are gratiously pleased to make him towards his Service & Charges in attending the said Expedition on our behalf.[5] And for so doing this with the proper Acquittances shall be as well to you, Our Master & Worker for making the said payments as to our Auditors & all others concerned in allowing thereof on Your Account a Sufficient Warrant, Given at our Court at St. James ye. first day of Nov[embe]r 1721 in the Eighth Year of Our Reign

By his Majesties Command

R. WALPOLE

GEO: BAILLIE

CHA: TURNER

H. PELHAM

To our Trusty and welbeloved Sr. Isaac Newton Knt. Master and Worker of our Mint & to all other Officers of our Mint herein concerned and to John Sydenham Esqr. Our Officer for inspecting the fishing [of] the wreck abovementioned

NOTES

(1) Mint/1, 8, p. 132.

(2) Jacob Rowe's primary interest was in the invention of instruments and devices for use at sea. He seems to have started his life at sea as a ship's schoolmaster (see P.R.O., Adm/2, 447, p. 543). In 1719 and 1720 a number of his inventions were treated favourably by the Admiralty, by whose order trials of them were made on board ship (see P.R.O., Adm/2, 453, pp. 442 and 468; Adm/2, 454, pp. 33 and 498; Adm/7, 339, p. 188; Adm/1, 4102, fo. 118). He took the expedition to Mayo as a further opportunity to try out his instruments, and petitioned the Admiralty for protection of the vessel on its voyage (see P.R.O., Adm/1, 4103, nos. 43 and 44).

In August 1725 other inventions proposed by Rowe were submitted to Newton for examination; see Letter 1474. To that year also is credited a short tract by Rowe called *Navigation Improv'd. Book I. The Fluid Quadrant for the Latitude. Book II. An Essay on the Discovery of the Longitude by a New Invention of an Everlasting Horometer founded on most unerring Principles*. (See E. G. R. Taylor, *The Mathematical Practitioners of Hanoverian England, 1714–1840* (Cambridge, 1966), p. 165).

(3) There is much more on this patent in P.R.O., T/1, 234, but none of these documents relate directly to Newton.

(4) See Letter 1383.

(5) At P.R.O., Mint/1, 8, p. 133 (top, is a copy of Sydenham's receipt for £97. 2s received from Newton as five per cent of the £1942. 7s received by the Crown as its share of the recovered bullion and Spanish coin.

1381   VARIGNON TO NEWTON

28 NOVEMBER 1721

From the holograph original in King's College Library, Cambridge[1]
Reply to Letter 1372

Nobilissimo Doctissimoque Viro
D.D. Isaaco Newton,
Equiti Aurato, nec non Regiæ Societatis Anglicanæ
Præsidi Dignissimo
S.P.D.
Petrus Varignon.

Vir Illustrissime,

Die 20. n.s. Octobris proxime elapsi literas tuas die 26. v.s. Septemb. ex-ar[a]tas accepi a Do. Arlaud[2] qui gratissima mihi de te narravit, & cum quo ad multam usque noctem honorificentissime de te sum collocutus. Harum literarum partem querelis D. Bernoullij respondentem, ei solam exscripsi die 21. ejusdem octob.[3] cum Addito quo vetabas ne in lucem edatur. Quod spectat conditiones a te requisitas ad tui picturam illi concedendam, eas reticui, ne ipsi molestæ essent. Porro ut te illi semper Amicum exhiberem, monui eum Optices tuæ prelo hic subditæ tria a te illi, filio ejus, & Nepoti exemplaria destinari: & vero in Responsione[4] sua Basileæ data die 22. n.s. ultimi mensis Novembris, rogat me ut ejus & eorum nominibus gratias de tot Donis Maximas tibi in antecessum agam. Quæ autem fuerit ejus Responsio ad ea quæ ex tuis ultimis literis ei exscripseram, non ausim tibi referre: satius duxi hanc transcribere Do. Moivreo qui eam tibi narret, ut scilicet ei dicere possis quod forsan scribere Nolles.[5]

Cl. Idem Moivreus nuper nunciavit mihi te accepisse literas quas ab Illustrissimo D.D. Cancellario tibi misi,[6] easque pergratas tibi fuisse: Gaudeo. Misit quoque schema a te excogitatum & delineatum ad ornandam tui libri frontem, quo nihil aptius ad id video, cum id clare referat experimentum (ni fallor) potissimum quo invicte Demonstras immutabilitatem colorum luminis, quæ ejusdem libri tui fundamentum est. Conquiro delineatorem solertem & acutum ad spectatores hac in Tabella rite disponendos, & ad accessiones ei congruentes excogitandos.[7] Ecce tibi mitto quatuordecim Folia, quorum

178

duodecim prima cum quatuor prioribus jam ad te missis, integram part. 1. Lib. 1. Optices tuæ comprehendunt. Alia duo sunt partis secundæ, cujus cætera folia simul pariter mittam tibi statim ac fuerit edita: Ec quando? prævidere Nequeo, cum lente admodum Editio ista procedat, licet de hac segnitie frequentissime conquerar. Velim persuasissimum tibi sit me hac in re nil prorsus negligere. Vale.

Dabam Parisijs
die 9. Decemb. 1721. n.s.

Translation

Pierre Varignon sends a grand salute
to the most noble and learned Sir Isaac Newton,
most worthy President of the English Royal Society

Most illustrious Sir,

On 20 October, N.S., last I received your letter written on 26 September, O.S., from Mr Arlaud,[2] who brought me most welcome news of you and with whom I have talked on and on about you, most respectfully, late into the night. That part alone of the letter answering the complaints of Bernoulli I copied from it for his benefit on 21 October,[3] with the addition in which you forbade its publication. As for the conditions you imposed before you would grant him a present of your picture, these I have kept secret lest they be hurtful to him. Further, so that you should always show yourself a friend to him, I have advised him that three copies of your *Opticks* printed at the press here are intended by you for him, his son, and his nephew: and indeed, in his reply[4] dated Basel, 22 November, N.S., last he asks me to offer you in advance the greatest thanks on behalf of himself and his [relatives] for so many gifts. What reply he made, however, to what I transcribed from your last letter [to me] for him, I would not dare to relate to you: I have preferred to transmit this to Mr De Moivre who will tell you of it so that, naturally, you can say to him what you would not, perhaps, wish to write.[5]

The same Mr De Moivre has told me recently that you had received a letter which I sent you from our most illustrious Chancellor [*Mr Daguesseau*],[6] and that it was very pleasing to you. I am delighted. He [*De Moivre*] also sent the figure you have devised and sketched as an ornament for the beginning of your book. I see nothing more appropriate for the purpose than this, since it clearly refers, if I am not mistaken, to that decisive experiment by which you demonstrate beyond doubt the immutability of the colours of light, which is the basis of that same book of yours. I am looking out for an ingenious and skilful draughtsman to arrange the observers properly in this plate and to devise additions suitable to it.[7] You see I am sending you fourteen sheets, of which the first twelve, together with the four earlier ones already sent to you, make up the whole of Part 1, Book 1 of your *Opticks*. The other two are from the second part, the remaining sheets of which I will send you altogether in the same way immediately after they have

179

been printed. But when [will that be]? I cannot foresee, as the printing proceeds extremely slowly, although I very often complain about this sluggishness. I wish you to be quite sure that I neglect nothing at all in this matter. Farewell.

Paris
9 *December* 1721, *N.S.*

NOTES

(1) Keynes MS. 142(I); microfilm 1011.26.
(2) See Letter 1400, note (1), p. 213.
(3) Letter 1374; see also Letter 1372, note (6), p. 165, on the presentation copies of *Opticks*.
(4) Bernoulli's reply was long and unrelenting; see Letter 1374, note (1), p. 169.
(5) This was done; see Letter 1385.
(6) Letter 1370.
(7) No observers were inserted, since the background was so dark that figures would not have been visible; see Plate I (upper).

1382 LITTLETON POWYS TO NEWTON & CLARKE
14 DECEMBER 1721
From the original in the Bodleian Library, Oxford[1]

To Sr Isaac Newton & Dr Clarke[2]

Sirs

I intended to have you two & Dr Halley to Eat a Commons[3] with me Here on next sunday. But Dr Halley being the remotest[4] I first Writ to him to know if he could comply with that day & I had his Answer last night (as by the Inclosed) that he will. I now therefore make it my Request that you two will Please to be here by 2 of Clock next Sunday, I name that Hour that Dr Clarke may be free from his Office.[5] I Hope it will be sutable to both Your Conveniences. You three will be all my Company.

<table>
<tr><td></td><td>Sirs I am</td></tr>
<tr><td>*Serjeants Inne*[6]</td><td>Your ever Honored Friend</td></tr>
<tr><td>*Dec:* 14. 1721</td><td>& Humble servant</td></tr>
<tr><td></td><td>LITTLETON POWYS</td></tr>
</table>

You need not Write only tell this Bearer. But Please to send back Dr Halleys Letter being I have it under his Hand & Seal that he will meet you here.

To the Hon[oure]d
Sr Isaac Newton & Dr Clarke

NOTES

(1) Bodleian New College MS. 361, II, fo. 55. Sir Littleton Powys (*c.* 1648–1732, called to the Bar at Lincoln's Inn in 1671, was created a Serjeant-at-Law and knighted in 1692. He had been a judge of the King's Bench since 1700, and was to be elected F.R.S. in 1724. His relations with Clarke, Halley and Newton are otherwise unknown.

(2) Presumably Dr Samuel Clarke (1675–1729), Leibniz's philosophical adversary, who was rector of St James's, Picadilly.

(3) A communal dinner at the Serjeants' Inn.

(4) Halley had been appointed Astronomer Royal on 9 February 1720, in succession to Flamsteed who had died on the last day of 1719.

(5) That is, morning prayer.

(6) The Serjeants had two Inns at this time (they united in 1758), one in Fleet Street, the other in Chancery Lane. Possibly the former (which survived in name till 1941, though the Serjeants had disappeared in 1877) is meant.

1383 NEWTON AND SYDENHAM TO THE TREASURY
c. 19 DECEMBER 1721
From a copy in the Public Record Office[1]
Reply to Number 1380

Sir Isaac Newton Master & Worker of His Majts. Mint,
and Jno: Sydenham, most humbly Represents

That Persuant to His Majestys Warrt: dated the 1st: Novr: 1721. We went to the House of Mr. Andr: Drummonds Goldsm[i]th at Charing Cross, where in the Presence of the Agent for the Proprietors of the Pattent granted to Jacob Rowe Esqr: for fishing on Wrecks, did see the Silver weighed, And Receiv'd for the Use of his Majesty 1/10 part thereof, and have converted the same into the Current coyn of this Kingdom, and pay'd the net Proceed unto Casper Frederick Henning Esq. Persuant to the Directions of the Said Warrt: An Account of which is Annexed.

Which is most humbly submitted

[The enclosed account[2]*]*

An Account of the Silver Monies proceeding from 618 lbs. 11 oz. of Silver Bullion and Spanish Silvere coyn melted down, being his Majes. Tenth part of the Treasure taken up from a Wreck in the Sea near the Isle of Mayo, pursuant to his Majes. Pattent granted to Jacob Rowe Esqr. for fishing of Wrecks, viz.

lbs. oz dwt	lb. oz	lb. oz dwt	
618. 11. 00.00	... 614. 03.00 ...	626. 07. 16.21.	£1942. 07s.00d.
before melting	After melted	Standard	In Tale

Out of which having deducted five per
cent paid John Sydenham Esqr. pursuant
to his Majes. Warrt . . . £97. 2. —. ⎫
and for the charges of   ⎬   99. 12. 00
melting into Ingotts . . . £ 2. 10. —. ⎭
Remains clear the sum of £1842. 15. 00
Paid to Casper Frederick Henning Esqr.
for the Use and Service of his Majts.
Privy purse.

5th. December 1721

Is. NEWTON
JNO. SYDENHAM[3]

NOTES

(1) T/1, 235, fo. 245, in a clerical hand, undated and unsigned. Both this report and the following account were read at the Treasury Board on 19 December 1721.

(2) *Ibid.*, fo. 247.

(3) The signatures are autograph.

1384 WARREN TO NEWTON
19 DECEMBER 1721
From the original in the Bodleian Library, Oxford[1]

Sr.

You are desired to meet the Rector & ye rest of the Trustees &c of His late Grace of Canterburys Charity to ye p[ar]ish of St. James,[2] At ye Chapel Vestry room, by ten of the clock in ye morning, On Thursday next ye 21 Instant, To Choose a Morning Preacher in ye room of the Ld. BP. of Glocester,[3] And 4 new Members to be Trustees, in ye places of ye 4 deceased, And you are desired not to fail, the surviving Trustees being very few.[4] I am

Sr
Your most dutyfull Servt.
AMB. WARREN

December 19
1721
To Sr Isaac Newton
Humbly

1385 VARIGNON TO NEWTON
?2 JANUARY 1722
From the holograph original in King's College Library, Cambridge[1]

Nobilissimo Doctissimoque Viro
D.D. Isaaco Newton,
Equiti Aurato, nec non Regiæ Societatis Anglicanæ
Præsidi Dignissimo
S.P.D.
Petrus Varignon

Vir Nobilissime ac Doctissime,

Ex nuper accepta D. Moivrei Epistola[2] intelligo per te mihi licere ac etiam D. Bernoullo, ut quod in ultimis tuis ad me literis respondebat[3] ad querelas ejus, privatim communicamus cum quovis dubitante num *Equitis erratici* Nomine D. Bernoullum æque ac D. Leibnitium illudere intenderis, quo suam publice læsam esse famam arbitratur ipse D. Bernoullus: inde postulat publicari pariter ea quibus mihi significasti Nomen hoc nihil ad illum attinere, te cum eo lites nunquam habuisse, contra eum nunquam scripsisse, nec in animo habere ut scribas.[4] Hæc quidem (quod moleste fert) vetuisti ne typis edantur, etiam numque vetas, ut monet D. Moivreus. Sed si loco interpretationis hujus tuis verbis editæ, permitteres eam meis æquivalentibus scribi in Epistola ad D. Bernoullum,[5] quæ tuo concessu prodiret in lucem, hac ratione forsan ille pacari posset, & pax inter vos ambos confi[r]mari; quod votorum meorum summa est. Velim igitur, Vir Nobilissime, aperias ipse mihi quam ultimo mentem Voluntatemque super hac re susceperis, ne scilicet inscio mihi quicquam facere contingat quod tibi displicere possit.[6]

Tuæ Optices editio per mensem & amplius languit, ac sæpius interrupta fere jacuit. Sed Illustrissimi Cancellarij jussu Frater ejus Do Abbas D'Aguesseau accersivit Montalanum ut illius edendæ privilegium ei adimeret; quod adeo concitavit infidum hunc Bibliopolam ut nunc isthæc editio longe celerius quam antea procedat. Ultimum hodie correxi folium part. 2. Lib. 1.[7] Unde brevi partem hanc alteram mittam tibi Ephemeridibus Academicis presentis anni 1722. ac etiam cum Exemplari figurarum ejusdem tuæ Optices, ut nimirum de ijs ipse judicare possis. Plurimum Vale.

Dabam Parisijs die 13. *an.* 1722. *n.s.*[8]

Translation

Pierre Varignon sends a grand salute
to the most noble and learned Sir Isaac Newton,
most worthy President of the English Royal Society

Most noble and learned Sir,

From a letter received recently from Mr De Moivre[2] I learn from you that I am permitted, as also is Mr Bernoulli, to communicate privately what you answered to his complaints in your last letter to me, to anyone doubting whether you intended to mock Mr Bernoulli as well as Mr. Leibniz with the name 'Knight-errant', by which Mr Bernoulli himself thinks his reputation is publicly injured. Whence he asks that those expressions should likewise be made public in which you have signified to me that this name was never applied to him, that you have never had quarrels with him, [and] that you have never written against him, nor have any intention of so writing.[4] These matters (and this vexes him) you have forbidden, and still forbid, to be printed, as Mr De Moivre tells me. But if instead of printing this explanation in your words, you will allow it to be written in equivalent [words] of mine in a letter to Mr Bernoulli[5] which will, with your agreement, be published, perhaps by this means he can be appeased, and peace be established between the two of you; which is the chief of my desires. I wish, therefore, most noble Sir, that you yourself would make known to me what opinion and resolution you have finally come to upon this issue, so that it does not happen that, unwittingly of course, I do anything which could displease you.[6]

The printing of your *Opticks* has been slowed down for a month or more, and being often interrupted has almost stopped. But by our most illustrious Chancellor's order his brother, the Abbé Daguesseau, threatened Montalant that he would take away his privilege of printing that [book], which has so stirred up this faithless bookseller, that now this same edition goes forward far more quickly than before. Today I have corrected the final sheet of Part 2, Book 1.[7] Whence I shall shortly send this next part to you with the *Connoissance des Temps* for the present year 1722, and also with examples of the figures for this same *Opticks* of yours, so that you can judge them yourself. A hearty farewell.

Paris, the 13*th day of* 1722, *N.S.*[8]

NOTES

(1) Keynes MS. 142(M); microfilm 1011.26. Although we place this letter here, we think that it was misdated by Varignon and was actually written in *February* 1722. See note (8) below.

(2) In a letter to Johann Bernoulli of 4 February 1722, N.S. (see the Bernoulli Edition) Varignon refers to this same letter as received only the previous day by himself. De Moivre wrote that he had never received a letter of (presumably) December 1721 in which Varignon had transmitted to him, for Newton, the contents of Bernoulli's letter to Varignon of 22 November 1721, N.S. In ignorance therefore of this reaction by Bernoulli to Newton's views, De Moivre had (in a letter of 28 December, that is, about a month before this 'most recent' one) reported Newton's downright refusal to allow any part of Letter 1372 to be printed; and this refusal, after some delay, was passed on to Bernoulli by Varignon (31 January 1722). However, earlier in January—presumably after receiving the letter of 28 December from De Moivre—Varignon again wrote to him, pressing De Moivre to make one final effort to compose the quarrel between Newton and Bernoulli, and it was to this appeal that De Moivre responded in the letter which reached Paris on 3 February. He had (as Varignon tells it) read to Newton the quotation from Bernoulli's letter of 22 November 1721, N.S., sent him by Varignon, and quoted (in French) Newton's answer: '"Dès ma jeunesse j'ay haï les disputes; mais presentement que j'ay quatre vingt ans je les abhorre. Je ne sçais point l'usage que M. Bernoulli pouroit faire de cette lettre imprimée ou d'un extrait de cette lettre: Ainsi je ne puis consentir à cette impression."' 'Vous pouvez bien juger, Monsieur,' added De Moivre, 'que cela étant prononcé avec un peu d'émotion, je ne pouvois pas argumenter contre lui pour prouver que sa crainte n'étoit peut être bien fondée. Je me contentay de lui dire: "du moins, Monsieur, ecrivez sur cela trois lignes à M. Varignon, affin qu'il sçache comme de votre propre bouche votre Resolution finale." "Non, non," me dist-il, "il suffira que vous ecrivez vous même," làdessus il me quita.' But De Moivre added that, although Newton adamantly refused to allow Letter 1372 to be printed, he agreed that it might be communicated privately to interested parties.

(3) *Read*: 'respondebas'.

(4) Varignon here summarizes the content of Newton's Letter 1372. De Moivre had been charged with transmitting verbally to Newton Bernoulli's reply to this letter, and with obtaining Newton's reply (see Letter 1381).

(5) Varignon drafted, and sent to Newton, his proposed letter to Bernoulli; see Letter 1390a.

(6) Newton does not seem to have replied to Varignon's request, despite frequent prompting; see Letters 1390 and 1391.

(7) Presumably Varignon meant to write Part 1, Book 2. Compare Letter 1390 *ab initio*.

(8) For various reasons it seems that this date must be erroneous (it is perfectly clear in the original). Letter 1390 suggests that the interval since Varignon last wrote was not so great as three months; the details of the printing of the *Traité d'Optique* there given indicate fairly precisely that the present letter was written at least three weeks before the fall of Chancellor Daguesseau on 28 February, N.S. A date about 7 February is consistent with the facts that (i) on 31 January Varignon still knew only of Newton's adamant refusal to allow Letter 1372 to be published; (ii) the present letter was written after 31 January, since now Varignon knows that Newton has so far weakened as to permit private communication of the contents of Letter 1372; and (ii) this relaxation of Newton's position was made known to Varignon on

185

3 February in a letter from De Moivre, to which he refers in the present letter (see note (2) above). It is possible that its true date is 13 February, but this seems to allow too little time for the events described in Letter 1390.

1386   WILLIAM BLUNDEL TO NEWTON
17 JANUARY 1722
From the original in the University Library, Cambridge[1]

By an Act[2] made in the 12th year of our Sovereign Lady
Queen Anne for providing a publick reward for such person
or persons as shall discover the Longitude at sea.
The prologue to the Act doth begin as followeth.

Whereas it is well known by all that are acquainted with the art of Navigation that Nothing is so much wanted & desired at sea as the discovery of the Longitude, for the safety & quickness of voyages, the preservation of ships, & the lives of men, And whereas in the judgment of able Mathematicians & Navigators severall methods have already been discover'd true in Theory tho very difficult in practice, some of which there is reason to expect may be capable of improvement, some already discover'd may be proposed to the publick, & others may be invented hereafter. And whereas such a discovery would be of particular advantage to the trade of Great Britain & very much for the honour of this kingdom but besides the great difficulty of the thing it self, & partly for the want of some publick reward to be setled as an encouragement for so usefull & beneficiall a work. And partly for want of money for tryalls & experiments necessary thereto, no such inventions or proposalls hitherto made have bin brought to perfection. Thus ends the prologue to the Act. By which prologue if carefully heeded the great benefit value & worth which the government is pleased to take notice of, is that there is nothing so much wanted & desired at sea as the discovery of the Longtitude.

In a concern of so great consequence as is mentioned in relation to the discovery of the Longitude: & likewise in a concern of so great consequence as is the finding out the variation of the Mariners Compas or Magnet Needle, as it is printed in a map of the world by Mr Henry Overton[3] at the White Horse without Newgate London which those that please to give themselves the trouble may read at pleasure the great value of finding out the variation as is aforesaid printed in the bottom of the map. and likewise the dilligent search that for a considerable time hath been made by severall great Mathematicians, the consideration of all which must needs make all intelligible men to believe the greatness of the value.

Whereas[4] in the beginning of the month of August last past 1721 the variation of the Mariners compas containing a circle of 360 degrees is most certainly found out by me Willm: Blundell living at the signe of the Bulls head in Bishop street Coventry to any degree of the said circle of 360 degrees on any parts of the surface of the globe of the whole Earth & Waters. By reversing the said circle of 360 degrees from North to South, from South to North, from East to West from West to East. This will prove it self & therfore needs no further demonstration. Note that what is here written relating to the finding out the variation by me Will[ia]m Blundell I did immediately spread abroad not for any pride or ostentation, but with a willing unwillingness. But the constant thoughts so run in my mind of the vast benefits & advantages that would accrew thereby cheifly to some men in particular, and not only so, but likewise to the whole race of mankind, and in regard of my old age & incapacity, was fearfull least a concern of so great importance should slip out of the world without discovery. Therefore I pushed on the discovery by degrees so fast that I thought it past danger of being lost, but not so willingly as to discover the whole process without being at some difficulty to others, who perhaps might have proved so disingenious as to have made my labours the ofspring of their own studies therefore I did give to as many as either asked or desired the aforesaid variation. Then I proceeded to write letters to others in particular as followeth.[5]

First to the Author of the Northampton Mercury Willm Dicey with whom I had some acquaintance, I writ after this manner Sr I desire you will enter the enclosed spot of news in your next mercury, God grant life & health I shall not be long before I return you a taste of what is much more wonderfull & surprizing in all the Mathematicall sciences, in this letter I inclosed the spot of news relating to the variation but from the middle of August to the 18th of Sept: 1721 he returned no answer but in his Mercury sept: 18th it came in print page 251.

At the same time I sent to the Author of the Stamford Mercury in the very same words in a letter with the variation inclosed in the said letter but he never returnd any answer, At the same time from the middle of August 1721 I sent a letter to the Author of the London Journall in the very same words as is aforesaid in the Northampton mercury and in it I sent the inclosed spot of news relating to the variation to be printed but came not printed till Saturday Nov. 11th it came printed in the advertisements in the London journall page 4th At the same time from the middle of August 1721 I sent a letter to the Author of the London Gazette in the very same words as is aforesaid in the

Northampton mercury & in it I inclosed the variation of the Mariners compas but no answer neither in print nor writing, I thought by writing to all these it migth have produced such amusements that would have produced the looking after a concern so valuable which I believe it did but not with any design to make me the author of the first discovery.

On thursday the 7th of Sept: 1721 I writ again to the Author of the London Gazette & in my letter the same as in my letter to the Northampton Mercury with the variation as is aforewritten enclosed & the same with what I aforewrit to the Author of the London Gazette, but with further instructions than I had before written relating to the variation as also some light of the taste of what I had written concerning that if it pleased God to grant me life and health I would give 'em a taste of what is more wonderfull and surprizing in all the Mathematicall sciences, but all in vain no answer came from him neither in word writing nor print.

Coventry on Thursday the 7th: of Sept: I writ an address to the right Worship-full the Maior and the Worshipfull the Aldermen, on Friday 8th. of Sept: it was deliver'd to Mr Collens the Maior and the same day communicated to the Aldermen the same as aforewritten.

On Wensday the 11th of Oct: I made sute to Mr Collens then Maior to deliver the honourable Coll[on]el Oughton member of parliament for the city of Coventry the variations of the aforesaid circle of 360 degrees in which I more largely discover the many performances that would accrew thereby.

But on Monday the 27th of November 1721 Mr Rilsey now Maior of Coventry did warn a full house of the Magistrates on purpose at the Maiors parlour there came Mr Maior & 8 Aldermen and one Common Councill making number 10 at which time I produced my papers relating to the variation and Longitude, as also the circle of variation at which time they all set their hands as witnesses to my papers to prevent any clandestine dealing not with any intent that they should approve or disapprove what I then laid before them for that could not be expected. And I may truely say that the finding out a way to discover my notions, & not to be wronged of them or to be fallaciously dealt with, and the finding but fit words whereby to explain my self in the discovery of my notions hath been more perplexity of mind, and hath occasioned sharper pains to my head than the finding out the Variation and Longitude. I have perplexed my thoughts. And allso used my Indeavors which way to make the Full Discovery, Therefore Honourable Sr I make my sute to you, And the members of your honourable Royall Society For Asistance in such a Method as I hope will not be unaceptable; And with pardon For my old Age and incapasity especially in this winter season that makes me very uncapaple [*sic*] to com to London to com and weat on you my selfe. Therefore in regards of my incapasity The First thing that I make sute for is that I may be put into such a method that my Discovery may com safely into your hands, And that honorab[le] Sr you, with so many members of the Royal Society as may be thought needfull will derect time and place for Receaving the papers of my Discovery together with the Circle of Variation and Reversion For I have allmost an Asured Confidence that you with the Rest of the intellidgeable and Apprehensive members of the Royall Society upon Reading the Discovery, And with a Few minuits sight and consideration of the Circle of Variation will Conclude with me in the truth and Asureance that they pleanly Demonstrate and prove them selves and will Admit of no obiection. Therefore in this concern Relating to the method afore-said, My second sute to you Honourable Sr. And to all the members of the Royall Society. It is that it may have your candid Reception, and Agreeable Aprobation no Farther then you will Finde by the truth of my proofes and Demonstrations And I Doubt not of your sutable sentiments acording to the Merits which they will Deserve. And my humble sute to you is that you will give me all the assistance you can in promoting my Interest so soon as you have Receaved my Demonstrations and perused them, I no waies doubting of your Findeing them in all Respects Agreeable to your sentiments. I had not been Capable of Discovering this valluable concern, had it not been for old papers which I have by me as is aforementioned Relating to the Spot of News which I call a Taste of what is more wonderfull surpriseing in all the mathematicall Sciences But I will forbear any Farther of this at present; And will make my Request that I may Receive an Answer of the Receipt of this Letter And I shall allwaies to my best Capasity Remain

Your most humble servant WILLIAM BLUNDEL

January the 17th: 1721

To
The Honorable Sir Isack Newton
President and to the rest of
the members of the Royall
       Society
             London

NOTES

(1) Add. 3972, fos. 3–4. Little is known of William Blundel (1647–1723), except that he lived in Coventry, and was an enthusiastic inventor of nonsensical methods for determining the longitude. He published a number of pamphlets on the method for finding the longitude which

he discusses in the present letter. Copies of these bearing notes in Blundel's hand, apparently printed separately, but bound together and paginated continuously to form one short tract, are in the British Library (534.m.23(3)). The first pamphlet, dated 19 December 1721, includes passages from the present letter (see notes below); this is followed by an appendix, printed 29 September 1723, and another pamphlet dated 22 April 1723; the fourth and final pamphlet is undated. All cover much the same ground, and all are incoherent (see E. G. R. Taylor, *Mathematical Practitioners of Hanoverian England* (Cambridge 1966), pp. 111–12).

(2) See vol. vi, p. 160, note (2).

(3) Henry Overton printed numerous maps and compiled many atlases, usually from engraved plates he obtained from other cartographers, adapted by himself before printing. We have not traced the particular map referred to here.

(4) This passage, down to 'needs no further description', is taken from the opening lines of Blundel's pamphlet of 19 December 1721 (see note (1) above). The copy in the British Museum has written in at the top of p. 1, 'To the President of the R.S.' and bears the printed heading, '*To all and Singular. the Honourable Commissioners, Constituted by Act of Parliament, for the Discovery of Longitude at Sea...*' After the passage quoted here, the pamphlet continues 'yet for the benefit of thoses that are not so Apprehensive of the Notions, let them please to take the following directions,' and then launches into incomprehensible detail about Blundel's invention. In so far as he may be understood, Blundel seems to suggest that the cardinal points are all in error by 45°, that the magnetic needle always orientates itself in the plane of the horizon, and that the magnetic variation varies through 360° as we go round the equator.

The idea of using magnetic variation as a means for measuring longitude was not a new one; Christopher Columbus apparently tried to make use of it (see E. Guyot, *Histoire de la détermination des longitudes* (La Chaux-de-Fonds, 1955), p. 88) and Giambattista della Porta (*Natural Magick* (London, 1658), Book vii, Chapter 38), claimed it as a useful method, stating that the line of zero variation passed through the Azores. William Gilbert, however, pointed out in 1600 (*De Magnete*, Book iv (London), Chapter 9) that the variation was not constant along each meridian, nor proportional to the longitude. However, he suggested it was still possible empirically to determine lines of equal variation on the Earth's surface. Edmond Halley constructed charts showing these for use in navigation, and wrote several papers on the subject of magnetic variation in the *Philosophical Transactions*. He hoped eventually to discover a law, presumably fairly complex, to describe the variation.

Blundel, in his pamphlets, openly criticized the work of Halley. Yet Blundel himself seems to have missed even the point that to measure magnetic variation we need to know the true north from celestial measurements; he claimed that his method was particularly useful when the sky was obscured.

(5) In the British Library (534.m.23(3a)) is a printed copy of one of these 'Advertisements' of Blundel's. In it he suggests the use of tide tables to determine longitude. On the back is a note in his own hand, in which he claims to have developed other methods using the diurnal revolution of the earth, 'tables of quadrantal circles and lines', clocks and watches, and sandglasses. But he adds that the method using magnetic variation was his first.

1387 J. BAYNES TO NEWTON
21 JANUARY 1722
From the original in the Bodleian Library, Oxford[1]

Sr

The bearer, Mr. Langbridge,[2] having been under great disappointmts. is an humble Suitor to You Sr. for Yr. favour in a particular, wch he desires to mention to Yr. Self: If it happen to be in Yr. power to comply wth his request, it will be a Seasonable relief to him at this Juncture

I most heartily wish You Sr. health & all prosperity, & am wth. the Greatest Respect

> Sr Your most Dutifull
> H. servt.
> J BAYNES

The 21. *Janry* 1722

To
 Sr. Isaac Newton
  These

NOTES

(1) Bodleian New College MS. 361, II, fo. 56. We know nothing of the writer. One of this name became a Serjeant-at-Law in 1724.

(2) Possibly Edmund Longbridge, who petitioned Newton for reinstatement as Deputy Controller of the Duties on Salt and Rock Salt; see Letter 1546.

1388 ROBERT WALPOLE TO THE MINT
8 FEBRUARY 1722
From a clerical copy in the Public Record Office[1]
For the answer see Letter 1389

Whitehal Trea[su]ry Chambers 8th: *Febry* 1721[/22]

The Right Honble The Lords Comm[ission]ers of His Majestys Treasury are Pleased to Refer this Petition and the Bill Annexed[2] to the Principal Officers of His Majestys Mint Who are to Peruse his Majestys Warrants Directing the Several publick seals to be prepared and the Respective Certificates of their being delivered pursuant thereto And Examine into the Reasonableness of the Prices set down for the same and thereupon make Report to their Lordships with their Opinion.

> R. WALPOLE

(1) Mint 1/7 p. 106, middle. Walpole had been appointed Chancellor of the Exchequer and First Lord of the Treasury on 3 April 1721.

(2) James Girard's petition and bill; see Letter 1389, note (2).

1389 NEWTON TO THE TREASURY
14 MARCH 1722
From the original in the Public Record Office[1]
Reply to Letter 1388

To the Right Honorable the Lords Commissioners
of his Majes. Treasury

May it please your Lordships

I have considered and examined the Vouchers of Mr. James Girard's Bill hereunto annexed,[2] for engraving of seales, which your Lordships were pleased to referr to the Principal Officers of his Majes. Mint the 8th of february Last,[3] and humbly certifie your Lordships that I find them right, the work good and the prices reasonnable, being the same heretofore allowed for work of the like sizes.

It likewise appears to me by proper receipts that the seales in the said Bill mentioned have been delivered to the respective Offices or Officers of State they were ordered for, and as they have been Weighed in the Mint that they are of the Weight expressed in the said Bill.

Which is most humbly submitted to Your
Lordships great Wisdom
Isaac Newton

Mint Office the 14th of March
1721[/22]

NOTES

(1) T/1, 239, fo. 131. The letter is in a clerical hand, but signed by Newton. A draft in Mint Papers, III, fo. 470, has the first paragraph in Newton's hand, undated, signed 'I.N.', with the second paragraph in another hand, unsigned, but dated 7 March 1721[/22].

(2) Girard's bill is at T/1, 239, fo. 135. In an introductory letter (fos. 133-4), Girard reports that he submitted a bill in May 1720, but it was not then examined, so he here submits it again, with additions for the period since then; the total sum for engraving seals is £341. A copy of the warrant authorizing the payment of this sum to Girard, dated 28 June 1722, is at T/53, 29, pp. 503-4. The list of seals Girard had engraved include seals for the Admiralty, for various Scottish government departments and one for the Dominion of South Carolina.

The unfortunate Girard had also suffered delays in receiving payment of his £50 salary; see P.R.O., T/53, 28, p. 207, and T/53, 29, p. 95.

(3) Letter 1388.

1390 VARIGNON TO NEWTON
24 MARCH 1722
From the holograph original in King's College Library, Cambridge[1]

Nobilissimo Doctissimoque Viro
D.D. Isaaco Newton,
Equiti Aurato, nec non Regiæ Societatis Præsidi Dignissimo
S.P.D.
Petrus Varignon.

Vir nobilissime,

In ultimis meis ad te literis[2] promiseram me paulo post fore tibi missurum
Optices tuæ folia quæ tibi desunt ad usque finem part. 1. lib. 2. sed cum illorum
ultimum, quod correxeram eo ipso die quo scribebam tibi, obtinere non
potuerim absolutam nisi post dies octo; prætereaque cum viderim ejusdem lib.
2. partem secundam brevissimam esse; ac proinde brevi post hac absolvendam,
Prælo nimirum tunc accelerato Minis D. Abbatis DAguesseau,[3] privilegij
Montalano adimendi: Fasciculum foliorum istorum distuli, ut nempe eadem
opera has ambas istius libri secundi partes simul ad te mitterem. Sed statim ut
prævisus est Illustrissimi D. Cancellarij casus quem ex Nuntijs publicis audisti,
prælum cœpit iterum elanguere ita ut loco duorum foliorum quæ per singulas
tres a superioribus Minis primas hebdomadas mihi venerant, unicum (ut
antea) per unamquamque duarum sequentium acceperim; ac deinde, postquam scilicet Illustrissimus idem D. Cancellarius ab Aula tandem semotus
est,[4] unicum etiam per duodecim dies expectatum mihi venerit. Tum hac
ultima Montalani Nequitia permotus, Vehementem Epistolam scripsi, non
huic inverecunde fallaci, sed Typographo, quæ felicem huc usque habuit
exitum: Hanc exscriptam ostendet tibi D. Moivreus, Typographique promissa
Narrabit, quorum effectu absolutam nunc habemus Optices tuæ partem 2.
lib 2. quam cum cæteris folijs ad ea retro quæ jam habes (quod totum compre-
henditur novis octodecim folijs) mitto tibi per unum ex discipulis meis
Londinum hinc petentem. His octodecim folijs in fasciculum Collectis adjunxi
Exemplar unum figurarum,[5] ut de ijsdem Judices, præterea Ephemerides
Academicas presentis an. 1722. Cum Exemplari[bus] duorum Operum quibus
adjudicata sunt præmia totidem Academica 2000. lb & 1500 lb.

Cæterum[6] ab aliquot jam diebus literas accepi a D. Moivreo, quibus me
monebat per te mihi licere ut Do. Bernoullo scribam Epistolam eam cujus ei
scribendæ copiam a te petieram,[7] teque ipsum copiam hanc mihi certam
facturum esse literis mox suas sequuturis. Has adhuc expecto ut ausim quicquam

hac de re D. Bernoullo scribere, ne scilicet incautus scribam quod ei scribi nolles. Rogo igitur, Vir Nobilissime & Humanissime, ut indices ipse mihi quousque facultatem des satisfaciendi ei super querelis ejus. Ut autem Clarius & enucleatius eam mihi explices, en Epistola quam ei mittendam paravi, si approbaveris:[8] Animadverte recte ne voluntatem tuam interpreter; Sin, hanc epistolam ipse corrigas, addendo, vel demendo, vel alia quacumque ratione immutando, ita tamen ut Do. Bernoullo placando sufficiat; Deinde sic a te correctam velim remittas mihi, ut ei parem scribere possim. Si vero hanc approbaveris qualem tibi mitto, poteris eam lacerare vel comburere, modo approbationis hujus tuæ me moneas, cum Autographa mihi restat. Vale.

Dabam Parisijs die 4. *Aprilis* 1722. *s.n.*

Translation

Pierre Varignon sends a grand salute
to the most noble and learned Sir Isaac Newton,
most worthy President of the Royal Society.

Most noble Sir,

In my last letter[2] to you I promised I would send you shortly afterwards the sheets of your *Opticks* which are lacking up to the end of Part 1, Book 2; but since I could not obtain the last of them completed until a week later (though I had corrected it that very day on which I wrote to you); and since besides I had seen that the second part of Book 2 was extremely short, and hence ought to be completed shortly afterwards, the press having actually been speeded up at that time by the Abbé Daguesseau's[3] threats of taking away the 'Privilège' from Montalant, I kept back the gathering containing these sheets with the object of sending you both of these parts of the second Book of that work at once. But immediately the fall [from power] of our most illustrious Chancellor was foreseen, which I heard of from public announcements, the press began to slow down again, so that in place of the two sheets which had come to me in each of the first three weeks after the above threats, I received one (as before) during each of the following two [weeks]; and after that, that is to say after the Chancellor was at last removed from court,[4] one [sheet] alone came to me after being expected for as much as twelve days. Then, stimulated by this last wickedness of Montalant, I wrote an angry letter, not to that shameless trickster [himself] but to the printer, which has worked wonders up to now. Mr Dr Moivre will show this letter to you, and will tell [you] of the printer's promise, the result being that we now have complete Part 2, Book 2, of your *Opticks*, which I am sending to you by one of my students making the trip from here to London, with the rest of the sheets going back to those you already have (which altogether comprise eighteen new sheets). I have added to these eighteen sheets, collected together into a gathering, one copy of the figures,[5] so that you may give your opinion of them, and furthermore the Academy's *Connoissance des Temps* for the present

194

year 1722, together with copies of two works to which the Academy has awarded prizes of as much as £2000 and £1500.

For the rest, [6] some days ago now I received a letter from Mr De Moivre in which he advised me that you permit me to write that letter to Mr Bernoulli, for the composing of which I had sought authority from yourself, [7] and that you yourself would inform me of this authority in a letter soon to follow this. I am still awaiting this before being so bold as to write anything about this matter to Mr Bernoulli, in case of course I should rashly write something which you do not wish to be written to him. I ask, therefore, most noble and kind Sir, that you yourself should indicate to me how far you would go in giving me the power to satisfy him concerning his complaints. However, so that you may explain this to me more clearly and precisely, here is a letter which I have prepared for sending to him, if you shall approve it: [8] see if I do not interpret your wish correctly; if not, you should correct this letter yourself, by adding to it, or shortening it, or changing it in any way whatever, so long (however) as it suffices to placate Mr Bernoulli; finally, please send [the letter] back to me when you have corrected it, so that I can write the like to him. If indeed you approve it in the form in which I send it to you, you can either tear it up or burn it provided that you advise me of your approval, as the original remains with me. Farewell.

Paris, 4 April 1722, N.S.

NOTES

(1) Keynes MS. 142(J); microfilm 1011.26. Varignon enclosed a holograph draft of his reply to Bernoulli in this letter. See Letter 1390a, following.

(2) Letter 1385.

(3) Chancellor Daguesseau's brother; see Letters 1365 and 1385.

(4) Daguesseau was dismissed, through the influence of Cardinal Dubois, on 28 February 1722, N.S., and retired to his estate at Fresnes immediately. He returned to office in 1727.

(5) The text figures probably, since the vignette is discussed in Letter 1391.

(6) Varignon now returns to the business of Letter 1385, repeating his request that Newton *personally* authorize in detail the terms in which Newton's disclaimers (Letter 1372) might be communicated to Bernoulli *for public use*. These were already known to Bernoulli, of course, from Varignon's Letter 1374 to him, but this last letter had been written under the proviso that Bernoulli was *not* to use its contents publicly. The difficulty was to arrive at a formula that Newton would allow to go before the public.

(7) In Letter 1385.

(8) Letter 1390a.

1390*a* VARIGNON TO J. BERNOULLI
MARCH 1722
From a copy of the draft in King's College Library, Cambridge[1]
Enclosed with Letter 1390

Clarissimo Doctissimoque Viro
D.D. Joanni Bernoulli, &c.

Vir Clarissime,

Cum mihi significaveris videri tibi Dm. Newtonum illudere te voluisse, dum, postquam dixerat te primas tenere inter D. Leibnitij discipulos, addidit hunc illis stipatum velle Equitis erratici more bellum gerere: Videri, inquam, tibi Dm. Newtonum jocoso hoc Equitis erratici nomine designare te quoque voluisse. Præterea suspicaris collectionem scriptorum pertinentium ad litem de primo calculi infinitesimalis Inventore (qua in collectione Irrisio hæc reperitur) Jubente aut saltem annuente Do. Newtono, recens editam fuisse a Do. Des Maizeaux. Tandem eundem Dm. Newtonum favisse Do. Kellio [*sic*] dum adversus te scriberet.

Ut autem ista credans, bona tua cum venia dixerim, Vir Clariss. nullo modo me adduci posse, tum quia nihil, eorum mihi videtur, tum maxime quia undequaque audivi Dm. Neutonum virum esse pacificissimum suavissimisque Moribus instructum; quod quidem ex ejusdem humanissimis ad me literis semper intellexi. Attamen veritus ne ex illis tuis suspicionibus dissidium aliquid suboriretur inter vos, quos ambos summopere colo, Do. Neutonij humanitatis fiducia sum ausus ab illo petere quid rei esset in illis tuis querelis.

Ille porro huic meæ quæstioni benigne respondit, Equitis erratici nomen ad te nihil attinere, Lites tecum nunquam habuisse, contra te nunquam scripsisse, nec in animo habere, ut scribat. Præterea testatur, ipso inscio, supradictam collectionem editam fuisse, donec schedæ quatuor vel quinque primæ tomi secundi[2] impressæ essent, ex Hollandia missæ, & ei ostensæ. Adde quod asserat collectionem hanc prope impressam esse cum primas tuas literas accepit. Denique tantum abest ut idem Dus. Neutonus faverit Do. Kellio adversus te scribenti, quin potius, ut stylum ejus cohiberet, has tuas literas amicas scriptas, cum eo communicavit; quas cum maligna hic sit interpretatus in Epistola dein ad te scripta, cum idcirco quasi liti studentem Dnus. Newtonus vehementer objurgavit.[3]

Hæc sunt quæ ex ejusdem ad me literis didici, quæque autumo tibi placitura: quam ob rem ea tecum communico de consensu Dni. Newtoni, per quem & per me tibi licet hac mea Epistola uti prout volueris. Vale.[4]

196

Translation

To the most famous and learned
Johann Bernoulli etc.

Most famous Sir,

Since you have indicated to me that it seems to you that Mr Newton meant to mock you when, after he had said he held you chief among Mr Leibniz's disciples, he added that Leibniz, surrounded by such men, wished to wage war in the manner of a knight-errant; it seems to you, say I, that Mr Newton meant to include you also under this jocular name of knight-errant. Further, you suspect that the collection of letters relating to the dispute over the first inventor of the infinitesimal calculus (in which collection this mockery occurs) was published recently by Mr Des Maizeaux by the order of, or at least with the assent of, Mr Newton. Finally, [you suppose] that the same Mr Newton encouraged Mr Keill when he wrote against you.

I would say, with your indulgence, most famous Sir, that I can in no way be induced to believe the same thing, not only because I can see nothing of the sort but also, and above all, because I have heard everywhere that Mr Newton is a man endowed with a most peaceable and sweet disposition; which I have indeed always understood from his most kind letters to me. But nevertheless, lest these suspicions of yours create some dissent between you, both of whom I honour very greatly, I have been so bold as to trust to Mr Newton's kindness in asking him what substance there is in these complaints of yours.

Further, he has kindly replied to this question of mine that the name of knight-errant in no way applies to you, that he has no argument with you, [and] that he has written nothing against you, nor has it in mind so to write. Besides, he testifies to his own ignorance that the above-mentioned collection was [about to be] published, until the first four or five sheets of the second volume [2] had been printed and sent from Holland, and shown to him. Add to this the fact that he asserts the collection to have been nearly printed when he received the first of your letters. Lastly, Mr Newton was so far from encouraging Mr Keill in writing against you, that instead, in order to restrain that man's pen, he communicated to him that friendly letter of yours written [to him]; which [Keill] interpreted maliciously in the *Epistola* he then wrote to you, although Mr Newton reproved him severely on that account as being zealous in the pursuit of wrangles. [3]

These are the matters I have learned from his letters to me, which I think will be pleasing to you; for this reason I communicate them to you with the agreement of Mr Newton; and both he and I allow you to use this letter of mine just as you wish. Farewell. [4]

NOTES

(1) Keynes MS. 142(N), microfilm 1011.26 (in the hand of Varignon). The version finally sent to Bernoulli differed in one or two minor details; see notes (2) and (3) below.

Varignon had first planned to write a public letter to Bernoulli in January (see Letter 1385)

197

but kept his plan secret from him until he had obtained Newton's permission. Eventually, on 4 June 1722, N.S., he sent him a copy of the present letter, enclosed with another more private one. In the latter Varignon explained how he had asked De Moivre to extract from Newton definite permission to have the present letter printed. De Moivre, after considerable difficulty partly due to Newton's periods of absence from London on account of sickness, had eventually done this, and had informed Varignon of his success—and of his determination to have nothing else to do with the matter.

Bernoulli's reaction to the letter is recorded in his reply to Varignon, dated 28 July 1722, N.S. (see Bernoulli Edition). Clearly he had expected a more apologetic letter; he declared that, out of courtesy to De Moivre and Varignon, he would not publish it, so long as 'Mr. Taylor ou quelqu'autre de sa bande' ceased harassing him. Varignon was horrified, and in a strongly worded reply of 30 October 1722, N.S., expressed his disappointment in Bernoulli's reaction: how could Bernoulli expect Newton to apologize; an apology implied culpability, whereas what Newton had intended was an *explanation* which left both parties their good names. As for the suggestion that the letter be suppressed out of consideration for De Moivre and Varignon, nothing in the letter implicated either man in the affair. Varignon declared himself very reluctant to inform either De Moivre or Newton of Bernoulli's reaction. Probably he did not do so, since Bernoulli's complaints were not discussed again until after Varignon's death, when Bernoulli himself eventually wrote to Newton (Letter 1404).

(2) De Moivre's draft, and the version sent to Bernoulli (see note (3) below) have 'Donec paginæ 88 primæ tomi secundi'.

(3) This paragraph is based on Letter 1372 from Newton. Newton seems to have instituted slight changes, in discussion with De Moivre, for in Keynes MS. 142(P) is another copy of the draft in De Moivre's hand and in Basel is the version sent to Bernoulli by Varignon, both of which here replace the draft text by a longer passage concerning Keill's ill-conduct, as indicated by Newton's draft in U.L.C. Add. 3968(42), fos. 618–19:

Denique Dm. Keillium suo Marte adversus te scripsisse, idque Oxonii; se lites illas neglexisse, literis tuis jam dictis fidem habuisse, eas Do. Keillio contradicenti [privatim] ostendisse ut eum convinceret, sed frustra; literas Dni. Keillij ad te subinde scriptas non vidisse antequam in lucem prodirent, & eundem Dnm. Keillium mox objurgasse quasi litibus novis studentem.

(Lastly [Newton writes] that Mr Keill wrote against you upon his own authority, and that from Oxford; that he himself had set aside these disputes, trusting in your aforesaid letter; and that as Mr Keill took the opposite view he showed it to him [privately] to convince him, but in vain; the letter that Mr Keill thereupon wrote he had not seen before it was published and he at once reproved Keill as being too eager to create fresh wrangles.)

The Bernoulli Edition omits the word 'privatim'.

The explanation of this second, fuller draft seems to be that Newton, feeling that Varignon's emphasis on Newton's blamelessness *vis-à-vis* Keill was insufficiently strong, drafted this new version which De Moivre sent to Varignon, of course leaving Newton a copy.

(4) The version sent to Bernoulli was dated 4 June 1722, N.S. (see Bernoulli Edition).

1391 VARIGNON TO NEWTON
17 APRIL 1722
From the holograph original in King's College Library, Cambridge[1]

Nobilissimo Doctissimoque Viro
D.D. Isaaco Newton
Equiti Aurato, nec non Regiæ Societatis Præsidi Dignissimo
S.P.D.
Petrus Varignon

Vir Nobilissime

Cum sæpesæpius circumspexissem cujus Pictoris ope uterer ad delineandam & ornandam libri tui Frontem, quam Argumento aptissimam ad me misisti,[2] qui nempe suæ Artis bene peritus & Intelligens tuam in ea mentem stylo rite posset exprimere: quacunque, inquam, me verterim ut hunc invenirem, Neminem tandem comperi eam ad rem & tibi magis idoneum quam Dm. Arlaud,[3] tum ratione peritiæ qua maxima pollet hac in arte, tum ratione observantiæ qua summa te colit. Itaque aliquot ab hinc diebus ipsum adivi, consilium & opem ab eo petiturus hac in re, quam ut tuam esse audivit, statim suam fecit, lætus hac occasione de te bene merendi suis erga te officijs; ac inde omnem operam & curam pollicitus est mihi se positurum ut schema quod optas, sit eleganter Delineatum ac sculptum ad tuam Mentem, quam ideo ipsi exposui, quamque penitius illico perspexit. Cum autem longe magis illi quam mihi sint noti Delineatorum & Sculptorum, quotquot hic adsunt, peritissimi, eundem rogavi ut eos ipse eligeret quos jam dirigendos benevole susceperat, ac cum illis de pretio conveniret, cujus æstimationem prorsus ignoro, quodque ad nutam ejus me soluturum esse promisi. At vero de hac mihi reddenda pecunia velim ne cures, nisi postquam persolutæ tum Montalano promissæ, tum solvendæ pro compactione libri tui exemplarium a te Donandorum, acceptilationes omnes ad te misero.

En schema paratum ad sculpturam, si arrideat tibi: quos in Archetypo delineaveras radios, quosque idcirco hoc in Exemplari delineatos vides, eos delendos esse sentit Dus. Arlaud, atque idem Ego; cum in spatio illuminato nulli seorsum discerni debeant nec revera possint: unde illuminatum spatium videtur album totum esse debere. Censuit præterea omittendos pariter esse spectatores & Genios quos ad hujus schematis ornatum finxeramus, utpote qui in obscuriore tenebricosi cubiculi loco positi, omnem oculorum aciem fugerent, aut saltem non viderentur nisi pullati;[4] quod lugubræ nimium foret. Porro tuam de hoc schemate sententiam expectamus ut ad mentem tuam corrigatur, si quid in eo desideres: utut sit, velim id prompte mihi remittas, cum ultima mea ad Typographum epistola (quam tibi D. Moivreus ostendit) Prelum adeo promoverit, ut tuæ Optices editionem brevi absolvendam fore sperem: Iam semel ultimum libri secundi folium correxi, & tertij primum expecto. Unde propediem istius libri secundi partes absolutas tertiam & quartam ad te mittendas dabo Do. Lauranzi amanuensi Excellentissimi D.D. vestri apud nos Legati. Ejusdem libri secundi partes primam & secundam tibi redditas fuisse confido ab uno ex meis discipulis hinc Londinum die sexta mensis hujus profecto, in quarum fasciculo specimen erat epistolæ D. Bernoullo destinatæ,[5] si eam approbaveris; Sin, velim eam corrigas ita ut in lucem edita possit illi placando sine molestia tua sufficere. Hanc sic voluntati tuæ accommodatam, & ejus illi mittendæ copiam in dies singulos a te expecto, rogoque ut gravi hoc officij onere quamprimum levare me digneris. Salutem plurimam tibi dicit Dus. Arlaud, ac Ego parem. Vale.

Dabam Parisijs
die 28. *Aprilis* 1722. *S.N.*

Translation

Pierre Varignon sends a grand salute
to the most noble and learned Sir Isaac Newton
most worthy President of the Royal Society.

Most noble Sir,

As I had over and over again looked round for an artist whose aid I might employ for drawing and perfecting the headpiece for your book, [the sketch] you sent me being most suitable for the business,[2] that is to say one well skilled in his art and intelligent, who can express your intention in it in a fitting style; wherever I turned (I say) to find this [man], I finally discovered no one better suited to the task and to yourself than Mr Arlaud,[3] both by reason of the great skill which he possesses in this art, and by reason of the respect with which he honours you most highly. Therefore several days ago I went to him, in order to seek advice and help from him in this matter, which, as soon as he heard it to be your [concern], he at once made his own, glad of this opportunity for obliging you by his services to you. Moreover he thereupon promised me to take the whole care and responsibility upon himself, in order that the figure you wish for may be elegantly drawn and engraved to your intention; therefore I showed it to him, and he there and then made a pretty thorough study of it. However, as he is far better acquainted with skilled artists and engravers than I am, such as there are of that sort here, I have asked him to select those whom he had now kindly undertaken to direct and to agree on a price with them himself; for I am absolutely ignorant of values in this context, and so I have promised to pay whatever he approved. But please do not for this reason take the trouble to send me money, until after what was promised Montalant has been paid, and also the cost of binding the copies of your book which you are to give as presents; I will send you all the receipts.

Here is the drawing made ready for the engraver, if you think well of it. The rays [of light] which you drew in the original sketch, and which you accordingly see drawn in the copy, should be deleted in the opinion of Mr Arlaud, and in mine also. For in the illuminated space nothing ought to be separately discerned, nor indeed can be; whence the illuminated space ought to appear wholly white [on the paper]. Moreover, he thought that the spectators and cherubs with which we had planned to adorn this drawing should likewise be omitted, especially as whatever figure is placed in the dark and shadowy part of the room must escape the sharpest sight, or at least be seen only as wholly black,[4] which would be too gloomy. We now await your opinion of this drawing so that it may be amended to your wishes, if you desire anything of the sort; however that may be, please return it to me promptly because my last letter to the printer (which Mr De Moivre showed you) so spurred on the press that I hope the edition of your *Opticks* will be completed shortly. I have now corrected the last leaf of the second book for the first time, and expect the first [leaf] of the third book. Hence I shall soon give Mr Laurenzy, amanuensis to His Excellency your Ambassador here, the finished third and fourth parts of that second book, for sending on to you. I feel sure that the first and second parts of Book II were delivered to you by one of my pupils who set out from here for London on the sixth of this month; in the package was a draft of the letter intended for Mr Bernoulli,[5] if you should approve it; if not, please correct it so that it can be made public with the object of placating him without trouble to yourself. Each day I expect to receive this from you adjusted to your wishes, and authority to send it to him, and I ask you to be so good as to relieve me of the burden of this onerous duty as soon as possible. Mr Arlaud sends you a grand salute, and I do the like. Farewell.

Paris, 28 April 1722, *N.S.*

NOTES

(1) Keynes MSS, 142(K); microfilm 1011.26.

(2) See also Letter 1369, note (5), p. 156.

(3) See Letter 1400.

(4) Because the room shown in the vignette is darkened, any human figures shown would have had to be depicted as completely black in order to be visible. See Plate I.

(5) Letter 1390*a*.

1392 NEWTON TO THE TREASURY
26 JUNE 1722
From the holograph copy in the Mint Papers[1]

To the Rt Honble the Lords Comm[issione]rs of
his Majties Treasury

May it please your Lordps

The Porter of the Mint being lately dead, I ordered one Robert Lowe to do the buisiness till a new Porter should be appointed. The Place was formerly a Patent place, but now the house belonging to it is taken away, & the Salary is only 20*lb* per annum. Which being too little for bearing the charges of a Patent, I humbly pray that a Porter may be appointed by a Constitution.[2] Robert Lowe hath hitherto behaved himself with great diligence, & I humbly pray that he may be appointed if your Lordps think fit

All which is most humbly submitted to your Lordps great wisdome

ISAAC NEWTON

Mint Office
June 26. 1722

NOTES

(1) I, fo. 224; there is a rough draft at fo. 220, top. This is one of three letters all bearing the same date, and all concerning the lesser-paid mint workers and their salaries. We place them in order of Newton's three rough drafts in Mint Papers, I, fo. 220; clearly Newton had been reviewing the salaries of his employees.

(2) Lowe was apparently constituted as Porter, but the Treasury omitted then to increase his salary; see Letter 1431. In May 1726 Lowe was also appointed Deputy Weigher and Teller under John Phillips (see P.R.O., Mint/3, 168, fo. 2).

1393 NEWTON AND BLADEN TO THE TREASURY
26 JUNE 1722
From the original in the Public Record Office[1]

To the Rt Honble the Lords Commissioners
of his Ma[jes]ties Treasury

May it please your Lordps

The salaries of the Clerks of the Mint, which were setled about sixty years ago, being now not sufficient for their maintenance, we humbly pray that they may be augmented by about a quarter;[2] so that the salaries of the Warden's Clerk, the Masters three Clerks, & of the Comptrollers Clerk, wch are forty pounds per annum each, may become fifty pounds each; and that those of the Assaymasters Clerk & of the Purveyor to the Mint wch are twenty pounds each per annum, may become twenty & five pounds each; & those of the Clerk of the Weigher & Teller & of the Clerk of the Surveyor of the meltings wch are ten pounds each per annum may become twelve pounds ten shillings each. All which is most humbly submitted to your Lordps great wisdome.

ISAAC NEWTON
MARTIN BLADEN

Mint Office.
June 26th 1722

NOTES

(1) T/1, 239, fo. 306, in Newton's hand throughout. Drafts in the Mint Papers, I, fo. 220, and in the Bodleian Library, New College MS. 361, II, fo. 39v, show that Newton originally drafted the letter as from himself alone.

(2) The Treasury accepted the Officers' recommendation and authorized the payment of these enhanced salaries by a Warrant to Newton dated 8 August 1722 (P.R.O. Mint/1, 8, p. 134; T/52, 32, pp. 160–1). This is a routine document of no intrinsic interest.

1394 NEWTON TO THE TREASURY
26 JUNE 1722
From the holograph copy in the Mint Papers[1]

To the Rt Honble the Lords Commissioners
of his Majties Treasury

May it please your Lordps

Mr Croker[2] Graver to the Mint having taken one Beresford[3] to be his Apprentice & the time of his Apprenticeship being lately expired, & the said Beresford now working under Mr Croker as Journeyman: I humbly pray yt Mr Croker may be allowed 30*lb* per annum for the maintenance of the said Beresford wth meat drink & cloaths, as in a former case of the like nature.

All wch is most humbly submitted to your Lordps great wisdome

ISAAC NEWTON

Mint Office
June 26. 1722.

NOTES

(1) I, fo. 151; there is a rough draft at fo. 220, bottom.
(2) See vol. v, p. 308, note (2), and Letter 1134, note (1) (vol. vi, p. 209).
(3) Francis Beresford; see Letters 923 and 927, vol. v, for his apprenticeship to Croker.

1395 NEWTON TO VARIGNON
?JULY 1722
From the holograph draft in King's College Library, Cambridge[1]
For the answer see letter 1396

Viro Celeberrimo Dno Abbati Varignon
salutem plurimam dicit
Is. Newton.

Vir Reverende

Mitto tibi exemplar Commercij Epistolici[2] hic iterum impressi una ejus Recensione Latine versa[3] & Judicio primarii Mathematici.[4] Hæc omnia diu ante obitum D. Leibnitii impressa fuerunt, sed Commercium illud in Bibliopolarum Officinis nondum venale extitit. Præmittitur Præfatio ad Lectorem[5] & subjungitur Annotatio,[6] quæ duæ novæ sunt, sed ex chartis veteribus collectæ. Et enixe rogo ut hasce duas legas: & siquid dictum inveneris quod dici non debuerit vel quod aliter dici debuerit, id mihi significes. Curabo enim ut corrigatur, si opus est, antequam Liber lucem viderit. Scopus Libri non est ut disputationes renoventur sed ut Quæstiones quæ ad rem nil spectant rejiciantur, et quæ de primo methodi, seu fluxionum & momentorum, seu differentiarum, Inventore dicta fuerunt ad posteros perveniant, & ad eorum judicium quiete referantur. Paulatim convalesco, & spero me salute cito fruiturum. Vale.

A Monsieur
Monsieur L'Abbé Varignon de L'Academie des Sciences
& Professeur Royal de Mathematiques au College Mazarin

a Paris[7]

Translation
Isaac Newton sends a grand salute
to the most famous Abbé Varignon

Reverend Sir,

I am sending you a copy of the *Commercium Epistolicum*,[2] printed a second time here together with the account of it translated into Latin,[3] and the opinion of the foremost mathematician.[4] All these things were printed long before the death of Mr Leibniz, but the *Commercium* has not yet been offered for sale in the booksellers' shops. A preface 'Ad Lectorem'[5] is prefixed and an 'Annotatio'[6] added, which two [items] are new, but taken from old papers. I earnestly entreat you to read these two, and if you should find anything said which ought not to be said, or which ought to be said otherwise, mention it to me. For I will take care it is corrected, if necessary, before the book is published. The object of the book is not to renew disputes but to set aside questions which have no bearing on the matter and to ensure that what has been said about the first inventor of fluxions and moments, or of differences, may be passed on to posterity and calmly referred to its judgement. I am getting well little by little, and hope I will soon enjoy good health. Farewell.

NOTES

(1) Keynes MS. 142(S); microfilm 1011.26. A translation of the letter is printed in Brewster, *Memoirs*, II, pp. 293–4. Its date can only be guessed from that of the reply.

(2) The second edition, *Commercium Epistolicum D. Johannis Collins Et Aliorum De Analysi Promota Jussu Societatis Regiæ in lucem editum: et jam Una cum ejusdem Recensione Præmissa, & Judicio primarii, ut ferebatur, Mathematici subjuncto, iterum impressum* (London, 1722). The edition was apparently produced wholly under the care of Newton himself. (For the original *Commercium Epistolicum* committee see vol. V, pp. xxiii–xxvi.) Apart from new introductory material and an appendix (see notes below) only minor changes were instituted throughout the text; these are discussed in detail by J. B. Biot and F. Lefort in the 1856 Paris edition of the *Commercium Epistolicum*. (A third edition of the *Commercium* (London, 1725) is simply a reissue of the second edition with a new title-page.)

(3) The *Recensio*, which had originally been published in English in *Phil. Trans.* for February 1715, 29, no. 342, pp. 173–225; see vol. VI, p. 242, note (2). It occupies pp. 1–59 of the *Commercium Epistolicum*.

(4) Johann Bernoulli; Newton refers to the 'Appendix' to the new edition (see note (6) below).

(5) The 'Ad Lectorem', which is unpaginated, precedes the *Recensio*, occupying six pages; it deals in brief with the *Charta Volans* (Number 1009, vol. VI) and subsequent correspondence between Leibniz, Conti and Newton. (The original 'Ad Lectorem', occupying pp. 64–5 of the main body of the book, is still retained.)

(6) The new edition ends with an 'Appendix' and 'Annotatio', occupying pp. 245–50. The 'Appendix' contains a printing of the letter extract of 7 July 1713 [N.S.], as printed in the *Charta Volans*, and the 'Annotatio' comments upon it, but makes no reference to Bernoulli, nor discusses who the author of the letter might be.

(7) The address is in De Moivre's hand.

1396 VARIGNON TO NEWTON
24 JULY 1722
From the holograph original in King's College Library, Cambridge[1]
Reply to Letter 1395

Nobilissimo Doctissimoque Viro
D.D. Isaaco Newton,
Equiti Aurato, nec non Regiæ Societatis Præsidi Dignissimo
S.P.D.
Petrus Varignon

Vir Illustrissime,

En absoluta tandem tuæ Optices editio, callidis ac fraudulentis artibus Montalani sæpe sæpius impedita: Hæc incœpta fuerat die prima Augusti 1721 [N.S.]. nec tamen fuit absoluta nisi die ultima Julij 1722 [N.S.]. licet ea sit duntaxat foliorum 66.[2] quorum unumquodque opus erat unius tantum modo diei, ut ipse mihi ultimo fassus est Typographus, hanc longam unius anni moram in tergiversantem ac sæpe mentientem Montalanum transferens. Porro hujus editionis ultima demum folia tibi mitto cum speciminibus litterarum capitalium,[3] initio cujusque partis addendarum Fronti cujus specimen misi tibi cum folijs aliquot alijs ad hæc usque pertingentibus. Quod spectat ista specimina, cum descripta Sculptaque sint præceptis & ducta Dni. Arlaud,[4] harum rerum peritissimi, & tuique observatissimi, non dubito quin tibi maxime probentur. Utinam & grata pariter tibi sit cura, qua summa novæ huic Editionis invigilavi ut emendatissima prodiret. Quam ad rem adjumento quidem mihi fuerunt Notæ Dni. Coste; sed Moivreanæ Longe majus attulere, utpote quæ ad longe plura quam illæ, libri tui loca pertinerent, ac sæpe Mathematica, in quibus Dno. Coste æque modeste ac merito cesseret Dno. De Moivre, cujus observationes huic Editioni Gallicæ multum lucis afferre mihi videntur. Quod sincere testari debeo; quodque Interpretationis hujus Autor ipse, magno maximeque laudando animo in sua Nova præfatione fatetur.[5]

Cæterum ubi in aliqua re vidi dissentientem eum a Dno. de Moivre, quod raro fuit, utriusque sententiam qua potui attentione & æquitate perpendi; ac si nihilominus anceps hærerem, editionem ultimam Latinam super hac re contulebam: contuluissem Anglicam, si sermonem hunc intellexissem. Tandem nonnunquam contigit ut inter illorum sententias fluctuans, eam fuerim sequutus quæ tuo sensui librique perfectioni accommodata magis visa est mihi; atque hic unus est mei Judicij usus per te mihi licitus. Utinam in his nil contra mentem tuam peccaverim, & mea quantulacunque opera tibi placeat, cum

206

nihil mihi sit antiquius & optatius quam ut me tibi gratum in omnibus præ-
beam pro tuis in me Beneficijs quorum Numerum recens auxissi dono *Commercij*
epistolici, quod mutuo petenti creditum amisi. De hoc novo dono pares ac de
præteritis ago tibi gratias, hoc est, maximas, quas ut a me tibi referrat; Nuper
Dum. Moivreum rogavi, ac etiam ut tibi nunciaret me hæsisse duobus in locis
primæ istius libri præfationis ad lectorem, quam pacis ergo jusseras ut expend-
erem una cum notis in Epistolam sine nomine datam 7. Jul. 1713. [N.S.] in
quibus nil mihi visum est paci noxium. [6]

At non ita in prædicta præfatione: nam in ejus pag. 5. duo mihi visa sunt
pacem dirimere posse, scilicet

1°. Lin. 12 & 13 legitur, *Jam velo sublato, ut Militem in hac rixa pro se inducere*:
mallem simpliciter, *Jam in hac rixa pro se inducere*, ne quis sub illo *velo* prius
latitantem putet Dum Bernoullum, cui Dus. Leibnitius Epistolam prædictam
ascripsit. Adde quod, *ut militem*, vilior est denominatio quam ut eundem Dum.
Bernoullum non offendat. [7]

2°. Ibidem Lin. 29. legitur de Do. Des Maizeaux, *& in lucem edidit*: Mallem,
&, me non consulto, in lucem edidit, [8] ut nimirum hæc loquendi ratio consilietur
cum Epistola quam ad Dum. Bernoullum tua cum Venia nuper scripsi. [9]

Hæc sunt quæ, te Jubente notavi in prædicti libri præfatione prima ad
lectorem. At in notis ad Epistolam sine nomine datam die 7° Julij 1713 [N.S.].
nil mihi visum est quod sic paci noxium esse possit, ut jam dixi.

Quod autem scribis te paulatim convalescere, id multa me lætitia afficit,
quæ summa erit ubi te pristina & integra salute frui audivero: Hoc ut faxit
cito deus, eum enixe precor. Vale.

Dabam Parisijs die 4. Augusti 1722. s.n.

P.S. Quod hic reperis Exemplar libri recens editi de Magnitudine Terræ, id
tibi mittitur a Dno. Cassino: [10] Opus hoc est collectio & usus Observationum
quas Pater ejus habuit in perquirenda Regni Gallici Meridiana transeunte per
observatorium Regium.

Meo nomine reddas quæso Dno. de Moivre pretium libri posthumi Dni.
Cottes, [11] quem nuper ad me misit: Illud de Impensis pro te hic a me factis &
faciendis deducam statim ac didicero quanti sit pecuniæ Nostræ.

Translation

Pierre Varignon sends a grand salute
to the most noble and learned Sir Isaac Newton,
most worthy President of the Royal Society

Most illustrious Sir,

Here, complete at last, is the edition of your *Opticks*, delayed so very many times by the cunning and deceitful machinations of Montalant: having been begun on 1 August 1721 [N.S.], it was nevertheless not completed until the last day of July 1722 [N.S.], although it has only 66 sheets,[2] each of which was the work of one day only, as the printer himself finally admitted to me, attributing this long delay of one year to the prevaricating, and indeed often lying, Montalant. Further, I am at last sending you the final sheets of this edition, together with specimen pulls of the capital letters,[3] to be added to the beginning of each part; I have [already] sent you an example of the head-piece [for each part] with a few other sheets extending as far as these. As for these specimens [of the letters] I do not doubt that they will be warmly approved by you, since they have been drawn and engraved under the guidance and management of Mr Arlaud,[4] who is most skilful in these matters and respectful of yourself. If only the great care with which I have watched over this new edition in order to publish it most correctly may equally deserve your gratitude! Certain notes from Mr Coste were of assistance in this matter; but those of De Moivre gave far greater help, both because they referred to far more places in your book than his [did], and because [they] often [related] to the mathematical places, in which Mr Coste, with equal modesty and rightness, gave way to Mr De Moivre, whose observations seem to me to give much clarity to this French edition. To which I must testify sincerely; and which the author of the translation himself acknowledges in his new preface, in a spirit of the highest admiration possible.[5]

For the rest, where I have seen in some place that he [*Coste*] differed from De Moivre, which happened rarely, I have considered with such care and equity as I could the opinion of each man, and if nevertheless I remained in two minds I brought into the issue the most recent Latin edition; I would have brought in the English edition, if I had understood that language. Finally, it sometimes happened that, vacillating between these opinions, I have followed that which seemed to me better suited to your meaning and to the perfection of your book; and this is the sole use of my judgement allowed me by yourself. If only I may have in no way offended in these points against your opinion and if only my labours, trivial as they are, may please you! For nothing is more important and more desirable to me than to deserve your thanks in everything, on account of your kindnesses to me, the number of which has recently been increased by a gift of the *Commercium Epistolicum*, which having loaned it out I believed lost. I give you like thanks for this new gift as for past ones; that is, the greatest thanks that may come from me to you; not long ago, I asked Mr De Moivre and he was to tell you, that I had stuck at two places in the first preface of your book, 'To the reader', which you ordered me to examine in the interests of the reconciliation together with the notes on the anonymous letter dated 7 July 1713, [N.S.], in which nothing seemed to me harmful to a reconciliation.[6]

But it was not so in the preface mentioned above, for on p. 5 of it two things seemed to me to be able to spoil a reconciliation, as follows:

1. In lines 12 and 13 we read, 'Now, after he has been disguised, he is brought forward as a soldier [fighting] on his own behalf in this quarrel': I would prefer more simply 'now he is brought forward on his own behalf', in case someone should at first

think it is Mr Bernoulli lying hidden under that disguise, to whom Mr Leibniz has ascribed the above-mentioned letter. In addition, 'as a soldier' is too vulgar an appellation not to offend Mr Bernoulli.[7]

2. On the same page, line 29, we read of Mr Des Maizeaux, 'and he published': I would prefer, 'and, without consulting me, he published',[8] as certainly this way of speaking is consistent with the letter which I wrote to Mr Bernoulli a little while ago, with your permission.[9]

These are the things which, in accordance with your instructions, I have noted in the above-mentioned first preface of your book, 'To the reader'. But in the notes on the anonymous letter dated 7 July 1713 [N.S.], there seemed to me to be nothing which could thus be harmful to a reconciliation, as I have already said.

Moreover, it gives me much pleasure that you write that you are getting well little by little, [a pleasure] which will be supreme when I have heard that you enjoy perfect health as formerly: that God may so make it speedily I earnestly pray to him. Farewell.

Paris, 4 August 1722, N.S.

P.S. The copy which you find with this of a recently published book on the figure of the Earth, is sent to you by Mr Cassini.[10] This work is a collection and application of observations which his father made in investigating the meridian of the French kingdom that passes through the Royal Observatory.

I beg you to pay Mr De Moivre, on my behalf, the price of the posthumous book of Mr Cotes,[11] which he recently sent me: I shall deduct the sum from the expenditure made and to be made by me on your account, as soon as I learn how much it is in our money.

NOTES

(1) Keynes MS. 142(L); microfilm 1011.26. Compare Brewster, *Memoirs*, II, p. 294.

(2) The book is numbered to page 596, but there are no pages 341–440; thus it has 496 numbered pages, or 62 sheets. Introductory material takes up a further $2\frac{1}{2}$ sheets. Possibly Varignon includes the figures in his count.

(3) The opening paragraph of each Part of each Book begins with an ornamental, engraved capital letter, depicting putti often engaged in some geometrical or optical activity.

(4) For Jacques-Antoine Arlaud, see Letter 1400, note (1), p. 213; Newton wrote this letter to thank Arlaud for his work. For Arlaud's handling of the vignette in the *Traité d'Optique*, see Letter 1391.

(5) Coste had written, at the end of his Preface, 'Au reste, quoique j'aye fait quelques corrections assez importantes dans cette seconde Edition, que j'ai revûë avec toute l'exactitude dont je suis capable; la reconnoissance m'oblige de faire sçavoir au Public, que M. *de Moivre*, non moins bon Critique, que sçavant Mathematicien, y a fait quantité d'excellentes corrections, tant à l'égard du sens, qu'à l'égard de l'expression, lesquelles donnent à la Traduction Françoise de cet Ouvrage, un degré de perfection, qu'elle ne pouvoit recevoir que de l'habileté, de la sagacité, & de la justesse d'un si bon Esprit.'

(6) The letter is printed as an Appendix to the second edition of the *Commercium Epistolicum* (London, 1722), pp. 245–6, and is followed by Newton's 'Annotatio'; see also Letter 1395, note (6), p. 205.

(7) The 1722 edition reads 'jam ut advocalum [*read* advocatum] in hac rixa pro se inducere' ('now he is brought forward as an advocate on his own behalf in this quarrel').

(8) This change was not made.

(9) Letter 1390*a*.

(10) Jacques Cassini, *De la grandeur et de la figure de la terre* (Paris, 1722); published as *Suite des Mémoires de l'Académie Royale des Sciences* for 1720. Cassini had in fact extended the measurements made by his father, Jean-Dominique Cassini, after his death in 1712. In his book he reaches the empirical conclusion that the earth is elongated in the direction of its axis of rotation, not flattened as predicted by Huygens and Newton.

(11) Cotes' *Harmonia Mensurarum*; see Letter 1310, note (2), p. 29.

1397 NEWTON TO THE SOUTH SEA COMPANY
8 AUGUST 1722
From a photocopy of the holograph original[1]

Sr

Pray pay to Dr Fauquier the three per cent Dividend due at Midsummer last upon twenty & one thousand six hundred & ninety six pounds six shillings & four pence south sea stock I am entituled unto in the Companies books; & his receipt shall be your sufficient discharge from

Your humble servant
ISAAC NEWTON

8 *Aug.* 1722
To the Accomptant
of the Southsea
Company.

NOTE

(1) We print the letter from a photocopy found amongst the papers left by Dr J. F. Scott; the location of the original manuscript is unknown to us. For Newton's dealing in South Sea Stock see Letter 1342, note (1), p. 96.

1398 SAMUEL BULL TO NEWTON
14 SEPTEMBER 1722
From the original in the Mint Papers[1]

To Sr. Isaac Newton Master and
Worker of his Maj. Mint in ye Tower
The Humble Memorial of Sam: Bull 2d. Engraver &c Sheweth.

That the sd. Sam: Bull has servd as an Engraver to the sd. Mint since ye Year 169$\frac{6}{7}$ being invited by ye late Mr Harris[2] in ye time of ye great Coynage, from a profitable business he was then engag'd in, with promises of good encouragm[en]t and accordingly recd. for ye 1st Years service above 100*lb*. wch. ye Officers were pleasd to declare he then very well deserv'd

That when ye Coynage was expiring, other Gravers, who had been Employd were discharg'd, but ye sd Sam Bull was invited by ye three principal Officers to continue, they promising to gett him an Establishmt. from ye Tre[asu]ry which was done accordingly, by a Warrt for an Allowance of 50*lb* per Ann to him, as a Probationer And that was afterwards advanced to 60*lb*[3] with an Allowance upon every Mony puncheon yt he made; and this continued to Mr. Harris his Death.[4]

That thereupon Mr. Croker, & ye sd: Bull were sent for[5] by Sr. John Stanley & your Honour &c, who directed them to go on with ye business promising at ye same time to gett a new Patent, for Mr. Croker to be first Graver at 200*lb* per annum, & for ye sd: Bull to be second Graver at 100*lb* per annum. But unhappily at this juncture, one Monsr. Le Clerc[6] insinuating himself into favour, was appointed & p[ai]d as third Graver for about 7 years, 80*lb* per annum tho he did not act or work for ye Mint during that whole Time, whereas ye sd: Bull in ye same time gravd all ye Arms & yet recd. no more than 80*lb* per annum of ye 100*lb*. which was offer'd to him at first by all ye three principal Officers

That ye said Bull since his Majesties accession has had more business than before, by ye Augmentation in ye Royal Arms, but without any consideration

That ye sd: Bull paying Taxes, & having very little perquisites by Medals, puncheons or otherwise, has less encouragm[en]t than at his first coming into ye Mint, tho he has now spent about 26 years in ye service (being ye best part of his life) which with all submission cannot but be thought an extraordinary hardship and discouragment

That during his whole service this Memorialist never recd. any complaint agst. him, but on ye contrary humbly hopes, that he has at all times acquitted himself in his Station, with skill, diligence & fidelity.

Therefore your Memorialist most humbly beseeches your Honour to whom his whole service & behaviour are well known, that you'd please to take his case into your serious & speedy consideration, that one, who modestly speaking might have gain'd a hansom Livelyhood in a private capacity may not after an unblamable service of 26 years, hardly gett his Bread in ye service of ye Crown who is

Your Honrs most Obedient
& most humble servt
SAM: BULL

The 14*th. Sepr.*
1722

NOTES

(1) I, fo. 166. For Samuel Bull, Second Engraver, see vol. v, p. 308, note (3). Bull died in 1726 and was replaced by the younger John Roos; see Letter 1494.

(2) See vol. v, p. 308, note (2).

(3) Newton appealed to Godolphin, then the Lord High Treasurer, for this increase in 1702; see Letter 656, vol. IV.

(4) Harris died in 1704; see vol. IV, p. 351, note (3).

(5) In August 1704; see Letter 674, vol. IV.

(6) Gabriel le Clerk's appointment was indeed a sinecure; see Craig, *Mint*, p. 202.

1399 NEWTON TO THE TREASURY
18 OCTOBER 1722
From the holograph original in the Public Record Office[1]

To the Rt Honble the Lords Commissioners
of His Majties Treasury

May it please your Lordps

In order to make up my Accounts of the Coynage of copper moneys & pay the ballance thereof into the Exchequer, I humbly pray your Lordps that a Tryall of the Pixes of that Coynage may be appointed.

Which is most humbly submitted to your Lordps great wisdome

ISAAC NEWTON

Mint Office
18 *Octob.* 1722

NOTE

(1) T/1, 242, no. 68, fo. 197. Endorsed 'Mr Powys to be present on the part of the Trea[su]ry at this Tryal'.

1400 NEWTON TO JACQUES-ANTOINE ARLAUD
22 OCTOBER 1722
From the holograph original in the Bibliothèque Publique
et Universitaire de Genève[1]

Vir Celeberrime

Gratias tibi debeo quam maximas quod Schema[2] Experimenti quo lux in colores primitivos & immutabiles separatur, emendasti, & longe elegantius reddidisti quam prius. Sed et me plurimum obligasti dum Schema illud in

Lamina ænea incisum, & inter imprimendum obtritum, refici curasti, ut impressio Libri elegantior redderetur. Gratias itaque reddo tibi quas possum amplissimas. Quod inventa mea de natura lucis & colorum viris summis Dno Cardinali Polignac[3] & Dno Abbati Bignon[4] non displiceant, valde gaudeo. Utinam hæc vestratibus non minus placerent quam elegantissimæ vestræ et perfectissime delineatæ picturæ nostratibus placuerunt. Ut Deus te liberet a doloribus capitis & salvum conservet, ardentissime precatur

Servus tuus humillimus et obsequentissimus
ISAACUS NEWTON

Dabam Londini
22 Oct 1722

Celeberrimo Viro Dno Arlaud

Translation

Most famous Sir,

I owe you the greatest thanks because you have corrected the diagram[2] of the experiment in which light is separated into its primitive and immutable colours, and have made it far more elegant than before. But you also greatly obliged me when you took care that that diagram, engraved upon a copper plate and worn out in the course of printing, was repaired, so that the impression of the book was rendered more elegant. And thus I offer you the greatest thanks I can. I am exceedingly delighted that the things which I have discovered about the nature of light and colours do not displease those great men, the Cardinal Polignac[3] and the Abbé Bignon.[4] Would that these things may please your countrymen no less than your most elegant and perfectly drawn picture pleased mine!

That God may free you from headaches and conserve your health is the very urgent prayer of

Your humble and most obedient servant,
ISAAC NEWTON

London, 22 October 1722

To the most famous Mr Arlaud

NOTES

(1) MS. lat. 136; printed in *Mémoires et documents de la Société d'Histoire et d'Archéologie de Genève*, **5** (1857), 366, and in Edleston, *Correspondence*, pp. 188–9. A holograph draft is in King's College Library (Keynes MS. 142(Q)). Jacques-Antoine Arlaud (1668–1743) was born in Geneva, and at the age of twenty went to Paris where he gained a high reputation as a miniaturist. Later the Duc d'Orleans became his patron. According to J. B. Descamps, *La vie des peintres Flamandes, Allemands et Hollandois*, IV (Paris, 1763), pp. 116–22, he visited England in 1721, bearing a letter of introduction addressed to Caroline, Princess of Wales, whose portrait

he later painted; he became very popular at the English court. Whilst in England he met Newton, and drafted the vignette for the second edition of the *Traité d'Optique* (see Letter 1391). He was apparently also responsible for drawing the figures. Newton had presented Arlaud with a copy of the book on 14 September 1722, now in the possession of the Bibliothèque Publique et Universitaire de Genève.

(2) See Letter 1391.

(3) For the Cardinal de Polignac, see Letter 1350, p. 117, note (5).

(4) Varignon had presented the Abbé Bignon (vol. VI, p. 5, note (2)) with a copy of the second English edition of the *Opticks* in 1718 (see Letter 1304).

1401   NEWTON TO VARIGNON

c. OCTOBER 1722

From a holograph draft in King's College Library, Cambridge[1]

Celeberrimo Viro Dno Abbati Varignon Matheseos Professori
et Regiæ Scientiarum Academiæ apud Parisienses Socio
Is. Newtonus
salutem plurimam dicit

Vir celeberrime

Impressionem Optices meae jam tandem completam esse gaudeo. Et gratias ago tibi quam maximas tum quod Editionem summa cum ἀκριβεία correctissimam reddidisti, tum quod exemplaria amicis meis meo nomine donari curasti.[2] D. Moyvræus Literas tuas novissimas mihi ostendit, per quas intelligo expensa tua varia circa hoc negotium ad summam Librarum Sterlingarum plus minus octodecim ascendere. Et remisi nummos aureos viginti quinque quos Guineas vocamus,[3] de quibus D. Moyvreus tibi fusius scribet. Ex nummis receptis expensa tua quæso primum solvantur; & quod superest nummorum Dno Montalantio Bibliopolæ vestræ dari cupio. Spero quod is Librum promptius vendet lucri gratia quam impressit. Spero etiam quod Experimenta Optica, coram Dno Cancellario tentanda, recte successerint. Exemplar Commercij Epistolici[4] correctum tibi mittendum Dno Moyvreo dedi ut et exemplar Algebræ[5] denuo impressæ. Siquid fuerit in quo tibi possim hic inservire, quæso mittas mihi mandata tua quam liberrime. Spero quod salutem Tuam recuperaveris, & summopere precor Deum O.M. ut te salvum in plurimos annos conservet. Vale.

Translation

Isaac Newton makes a grand salute
to the most celebrated Abbé Varignon, professor of mathematics
and member of the Royal Academy of Sciences at Paris

Most famous Sir,

I am pleased that the impression of my *Opticks* has now at last been completed. And I give you my very best thanks, not only because you have with the greatest precision rendered this edition extremely correct but also because you have taken care to give copies to my friends on my behalf.[2] Mr De Moivre has shown me your most recent letter, by which I understand that your various expenses over this business rise to round about the sum of £18 sterling. And I have sent you back twenty-five gold coins, which we call guineas,[3] of which Mr De Moivre writes to you at greater length. I beg you first to recoup your expenses from the coins received; and what remains of the money, I desire to be given to Mr Montalant your bookseller. I hope that for the profit's sake he will sell the book more quickly than he printed it. I hope also that the optical experiments which were to be tried in the presence of the [then] Chancellor, Mr [Daguesseau], have succeeded properly. I have given Mr De Moivre a corrected copy of the *Commercium Epistolicum* for sending to you,[4] and also a copy of the *Algebra*[5] recently printed. If there is any way in which I can serve you here I ask you to send me your commands as freely as possible. I hope that you will have recovered your health, and above all I pray God to preserve you safely for many years. Farewell.

NOTES

(1) Keynes MS. 142(R); microfilm 1011.26. There is a rougher partial draft of the letter at Keynes MS. 142(Q), following a draft of Newton's letter of 22 October 1722 to Arlaud; hence we assume the present letter to have been written about the same time. A comparison of the two drafts to Varignon is interesting. The rougher draft ends at the reference to the optics experiments performed before the former Chancellor, Daguesseau, and so originally did the draft we print, that is with the word 'successerint'. Then, evidently, De Moivre wrote below Newton's draft the words: 'I gave Mr De Moivre ye Commercium corrected & ye new Edition of ye Algebra to be sent you, I wish it were in my power to serve you'. Newton then added round this, and round a money computation by De Moivre, the final sentences of his letter. Thus, it appears, Newton relied on De Moivre even to the suggestion of his closing compliment.

Possibly no letters had passed between Newton and Varignon since Varignon's letter of 24 July (Letter 1396) but clearly De Moivre and Varignon were in continuous correspondence, and De Moivre kept Newton informed of the progress of *Traité d'Optique*.

(2) See Letter 1372, p. 160.

(3) The rougher draft adds here '& qui faciunt 771 Livres Tournois in pecunia vestra' ('which makes 771 French *livres* in your money'). The phrase was probably omitted in the next version because De Moivre disagreed with Newton's calculation. His own calculation appears at the bottom of the draft at Keynes MS. 142(R); according to him an exchange rate of $23\frac{1}{4}d$ = 3 livres gives 25 guineas = 812 livres (actually the calculation gives a figure nearer 813 livres).

The edition seems to have been less expensive than originally expected (compare Letter 1372); perhaps Montalant's payment was to be reduced because of the delays he had caused.

(4) Compare Letter 1396, notes (6) and (7), pp. 209–10.

(5) Newton means the second edition of his *Arithmetica Universalis* (London, 1722); compare p. 101, note (13).

1402 FONTENELLE TO NEWTON
11 NOVEMBER 1722
From the original in the Royal Society[1]

Monsieur

L'Academie Royale des Sciences m'a chargé de vous remercier trés humblement de la Traduction françoise de votre Optique:[2] qu'elle reçue hier par M. Varignon. vous savés ce que toute l'Europe savante pense d'un Ouvrage si original, si ingenieux, si digne de vous, mais l'Academie, qui vous conte pour un de ses membres, en sent le merite, et le loüe avec un interest plus particulier. Je suis

Monsieur
Votre très humble et très
obeissant serviteur
FONTENELLE
Sec. perp. de l'Ac. Roy. des Sc

de Paris ce 22 Nov. 1722 [N.S.]

Trouvés bon, Monsieur, qu'aux remerciements de l'Academie je joigne aussi les miens pour l'Exemplaire que j'ai reçü de votre part. je ne puis assés vous exprimer combien je suis sensible a l'honneur que me fait un homme tel que vous, lors qu'il se souvient de moi d'une maniere si obligeante. quand vous ne feriés que savoir mon nom, j'en serois très glorieux, et conterois pour un extrème bon heur qu'il eust été jusqu'a vous. j'ai été aussi infiniment touché de l'avoir trouvé dans la Preface de M. Coste,[3] il faudra donc qu'on le connoisse, puis qu'il est dans un Ouvrage de grand M. Neuton. j'en ai une très vive reconnoissance pour M. Coste, qui ne pouvoit jamais me faire un plus grand honneur, mais je sens aussi que je vous doi beaucoup, Monsieur, de ce que vous avès eu la bonté d'y consentir.

NOTES

(1) MS. MM 5, 48.

(2) In Letter 1372 Newton had mentioned his intended gift of the *Traité d'Optique* (Paris, 1722).

(3) Coste wrote in his Preface, 'J'aurois souhaité pouvoir joindre à l'exactitude [de ma Traduction] ce tour vif & delicat du Secretaire de l'*Academie Royale des Sciences*, qui a trouve l'art de donner de l'agrément à la Solidité, sans lui rien ôter de son poids.' A footnote specifies Fontenelle as the secretary.

1403 NEWTON TO THE TREASURY
21 JANUARY 1723
From the holograph draft in the Mint Papers[1]

To the Rt Honble the Lords Comm[ission]ers of
his Ma[jesty]'s Trea[su]ry

May it please your Lordps

The Corporation of the Moneyers represent that they have been brought up Apprentices to the trade of coining & that to set up new Mints without them diminishes the right of their Apprenticeship.[2] They represent also that the multiplying of Mints tends to promote the skill of counterfeiting the gold & silver moneys, as happened in the coinage of Tin half pence & farthings in the beginning of the reign of King William & Queen Mary. For obviating these objections, & that of the insignificancy of a Comptroller of such a Mint, I humbly propose that Mr Wood prepare the blanks of fine Copper & make them fit to be stamped & then send them to the Mint in the Tower to be delivered there by weight & stamped & delivered back by the same weight.[3] This may be done by a Signe manual appointing the Assay & the number of pieces in the pound weight & the stamp & yearly quantity, & what shall be allowed to the Moneyers Graver Smith & Assayer & entring all receipts & deliveries in books, & acquainting me with what he finds amiss. the whole charge will not exceed two pence half-penny per pound weight. For I reccon nothing for myself. This I propose as safest for the governement & to be done by vertue of the power reserved in his Ma[jes]ty & your Lordps of controlling Mr. Wood.

All wch is most humbly submitted to your Lordps great wisdome

ISAAC NEWTON.

Mint Office
21th[4] *Jan.* 172⅔

NOTES

(1) II, fo. 464. On 8 May 1722 a draft Warrant had been submitted by Lowndes to the Attorney and Solicitor General empowering William Wood (see Letter 1319) to coin copper farthings and halfpence for Ireland (see P.R.O., T/27, 23, p. 272). In August a constitution was prepared by the Treasury 'appointing Sir Isaac Newton Compt[rolle]r of this Coynage & that for the West Indies to act by himself or Deputy. & then my Lords will give the proper directions & powers to [Mr. Wood] to coin a certain quantity of Copper Money at Bristol according to his request'. Newton's copy of the Treasury Warrant to Wood (dated 23 August 1722) authorizing him to establish his Mint at Bristol is at Mint Papers, II, fo. 460.

The Treasury was continually plagued by complaints about the scheme. The Commissioners of Revenue in Ireland were first to object, saying that the private coinage of copper would be prejudicial to trade (see their representation to the Treasury of 19 September 1722, P.R.O., T/1, 244, no. 50, fo. 276). Objections from the Corporation of Moneyers followed (see below). The coinage went ahead, but the Irish renewed their complaints in 1724 (see Letter 1430), demanding fuller assays of the coin.

(2) The petition (undated) from the Provost and Corporation of Moneyers to the Treasury is at P.R.O., T/1, 244, fo. 17.

(3) Wood must have had the first blanks ready by 29 July 1723, for he then submitted a petition requiring the Controller of the coinage (Newton) to report on assays of the coin (see P.R.O., T/1, 244, fo. 50). It was Matthew Barton, appointed by Newton as inspector of the Irish coin, not Newton himself, who submitted a favourable report on the assay of the coin on 2 August 1723 (P.R.O., T/1, 244, no. 13, fo. 48).

Curiously, the assays of the coin Wood prepared for the West Indies were not so successful; in this case not only the coin was assayed, but also the uncoined ingot and fillet. Wood objected to this (13 December 1723; see P.R.O., T/1, 244, fo. 222, no. 49) and Barton accepted his objections, but later further complaints against Wood were made, suggesting that he had made improper use of his coining machines, and did not keep proper accounts (P.R.O., T/1, 252, fo. 27).

Newton's proposal of bringing the blanks to London for minting was not adopted; they were minted at Bristol.

(4) The date is difficult to read; apparently Newton first put '19', and then wrote '21' over it.

1404 J. BERNOULLI TO NEWTON
26 JANUARY 1723
From the original in Yale University Library[1]

Illustrissimo atque Nobilissimo Viro
Isaaco Newtono
S.P.D.
Joh. Bernoulli

Ad te iterum venio Vir Inclyte, ut iteratas persolvam gratias pro novo munere quo me beasti, nec me tantum sed et filium meum atque Agnatum. Accepi nimirum tria inter nos tres distribuenda Exemplaria nitidissime compacta Optices Tuæ Parisiis nuper editæ,[2] quæ Cl. Varignonius, paulo ante obitum suum,[3] cunctis qui sinceritatem cum eruditione conjunctam amant vehementer lugendum, mihi nomine Tuo transmiserat. Etsi nesciam quid sit quo hanc Tuam erga me meosque munificentiam demeruerim aut postea demereri possim; id saltem persuasum Tibi habeas Vir maxime, neminem esse qui immortalia Tua inventa ex vero rerum pretio pluris æstimet et simul

sincerius quam ego: Hoc cumprimis quod de Lumine et Coloribus systema pro
ingenii Tui sagacitate felicissime eruisti me summum habet admiratorem:
inventum sane quovis ære perennius et a posteritate magis quam nunc fit
suspiciendum. Sunt enim qui illud partim ex invidia partim ex imperitia
obtrectare non verentur, quin et cum nihil habent quod pretium ejus im-
minuat, audent inventionis laudem Tibi surripere eamque sibi arrogare: En
exemplum in quodam Hartsoekero homine inepto et in Geometria prorsus
hospite, qui in opusculo [4] aliquo in lucem protruso perfricta fronte sustinet
novam tuam Colorum Theoriam eorumque diversam refrangibilitatem sibi
dudum notam variisque experimentis perspectam fuisse, antequam quicquam
ea de re inventum a Te aut evulgatum fuisset, quod apud me summam excita-
vit indignationem sicut et hoc quod reliquas Tuas rerum physicarum explica-
tiones utut ingeniosissimas, præsertim quæ ad systema planetarium spectant,
ubi cum phænomenis tam mirifice consentiunt, admodum salse et scoptice
traducit, quamvis de rebus istis non aliter argumentetur quam cæcus de
colore, nec mirum, siquidem homo sit ἀγεωμέτρητος et omnis humanitatis
expers, nemini parcens, imo summorum virorum atque de re mathematica ac
philosophica optime meritorum famam arrodere non dubitans. Ita ut mirer,
neminem ex Vestratibus adesse, qui Tuam Vir Illustrissime, existimationem
vindicet contra rudem et barbarum hominem. Ad me quod attinet, fateor, me
ab illo tractari multo acerbius, nihil enim injuriarum est, nihil aculeorum
quod in me non sparserit, idque non aliam ob causam quam quod aliquis ex
meis discipulis phosphorum meum mercurialem ab illius morsibus defenderit; [5]
Licet indignus sit homo cui ego respondeam, unum tamen est quod me magno-
pere urit; scilicet ut me omnium risui exponat, impudentissime comminiscitur,
me mihimet ipsi tribuisse titulum *excellentis mathematici*, et ut calumniæ crimen
a se amoliatur Te Vir Illustrissime ejus Auctorem facit, dum locum citat ex
Tomo 2. Collectaneorum Di. Desmaiseaux, (*Recueil de diverses pieces,*) p. 125.
l. 32. ubi loqueris de epistola illa 7 Junij 1713 [N.S.], quam Leibnitius a me
scriptam esse contenderat, et in qua prout erat impressa in scheda illa volante
29 Julij 1713 [N.S.], elogium illud, sed quod parenthesi includebatur, mihi
erat adscriptum. Hinc malitiose colligit calumniator, quasi insinuare volueris,
me eo arrogantiæ processisse ut hunc mihi titulum sumpserim, cum tamen te
voluisse contrarium dicere luculentissime pateat ex verbis quæ locum citatum
immediate sequuntur, quibus nimirum fateris in eadem illa epistola per
Leibnitium altera vice edita in Novis Litterariis citationem parenthesi inclusam esse omissam; unde sponte fluit Auctorem epistolæ non fuisse Auctorem
parentheseos, sed hanc fuisse insertam ab eo qui schedam volantem 29 Julij
edidit; possum itaque haberi pro Auctore Epistolæ et tamen non haberi pro
scriptore elogii parenthetici. Interim quicquid sit calumnia Hartsoekeri in te

magis quam in me redundat, eam enim ex verbis tuis maligne detortis elicere conatur.[6] Quid igitur faciendum statuas, ut innocentia in tutum collocetur apud eos qui Collectanea Desmaisavii non viderunt, libentissime equidem ex te ipso intelligerem, si qua responsione me dignari volueris. Quod superest te Vir Nobiliss[ime], rogatum volo nomine Celeberrime nostri Scheuchzeri,[7] vestræ Societatis Regiæ Socii, ut Filio ipsius, qui nunc Londini agit te accedendi alloquendique copiam indulgeas; id namque in maxima laude sibi ponet, quod viderit summum Philosophorum et Mathematicorum Principem. Vale, et me Nominis tui Cultorem perpetuum amare perge.

Dabam Basileæ, a.d. VI. *Febr.* MDCCXXIII [*N.S.*]

Translation

Johann Bernoulli sends a grand salute
to the most famous and most noble Isaac Newton

I write to you again, famous Sir, in order to offer you repeated thanks for the new gift with which you have favoured me, and not only me but also my son and my nephew. For I have actually received three elegantly bound copies of your *Opticks*,[2] recently published in Paris, for distribution amongst us three, copies which Mr Varignon had sent to me in your name a little while before his death,[3] which is mourned by all those who ardently love sincerity joined with wisdom. And yet I do not know how it is that I have deserved this generosity of yours in relation to myself and my [relatives], or how I can deserve it hereafter; of this at least you may be certain, greatest Sir: that there is no one who values your immortal discoveries more greatly, and at the same time more sincerely according to the true worth of things, than I do. In particular, what you have most successfully discovered, thanks to the shrewdness of your talent, concerning light and the system of colours, has in me a very great admirer; indeed, it is a discovery more enduring than any bronze [monument], and one to be more greatly prized by posterity than it is now. For there are those who, partly from envy and partly from ignorance, do not fear to disparage that [discovery], but rather, since they have nothing which threatens the value of it, dare to steal from you the praise of your discovery and claim it for themselves: of this there is an example in a certain Hartsoeker, a foolish man and a complete stranger to geometry, who has with great effrontery maintained in a certain little treatise thrust upon the world[4] that your new theory of colours and of their varying refrangibility had been known to himself some while before, and had been investigated by various experiments before the publication of anything discovered by yourself concerning that matter, [a claim] which aroused in me the highest indignation; and in the same way he has also handled your other explanations of physical phenomena in a caustic and sarcastic manner, although they are most ingenious (especially what concerns the planetary system, where everything fits so remarkably with the phenomena). And yet he adduces no more evidence about these matters than a blind man can do concerning colours, nor is this surprising since the man is deficient in geometry and lacking in all humanity, sparing none and never hesitating to derogate from the fame of the greatest men, highly meritorious in questions of mathematics and philosophy. So that I am astonished that no-one comes forward from your fellow-countrymen to defend your reputation, most illustrious Sir, against this rough and barbarous man. As for myself, I admit that I have been much more sharply treated by him; for there is no wound or sting which he would not have inflicted on me, and for no other reason than because one of my students defended my [theory of] the phosphorescence of mercury from his attacks.[5] Granted that the man is not worthy of my reply, nevertheless there is one thing which irritates me greatly, namely, that in order to expose me to everyone's laughter, he pretends that I attributed to my own self the title 'excellent mathematician', and, in order to remove the odium of calumny from himself, made you, most illustrious Sir, responsible for it, by citing the place in volume II of Des Maizeaux's collection (*Recueil des diverses pièces*) p. 125, l. 32, where you speak of that letter of 7 June 1713, [N.S.], which Leibniz maintained was written by me, and in which, as it was printed in the *Charta Volans* of 19 July 1713, [N.S.], that word of praise was ascribed to me, but enclosed in parenthesis. Whence the liar has maliciously inferred that you meant to imply that I had behaved with such arrogance that I had taken this title upon myself, when on the contrary it was most abundantly clear from the words which immediately follow the place cited, that you meant to say the opposite; for in these words you do indeed allow that when that same letter was published by Leibniz a second time in the *Nouvelles Litteraires* the quotation enclosed in parentheses was omitted; whence it follows at once that the author of the letter was not the author of the parenthesis, but that this was inserted by whoever published the *Charta Volans* of 29 July, [N.S.]. Therefore I may be taken as the author of the letter, and nevertheless not be taken as writer of the parenthetical eulogy. Meanwhile, whatever the case, Hartsoeker's accusation redounds more on you than on me, for he tries to produce it from your words, wickedly distorted.[6] If you should consider me worthy of a reply, I would very gladly learn from your own pen what you decide to do, in order to defend your innocence in the eyes of those who have not seen Des Maizeaux's *Recueil*. For the rest, most noble Sir, I wish to ask you on behalf of our most celebrated country-man Scheuchzer,[7] a Fellow of your Royal Society, that you should so far indulge his son, who is now living in London, as to allow freedom of access and conversation with you; for thus he thinks he will be most deserving of praise, because he will have seen the great prince of philosophers and mathematicians. Farewell, and hold me bound to a perpetual esteem of your name.

Basel, 6 February 1723 [*N.S.*]

NOTES

(1) Newton Papers, no. 93; printed in Brewster, *Memoirs*, II, pp. 507–8, and partly translated, *ibid.*, p. 74.

(2) The second French edition; see Letter 1338, note (7), p. 91.

(3) Varignon died on 22 December 1722, N.S. The gift of the *Traité d'Optique* is mentioned

in letters passing between Bernoulli and Varignon in the summer of 1722; see Bernoulli Edition.

(4) Nicolas Hartsoeker (1656–1725), a convinced cartesian and correspondent of Leibniz, criticized Newton strongly in his *Recueil de Plusieurs Pieces de Physique où l'on fait principalement voir l'Invalidité du Systeme de Mr. Newton* (Utrecht, 1722). The *Recueil* began with a reprinting of a letter from Hartsoeker to Le Clerk criticizing Newtonian gravitational theory; this was followed by Le Clerk's remarks, and Hartsoeker's reflections upon these. In his reflections, Hartsoeker digressed into other aspects of physics, including optics. On pp. 103–4 he claimed that he did experiments on refraction before hearing of Newton's optical researches, and that in his *Essai de Dioptrique* of 1694 he had postulated that colour depended on two factors inherent in a ray of light, which he called 'force et vigueur' and that these were not changed by refraction or reflection. (In his *Essai de Dioptrique*, pp. 32–3, where he had indeed put forward a vague qualitative theory relating colour to refrangibility, he labelled these two factors 'vitesse et epaisseur', and mentioned experiments he had done, but without describing them in detail.) In the *Recueil* he added that

> ...j'en ai conjecturé depuis, que les rayons de lumiere ne sont ainsi differens entre eux, que parce qu'ils coulent au travers de differens corps ou tuyaux, qu'on peut appeller corps cilindriques ou tuyaux à lumiere, & que ces tuyaux sont parfaitement durs, immuables, & aussi anciens que l'Univers, comme tous les corps premiers qu'on appelle atomes.

(5) Johann Bernoulli had written a number of articles on his 'Nouveau Phosphore' in the *Mémoires de l'Académie Royale des Sciences* for 1700 (pp. 178–90) and for 1701 (pp. 1–9 and 137–48). There he had claimed that the luminescence observed on agitating a partially evacuated vessel of mercury is caused by the pulling and pushing of a subtle matter through the walls of the tube. Hartsoeker, in his *Eclaircissemens sur les conjectures physiques* (Amsterdam, 1710), described Bernoulli's theory as 'embarrassée et defectueuse'. Bernoulli retaliated by persuading one of his pupils, Wilhelm Bernhard Nebel (1699–1748) to compose a *Disputatio de Mercurio Lucente in Vacuo sub Præsidio J. Bernoulli* (Basel, 1719), where Hartsoeker's objections were criticized in their turn. (Hartsoeker discusses the whole dispute in the 'Avertissement' to his *Recueil*, p. vii.)

In his *Recueil*, pp. 268–88, Hartsoeker published an article 'Remarques sur deux passages d'une these que Mr Bernoulli a fait soutenir à Bâle par un certain Nebel son ecolier'. There he begins by quoting from Nebel's *Disputatio* passages where Nebel strongly accuses Hartsoeker of deriding the work not only of Bernoulli, but also of Huygens, Leibniz and Newton. He then deals with these accusations one at a time, and in particular deals at length with the Newtonian system. On pp. 277–8 he suggests that 'Mr. *Bernoulli, tout excellent Mathematicien*' should do the necessary calculations to save it. This leads him to ask who gave Bernoulli the title 'excellent Mathematician', and to bring up briefly the whole question of the *Charta Volans* (see Number 1009, vol. VI); he suggests that it was Bernoulli himself who was responsible for the appellation. He goes on to criticize Bernoulli's theory of mercurial phosphor at length, and finally, pp. 286–8, arraigns him for his rudeness and method of attacking his critics.

(6) Bernoulli builds the whole of this discussion on one short paragraph in Hartsoeker's *Recueil*:

> Mais de qui fait-on, dira peut-être quelqu'un, que ce titre pompeux d'*excellent Mathematicien*, appartient à Mr. Bernoulli? De qui? De lui même, ou d'une lettre, qu'il écrivit le 7me Juin 1713 [N.S.]. à Mr. le Baron de *Leibnitz* a l'occasion d'une dispute que ce Baron eut alors avec Mr. le Chevalier *Newton* sur le droit de l'invention des *Fluxions ou du Calcul Differentiel*, &

motibus a posteriori, et ex observationum suarum collatione deprehendisset
Keplerus, a Te fuisset a priori et Geometrice demonstrata, nimirum, quod □ta.
temporum periodicorum sint in iis cunctis ut □i distantiarum a sole mediarum;
atque in nobilissimo illo Theoremate, quandam illico demonstrationem quasi
viderer mihi cernere, universi tui Systematis, tum impotenti equidem desiderio
flagrabam, intelligendi demonstrationem saltem hujus Theorematis, idque
demum me tandem impulit, Vir Illustrissime, ut Incomparabile illud tuum
opus, quod merito summum quendam humani acuminis apicem dixerim,
quodque Geometras Astronomosque ævi nostri peritissimos a triginta retro
annis torsit exercuitque, puta Librum tuum *Principiorum* legere tentarem, quod
equidem opus aggredi nondum fueram ausus. At ibi vix effabili voluptate, non
modo dicti illius Theorematis demonstrationem, quam tradis Prop: 15. lib: 1
sed et tertiam circiter operis partem intellexi, atque sexcentas proprietates
omnino novas stupenda universalitate patentes, immanique Geometriæ vi
erutas ibi cernere mihi contigit. Hoc facto, resumpta Dni. Keilij Astronomia,
cum pervenissem Pag[in]a 375 ad solutionem Kepleriani Problematis[9] Geometricam; scilicet de invenienda Anomalia Planetæ vera ex media data; seu
quod eodem redit; de secanda, in data ratione, Ellipsi, Circulove, per rectam e
dato axis puncto ductam; hanc equidem solutionem non capiebam, eo quod
ipsum ibi arcum viderem exprimere per seriem infinitam; quare cum circa hoc
etiam solutionem quam Clar: Gregory Pag. 211. Astron[omi]æ[10] suæ tradidit,
consuluissem, ipsumque ibi Lectorum, pro doctrina serierum, mihi tum adhuc
incognita ablegare viderem ad Exercitationem Geometricam de Dimensione
figurarum[11] a se traditam Edinburgi Ao 1684. et ipsum avidus legi, atque etiam
intellexi, quandoquidem mihi jam innotuerat primum ejus Lemma cui super-
structa sunt omnia; Est enim illud Lemma[12] unicum, sed quod mira patet
universalitate Theorema demonstratum a Clarissimo vestrate Wallisio in
Arith[meti]ca sua *Infinitorum*,[13] et Clarissimo etiam Barrowio adhibitum in
demonstrationibus perelegantibus Theorematum Archimedeorum tam circa
superficiem quam soliditatem sphæræ partiumque suarum, quas ad calcem
Euclidis[14] sui subtexuit; Tum demum intellexi, qua ratione ex dato arcu posset
per infinitam seriem exprimi ejus sinus rectus et versus, Cosinus, Tangens;
atque vicissim etiam, quomodo ex horum aliquo dato, arcus posset exprimi, nec
non per easdem darentur Curvarum Geometricarum tam areæ quam soliditates
ab arearam illarum circa axes gyratione genitæ; ut et quomodo etiam per infinitas series exprimeretur valor radicis cujuscunque dignitatis e Binomio
quovis. At cum nec ista sufficerent ad intelligendam penitus illam Kepleriani
Problematis solutionem, sed ipsum ad id tandem adhibere etiam viderem
mirabilem tuam Reversionum Methodum, illud me demum compulit, ut
celebratissimas illas binas a te Vir Illustrissime ad Dnum Leibnitium Ao. 1676

datas Literas adierim, in Tomi 3^{tij} Wallissij *Commercio Epistolico*,[15] ubi non modo aperuisti, ipsi Leibnitio tum nunquam visa, sed quod facile crediderim talia, quæ ille funditus penitusque nunquam introspexit. Ægerrime enim adducor, ut credam, illum omnino perspexisse demonstrationes mirabilium illorum Theorematum, quorum ope stupendum illum Regressum doces, atque invenis seriem quæ exprimat Curvæ Aream ex data illius Ordinata, et vicissim seriem reperis quæ ex data Curvæ Area, ejusdem exprimat Ordinatam, neque facilius mihi persuasero, Dnum Leibnitium intellexisse, qua mirabili arte tibi Vir Illustrissime innotuerit illa genesis numerorum Coefficientium istius seriei Pag[in]æ 626 *Commercij Epistolici* pro arcu Elliptico, et ejus quæ Pag[in]a 627 occurit, qua exprimis soliditatem portionis Sphæroideos Elliptici. Enimvero Clar[issim]us Gregorius easdem binas series equidem Pag[ini]s 33 et 44 dictæ illius suæ Exercitationis Geometricæ etiam invenit; at ipse quidem illam Coefficientium genesin non detexit, qua levissima opera produci series, magisque ad veritatem appropinquari potest, quoad libuerit;[16] quod quidem facere in omnibus seriebus solenne est tibi soli.

Quod autem asseram, Dnum Leibnitium inventa mirabilia binis illis Epistolis contenta omnino non perspexisse omnia, ejus documentum mihi suppeditavit Doct: Keilius. Etenim Dno. Bernoulli Dni. Leibnitij discipulus, ab eo didicit Calculum quidem Differentialem, at non item difficiliorem Integralem, unde etiam hunc sibi arrogare voluit Dno Bernoulli, sibique magni quid præstitisse est visus, quod quantitatis differentialis $dy\sqrt{bby-a^3}|$ detexisset Integralem.[17] At vero, res ipsa clamat, quod si harum literarum rationes ususque penetrasset intime Dno. Leibnitius, tum sane vidisset, Integralem non modo quantitatis $dy\sqrt{bby-a^3}|$ sed et infinitarum aliarum longe magis compositarum restitui ope solius tui 1^{mi} Theorematis, quod ipsi aperueras; scilicet; quod si Curvæ Ordinata exprimeretur per $dz^\theta \times \overline{e+fz^\eta}|^\lambda$ tum ipsius Area exprimeretur per seriem[18] infinitam, sed tum convergentem;

$$Q \text{ in } \frac{z^\pi}{s} - \frac{r-1}{s-1} \times \frac{eA}{fz^\eta} + \frac{r-2}{s-2} \times \frac{eB}{fz^\eta} - \frac{r-3}{s-3} \times \frac{eC}{fz^\eta} \text{ \&c:}$$

si autem istud vidisset, sane sibi id lucro fecisset Dnus Leibnitius, atque tum etiam ægre hoc discipulum ejus Dnum. Bernoulli latuisset. Etenim, ut quod res est dicam, vix mihi sit credibile, Dnum Bernoulli fidei esse usque adeo sublestæ subdolæve, quam ut de re tantilla quanta est restitutio Integralis istius quantitatis $dy\sqrt{bby-a^3}|$ sibi fuisset tantopere gratulatus, ut etiam se pro Integralis Calculi inventore voluisset venditare, si dicti hujus tui Theorematis usum penitus perspexisset; et multo quidem adhuc minus eum id ausum fuisse mihi persuaserim si illam Integralem novisset restituere, eo quo docuit Doct: Keilius modo, Pag[in]a 265 Hagensis Diarij Gallici Anni 1719;[19] nimirum ad id adhibendo mirabiles illas formulas Tabulæ 1^{æ}[20] tui Tractatus de Quadratura

Curvarum. At candide fatear, solum illud a Dno. Keil ibidem allatum exemplum exquisitissima sane me suffudit voluptate; eo enim exemplo me docuit, συνώνυμα tibi esse, Curvam quadrare et Integralem quantitatis restituere, quæ vere tua est, eademque simplicissima totius negotii idea, mihi nondum percepta.[21] Hac autem clave mihi Autor fuit Dnus. Keil ut Tractatum illum de Quadratura Curvarum, qui Differentialem pariter ac Integralem complectitur Calculum, et legerim et pene integrum commentatus fuerim; atque verissime sane affirmaverim, quod etiamsi tuus ille Quadraturarum seu Integralis Calculus non vera foret clavis, qua 2do. tuo *Principiorum* Libro solutiones detexisti Problematum Physico-Mathematicorum subtilissimorum et omnibus aliis ante te imperviorum; etiamsi, inquam, hic non esset fructus longe mirabilissimus istius tui Calculi Integralis, et licet etiam nullius plane foret usus, minime tamen dixisse verebor, nunquam sane ullum ab humano ingenio profectum profecturumve esse divinius inventum; neque fieri potest, quin exultabundus in summam mentis rapiatur admirationem, quicunque intellexerit, mirabiles illas tuas series[22] Theorematum 3^{tij} et 4^{ti}. videritque illas eminenter continere Integrales, omnis quantitatis differentialis compositæ ex uno vel duobus Polynomiis, cujusvis gradus et quodlibetcunque terminorum numero complexis; Sive etiam, si quis rite ceperit eam, pulchritudinis tantum non divinæ transmutationem Curvarum, qua condidisti immensæ universalitatis formulas illas nunquam satis demirandas 2^{dae}. tuæ Tabulæ Curvarum nonquadrabilium, sive quantitatum, quarum Integrales nullo finito Algebraicorum terminorum numero possunt exprimi; Certo enim est certius, mihi sane de nulla unquam re, quam tota vita viderim gratulari impensius, quam quidem de hisce mihi intellectis. Etenim Vir Illustrissime, ad ista Theoremata quod attinet, illa equidem eatenus perspexi, ut cum te asserere viderem, sub finem Theor[ema]tis 4^{ti}. quod *series talium Propositionum pergeret in infinitum*; hinc ad tui imitationem mihimet seriem infinitam adornavi quæ exprimit Ordinatam Curvæ cujus Area esset $z^{\theta}R^{\lambda}S^{\mu}T^{\omega}$; atque deinde etiam alteram seriem, quæ exprimat Aream Curvæ cujus Ordinata sit

$$z^{\theta-1}R^{\lambda-1}S^{\mu-1}T^{\omega-1} \quad \text{in} \quad \overline{a+bz^{\eta}+cz^{2\eta}} \quad \&c$$

per eamque seriem Integralem hujus quantitatis restitui;[23]

$$\frac{\frac{1}{3}a^3b^4\,dy - \frac{1}{3}a^2bc^2d^2\,dy + \frac{2}{3}a^2b^4y\,dy - \frac{5}{2}a^2b^3cy\,dy - \frac{2}{3}abd^2c^2y\,dy}{}$$

$$+\frac{5}{2}ad^2c^3y\,dy - \frac{7}{2}ab^3cy^2\,dy + \frac{7}{2}c^3d^2y^2\,dy - \frac{13}{6}a^2b^2y^3\,dy$$

$$\frac{-\frac{17}{6}ab^2y^4\,dy + \frac{19}{2}abcy^4\,dy + \frac{23}{2}bcy^5\,dy}{\sqrt{ab-3cy}\mid \times \sqrt[3]{\overline{ay+yy}^2} \times \sqrt[4]{\overline{ab^3-ccdd-2by^3}^3}}$$

226

Idque quidem præstiti, servando valorem ipsius η positivum, sex variis modis, prout scilicet Polynomiorum $R^{\lambda-1}$, $S^{\mu-1}$, $T^{\omega-1}$ valores varie combinari possunt; atque sex itidem adhuc aliis modis, dando scilicet huic quantitati expressionem, in qua valor ipsius η in formula sit negativus, quod sane Bernoullios Hospitalliosque ad unum omnes frustra tentare videas. Postea etiam expertus sum, plurimas quantitates differentiales, quæ prima fronte vix videantur, neque ad hæc tua Theoremata neque ad ullum Tabulæ tuæ 1^{mæ} formulam posse referri, levi quadam alterius incognitæ substitutione eo reduci; Ita enim harum quantitatum Integrales ope formularum tuarum repperi.

$$\frac{\frac{1}{2}abc\,dy\,\sqrt{ay}}{b^2y\,\sqrt{ac}-2aby\,\sqrt{cy}+ay^2\,\sqrt{ac}}; {}^{(24)} \qquad \frac{6ax^4\,dx+4bbx^3\,dx}{4a^2x^2+4abbx+b^4} {}^{(25)}$$

$$\frac{-2aabcx\,dx-c^4\,dx}{a^4bb-2aabx^3+x^6}; {}^{(26)} \qquad \frac{axx\,dx+abbx\,dx}{aaxx+2abbx+b^4}; {}^{(27)}$$

imo etiam hæc quantitas $\dfrac{\frac{3}{2}\,abbxdx+2acxdx\,\sqrt{ac-xx}}{\sqrt{a^4bc+2a^2c^2x^2-a^3bx^2-2acx^4-3ab^2\,\sqrt{ac-xx}^3}}$

Polynomio complexo constans substituendo, atque vocando $\sqrt{ac-xx}=\sqrt{y}$ abiit in aliam expressionem, induitque formam cum Polynomio incomplexo, quod quia tum referri poterat ad formulam tui Theor[ema]tis 3^{tij} series convergebat et Integralem quæsitam ita mihi restituebat.$^{(28)}$

Hisce tandem intellectis, secundam etiam tuam Tabulam Curvarum cum Ellipsib[us]que Hyperbolisve comparabilium sum aggressus, cui condendæ adhibita fuerunt mira tua Theor[ema]ta 5.6.7. cum suis Corollarijs atque etiam Problema 3^{tium} estque hæc ipsamet mirabilis illa transmutationis Methodus præ qua se Craigius pauci facere confitetur quicquid super hoc argumento elaboraverat, nec id injuria dicit Pag[in]a 60 Tractatus sui de *Calculo Fluentium*.$^{(29)}$ Ea vero Tabula 2da. mihi equidem initio plus aliquanto laboris creavit, eam tamen eatenus intelligo, ut in illa videam quod quæcunque mihi etiam illius 2^{dæ}. Tabulæ formula proponatur, v.g. Casus 1^{mus} Formæ 8. $\dfrac{dz^{\eta-1}}{\sqrt{e+fz^\eta+gz^{2\eta}}}$ si$^{(30)}$ pro illius transmutatione supponatur, ut jubes, z^η valere abscissam x curvæ novæ per æquationem $\sqrt{e+fx+gxx}=v$ determinatæ, tum equidem adhibendo tuum Theor: 7 quod evidens est, facile demonstrare possum quantitatis

$$\frac{dz^{\eta-1}}{\sqrt{e+fz^\eta+gz^{2\eta}}}$$

aream suam Integralem esse plane ut dicis, $\dfrac{8dgs-4dgxv-2dfv}{4\eta eg-\eta ff}$; quod levi etiam opera æquivalere ostenditur ipsi $\dfrac{8dg}{4\eta eg - \eta ff} \times \alpha GDB \pm DBA$; scilicet

$$\frac{8dg}{4\eta eg - \eta ff} \times \alpha GDB + DBA$$

si abscissæ x versus Curvæ centrum tendant, at ipsi $\dfrac{8dg}{4\eta eg - \eta ff} \times \alpha GDB - DBA$
si abscissæ x tendant ad partes contrarias in tuis Figg. 6 vel 7. vel 8; namque
æquatio $\sqrt{e + fx + gxx} = v$ Locus esse potest vel ordinataram ad primas diametros innumerarum Hyperbolarum ut in Fig: 7 vel ad totidem secundas
diametros ut in fig. 8 vel ad Ellipsum Circulorumve diametros ut in Fig: 6. prout
scilicet datarum quantitatum e; f; g valores erunt vel positivi, vel nulli vel
negativi. Unde liquido sequitur, omnem ordinatam, quantitatemve differentialem, cujus Integralis nullis numero finitis Algebraicis terminis exprimi potest,
si ad hujus Tabulae aliquam formulam referre se patiatur, tum utique illico
ipsius Area seu quæsita Integralis Geometrice hic exhibebitur, exprimeturque
per datæ Ellipseos aut Hyperbolae Conicae altiorisve etiam cujusvis gradus
portionem determinatam. Inventum sane usque adeo alte erutum, cujusque
usus tam late patet, ut illud, me judice, merito habendum sit revera pro omnium
mirabilium mirabilissimo. Namque si ex: gr: proponeretur invenienda Inte-
gralis quantitatis $\dfrac{4b\,dy\sqrt{a^2c^2 + a^3b - b^2y^2 - acy^2}}{ayy}$ sublato primum $dy = 1$;[31] video
eam quantitatem sub hac quidem forma in casum Tabulae neutrique cadere
posset, illa vero forma cum modis mutari possit infinitis, divido hujus fractionis
utrumque terminum per ay; et evadit expressio $\dfrac{4b}{y}\sqrt{\dfrac{-bb-ac}{aa} + \dfrac{cc+ab}{yy}}$ quam
nunc utique referre posse in Tabula 2da. video, ad Formæ $3^{\text{iæ}}$ casum 1^{mum} $\dfrac{d}{z}\sqrt{e + fz^{\eta}}$
ponendo nimirum $z = y$; $d = 4b$; $e = \dfrac{-bb-ac}{aa}$; $f = cc + ab$; $\eta = -2$. Atque
cum fuerit ibi posita $xx = \dfrac{1}{z^{\eta}}$; valebit $xx = \dfrac{1}{y^{-2}} = yy$ unde cum in $\sqrt{f + exx} = v$
obtineat hic e valorem negativum, f vero positivum, referet æquatio Ellipsin,
quæ sit tua Fig: 6 cujus diameter Aa esset $\sqrt{\dfrac{f}{e}}$ illiusque ad suam parametrum
ratio illa, quæ est 1 ad e.[32] Cumque hic $\sqrt{f + exx}$ valeat $\sqrt{cc + ab - \dfrac{bbyy - acyy}{aa}}$ id
nobis designat, Locum etiam ad Ellipsin, cujus diameter sit $a\sqrt{\dfrac{ab+cc}{ac+bb}}$ ejusque
ad latus rectum[33] ratio, quæ est a^2 ad $ac + bb$. Sed juxta formulam Tabulae area
seu integralis est $\dfrac{4de}{\eta f} \times \dfrac{v^3}{2ex} - s$ sive $\dfrac{2dv^3}{\eta fx} - \dfrac{4de}{\eta f} \times s$ quod pro nostris valoribus docet,

quæsitam propositæ quantitatis aream seu integralem esse

$$\frac{-4b}{ccy+aby} \times \sqrt{\overline{\frac{-bbyy-acyy}{aa}+cc+ab}^{\,3}}$$

demto producto $\frac{8b^3+8abc}{a^2c^2+a^3b} \times s$ id est producto $\frac{8b^3+8abc}{a^2c^2+a^3b}$ per Ellipticum Trape-
zium *ABDP* fig[ur]æ tuæ 6^{tæ} modo jam dicto determinatæ. Responsum, cujus
veritatem a posteriori per Calculum Differentialem demonstrare proclive esset,
at quod tantum abest, ut ipse, si viveret Dnus Leibnitius nec Bernoulli nec
omnium ejus discipulorum ullus alius unquam retegere potuissent, ut universis
ipsis potius illud e tripode dictum aut ab quodam Œdipo profectum esse, videri
debeat. Porro mihi etiam constat, Vir Illustrissime, mirabiles illos formulos
similiter inservire pro areis inveniendis, si dentur fluxiones fluxionum; imo
fluxiones 3^{tiæ}. 4^{tæ}. &c. in infinitum, ita ut parvulo illo Tractatu sane sis
dicendus Infinitum te vere exhausisse atque complexum esse infinities. Facile
etiam crediderim, paucissimos dari qui te Vir Illustrissime omnino assequi
valeant, et qui te ubique adæquate capiant intelligantve perinde ac tu teipse.
Etenim licet hic mirabilis hujus 2da. Tabulæ, transmutationisque usam videam,
pro invenienda Area quantitatum quæ cum Ellipsi et Hyperbola possunt comparari, attamen, circa modum quo illa ipsa Tabulæ 2^{dæ}. transmutatio facta
fuerit nil video, præterquam quod id ope Theorematum 5.6.7. ejusque Corol-
lariorum nec non tui Probl[ema]tis 3^{tij}. idque quidem, ni valde fallor, improbo
calculo absolveris. Enimvero ubi proposita tibi est quædam Tabulæ 2^{dæ}.
Ordinata; ex: gr: Casus 1. Formæ 2^{dæ}. $\frac{dz^{\frac{1}{2}\eta-1}}{e+fz^\eta}$; qua quæso mirabili arte, Vir
Incomparabilis detexisti, ad faciendam illam transmutationem ponendum tibi
esse valorem x Abscissæ in Curva nova æquale $\sqrt{\frac{d}{e+fz^\eta}}$ et vel hoc etiam facto,
unde etiam tum tibi innotuit veram illius Aream fore $\frac{2xv \div 4s}{\eta}$;[34] Estque
ipsamet illa ars mirabilis quam Probl[ema]te 3^{tio}. Prop: 10 vocas, *Curvam
comparare cum simplicissimis.* Quod ibidem etiam dicis, transmutationem illam
usui fore, pro quantitatibus quadrabilibus, quæ statim neque ad Tabulæ 1^{mæ}.
neque ad Theor[ema]tis 3^{tij} vel 4^{ti} formulas referri poterunt, atque ipsas
beneficio mirabilis illius artis in aliis mutatum iri, quas deinde ad formulas
referre possimus, id equidem se ita habere debere cerno, sed nec ejus exemplum
video nec ipse præstare valeo. Illam etiam mihi subolet esse causam, cur acci-
derit mihi, ut nonnullas differentiales quantitates a priori formatas mihi pro-
posuerim, quæ sola incognitæ substitutione nulla ad formulas referri poterant,
proindeque etiam earum Integrales restituere non potuerim; nimirum, dum

pluribus compositæ sunt incognitis, sive Polynomio complexo constantes, idque mihi procul dubio minus ex voto successit, ut suspicor, quoniam Corollaria 7.8.9 et 10 Prop:nis 9æ et Problema 3^{tium} non plane penitusque intellexi, et quia scilicet pro ejusmodi exemplis necessario venit adhibenda ipsa mirabilis illa tua Curvarum transmutatio et inventio seu reductio ad curvas simplicissimas. Ita enim sequentium quantitatum Integrales reperire non potui.[35]

$$\frac{3x^4\,dx + 6abcx\,dx - 3abxx\,dx - 2ax^3\,dx - a^2bc\,dx}{x^6 + 2abx^4 - 2abcx^3 + aabbxx - 2aabbcx + aabbcc}\,;$$

$$\frac{2axy^2\,dx + cxy^2\,dy - cy^3\,dx - 2c^2y^2\,dy - 4ax^2y\,dx - 2ax^2y\,dy + 4ax^3\,dy + 8ccyy\,dx}{y^4 - 8xy^3 + 16xxyy}\,;$$

$$\frac{10cxxdx - \frac{4}{3}acx\,dx - 3aab\,dx + \frac{5}{2}b^3\,dx\,\sqrt[3]{aab - 2cxx}^{\,2}}{\sqrt{5b^3x + \overline{2a - 6x}^3\,\sqrt{aab - 2cxx}} \times \sqrt[3]{aab - 2cxx}^{\,2}}\,;$$

Interim tamen acquiesco; etenim quis quæso Lector unquam, mentem Autoris Infinitorum Infinita tantum non exhaurientis, se penitus funditusque capere glorietur; atque in eo nil nisi humani quid patior, imo vero nil quod mihi non commune ducam cum ceteris Mathematicis universis, Atque eo ipso Tractatus ille de Quadratura Curvarum suum apprime sapit Autorem. quippe cui operi accidat, idem, quod proprium sane tuis operibus omnibus esse solet, nimirum ut illa opera vel summi etiam Mathematici nunquam absolute, sed secundum quid modo capere valeant.

Cæterum Vir Admirande, obsecro te atque obtestor, ne qua nimiam forte luxuriantis epistolæ verbositatem mali consulas; id enim debitum quoddam est, quod tibi persolvendum dixi, quo tibi melius scilicet constaret, atque ipse gratissima devotissimaque mente profiterer, quo me nulli mortalium imitanda tua scripta evexerint. Quod autem ægerrime fero unum est, nimirum hocce; quod pro tot tantisque sublimioris Matheseos accessionibus quibus a te ditatum me video, nihil suppetat præter verba quæ redhostimenti loco offeram; illa tamen, utpote candidissimo e pectore tibi manantia non omnino te aspernaturum esse, pro æquitate, et ceteris tuis donis pari illa humanitate tua confideo. Imo obsecro V.I. ut te penes antiquum illud obtinere sinas; *Mola litet qui hecatomba nequit.*[36] Cæterum si cogitaveris his in terris pene jacere nobillissimas hasce artes, versatissimumque in sublimiori Mathesi hic esse neminem, tum equidem id tibi fidem faciet, nec ægre adduceris ut credas mihi, dum dixero; quod hac de causa, solo impotenti scientiarum illarum amore ductus, ἀυτοδίδακτος esse coactus sim, unde tibi minime videbitur mirum, si sub lectione operum tuorum, modo nondum intellecta anxius rimando, modo jam perspecta attonitus suspiciendo, sexcenties exclamaverim:

O felicem qui Newtoni pendet ab ore!

Quam felix horum potuit qui noscere causas ![37]

Et revera Vir Illustrissime tantam animo concepi tui, veri Philosophorum, Mathematicorumque Phœnicis, admirationem atque eo usque Te, acuminis monstrum veneror, ut, cum per invida fata plura negentur, majorem nullam nunc exoptaverim voluptatem, nilque mihi jam in ardentioribus sit votis præter hocce unicum; scilicet, quandoquidem Vir Incomparabilis; vultum tuum in depicta imagine, quæ de die nocteque mihi latus claudit, jam aspicio, quia etiam mentem animumque tuum in operibis tuis, mihi manibus continuo tritis jam suspicio et contemplor, enixissime rogaverim, ut et manum duntaxat tuum deosculari liceat, etiamsi illud solummodo foret in autographa aliqua tua schedula, qua scilicet me saltem faceres certiorem, utrum adhuc Londini veneat solum quod mihi deest tuum Opus, quodque hic terrarum nunquam appulit, cuique est Titulus. *Analysis per quantitatum series fluxiones ac differentias, cum Enumeratione linearum tertii ordinis Londinii ex Officina Pearsonij* 1711.[38] atque etiam an adhuc prostet eleganter a Maclaurino scoto conscriptus ille Commentarius[39] in tuum ultimum illum de Enumeratione Curvarum tertii ordinis Tractatum; denique numquid etiam vera sit illa quæ hic percrebuit fama, necdum te quieti indulgere, maxime alias provectæ illæ tuæ ætati competenti sed adhuc te esse occupatum in elaborando Opere quodam circa Chronologiam sacram,[40] teque etiam illud mare scopulosissimum fere esse emensum, est enim materia etiam illa quæ plurimis scatet adhuc nodis, tali vindice dignissimis, cui propterea merito inhiant atque intenti sunt, quotquot illa spe lactantur Eruditi. Hæc inquam, si mihi venia daretur, ut horum abs te fierem certior, ejus profecto beneficij memoria me penes non esset nisi cum ipso immortalis Newtoni nomine intermoritura. At denique merito ego vereor, ne tandem nimiæ sis meæ prolixitatis pertæsus, nec tamen prius ego figam pedem, quam te fuerim obtestatus, animitus me ad fata usque credas

Vir Illustrissime

Summi Tui Nominis

Observantissimum Studiosissimumque
Cultorem

PHILIPPE NAUDÉ JUNIOR.

P.S. Si aliqua me exoptatissimo cujusdam responsi mactare libeat honore, posset illud inscribi. *A Phillipe Naudé le fils, Professeur en Mathematiques au College Royal de Joachim et membre de la societé Royale des sciences de Berlin*

à Berlin

Translation

Berlin, 6 February 1723 [N.S.]

Most illustrious Sir,

I must assuredly confess myself deeply indebted to the very excellent and noble Mr Schott,[2] Ambassador at our Court from his Britannic Majesty, because I owe to his kindness the assurance he gave me that because of the extreme affability in manners you exhibit you would not take ill my addressing this letter to you; as also because he undertook to have the letter safely delivered to you. Hence it is clear that I shall most eagerly seize the opportunity of that honour (which I anxiously sought some time ago) now that I enjoy it, of expressing to you my sentiments of gratitude and devotion towards you on account of so many and so great additions to the higher mathematics with which I have been enriched through [reading] your never sufficiently admired works, and at the same time of signifying to you my supreme mental veneration, which throughout my life I shall devote to you and those remarkable gifts that God has pleased to bestow upon you, greater than those of any mortal of any age.

However, so that I may the more graphically describe how much I owe to you, most illustrious Sir, it should at once be stated that because my unhappy fate has never permitted me to visit your island (truly, for the reason that it rejoices in yourself, fortunate) it thence came about that as I was denied access to the actual discoverer and originator of the most recondite and mysterious branches of mathematics I was forced to draw [my knowledge of] the calculus of fluxions from a foreign spring, that is, the book[3] of the Marquis de l'Hospital, [a calculus] wholly owed to yourself and of which Leibniz attached to himself the credit by denominating it *differential*.[4] After this I saw and understood the laws of motion also, and those of the impact of bodies, demonstrated *a priori* by the learned Keill of pious memory in his *Lectiones Physicæ*[5] but by 'sGravesande and Mariotte proved *a posteriori*.[6] Thence as I was most eager to understand your astronomy as well, I first read the *Introduction*[7] to it written by the said Mr Keill so that thus it would be possible to gain access to that same astronomy of yours as explained at greater length by the famous Mr Gregory.

But when in reading Keill's *Astronomy*[8] I saw him to assert on p. 36 that that marvellous harmony, which Kepler discovered in the motions of all the planets *a posteriori* and by the collation of his observations was demonstrated by you *a priori* and geometrically, namely [Kepler's Third Law] that the squares of the periodic times are in all of them as the cubes of the mean distances from the Sun; and when in that most noble theorem I seemed to discern as it were some demonstration there of your whole system, then truly I burned with an impotent desire to understand the demonstration of this theorem at least and that drove me at last, most illustrious Sir, to attempt to read that incomparable work, which I have justly called the climax of human intelligence and which has tormented and exercised the most skilful geometers and astronomers of our age for thirty years, I mean your book of the *Principia*, which book I had indeed not yet dared to tackle. But there with indescribable joy I understood not only the demonstration of that theorem (which you discuss in Book I, Proposition 15) but also about one-third of the work, and I happened to discover there countless wholly new properties displayed with stupendous universality and worked out by immense geometrical power. This done I returned to Keill's *Astronomy*, and when I had reached on p. 375 the geometrical solution of Kepler's Problem,[9] of finding the true anomaly of a planet from the given mean anomaly (or, what is the same thing, the problem of cutting an ellipse or circle in a given ratio by means of a line drawn through a given point on the axis) I did not understand this solution because I found him to express the arc [of the anomaly] by means of an infinite series; for which reason when I had examined what the famous Gregory has to say on p. 211 of his *Astronomy*[10] concerning this same solution and found that he there directed the reader, seeking for the doctrine of series (which was hitherto unknown to me), to the *Exercitatio Geometrica de Dimensione Figurarum*[11] published by him at Edinburgh in 1684, I read that eagerly also and understood it too, as soon as I had become acquainted with its first Lemma[12] on which everything is based; for that is a unique Lemma, but one that with a wonderful universality makes plain the theorem demonstrated by your most celebrated Mr Wallis in his *Arithmetica Infinitorum*[13] which is also provided by the very famous Mr Barrow in his exceedingly elegant demonstrations of the theorems of Archimedes concerning both the surface and the volume of the sphere and its parts, which he annexed to the end of his *Euclid*;[14] then at length I understood how when the arc is given, its right and versed sines, cosine and tangent may be expressed by an infinite series, and conversely how, when each of these is given, the arc may be expressed, and how, through the same, both the areas of geometrical curves and their volumes generated by revolving the areas about axes of revolution are obtained; and also how the value of any root whatever of any power from any binomial may be expressed in an infinite series. As these things are not sufficient for a complete understanding of that solution of Kepler's Problem, but I saw that this was to be obtained at last from your wonderful inverse method, that drove me to tackle those two famous letters of yours, most illustrious Sir, addressed to Mr Leibniz in 1676, in the third volume of Wallis's *Commercium Epistolicum*[15] where you disclosed such things as were not only then unknown to Leibniz himself, but such as I can readily believe he would never have investigated thoroughly to their foundations. I can only with great difficulty be brought to believe that he thoroughly understood the demonstrations of those marvellous theorems by means of which you teach that stupendous inversion and find the series expressing the area of a curve from its given ordinate and, conversely, find the series expressing the ordinate of a curve from its given area; nor can I more easily convince myself that Mr Leibniz appreciated by what wonderful art you, most illustrious Sir, became master of the genesis of coefficient numbers for that series for the elliptical arc (*Commercium Epistolicum*, p. 626) and for that which occurs on p. 627 by which you express the volume of a portion of an elliptical spheroid. It is true that Mr Gregory also discovered both those same series (pp. 33 and 44 of his aforesaid *Exercitatio Geometricae*) but he did not disclose that genesis of the coefficients by which with trifling labour the series is formed, and can be made to approximate as closely to the truth, as one wishes.[16] The doing of this in all series is in the most solemn sense yours alone.

As for my assertion that Mr Leibniz did not thoroughly grasp all the wonderful discoveries contained in those two letters, Dr Keill furnished me with the evidence for it. For Mr Bernoulli, Mr Leibniz's pupil, learned the differential calculus from him, but not the more difficult integral calculus, whence Mr Bernoulli wished to claim this for himself also, and seems to make a great fuss because he discovered the integral of the differential quantity $(b^2y - a^3)^{\frac{1}{2}} . dy$.[17] But it stands to reason that if Mr Leibniz had thoroughly assimilated the methods and practices of these letters, he would surely have perceived that the integral not only of the quantity $(b^2y - a^3)^{\frac{1}{2}} . dy$ but of many others much more complex may be recovered by means of your first theorem alone, which you had disclosed to him. Because if the ordinate of a curve may be expressed by $dz^\theta(e + fz^\eta)^\lambda$ then its area is to be expressed by an infinite though converging series[18]

$$Q\left(\frac{z^\pi}{s} - \frac{r-1}{s-1}\cdot\frac{eA}{fz^\eta} + \frac{r-2}{s-2}\cdot\frac{eB}{fz^\eta} - \frac{r-3}{s-3}\cdot\frac{eC}{fz^\eta}\cdots\right)$$

However, if he had seen that, surely Mr Leibniz would have turned that to his own profit and then it would hardly have been concealed from his pupil Mr Bernoulli. Moreover, to speak the truth, it is hardly credible to me that Mr Bernoulli is in matters of trust either so heedless or so cunning as to congratulate himself so much on so slight a thing as the recovery of the integral of that quantity $(b^2y - a^3)^{\frac{1}{2}} dy$, in order to claim the discovery of the integral calculus for himself, if he had thoroughly perceived the usefulness of this first theorem of yours; and I am convinced he would have been the less bold so to do if he had known how to recover that integral in the manner taught by Dr Keill in the French *Journal* [*Literaire*] *de la Haye* for 1719 p. 265;[19] that is, by applying to it the remarkable formulae of the first table in your *De Quadratura Curvarum*.[20] To make a candid admission, that one example there brought forward by Dr Keill fills me with exquisite delight for by that example he teaches me that it is for you the same thing, either to give the quadrature of a curve or to recover the integral of a quantity, which I had never yet known to be truly your idea, and is the very simplest notion of the whole business.[21]

Having this clue, I was inspired by Mr Keill to read that treatise *De Quadratura Curvarum*, which embraces both the differential and the integral calculus, and to annotate virtually the whole work, and I would most truly have asserted that even if that integral calculus or calculus of quadratures of yours were not the actual key with which in the second book of your *Principia* you unlocked the solutions of most subtle mathematical–physical problems, unyielding to all others before yourself; even though, as I say, this were not by far the most remarkable fruit of that integral calculus of yours, and though it were even of no utility at all, still I should not hesitate to say that no more divine discovery has been or will be vouchsafed to human wit. Nor can it be, but that whoever understands those two wonderful series[22] of Theorems 3 and 4 will find his mind exalted in high admiration; and he will see that they do eminently contain the integrals of any differential quantity composed of one or two polynomials of any degree and including any number of terms. Or again, when one has rightly understood that transformation of curves (of almost superhuman elegance) upon which you have founded those formulae of immense universality, never sufficiently to be praised, of your second table of non-quadrable curves, or of quantities whose integrals can be expressed in no finite number of algebraic terms. It is a most certain thing, that I have never in my whole life seen anything for which I have felt more deeply grateful, than for these things that I have understood [from you].

234

For, most illustrious Sir, as to that theorem, so far as I have examined it, as when I have seen you make the assertion (at the end of Theorem 4) that 'the series of such propositions might be continued to infinity'; hence, in imitation of yourself, I have developed an infinite series expressing the ordinate of a curve whose area would be $z^\theta R^\lambda S^\mu T^\omega$ and then also another series which expresses the area of a curve whose ordinate is

$$z^{\theta-1}R^{\lambda-1}S^{\mu-1}T^{\omega-1}(a+bz^\eta+cz^{2\eta}\ldots)$$

by means of the same series I have restored the integral of this quantity;[23]

$$(\tfrac{1}{8}a^3b^4.dy - \tfrac{1}{8}a^2bc^2d^2.dy + \tfrac{2}{3}a^2b^4y.dy - \tfrac{5}{2}a^2b^3cy.dy - \tfrac{2}{3}abd^2c^2y.dy + \tfrac{5}{2}ad^2c^3y.dy$$

$$-\tfrac{7}{2}ab^3cy^2.dy + \tfrac{7}{2}c^3d^2y^2.dy - \tfrac{13}{6}a^2b^2y^3.dy$$

$$-\tfrac{17}{6}ab^2y^4.dy + \tfrac{19}{2}abcy^4.dy + \tfrac{23}{2}bcy^5.dy).\,[(ab-3cy)^{\tfrac{1}{2}}.(ay+y)^{\tfrac{2}{3}}.(ab^3-c^2d^2-2by^3)^{\tfrac{3}{4}}]^{-1}.$$

This I have accomplished, actually, while maintaining a positive value of η, in six different ways, that is to say according as the value of the polynomials $R^{\lambda-1}$, $S^{\mu-1}$, $T^{\omega-1}$ may be variously combined; and likewise in six different ways more, that is to say by giving an expression for this quantity in which the value of η in the formula is negative, which you will see the Bernoullis and the L'Hospitals together have all attempted in vain. Afterwards too I found that many different quantities, which at the first glance seemed hardly reducible to this theorem of yours nor to any formula of your first table, could be reduced in that way by means of a trivial substitution of another unknown [quantity]; for thus I discovered the integrals of these quantities by means of your formulas:

$$\frac{\tfrac{1}{2}abca^{\tfrac{1}{2}}c^{\tfrac{1}{2}}.dy}{b^2ya^{\tfrac{1}{2}}c^{\tfrac{1}{2}}-2abyc^{\tfrac{1}{2}}y^{\tfrac{1}{2}}+ay^2a^{\tfrac{1}{2}}c^{\tfrac{1}{2}}},\;{}^{.(24)} \qquad \frac{6ax^4.dx+4bbx^3.dx}{4a^2x^2+4abbx+b^4}\;{}^{(25)}$$

$$\frac{-2a^2bcx.dx-c^4.dx}{a^4b^2-2a^2bx^3+x^6}\;{}^{.(26)}, \qquad \frac{ax^2.dx+ab^2x.dx}{a^2x^2+2ab^2x+b^4}\;{}^{.(27)},$$

and also this quantity

$$\frac{\tfrac{3}{2}ab^2x.dx+2acx(ac-x^2)^{\tfrac{1}{2}}.dx}{(a^4bc+2a^2c^2x^2-a^3bx^2-2acx^4-3ab^2(ac-x^2)^{\tfrac{3}{2}})^{\tfrac{1}{2}}}$$

containing a complex polynomial is made by substituting $y^{\tfrac{1}{2}}$ for $(ac-x^2)^{\tfrac{1}{2}}$ to become another expression, there taking on a form not having a complex polynomial; because this could then be reduced to the formula of your third theorem, the series converged and the sought-for integral was thus restored for me.[28]

Having understood all this I tackled also the second of your tables of curves comparable with ellipses and hyperbolas, for the preparation of which you supplied those wonderful Theorems 5, 6 and 7 with their corollaries and also the third problem. This is actually that marvellous method of transformation towards which Craig confesses he has achieved little, whatever his achievement in this area, nor does he speak unjustly on page 60 of his *De calculo fluentium*.[29]

That second table did at the first create more than a little trouble for me, but I understand it so far in the end as to see in it what each formula of that second table also puts before me, for example the first case of Form 8. $dz^{\eta-1}/(e+fz^\eta+gz^{2\eta})^{\tfrac{1}{2}}$ if,[30] as you instruct

z^η is supposed for the sake of the transformation of the form to be equal to the abscissa x of the new curve determined by the equation $(e+fx+gx^2)^{\frac{1}{2}} = v$, then by applying your Theorem 7 as is obvious, I can easily demonstrate that the integral area of the quantity $dz^{\eta-1}/(e+fz^\eta+gz^{2\eta})^{\frac{1}{2}}$ is clearly (as you say), $(8dgs-4dgxv-2dfv)/(4\eta eg-\eta f^2)$, which is by a trifling effort shown to be equal to $(8dg/(4\eta eg-\eta f^2)) \cdot (\alpha GDB \pm DBA)$; that is

$$(8dg/(4\eta eg-\eta f^2)) \cdot (\alpha GDB + DBA)$$

if the absiscissæ x are in the direction of the centre of the curve, but equal to

$$(8dg/(4\eta eg-\eta f^2)) \cdot (\alpha GDB - DBA)$$

if the abscissae x are in the opposite direction in your Figures 6, 7 or 8. For the equation $(e+fx+gx^2)^{\frac{1}{2}} = v$ can be the locus of the ordinates either to the first diameters of innumerable hyperbolas as in Figure 7 or to as many second diameters as in Figure 8 or to the diameters of ellipses or circles as in Figure 6, according to whether the values of the given quantities e, f, g are positive, zero, or negative. Whence it plainly follows that every ordinate or differential quantity whose integral can be expressed in no finite number of algebraic terms, if it suffers itself to be reduced to any formula of this table, then certainly its sought-for area or integral will be shown there geometrically, and expressed by a determinate portion of a given ellipse, hyperbola or other conic of any higher order whatever. This is so sublime a discovery and of so wide-ranging utility that in my opinion it may be justly held to be the miracle of miracles. For if, for example it is proposed to find the integral of the quantity $4b\,dy(a^2c^2+a^3b-b^2y^2-acy^2)^{\frac{1}{2}}/ay^2$; allowing first that $dy = 1$,[31] I see that this quantity can nowise under this form fall under any case of the table; but as this form can be changed in innumerable ways, I divide both [top and bottom] terms of this fraction by ay, and the expression becomes $\dfrac{4b}{y}\left(\dfrac{-b^2-ac}{a^2}+\dfrac{c^2+ab}{y^2}\right)^{\frac{1}{2}}$, which I see can now certainly be related to the second table; [that is], to the first case of the third form $(d/z)(e+fz^\eta)^{\frac{1}{2}}$ by taking $z = y$; $d = 4b$; $e = (-b^2-ac)/a^2$; $f = c^2+ab$; $\eta = -2$. And when it is there postulated that $x^2 = z^{-\eta}$, it appears that $x^2 = 1/y^{-2} = y^2$ and since e in $(f+ex^2)^{\frac{1}{2}} = v$ here receives a negative value, and f is positive, the equation describes an ellipse, which is your Figure 6 whose diameter Aa would be $(f/e)^{\frac{1}{2}}$ and the ratio of that to its parameter will be as 1 to e.[32]

And since $(f+ex^2)^{\frac{1}{2}}$ is here equal to $(c^2+ab-(b^2y^2-acy^2)/a^2)^{\frac{1}{2}}$, this shows us that the locus is also to an ellipse whose diameter is $a((ab+c^2)/(ac+b^2))^{\frac{1}{2}}$ and the ratio of its diameter to the latus rectum[33] is as a^2 to $(ac+b^2)$. But from the formula of the table the area or integral is $(4de/\eta f)(v^3/2ex-s)$ or $2dv^3/\eta fx-4des/\eta f$ which teaches us that for our values the sought-for area or integral of the quantity proposed is

$$\frac{-4b}{c^2y+aby}\left(\frac{-b^3y^2-acy^2}{a^2}+c^2+ab\right)^{\frac{3}{2}}$$

after the product $((8b^3+8abc)/(a^2c^2+a^3b)) \cdot s$ (that is the product $(8b^3+8abc)/(a^2c^2+a^3b)$ by the elliptical trapezium $ABDP$ of your Figure 6, determined in the way already mentioned) has been subtracted. A solution, whose truth it was easy to demonstrate *a posteriori* by the differential calculus, but it is far from being the case that Mr Leibniz himself (if he had lived), nor Bernoulli nor any other of all his disciples could ever have disclosed it, so that it should seem to every one of them rather a pronouncement of the Delphic oracle or the work of some Œdipus. Moreover, it also appears to me, most illustrious Sir, that those wonderful formulas are likewise useful for finding areas when the second fluxions are given, not to say the third and fourth fluxions *ad infinitum* so that in that little treatise you may surely be said to have truly exhausted infinity and to understand infinite numbers. I would readily have believed that there are few who are altogether able to follow your steps, most illustrious Sir, and who everywhere take and understand your meaning as you do yourself. For although I see here the usefulness of this marvellous second table and of transformation for finding the areas of quantities which can be compared with the ellipse and hyperbola yet I see nothing of the way in which that transformation itself of the second table was performed, except that it [was done] by means of Theorems 5, 6 and 7 and their corollaries as well as your third problem, and that if I am not much mistaken you did it by a monstrous computation. For when you addressed yourself to certain ordinates of the second table, for example Case 1 of Form 2, $dz^{\frac{1}{2}\eta-1}/(e+fz^\eta)$, by what extraordinary art, I ask, did you find out, incomparable man, that to make that transformation you had to take the value of the abscissa x in the new curve to be equal to $(d/(e+fz^\eta))^{\frac{1}{2}}$ and even when this was done how did you then know that the true area of that [curve] will be $\left|\dfrac{2xv-4s}{\eta}\right|$?[34] That is the very same wonderful art which in the third problem, Proposition 10, you call 'To compare a curve with the most simple ones'. What you also affirm in the same place, that that transformation will be of use for quadrable quantities which can be immediately related neither to the formulas of Table 1, nor to those of the third and fourth theorems, and that these with the aid of that marvellous art can be changed into others which we are then able to relate to the formulas, I see as something that ought to be so, but I do not see an example of it nor can I furnish one myself. This suggests to me the reason why it has happened to me to propose to myself several differential quantities formed *a priori* which could by no single substitution of an unknown be related to the formulas, and hence I could not recover their integrals either. Truly, when they are composed from many unknowns, or constitute a complex polynomial, doubtless the process has gone badly for me (or so I suspect) because I have understood neither completely nor clearly Corollaries 7, 8, 9 and 10 of Proposition 9 as well as Problem 3, and because for instances of this sort it becomes necessary to apply that marvellous transformation of curves of yours, and the discovery of or reduction to the most simple curves. For thus I could not recover the integrals of the following quantities[35]

$$\frac{3x^4.dx+6abcx.dx-3abx^2.dx-2ax^3.dx-a^2bc.dx}{x^6+2abx^4-2abcx^3+a^2b^2x^2-2a^2b^2cx+a^2b^2c^2};$$

$$\frac{2axy^2.dx+cxy^2.dy-cy^3.dx-2c^2y^2.dy-4ax^2y.dx-2ax^2y.dy+4ax^3.dy+8c^2y^2.dx}{y^4-8xy^3+16x^2y^2};$$

$$\frac{10cx^2.dx-\frac{4}{3}acx.dx-3a^2b.dx+\frac{5}{2}b^3(a^2b-2cx^2)^{\frac{2}{3}}.dx}{(5b^3x+(2a-6x)^3.(a^2b-2cx^2)^{\frac{1}{2}})^{\frac{1}{2}}.(a^2b-2cx^2)^{\frac{2}{3}}}.$$

Meanwhile, however, I accept the situation; for, I ask, what reader can boast of a complete and fundamental understanding of the mind of a writer who has all but exhausted the infinity of infinities, and in this respect I suffer nothing that is not common to all humanity, indeed nothing which I do not share with the whole race of mathematicians. And to that extent that treatise *De Quadratura Curvarum* absolutely takes after its author. Indeed, it happens with that work as with all your works, that is to say that that book is never completely understood even by the finest mathematicians, but each one is able to understand it to some extent.

For the rest, I beseech and supplicate you, excellent Sir, not to take amiss the verbosity of this perhaps overgrown letter; for this in a measure arises from my laying before you matters to be resolved, in trying to make them clearer to you, and from my making profession of the grateful and devoted feelings inspired in me by your writings, which no mortal man can imitate.

One thing only is a cause of grief to me, namely this: that I can have no recourse but to words in offering a return to you for so many and so great advances in the higher mathematics by which I find myself enriched thanks to you; these, however, as swelling towards you from the sincerest of hearts you will not wholly spurn, because I am confident that fairness is as much among your gifts as kindness. I especially conjure you, most illustrious Sir, to cling dearly to the proverb: 'He offers grain as sacrifice who has no oxen'.[36] Further, if you have supposed that these noble arts had almost perished in these lands, and that no-one here was really skilled in higher mathematics, then indeed it will make you believe me, nor will it displease you to believe me, when I say that for this reason, spurred on solely by an impotent love of those branches of learning, I was compelled to be self-taught; whence to you it will seem no miracle if in reading your work, now anxiously delving into what is not yet understood, now regarding with astonishment [the truths] I had perceived, I have exclaimed a thousand times:

O happy he who hangs on Newton's words!

How fortunate to be of those who know the causes![37]

And indeed, most illustrious Sir, I have formed in my mind so great an admiration of you as the truest of philosophers, the Phoenix of mathematicians, that I venerate you as a prodigy of intelligence, to the point where (though many things are denied by envious destiny) I could choose no greater delight and none more ardently longed for, than this that, in as much as I now behold your face, incomparable Sir, in a painted likeness, which I keep close by me day and night, and because I glimpse and contemplate your mind and spirit too in your works, continually turned over by my fingers, I most earnestly beg you that I may at least kiss your hands even though it be merely by some autograph note by which you may at any rate inform me whether the only one of your works that is lacking to me and which has never been brought to these lands, is still on sale in London; this is its title: *Analysis per Quantitatum Series* [etc.] *1711*.[38] And also whether the elegantly composed *Commentary* upon that most recent treatise of yours, *Enumeratio curvarum tertii ordinis*,[39] by the Scott Maclaurin, is still on sale; and lastly whether the rumour which has been spread abroad here is true, namely that you have not yet retired to rest but that far extending your working life into another age, you are still engaged in perfecting some work on the sacred chronology[40] and that you have also almost traversed that most perilous ocean, for this is a subject too that abounds in many puzzles, worthy of such an arbiter, upon whom moreover those scholars rightly gaze who are flattered by that hope. This I say, if I may be forgiven, in order to have information from you on these points, surely the memory of the kindness of it will only desert me when the name of the immortal Newton shall have passed away. But lastly I have reason to fear that I have at last fatigued you with my excessive prolixity, yet I cannot take my leave before I have assured you that you may heartily believe me, till death, most illustrious Sir, your most devoted and zealous admirer, Philippe Naudé Jnr.

P.S. If you choose to honour me with a much-desired reply of some sort, it may be addressed thus:...

NOTES

(1) MS. MM 5, 50. Philippe Naudé (1684–1745), son of the Berlin mathematician and theologian of the same name, was educated at the Joachim College in Berlin, and intended for the Church. However, it became clear that his abilities lay in the direction of mathematics, and he succeeded his father as Professor of Mathematics, first at the Langerfeld Academy of Arts (1707) and then at the Joachim College (1708), where he remained until his death. In 1711 he was elected to the Académie Royale des Sciences et Belles-Lettres of Berlin, and in 1738 he became a Fellow of the Royal Society. He published a number of mathematical papers in the *Mémoires* of the Berlin Académie, and left in manuscript a *Commentaire sur les Principes de Newton*.

The function of the present prolix letter is not altogether clear; Naudé merely praises Newton's mathematical prowess and contrasts it to his own poorer talents. The mathematical examples he gives in the letter show a very meagre understanding of Newton's work.

(2) Possibly George Scott of Bristo in Scotland, who held diplomatic offices at a number of German Courts, and was a great friend of George I, after whom he named his more famous son, George Lewis Scott.

(3) *Analyse des infiniment petits* (Paris, 1696); see Letter 1008, note (4) (vol. VI, p. 15).

(4) Later in his Letter Naudé mentions that he has seen Keill's 'Lettre à Bernoulli', *Journal Literaire de la Haye*, **10** (1719), 261–86 (see vol. VI, p. 386, note (4)). It seems likely that this was the source of his present comment.

(5) *Introductio ad Veram Physicam, seu Lectiones Physicæ...* (Oxford, 1702, 1705). Lectures XII–XIV deal with the impact of bodies.

(6) Edmé Mariotte, in his *Traité de la percussion ou choc des corps*, reported to the Académie Royale des Sciences at Paris in 1671, first published in Paris in 1673, and modified in later editions, and W. J. 'sGravesande, in his *Physices Elementa Mathematica, Experimentis Confirmata* (Leyden, 1720, 1721), both discussed the impact of bodies with detailed reference to experimentation.

(7) Naudé refers to Keill's *Introductio ad Veram Astronomiam seu Lectiones Astronomicæ* (Editio Secunda, multo Auctior & Emendatior, London, 1721).

(8) Keill's *Introductio ad Veram Astronomiam*; see note (7) above. On p. 37 (not p. 36) Keill wrote of Kepler's Third Law, 'Gloriam illam a priori investigandi & illius causam ex necessitate Physica monstrandi, magno Newtono nostro reservata fuit, qui demonstravit salvis naturæ legibus, aliam regulam in mundo locum obtinere non posse.' ('The glory of investigating it *a priori*, and of showing the cause of it from physical necessity, has been kept for our great Newton, who demonstrated that no other relationship could have place in the Universe, given the laws of nature.')

(9) Keill dealt with Kepler's Problem on pp. 363–80, first giving his own solution, and then one derived from Newton. *Inter alia*, on p. 377 he stated the expansion

$$\sin(\theta + x) = \sin\theta + x\cos\theta - \frac{x^2}{2!}\sin\theta + \ldots$$

without offering any explanation, but referring the reader to his *Trigonometriæ Planæ et Sphericæ Elementa* (Oxford, 1715). He then stated x in terms of sine θ, mentioning that Newton's method of reversion was the method by which this was obtained.

(10) David Gregory, *Astronomiæ Physicæ & Geometricæ Elementa* (Oxford, 1702). Keill had strongly recommended this work in the Preface to his *Introduction ad Veram Astronomiam*.

(11) David Gregory, *Exercitatio Geometrica de Dimensione Figurarum*... (Edinburgh, 1684).

(12) The Lemma, on p. 6, stated that 'Quavis recta in partes æquales innumeras discerpta, summa quarumvis dignitatum, ab innumeris istis rectis ab extremitate propositæ rectæ continuo incipientibus, genitarum, æqualis est rectæ propositæ potestati, quæsitis potestatibus proxime superiori, divisæ per suum exponentem.' ('Any straight line having been divided into innumerable equal parts, the sum of any powers generated from those innumerable straight lines, beginning successively from the end of the given line, is equal to the power of the given straight line, next above the sought for power, divided by its exponent.') He then explains this geometrically, equating the sum of the powers with the area under the graph (but giving no proof) and gives a number of examples.

(13) First published in 1655. Naudé probably refers to the version published in John Wallis' *Opera Mathematica*, I (Oxford, 1695), pp. 355–478.

(14) *Euclidis Elementorum Libri xv Breviter Demonstrati*, edited by Barrow, was first published in 1655. In 1657 a separately paginated section entitled *Euclidis Data Succincte Demonstrata* was added. The whole was republished in 1678 together with a third section, that referred to here, entitled *Lectio Reverendi et Doctissimi Viri D. Isaaci Barrow...in qua Theoremata Archimedis De Sphæra & Cylindro per Methodum Indivisibilium Investigata Exhibentur*.

(15) Wallis' *Commercium Epistolicum* was first published in 1658, and then again in his *Opera Mathematica*, II, pp. 757–860, but this does not contain the letters Naudé refers to here. Clearly he means the section 'Epistolarum quarundam collectio, rem Mathematicam Spectantium, Nunc primum edita', occupying pp. 615–708 of Wallis' *Opera Mathematica*, III. The Epistola Prior, the letter to which Naudé refers here, is printed on pp. 622–9. (See also vol. II, Letter 188.)

(16) Gregory gave, on pp. 20–1 of *Exercitatio Geometrica*... (see note (11) above), a method of extracting the root of $c^2 + x^2$ from first principles, and stated, without giving further examples, that the method could be extended to other roots. The series Gregory gives on pp. 33 and 44 are simply the integrals of other series which Gregory has formed. Hence it is not quite clear what Naudé means here. Either he intends to say that he could not generalize Gregory's method, but could only use Newton's algorithm for the binomial expansion when it became available to him, or he has missed the point completely, not realizing that the series here given were produced by simple integration, using the rule given in the Lemma.

(17) See Keill's 'Lettre à Bernoulli' (see note (4) above). On p. 264 of the paper Keill discusses a point raised by Johann [I] Bernoulli in his 'Epistola pro Eminente Mathematico' (see Letter 1196, vol. VI), concerning $\int (b^2 y - a^3)^{\frac{1}{2}} \cdot dy$ which had been treated by Jakob Bernoulli in a paper in the *Acta Eruditorum* for 1960, p. 218, using a method borrowed from his brother Johann.

(18) Naudé quotes the series from the Epistola Prior; see vol. II, p. 115.

(19) In Keill's 'Lettre à Bernoulli'; see vol. VI, p. 386, note (4).

(20) The integral is the special case of Form 3 of the first table of integrals in Newton's *De Quadratura Curvarum*, when $\eta = 1$.

(21) Keill wrote of the integral, 'l'on aura l'Aire de la Courbe, ou selon vous [Johann [I] Bernoulli], l'Integrale de votre quantité, telle que votre frere ou vous l'avez trouvée.'

(22) Naudé refers again to Newton's *De Quadratura Curvarum*.

(23) The expression may be integrated in the case of $\eta = 1$ by putting

$$z = y; \quad R = ab - 3cy; \quad S = a + y; \quad T = -2b + (ab^3 - c^2d^2)y^{-3}; \quad \theta = \tfrac{1}{3}; \quad \lambda = \tfrac{1}{2};$$
$$\mu = \tfrac{1}{3}; \quad \omega = \tfrac{1}{4}.$$

Alternatively, if $\eta = -1$, then

$$z = y; \quad R = -3c + aby^{-1}; \quad S = 1 + ay^{-1}; \quad T = -2b - (ab^3 + c^2d^2)y^{-3};$$
$$\theta = \tfrac{23}{12}; \quad \lambda = \tfrac{1}{2}; \quad \mu = \tfrac{1}{3}; \quad \omega = \tfrac{1}{4}.$$

It is not quite clear what Naudé means by 'sex variis modis'; perhaps he means the six different powers of z (z^0, z^η, $z^{2\eta}$ ... $z^{6\eta}$) which appear when $Z^\theta R^\lambda S^\mu T^\omega$ is differentiated.

On integration, the expression $y^{\frac{1}{3}}(ab - cy)^{\frac{1}{2}}(a+y)^{\frac{1}{3}}(ab^3 - c^2y^2 - 2by^3)^{\frac{1}{4}}$ is obtained. It seems reasonable to suppose that Naudé proceeded by choosing this function, differentiating it to obtain the expression given, and then testing the method by integrating. Of course, the fact that the integral is not multiplied by a complicated polynomial in y makes the case a very simple one, lacking the generality of Newton's Propositions 5 and 6 in *De Quadratura Curvarum*, of which this is intended as an extension.

In this and the mathematical expressions which follow, Naudé uses a greek δ and roman d interchangeably. He draws no distinction between d as signifying differential, and d used as a constant.

(24) The expression reduces to $\tfrac{1}{2}abc^{\frac{1}{2}}.dy/y^{\frac{1}{2}}(a^{\frac{1}{2}}y^{\frac{1}{2}} - b)^2$ which may be simply integrated by putting $R = a^{\frac{1}{2}}y^{\frac{1}{2}} - b$, $\lambda = -1$ and $\theta = \tfrac{1}{2}$, and proceeding as usual, to give $- ac^{\frac{1}{2}}y^{\frac{1}{2}}(a^{\frac{1}{2}}y^{\frac{1}{2}} - b)^{-1}$.

(25) Again this is a very simple case; the integral is $x^4/(2ax + b^2)$.

(26) Naudé probably intended to write

$$\frac{- 2aabcx\,dx - cx^4\,dx}{a^4bb - 2aabx^3 + x^6}.$$

On integrating this gives $cx^2/(x^3 - a^2b)$.

(27) The second term in the numerator is clearly wrongly copied, as it has too few dimensions. Integration of this function produces an infinite series, which is clearly not what Naudé intended. It is possible that he meant to write,

$$\frac{axx\,dx + 2bbx\,dx}{aaxx + 2abbx + b^4}$$

which integrates to give $x^2/(ax + b^2)$.

(28) The substitution Naudé suggests does indeed lead to a function of the type Newton integrates in Theorem 3, and the series yields an exact integral after the substitution has been made.

(29) John Craige, *De Calculo Fluentium Libri Duo* (London, 1718). On p. 60 Craige explains the usefulness of transforming curves in order to integrate. He notes that he himself in his *Tractatus Mathematicus de Figurarum Curvilinearum Quadraturis et Locis Geometricis* (London, 1693) gave a few simple examples, but did not achieve a general method of the kind Newton gives.

(30) Naudé here, rather pointlessly, simply paraphrases Form 8 from Newton's *De Quadratura Curvarum*.

(31) *dy* is, of course, the Leibnizian differential, here made a unit-indivisible.

(32) *Aa* is the *semi*-diameter, strictly speaking equal to $(f/-e)^{\frac{1}{2}}$. Naudé means by the 'parameter' of the ellipse its latus rectum.

(33) The semi-diameter, note (32) above.

(34) Naudé, following Newton, uses the sign ÷ to indicate that the lesser quantity should be subtracted from the greater.

(35) There seems to be no good reason why Naudé should choose to attempt these particularly unwieldy integrals.

(36) Apparently a variation on Pliny's proverb, 'Mola tantum salsa litant, qui non habent tura' ('Those who have no incense make sacrifice of grain and salt'); see C. T. Lewis and C. Short, *A Latin Dictionary* (Oxford, 1879), under *lito*.

(37) The second line echoes Virgil, *Georgics*, 2, 490.

(38) The *Analysis per Quantitatum Series, Fluxiones, ac Differentias*, edited by William Jones, was first published in 1711, and was to be appended later in this same year, 1723, to the pirated Amsterdam reprint of the second edition of the *Principia*.

(39) Naudé mistook the name, meaning no doubt Stirling's *Lineæ Tertii Ordinis Newtonianæ* of 1717 (see p. 55, note (1), above).

(40) Presumably Naudé had heard the rumours spread by Conti, and perhaps even seen a manuscript copy of the pirated *Abrégé de Chronologie* (see Letter 1436, note (2)).

1406 AUGUSTINE TAMPYAN TO NEWTON
3 APRIL 1723
From the original in the Bodleian Library, Oxford[1]

Aprill: 30: 1723

Hon[ou]red Sr

According to your desire, these waits on you by Cou[sin] Hurst with an Acc[oun]t of what money I lent to Mr. Pilkinton:[2] vizt: twenty pound, of wch I have Received five pound—A Favour wch I could never have asked nor should have ever writ about, had not your Goodness proposed it, but I should have sat down with ye same thoughts as I allwayes had with patience—neither durs[t] I have presumed to have mention'd again my Tobacco, had not you been pleased to have made enquiry after [it] wch indeed grows now very low. I have only to add my unspeakable Thanks, wth my sincere wishes for your Health & continuation of it, and Remain,

Honred Sr, Your most obedient
humble servt & Affectionate
kinsman—AUG: TAMPYAN

NOTES

(1) Bodleian New College MS. 361, II, fo. 65. The writer was presumably Augustine Tampyan, husband of Newton's niece Elizabeth, the daughter of William Ayscough, his mother's eldest brother (see Appendix II, Newton's Genealogy).

(2) Presumably Thomas Pilkington, Tampyan's cousin.

1407 MARY AND ANN DAVIES TO NEWTON

28 APRIL 1723

From the original in the University Library, Cambridge[1]

Cambridg. April ye 28. 1723

Honoured Sir

I have made bould to troubel your Honour with these few lines to return your Honour thankes for the too ginnes[2] that your Honour whas pleas'd to send ous by the gentelman that whated[3] upon your Honour with the letter[.] Honoured Sir we hope your Honour will pardon our rudeness in not riting before but my Mother and I have bin very bad and that was the caus of our not riting to return your Honour thankes before now[.] this being all at presant but our duty to your Honour and our continual prayers to God of your Honours health and long life in which we remain your Honours humbel Obedient and dutiful sarvants to command

MARY and ANN DAVIES

For
ye Honoured Sir Isaac Newton at
his Hous in Saint
Martens Streat near
Lester Squair
London

NOTES

(1) Add. 3965(17), fos. 650 and 652. A possible hypothesis is that Mrs Mary Davies (if she was the mother), or else her husband, had been Newton's servant at Trinity.

(2) *Read:* Two guineas.

(3) *Read:* Waited.

1408 BURCHETT TO NEWTON
2 JULY 1723
From a copy in the Public Record Office[1]

2 *July* 1723

Sr:

The Bearer hereof Mr: Zachariah Williams[2] having represented to my Lords Commiss[ione]rs of the Admiralty that he hath something to offer relating to the Discovery of the Longitude at Sea. Their Lordships desire you will please to give him an Opportunity of communicating his scheme to you, and they may have your Opinion of his Project.[3] I am

Sr:

Your most &c

J BURCHETT

Sr: Isaac Newton.

NOTES

(1) Adm/2, 456, p. 412.

(2) According to E. G. R. Taylor, in *The Mathematical Practitioners of Hanoverian England* (Cambridge, 1966), pp. 148–9, Zachariah Williams (1672–1753) came to London about 1726 to try for the longitude prize; the present letter indicates that his visit was somewhat earlier. Williams' account of his method was published posthumously, under the care of Samuel Johnson, as *An Account of an Attempt to ascertain the Longitude at Sea, by an exact Theory of the Variation of the Magnetic Needle, with a Table, etc.* (London, 1755). In it Williams describes his application to the Board of Longitude, although the date he gives for the event leads us to suspect his accuracy:

> About the year 1729, my Subscribers explained by Pretensions to the Lords of the Admiralty, and the Lord *Torrington* declared my Claim just to the Reward assigned in the last Clause of the Act to those who should make Discoveries conducive to the Perfection of the Art of Sailing. This he pressed with so much Warmth, that the Commissioners agreed to lay my Tables [of magnetic variation] before Sir *Isaac Newton*, who excused himself, by reason of his Age, from a regular Examination: But when he was informed that I held the Variation at *London* to be still encreasing; which he and the other Philosophers, his Pupils, thought then to be stationary, he declared that he believed my System visionary.

We have found no account of these proceedings in the Admiralty papers. Williams also tells us that Halley later encouraged him to publish his table of variations, and that he exhibited his experiments at the Royal Society. Williams also assisted Stephen Gray in his electrical experiments, with the help of his daughter, who was to become the 'blind Miss Williams' of Dr Johnson's household. (See also E. G. R. Taylor's article, 'A Reward for the Longitude', *Mariners' Mirror*, **45** (1959), 59–66.)

1409 TILSON TO NEWTON
3 AUGUST 1723
From the original in the Public Record Office[1]
For the answer see Letter 1411

Whitehall Treasury Chambers 3d *August* 1723

The Right Honoble the Lords Comm[issione]rs of his Ma[jesty']s Treasury are pleased to referr this Memorial[2] to Sr: Isaac Newton Knt: Master and Worker of his Ma[jesty']s Mint who is to consider the same and Report to their Lordp's his Opinion what is fit to be done therein Which in the absence of the Secretarys is signifyed by their Lordp's Command from

CHRIS TILSON

Earl of Lauderdale referred to Sr: Isaac Newton

NOTES

(1) T/1, 244, fo. 94. A copy of the Royal Warrant to Lauderdale of 17 November 1718 follows (fo. 96). Compare Letter 1375 and Newton's reply, Letter 1378.

(2) Lauderdale's memorial is at P.R.O. T/1, 243, no. 32, fo. 10. It points out that since Lady Day (25 March) 1721 no payment has been made for charges of the Mint in Scotland. Lauderdale now requests payment of charges up to Lady Day 1723, together with an additional £2144. 8s. 10⅜d for the payment of salaries.

1410 TILSON TO NEWTON
12 SEPTEMBER 1723
From the original in the Public Record Office[1]
For the answer see Letter 1411

Whitehall Treasury Chambers 12 *Sept* 1723

The Right Honoble the Lords Comm[issione]rs of his Ma[jesty']s Treasury are pleased to refer this Memorial[2] to Sr: Isaac Newton Knt: Master and Worker of his Ma[jesty']s Mint who is to consider the same and Report to their Lordp's his opinion what is fit to be done therein Which in the absence of the Secretarys is signifyed by their Lordps Command from

CHRIS TILSON

Mr. Montgomerie referred to Sr: Isaac Newton

NOTES

(1) T. 1, 244, fo. 97. The wording is identical with that of Letter 1409.

(2) The memorial from John Montgomerie, Master of the Edinburgh Mint, requesting £2100 to pay wages and expenses for the Scottish Mint precedes the reference at fo. 97. Montgomerie's plea is essentially a repetition of that of Lauderdale.

1411 NEWTON TO THE TREASURY
13 SEPTEMBER 1723
From the holograph original in the Public Record Office[1]
In reply to Letters 1409 and 1410

To the Rt Honble the Lords Comm[issione]rs
of his Ma[jesty]s Treasury

May it please your Lordps

In obedience to your Lordps Order of Reference (dated 3 Aug. & 12 Sept) upon the annexed Memorials of the General & Master of his Ma[jes]ties Mint at Edinburgh, I humbly represent that by the Coynage Act wch passed last Sessions of Parliament his Majties Signe Manual directed to the General 17 Novem. 1718, is become voyd.[2] And a new one may be procured to the Master for putting in force the Indenture made between the late Queen Ann & the said Master untill a new Indenture shall be made. And that the summ of two thousand & one hundred pounds be imprest upon account to the said Master for the use & service of that Mint, as is desired.[3] Which summ together with the moneys formerly imprest, I conceive sufficient for defraying all the charges of that Mint till Lady day last, & also for keeping five or six hundred pounds in his hands for defraying all the charge of coynage till more moneys shall be issued unto him.[4]

All wch is most humbly submitted to Your Lordships greate Wisdome

Is. NEWTON

Mint Office
Sept. 13. 1723

NOTES

(1) T/1, 244, fo. 91, no. 25. There are drafts in Mint Papers, III, fos. 145 and 198.

(2) See Letter 1297, note (2), p. 2.

(3) An Order for the payment of £2100 to John Montgomerie was made on 19 August 1724, for payment of the Edinburgh Mint salaries to 25 March 1723.

(4) Since the Scottish Mint had been last paid to Michaelmas 1721 (see Letter 1309 and P.R.O., T/53, 30, pp. 148–9) a further imprest of £1800 was due for its salaries to Lady Day 1723. Montgomerie was in fact advanced only £300 more than this. However, when the next order for payment was made on 19 October 1725 (P.R.O. T/53, 32, p. 199), £1800 was again imprest to Montgomerie for the one and a half years from Lady Day 1723 to Michaelmas 1724.

1412 TILSON TO CRACHERODE
18 SEPTEMBER 1723
From the copy in the Public Record Office[1]

Sir

Sr: Isaac Newton being threatened with a prosecution[2] at Law by one of the Hamiltons who was sent some time since with Monsr. Brandshagen to make a tryall of Silver Mines said to be found in Scotland The Lords Comm[issione]rs of his Ma[jesty']s Treasury are pleased to direct you (in case of such a prosecution) to appear and Defend Sr. Isaac at his Ma[jesty']s charge[3] this in the absence of the Secretarys is Signifyed by their Lordps Command from

Sr. Your &c 18 Septr: 1723
C. TILSON

NOTES

(1) T/27, 23, p. 386.

(2) It was James Hamilton who had made the 'discovery' of Sir John Erskine's silver mine at Alva in April 1716 (see Letter 1200, vol. VI). Newton—who was deputed to make the detailed arrangements for the survey of the mine—agreed with Hamilton that he should be paid the large sum of 10s per day while employed on this government service. Hamilton, who some six years later was in so bad a condition that his lodging was the Marshalsea Prison, claimed that he was entitled to be paid for 145 days' service and various undefined 'extras'; it is hard to understand the reckoning but it is quite clear that his story was that Newton owed him £52. 10s and refused to pay. Therefore Hamilton proposed to sue him.

As usual, Newton prepared several papers touching this business. One is a history of the Alva mine, in which Newton blamed the 'experts' (that is, Justus Brandshagen, James Hamilton, and his brother William) who were sent to Scotland to explore and survey the mine for wilfully and unconscionably protracting the whole affair, and disclaimed all personal responsibility: 'Sr. I. Newton neither received nor paid any moneys for or to James Hamilton or on account of the Princes Warrant nor had any benefit by that business. He acted all along by order and had nothing to do in that matter after June 27 [1717] the day on which he gave in his final report.' He further pointed out that the Treasury itself had later dealt with the outstanding claims of the Scottish party (Mint Papers, III, fos. 246–7). Newton also drafted a reply to Hamilton's writ, in Latin, on fos. 240–1 of the same volume.

(3) It appears from a note on the Latin draft that counsel for each side was selected but it is not known whether the action was brought to court.

1413 HENRY PEMBERTON TO NEWTON
?OCTOBER 1723
From the holograph original in the University Library, Cambridge[1]

Sr.

I have reviewed this second proof of your second sheet.[2] And I beg leave to ask, if the expression in p. 10. lin. 16. would not be a little clearer, if the word *vas* in that line were removed from the place where it now stands, and put immediately after *postquam*.[3]

I am
Your most humbl.
and most obednt. servt.
H. PEMBERTON

P.S. I have received the corrollary you left with Mr. Innis,[4] and will take care to insert it in its proper place.

NOTES

(1) Add. 3986, no. 1. Henry Pemberton (1694–1771) has already been mentioned in vol. VI, p. 284, note (6), in connection with John Keill and the Leibniz–Bernoulli challenge problem. At that time Pemberton had neither taken his M.D. degree at Leyden nor published the thesis *Dissertatio...de Facultate Oculi qua...se Accommodat* (Leyden, 1719) by which it was justified— his major original work in science. He was introduced to Newton by Richard Mead, on the basis of an anti-Leibnizian exercise in mechanics printed in the *Philosophical Transactions* (**32**, no. 371 (1722), 57). Why Newton chose Pemberton as the 'editor' of the third edition of the *Principia* remains a mystery, for he probably had been acquainted with him for less than a year when the work began.

Again, the date at which Newton positively decided to embark on this new edition is unknown. He had been collecting emendations for many years—indeed, almost as soon as the second edition was on sale, Newton had decided that a third was necessary. Possibly the ease and elegance of the production of the second French *Traité d'Optique* by the team of Coste, De Moivre and Varignon persuaded him to return to the *Principia*, once a pliant 'editor' was available. Probably the printing began in October or early November 1723 (Cohen, *Introduction*, p. 265).

Newton clearly placed at Pemberton's disposal an annotated copy of the second edition of the *Principia*. Pemberton seems to have dealt with small portions of the book at a time, first sending any major suggestions for alteration to Newton, then sending the amended copy to the printer, and finally sending the corrected proofs to Newton, together with any further suggestions for amendment. Almost all Pemberton's letters to Newton are undated. Rather than print them in unbroken sequences, we have assigned approximate dates to them, on the assumptions that each portion was brought to completion before the next was examined and that the average rate of progress was about 24 pages a month; this is borne out by the few dated letters, from which it appears that the printing of the first 450 pages required rather more than 18 months. Hence none of the letters is likely to be very far from its true place in time. We have included in the series a number of pages of queries which Pemberton sent to Newton, presumably enclosed with the relevant sections of copy. No replies from Newton are extant and (for the most part at least) probably none were written. Newton simply indicated his decisions on the copy or proof he returned.

Many of the most important passages from Pemberton's letters have already been printed by Professor I. Bernard Cohen in an analysis of Pemberton's effect upon the *Principia*—see Koyré and Cohen, *Principia*, ii, pp. 828–47.

(2) The third edition of the *Principia* is in quarto; Pemberton now sends Newton signatures B and C (there is no signature A); that is, the pages of text numbered 1 to 16.

(3) The change was made; see Koyré and Cohen, *Principia*, i, p. 51.

(4) John Innys, one of the printers to the Royal Society and of this book.

1414 PEMBERTON TO NEWTON

?NOVEMBER 1723

From the holograph original in the University Library, Cambridge[1]

Sr.

I here enclose a proof of your third sheet,[2] which I send you before the printer has corrected it, to ask your opinion of the passage in pag. 23 from l. 17 to 22. which contains in it some expressions, that you have thought proper to change in other places. Suppose it were put in some such manner as this.[3] Tandem, quemadmodum motus corporis A proxime ante reflexionem erit ad ejusdem motum proxime post reflexionem ut chorda arcus TA ad chordam arcus tA; ita motus corporis A proxime ante reflexionem, erit ad motum corporis B proxime post reflexionem in ratione composita ex rationibus corporis A ad corpus B, et chordæ arcus TA ad chordam arcus Bl; item motus corporis A proxime post reflexionem ad motum corporis B proxime post reflexionem, erit in ratione composita ex rationibus corporis A ad corpus B, et chordæ arcus tA ad chordam arcus Bl. Et simili modo &c But in l. 23. instead of inveniendi sunt motus, should it not rather be inveniendi sunt rationes motuum, or some such expression?[4]

In this page 23d. l. 6 from the bottom, instead of corpus B, should it not rather be, corpus B quiescens: because it is said presently after, si corpora obviam ibant.[5]

In pag. 22. l. 15. How do you approve of the expression hujus ætatis now at this distance from the time, that these Geometers florished in?[6]

When this sheet returns again to me from the printer I shall send it two you again that you may see it before it is wrought off.

I am

Your most humbl.

and most obedient servt.

H PEMBERTON

P.S. In page 19. l. 11. I have added after Lemmate XXIII, the words *ejusque corollario*, in regard to the corollary you have ordered to be subjoined to that lemma.[7]

NOTES

(1) Add. 3986, no. 2.

(2) Signature D, p. 17–24, *Principia*, 1726.

(3) Pemberton objects to Newton's wording in Corollary 6 to the Laws of Motion (see Koyré and Cohen, *Principia*, I, p. 67) because Newton *equates* the chord of the arc through which a pendulum falls to its velocity, whereas Pemberton wishes him to make it *proportional* to the velocity, using ratios. Newton rejected the revision proposed by Pemberton, only adding 'ut ita dicam' (so to speak) to the text.

(4) Also rejected.

(5) Newton agreed.

(6) Newton altered *hujus* ('of this') to *novissimæ* ('most recent')—but see Pemberton's next letter (Letter 1415).

(7) This was done.

1415 PEMBERTON TO NEWTON
?NOVEMBER 1723
From the holograph original in the University Library, Cambridge[1]

Sr.

In this revise you will see I have made the insertions, you directed me in the 23d page;[2] and in page 22d I have taken the liberty to change ætatis novissimæ into ætatis superioris,[3] the latter phrase being directly an expression in Cicero. If you approve not of this change the printers can alter it to [the] former phrase without any difficulty.

I am

Your most humbl.

and most obednt. servt.

H PEMBERTON

NOTES

(1) Add. 3986, no. 3.

(2) Almost certainly the parenthesis '(ut ita dicam)' on lines 17–18, since the other altera-

tion from the second edition would have been in Pemberton's copy. Compare Letter 1414, notes (3) and (5), p. 250.

(3) Line 15; Pemberton prefers to say that Wren, Wallis and Huygens were the leading geometers of the *last age*, rather than *the most recent age*. Compare Letter 1414, note (6).

1416 PEMBERTON TO NEWTON
?DECEMBER 1723
From the holograph original in the University Library, Cambridge[1]

Sr.

In this sheet occurrs the expression concentrice tangit,[2] which I have once already presumed to make a query upon. But by what you were pleased to answer, I am in doubt, whether I sufficiently explained my meaning. What I scrupled was this: That in the case where the circle does not fall on the same side of the curve on each side of the point in the curve, through which it is described; but does really intersect the curve, although the angle of intersection be less than what any two circles can make, when they touch each other; and though every circle either greater or less than this would actually touch the curve in this point, and intersect it in another in such manner, that in this concentrical circle these points of contact and of intersection may properly be considered as coalescing; yet when this circle crosses the curve in this point, which in all other circles is a point of contact; should not this point be considered as really converted from a point of contact into a point of intersection by the coalescence of the forementioned point of intersection with it? and if so, should not this concentrical circle receive a name strictly consistent with the present condition of this point; rather than such a denomination, that has reference to that condition, which here this point has, as it were, laid down; so that in the exactest propriety of speech it does not really belong to it. Methinks if Circulus concentricus, Circulus concentrice descriptus, Circulus qui concentrice describitur, or some expression of like sort, would be sufficient to represent fully, what is here intended, there would be less room for any cavil of this nature.

I am
Your most humble
and most obednt. servt.
H PEMBERTON

NOTES

(1) Add. 3986, no. 7. Cohen, *Introduction*, pp. 273–4, supposes that this undated letter is subsequent to Pemberton's letter of 9 February 1726, the latter part of which comments upon Proposition 6, Corollary 3 of *Principia* Book I, p. 47, as does the present letter. He reads 'which

I have once already presumed to make a query upon' as referring back to that 1726 letter, which, of course, was written long after p. 47 was printed off. We think Professor Cohen to have been mistaken on this point for the following reasons. First, the opening of the present letter 'In this sheet occurrs...' suggests that Pemberton was sending the sheet (whether of the revised second edition text, or of the third edition text printed from it makes little difference, but we presume the latter) to Newton for his examination; this he did not do with his 1726 letter noted above, and *a fortiori* would have had no occasion to do thereafter. Secondly, the discussion in this letter is general, and does not follow logically from the content of the 1726 letter. Thirdly, in the present letter Pemberton makes no allusion to the form of words he was to propose in 1726 for Corollary 3 and which Newton did not adopt; instead, he discusses different phrases. It seems evident that if the present letter followed that of 1726, it would show some unmistakable link with it. It is our hypothesis that Pemberton 'presumed to make a query upon' p. 42 of the second edition as revised by Newton for the new edition (changing the wording of Corollary 3), and returned to the charge again when transmitting the freshly printed sheet of the third edition with the present letter. The letter of 1726 is, of course, a late afterthought of Pemberton's.

(2) If the sheet was of the 1726 text, it was sheet G, Pemberton referring particularly to p. 47. At this stage, probably, Newton made no further modification to the revised Corollary 3; indeed, Pemberton's proposal here seems quite trivial. Compare Cohen, *Introduction*, p. 273 for the probable text at this time.

1417 CHARLES RAWSON TO NEWTON
31 DECEMBER 1723
From the original in the Bodleian Library, Oxford[1]

Hond. Sr.

I make bold to present you with a new Almanack wishing you a happy new Yeare with my humble service to you I remaine your most humble servant

CHA. RAWSON
stationar to ye Mint

Decemb. 31 1723
The Honble. Sr. Isaac
Newton Master of his
Ma[jes]ties Mint wthin ye Tower
of   London
Present.

NOTE

(1) Bodleian New College MS. 361, II, fo. 63. The writer was not also a publisher of books.

1418 J. C. F. VON HATZFELD TO NEWTON

[1723 or 1724]

From a copy in the University Library, Cambridge[1]

Sr.

Ther is nothing seems to be more strange to me then to observe amongst the learned of this age, that they look upon a person who pretends to something more than what everion [*read:* everyone] is capable of to be nothing but a fool and a mad man, just like as the vulgar uses to look upon a dexterous man to deal whit the devil or some of his imps, which to me is an evident demonstration that most of the learned knows no more of ingenuity then the vulgar knows of dexterity. Wherefor it is absolutely impossible for arts and sciences to arrive at their point of perfection, so long as they have the misfortune of depending upon the discretion of such like men, who not only are incapable of discovering any truth but are likewis incapable of coming in to it when they are put in the way that leads to it, for what I have actually experienced, show'd and explained to several of those I know to be most famas in mechanicks proves most evidently that ther is such a thing as a perpetual motion to be made; as well as the description we have of Dr Orfireuses model proves it to be what his S. H. the Landgrave of Hesse Cassel has been pleased to give his word and honnor for[2] Nevertheless that Eminent and Ingenious Prince and Admirable incourager of all arts and sciences is laughed att as well as I am for pretending to confirm his assertion by an ocular demonstration, and what can all this be imputed to but an insurmountable ignorance and stupidity or to such like jealousie in all thos who ar so violently against that and all other points of Ingenuity. As soon Sr as I had accomplished the first experiment mentioned in the enclosed to the Society you represent the head of I thought it to be a due in me to offer you the sight of it, but to my surprise I was told by your servant that without a Recommendation he had no power to introouce me which I most confess was a very dis-agreeable answer to me, for I have had the honnor to be introduced to sovereign Princes without a Recommendation and much less did I expect to be asked for one at your House when I came for nothing else but to invide you to see a thing, which I thought and yet think it worth your while to see and show you Sr. that I have more power over my passions then they have over me I do hereby for the sake of your own satisfaction invite you again to see and examine into the experiments I have made towharts the subject above mentioned, and if you deny me upon the second Invitation I shall take it for granted that you ar as unwilling of being convinced of your errors as most men uses to be; which I look upon to be a very unpardonable thing in any body, but chiefly in a philosopher, whose business it is to endevor to come into the truth at any rate, and to be at all times ready to exchange the most important of his errors for the most inferior point of truth that can be offered to him. [3]

I do not doubt Sr but you will think it very bold in me as a novice to pretent to give you more light then you have seen your selves in [a] thing, wherein you have spent the best part of your life, but notwithstanding, I can assure you that I have actually discovered more important errors of your production then I have yet met with in any philosophers writings. I am. Sr.

Your most obedient humble servant

JOHN CONRAD FRANCIS DE HATZFELD

NOTES

(1) Add. 4007, fo. 750. This is one of the copies of letters made in the 1870s at Cambridge from originals in the Portsmouth Collection, which were subsequently returned to the Earl of Portsmouth.

The writer was perhaps a member of the German noble family of this name, which may explain the hectoring tone of his letter.

He published in 1724 *The case of the Learned represented according to the merit of the ill-progress hitherto made in Arts and Sciences, shewing I. The cause of gravity and attraction. II. What Nature is and the effects it is capable of etc...Contained in two Letters to the Royal Society*. Probably this letter to Newton was concerned with the presentation of this work to the Royal Society. As to its date, the first *Letter* contains the date 6 May 1723; but the two *Letters* were only read at the Royal Society on 27 May and 21 October 1725.

(2) Von Hatzfeld's first *Letter*, critical of the Royal Society, is concerned with the reality of perpetual motion; for the machine of Orffyræus see Letter 1364, note (3), p. 146.

(3) Von Hatzfeld's second *Letter* shows 'the true Way of proving the Existence of God, and that Sir Isaac Newton, instead of having proved that most important Point, and established Natural Religion better than ever any Philosopher did (as he and his Disciples pretend) they have together with their Author, and his like Predecessors disproved it, and overturned both Natural and reveal'd Religion.'

1419 NEWTON TO JOHANN BURCHARD MENCKE

1724

From a holograph draft in private possession [1]

Sr

I perceive you have been misinformed. The Royal Society leave it to their Clerk to print a list of their Fellows Annually. And if any mistake happens he has power to correct the List without troubling the R. Society about it. Accordingly he corrected it in November last, as you will find by your own & Mr Wolfius's names in the List then printed, a copy of wch I ordered Mr Innys to send you.[2] Dr Vater[3] is a perfect stranger to me. I thank Mr Richter for his love to the truth. I meddle not with disputes between England & Germany. I honour Ticho Brahe, Kepler, Hevelius, Otto Gueric, Galilæo, Torricellius, Hugenius, Cassini, Bianchini & such others of all nations as were candid promoters of truth: but men that are contentious & noisy for the sake of fame or preferment hinder the growth of truth by writing & disputing for something else. I am

NOTES

(1) Another draft, differently worded but apparently of the same letter, is in King's College Library, Cambridge, Keynes MS. 111. The opening sentence implies that Mencke may have suggested a correspondence with Newton: 'Its now about fifty years since I began for the sake of a quiet life to decline correspondencies by Letters about Mathematical & Philosophical matters finding them tend to disputes and controversies...'

Johann Burchard Mencke (1674–1732), German philosopher and theologian, was born at Leipzig and later became Professor of History there. In 1708 he was appointed Historiographer of the Elector of Saxony, and rapidly gained a high reputation at court. He was editor of the *Acta Eruditorum* from 1707 to 1732, and in the issue for May 1721, published his 'Epistola ad Taylor', in defence of Johann Bernoulli (see Letter 1303, note (3)). Mencke's support of Bernoulli was probably responsible for the somewhat terse tone of Newton's note.

The final sentence of the King's College draft makes it clear that the intended recipient is Mencke. Newton writes, 'I thank Mr Richter heartily for defending me, & I thank you as hartily for the several civilities you mention shewed unto me in your Acta Eruditorum.' We have not been able to identify Mr Richter.

(2) The names of Christian Wolf and of Mencke had been omitted from the lists of members published by the Royal Society for the years ending 30 November 1720, 1721, 1722 and 1723. They were reinstated in the list for 1724.

(3) Probably Richter had noticed the omission in the list of members of the Royal Society which Abraham Vater appended to his short tract, *Dissertatio Regiæ Magnæ Britanniæ Societati Dicata qua Ductus Salivalis in Lingua...* (Wittenburg, 1723).

1420 PEMBERTON TO NEWTON
?JANUARY 1724
From the holograph original in the University Library, Cambridge[1]

Sr.

That the press may not be retarded I make bold to send you two more of your sheets with a few observations upon them.[2] You will see one or two q[ue]r[ie]s marked on the sheets, which I do not here take notice of; because they are of the same kind with some of the remarks in my last paper: so that by your opinion on them I shall be guided in respect to these.

1st qr in pag. 49.[3] Whether after the word diametri, the word aliæ may not
   properly be inserted; since the Axes are diameters as well as any others.

qr in pag. 51.[4] Whether the word conjugata should not rather be opposita:
   for the two hyperbola's belonging to the same transverse axis are never
   called conjugate hyperbola's by Apollonius, or by any other conic writer
   of authority that I can call to mind, but are always called, opposite
   hyperbolas, or rather opposite sections. If to these two opposite sections
   two other hyperbolas are added, having for their transverse axis the conjugate axis of the other, and the transverse axis of the other sections for
   their conjugate, these four hyperbolas are called conjugate by Apollonius.

1st qr in pag. 57.[5] Whether conisectionis should not be two distinct words.

2d qr ibid.[6] Does not the sentence here added in the margent, in order to be
   understood, seem to require something, that has not been demonstrated
   before. As it has not been shewn from what height a body must directly
   fall to acquire the velocity, by which it may move in any given conic
   section.

qr in page 53.[7] Instead of eadem vi centripeta, might it not rather be eadem
   vi centripeta eademque velocitate?

qr in pag. 59.[8] In this 18th proposition, which teaches the description of the
   conic section having one focus given, and the transverse axis given in
   magnitude; by means of the other data specified in this proposition. If
   you do not intend to limit the proposition in the hyperbola to that case
   when all the given points and lines touch that hyperbola, which contains
   within it the focus S; but do design to give the proposition its full extent;
   so that the given lines and points shall touch either of the opposite
   sections: then when SP exceeds AB in the hyperbola the radius of the
   circle to be described to the center P may be either $AB+SP$, or $SP-AB$,

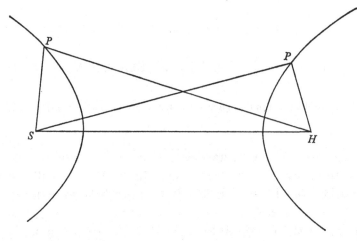

If P be in the hyperbola to which the focus S is nearest, then is $PH-SP = AB$, and $PH = AB+SP$; but if the point P be in the opposite section, then is $SP-PH = AB$, and $PH = SP-AB$. On this ground in page 60. l. 9. instead of $PH-SP$, it should be more generally the difference between PH and SP.

qr in pag. 64.[9] In this second case I think the two lines, by whose intersection you seek the point Z, do coincide and become one line, so that the point Z cannot be determined, by their means. Though in the first and most general case, I find that the two loci, you assign of the point Z, will be different. I have taken the freedom to inclose papers,[10] wherein I have endeavoured to demonstrate these remarks; in the last page of which you will find the figure belonging to the papers.

The solution of the first case will agree to the second likewise, if it be adjusted in some such manner as this. The position of the line TZ being determined, the ratio of RZ to ZT will be given, and therefore the ratio of AZ to ZR being given, the ratio of AZ to ZT will be given. Whence if a right line were drawn from A to T the triangle AZT would be given in species [per Eucl. dat. prop. 44] for the angle under the lines AT and TZ will be given. But hence the ratio of AT to TZ will be given, and consequently the point Z given.

To give the proposition which follows it's full extent as I said of prop. 18. it is required to solve this lemma not only when the differences of the

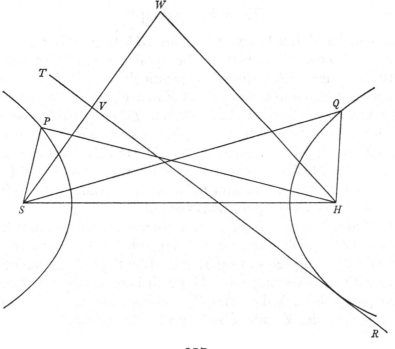

lines are given; but also when the sum of two of them shall be given, and either the sum or difference of one of these two and the third line; which is done in the same manner by using an ellipsis where the sum is given. [11] For suppose two points P, Q to be in opposite sections; then $PH-SP$ will be $= SQ-QH$, and $PH+QH = SQ+SP$; so that the sum of PH and QH will be given. So likewise if the tangent TR touch either of the sections, drawing SVW perpendicular thereto that VW be $= SV$, and drawing WH, $PH-SP$ will be $= WH$, and $PH-WH = SP$; so that the difference between PH and WH will be given; but $PQ-QH$ will be $= WH$, and $PQ = WH+QH$; in so much that the sum of WH and QH will be given.

In pag. 51. [12] I intend according to your directions to retrench the figure of all its superfluous lines; and draw such a one as I have roughly delineated in the margent.

In pag. 52. [13] I design to retrench those lines I have blotted out as of no use in this proposition.

In pag. 53. [14] I shall draw, if you so approve, such a figure as I have delineated in the margent, for the use of this 14th prop. and of the same size with the ellipsis in page 58th.

In pag. 54. [15] I intend for this 16th. prop. to draw the figure represented in the margent, drawing it to the same size with the forementioned figure in page 58th.

[*The enclosed papers*] [16]

The construction of this Lemma [17] is this. MN being taken in AB equal to the excess of BZ above AZ, so that AM be equal to BN; PM is taken to AM as MN to AB, and then TPR is drawn perpendicular to AB. In like manner mn being taken in AC equal to the excess of CZ above AZ, so that Am be equal to Cn; mQ is taken to Am as mn to AC, and the line TQS is drawn perpendicular to AC. Then it is shewn that ZR being perpendicular to TR, ZR is to ZA as MN to BA; and ZS being perpendicular to TQ, AZ is to ZS as AC to mn: in so much that the ratio of ZR to ZS is compounded of the ratio of MN to BA and of the ratio of AC to mn: and by this ratio of RZ to ZS is determined the position of the right line TZ, in which the point Z is posited.

Now if BZ and ZC are to be equal it is observed, that the point Z is also in the right line, which divides the line BC perpendicularly into two equal parts; and the point Z is supposed to be determined by the intersection of this perpendicular with the foregoing line TZ. But it here so falls out, that this perpendicular passes through the point T, and coincides with the line TZ. for which reason the point Z cannot be found by this means.

This appears in the following manner.

If AB be divided into two equal parts in D, MN will be likewise divided into two equal parts in the same point. And PM being to MA as NM to BA, PM will be to MA as DM to DA; and therefore DP, the excess of DM above PM, will be to DM, the excess of DA above MA, as the same DM to DA. After the same manner if AC be divided into two equal parts in E, Em will also be a mean proportional between AE and EQ. But if BZ be equal to CZ, DM, equal to half the excess of BZ above ZA, will be equal to Em, which is half the excess of CZ above ZA: therefore DM the mean proportional between DA, DP being equal to Em the mean proportional between EA, EQ, the rectangles under AD, DP and under AE, EQ are equal, and as AD to AE, that is as AB to AC so is EQ to DP.

Now erect the perpendiculars DF, EF, draw TGF; as also TH parallel to AB, and TI parallel to AC, connecting HI with a right line. Then will TH be equal to DP, and TI equal to EQ. therefore since AB was to AC as EQ to DP, AB will be to AC as IT to TH, and the triangle ITH similar to the triangle BAC; so that the angle under HIT in particular will be equal to the angle under CBA. But the angles under FHT and under FIT are both right, and therefore the four points F, H, T, I are in the circumference of a circle; whence the angle under HIT is equal to the angle under HFT: consequently this angle under HFT is equal to the angle under CBA. And therefore DF and BG cutting each other in V, the triangles BVD, FVG will be similar, the angle under VGF equal to that under BDV, and DF being perpendicular to BD, GF will be perpendicular to BG. But it is moreover manifest that BC is divided into two equal parts in G. for DF and EF dividing perpendicularly AB and AC into two equal parts, F is the center of the circle passing through B, A, C. therefore FG a perpendicular from this center to the chord BC will divide this chord into two equal parts.

I say in the next place FT and TZ are one continued line. It was observed above, that the ratio of RZ to ZS is compounded of the ratio of MN to BA and of the ratio of AC to mn. therefore when MN and mn are equal, RZ will be to ZS as AC to BA, or as HT to TI for as much as the triangles CAB, HTI are similar. Hence the lines DF, FE being parallel to the lines PT, TQ; and also the lines HT, TI parallel to RZ, ZS, FT and TZ will be one continued right line.

But I find it to be in this case only, where this solution fails. for in all other cases the two lines found by this method are different; and consequently the point Z is determined by their intersection.

This may be proved as follows.

The construction of this Lemma, when all the lines are unequal, amounts to this. Find the line *PT* by taking *DM* equal to half the excess of *BZ* above *ZA*, and making *DP* a third proportional to *AD*, *DM*. Then taking *EL* equal to half the excess of *CZ* above *ZA*, and making *Eq* a third proportional to *AE* and *EL*, erect *qXt* perpendicular to *AC*, that it may meet *PT* in *t*. Which being done one of the lines in which *Z* is found passes through the point *t*. Again if *GK* be taken equal to half the excess of *BZ* above *CZ* and *GY* be made a third proportional to *GC* and *GK*; a perpendicular erected from the point *Y* upon *BC*, will meet either of the perpendiculars *PT*, *qt* in a point from whence a line also may be drawn, in which the point *Z* shall be posited. Now if this perpendicular passes not through the point *t*, then this line, in which the point *Z* is posited, will not be the same, with the forementioned line passing through *t*, in which *Z* is also posited.

But that this perpendicular does not pass through *t* is thus shewn. Draw *ty* perpendicular to *BC*, and I say *Gy* is greater than *GY*: which appears thus.

The excess of *BZ* above *CZ* is equal to the difference between the excess of *BZ* above *ZA* and the excess of *CZ* above *ZA*. therefore *GK* is equal to the difference between *DM*, or its equal *Em*, and *EL*; that is *GK* is equal to *mL*. But the rectangle under *AEQ* being equal to the square of *Em* and the rectangle under *AEq* equal to the square of *EL*, the rectangle under *AE*, *Qq* will be equal to the rectangle under *mLn*, and therefore greater than the square of *mL*, or

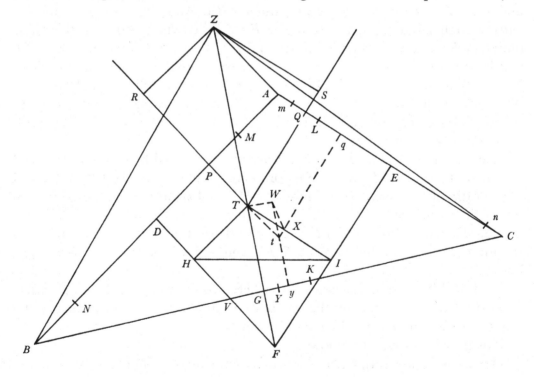

than the square of GK, or lastly than the rectangle under CGY. In the next place produce yt, let TW be parallel to BC, and draw WX. then the angle under TWt will be a right one, and the angle under TXt also right, because TX is parallel to AC. hence the points T, W, X, t are in a circle. therefore the angle under TXW is equal to that under TtW. but the angle under TtW is equal to that under VFG; for Tt is parallel to VF and Wt parallel to FG. but it has been likewise shewn that the angle under VFG is equal to that under ABC. therefore the angle under TXW is equal to that under ABC. besides the angle under WTX is equal to that under BCA, TW being parallel to BC and TX parallel to AC. hence the triangles XTW and BCA are similar; so that TX, or its equal Qq, is to TW, or its equal Gy, as BC to CA, or as CG to AE therefore the rectangle under AE, Qq which was proved above to be greater than the rectangle under CGY, is equal to the rectangle under CGy. whence it appears that Gy is greater than GY.

NOTES

(1) Add. 3986, no. 4. Corrigenda relating to this letter are listed by Newton in U.L.C., Add. 3965, fo. 499. This document is printed in Koyré and Cohen, *Principia*, II, pp. 827–8.

(2) Sheets H and I of the second edition, pp. 49–64.

(3) Pemberton here and below employs the page numbers of his second-edition printing copy; he refers now to Proposition 12 (*Principia*, 1726, p. 56). *aliæ* was inserted.

(4) *Ibid.*, p. 58. The correction was made.

(5) *Ibid.*, p. 63. This was also done.

(6) *Ibid.*, p. 63. Near the end of Proposition 17 Newton proposed to add a few words (see Koyré and Cohen, *Principia*, p. 132, line 17) which, at Pemberton's instance, he later omitted.

(7) *Ibid.*, p. 59, end of Corollary 1. The alteration was made.

(8) *Ibid.*, pp. 66–7. Newton did not accept this criticism, although Pemberton's argument is correct. Possibly he felt that in all situations involving real bodies in motion, his concern would always be with the nearer focus; compare note (11) below. Neither Newton nor Pemberton shows any concern over the fact that the problem has a double solution, giving two possible hyperbolas or ellipses. (Note that AB is a fixed magnitude, the conic hyperbola's diameter, not marked by Pemberton.)

(9) *Ibid.*, p. 71. Here Newton altered the wording of Lemma 16, case 2 to meet Pemberton's objection. See note (16) below.

(10) We print the enclosed paper at the end of the Letter.

(11) Newton obviated this objection by the addition of a comment, at the beginning of the Scholium following Proposition 21, to the effect that since real bodies in motion could not pass from one branch of the hyperbola to the opposite one, he was not concerned with cases where there were points lying on both branches.

(12) *Principia*, 1726, p. 58; Lemma 14. The figure has been simplified.

(13) *Ibid.*, p. 59, Prop. 13. Again, a simpler diagram is given.

(14) *Ibid.*, p. 60; a figure has been added.

(15) *Ibid.*, p. 61; a figure has been added.

(16) Now at Add. 3986, no. 25.

(17) Book I, Lemma 16. Pemberton adds additional points to Newton's figure (compare

Koyré and Cohen, *Principia*, ɪ, pp. 140–1). We have copied Pemberton's sketch, which, it will be observed, does not exactly match all the prescriptions of the text. The problem is to find a point Z, given three points A, B, C and the quantities $BZ - AZ$ and $CZ - AZ$; the argument is based on properties of hyperbolæ discussed in the preceding propositions.

Pemberton shows here that in the special case where $AZ = CZ$ the lines thus constructed are coincident, whichever pair of hyperbolæ is treated; hence the intersection Z cannot be determined. He might equally have quarrelled with Newton's general solution in Case 1, which suffers from the same frailty that Pemberton demonstrates for the particular Case, 2, where $AZ = CZ$. The fact is that this problem of 'three-circles tangency'—here posed in Vieta's simplification of his *Apollonius Gallus* (1600)—has always a pair of solutions and so cannot possibly be constructed by the intersection of straight lines alone. Whether Newton saw this point or merely acquiesced in Pemberton's urging that he give a common construction for both cases, he amended his general solution of the problem (*Principia*, 1726, p. 71) to the form Pemberton had suggested in his previous letter, where the two points Z are constructed effectively as the meet of the straight line TZ with the Apollonian circle determined by the constancy of the ratio of TZ to AZ. For further elucidation see Whiteside, *Mathematical Papers*, ᴠɪɪ, p. 253, note (16).

Corrigenda relating to this are given in U.L.C., Ad. 3965, fo. 499; see note (1) above.

1421   PEMBERTON TO NEWTON
11 FEBRUARY 1724
From the holograph original in the University Library, Cambridge[1]

Sr.

Be pleased to take particular notice of the two last lines in page 63;[2] because I have made a small alteration in them; the words Sit istud L, which were in the last line but one, appearing to me ambiguous; it not being thereby expressed whether this line L be the latus retum of the conic section, or of the other orbit.

I am

Your most humbl.

and most obednt. servt.

H Pᴇᴍʙᴇʀᴛᴏɴ

Feb. 11. 1723–4

NOTES

(1) Add. 3986, no. 5.

(2) Pemberton now refers to the proof, that is to *Principia*, 1726, p. 63, where he has written (line 34), 'Sit L coni sectionis latus rectum'.

1422 PEMBERTON TO NEWTON

18 FEBRUARY 1724

From the holograph original in the University Library, Cambridge[1]

Sr.

Please to take particular notice of a small alteration I have directed to be made in page 71.[2] If you approve not of it; by striking out the correction in the margent and the bottom of the page, the printers will leave it as it is.[3]

I am

Your most humbl.

and most obednt. servt.

H PEMBERTON

Feb. 18. 1723–4

NOTES

(1) Add. 3986, no. 6.

(2) Pemberton refers to *Principia*, 1726, p. 71. If accepted by Newton, it must have been one of the changes in lines 16–25, already discussed in Letter 1420.

(3) At about the same time as this letter was written, presumably, Pemberton drafted and sent to Newton (U.L.C., Add. 3986, nos. 26/7) a Lemma followed by two Corollaries and new versions of Book I, Propositions 25 and 26 (*Principia*, 1726, pp. 89–91 correspond). As there is no significant difference at this point between the second and third editions nor any discussion of Pemberton's abortive proposals, we have not reproduced his text here.

1423 LAUDERDALE TO NEWTON

27 FEBRUARY 1724

From the original in the Mint Papers[1]
For the answer see Letter 1425

Edinburgh feb: 27: 1724

Sir

My accounts for the money impressed to me since I was concerned for the mint in Scotland were layed before the Exchecquer here some time ago and would have passed this last terme had not my agent at London neglected to send me down a controlment certificate from the Exchecquer in England of what I had recieved, thinking that extracts from his own books would have been sufficient[.][2] I have ordered a friend of mine one Mr Robertson to wait of [*sic*] you and I hope you will be so good as to allow him to converse with you upon our affair's

I know no other difficulty our Baron's of Exchecquer cane [*sic*] have about my accounts unless it is my stating the Treasury and Exchecquer fees according to the double of a fitted account of your's, which at my desyer you were pleased to send me some time ago and you know that before the Union the collector of the Bullion here payed all the officer's sallary's without any manner of deduction so if you will be so good as to signify to me by a letter that these fee's are allowed to you I believe it may help to clear our Scots Baron's[3]

Mr Robertson will shew you a couble of my accounts and as to any other particular's I beg leave to refer you to him

I am

Sir

Your Most Obedient and

Most humble servant

LAUDERDALE

NOTES

(1) III, fos. 200–1.

(2) The Barons of the Exchequer of Scotland had written to the Lords of the Treasury about this difficulty on 17 February (P.R.O., T/1, 247, no. 22).

(3) Lauderdale succeeded in clearing his accounts for the £6885 imprest to him between Christmas 1714 and Lady Day 1721 on 25 July 1724 (P.R.O. T/1, 251, fo. 176). Compare Newton's Letter 1378.

1424 LUIGI FERDINANDO MARSIGLI TO NEWTON

29 FEBRUARY 1724

From the original in King's College Library, Cambridge[1]

Ill[ustrissi]mo Præsidi Regiæ Societatis Britannicæ
Aloysius Ferdinandus Marsili
S.P.D.

Cum frequentes literas ad te scriberem, tuas vere numquam acciperem, nullas ego deinceps decreveram ad te scribere; hactenus ergo silui eiusque propositi tenax fuissem, nisi ab eo me removeret præsens occasio. Transitus enim per istam Patriam D. Pirardi[2] fratris nostri Consocij eximij Botanici me allicit, ut certiorem te faciam ineunte proximo majo danubiale opus[3] mandandum esse typis quorumdem Bibliopolarum Amsterdamiensium, quia paucos infra dies opus integrum absolutum cum æneis laminis incisis illuc mittendum. Equidem vix putavi me vivo luce publica donandum fore, sed meum recens, novissimumque in Olandiam iter, cur id contingerit, in causa fuit. Mox aliud

264

opusculum huic inclytæ Societati inscribam de generatione tuberum terræ, quod prodibit cum secunda editione de generatione fungorum[4] Societati iampridem nota. Vetus Auctor Italus de sæculo quingentisimo, nomine Caccarellus,[5] super materia tuberum non pauca scripsit, methodum sequutus illorum temporum, et opusculum imprimatur curabo. Circa eadem terræ tubera varias observationes conserere mihi contigit, ignotas hactenus, et, quod natura propagat omnia p[er] semina, sed ea diversæ naturæ formæ ac substantiæ a seminibus regularium plantarum. Fungarius lapis vobis abunde notus, lapis vere non est, sed unum terræ tuber, et in eo crescentes fungi, non sunt plane filij quorumdum lignorum, foliorumve, in substantia d[icti] lapidis existentium, sed ipsiusmet substantiæ tuberis, quod omnino discrepat a sententia Bucconis[6] contrarium aperentis. Me, atque hoc scientiarum, et artium Institutum, studia omnia, et officia nostra in te collaturos, pro certe habe. Idem sibi suadeant D.P., Slon, Mid, Deram, cæterique, quibus meis verbis salutem nuncia. Vale.

Bononiæ 11 *Mart.* 1724 [*N.S.*]

Translation

Luigi Ferdinando Marsigli presents a grand salute
to the most illustrious President of the British Royal Society

As I wrote frequent letters to you without ever receiving yours, I decided to write to you no more; accordingly I have remained silent hitherto and I would have held fast to my resolution had not the present opportunity caused me to depart from it. For the journey through that country of Mr Pirardi,[2] brother of our eminent botanical colleague, entices me to inform you that at the beginning of next May a work on the Danube[3] is to be committed to the press of certain booksellers in Amsterdam, because the whole work complete with engraved copper plates is to be sent thither within a few days. I had indeed little thought that this work would be laid before the public in my lifetime, but my late and most recent trip to Holland was the cause why this will come about. I shall soon dedicate to this distinguished Society another little tract on the generation of tubers in the earth, which will appear with the second edition of *The Generation of Fungi*[4] already known to the Society for some time. An ancient Italian author of the fifteenth century, named Caccarellus,[5] has written quite a lot on the subject of tubers according to the method of that age and I shall see to it that the work is printed. It happens that I have compiled a variety of observations concerning these same tubers of the earth, neglected hitherto, and [to the effect] that nature propagates all things by seeds, though these are of a different nature, form and substance from the seeds of ordinary plants. The 'fungus-stone' well known to you is not really a stone but a tuber of the earth and the fungi growing in it are obviously not the threads of certain wood or leaves existing within the substance of the said stone but are of the very substance of the tuber itself, which is completely unlike the opinion of Boccone,[6] who pronounces the contrary.

Be sure that I and this Institute of Sciences and Arts offer you all zeal and our services. We also recommend ourselves to Mr President, Sloane, Mead, Derham and the rest, to whom give my greetings. Farewell.

Bologna, 11 March 1724 [*N.S.*]

NOTES

(1) Keynes MS. 94(E); microfilm 931.3. Luigi Ferdinando, Count Marsigli (1658–1730), a student under G.-A. Borelli and Marcello Malpighi, embarked on a military career which resulted in his capture by the Turks at the siege of Vienna (1683) and his court-martial for the loss of the Imperial fortress of Breisach (1703), of which he had been second-in-command.

The rest of Marsigli's life was devoted to learning and collecting. He was elected F.R.S. between 30 November 1692 and 30 November 1693, and became *associé étranger* of the Académie Royale des Sciences in 1715. In 1712 he presented to his native town of Bologna a vast collection which became the basis for an Institute of Arts and Sciences (see Letter 1440, note (8)). In 1721 he visited London, where he was warmly received by Newton, who introduced him at a meeting of the Royal Society on 14 December (see G. Fantuzzi, *Memorie...[di L. F.] Marsigli* (Bologna, 1770), pp. 248–51). He travelled to Paris in the same year. Apart from his writings noted below, his *Histoire physique de la mer* (Amsterdam, 1725) is regarded as the first work on oceanography.

(2) The reading is doubtful, and we have been unable to identify an Italian botanist of this name.

(3) Marsigli had studied the Danubian basin attentively in his military days, and published *Danubialis operis Prodromus* (Nuremburg, 1700). The present work was *Danubius Pannonico-Mysicus Observationibus Geographicis, Astronomicis, Hydrographicis, Historicis, Physicis Perlustratus* (Amsterdam, 1726), in six folio volumes.

(4) According to the Journal Book of the Royal Society (14 December 1721) Marsigli had already presented to the Society a copy of his *Dissertatio de Generatione Fungorum* (Rome, 1714); and a letter of his was read on 15 March 1722. The second edition, mentioned here, was never published.

(5) Alfonso Ciccarelli (d. 1580), *De Tuberibus* (Padua, 1564).

(6) Possibly Paulo Boccone (1633–1703), a Cistercian monk from Palermo who became Botanist to Cosimo III of Tuscany. (See P. A. Saccardo, 'La Botanica in Italia', *Memorie del Reale Istituto Veneto di Scienze, Lettere ed Arti*, **25**, no. 4 (Venice, 1895), and **26**, no 6 (Venice 1901).) Boccone travelled widely over Europe and was a prolific writer.

1425 NEWTON TO LAUDERDALE

18 MARCH 1724

From the holograph original in the Scottish Public Record Office[1]
Reply to Letter 1423

London March 18th
172¾

May it please your Lordp

I had the honour of your Letter of Feb. 27th last & thereupon ordered a copy to be made of one of my Bills of the Fees & charges of the receiving the money for the Mint Office in the Tower of London out of his Majties Exchequer that I might shew it to Mr Robertson. But having not yet seen him, I send it to your Lordship enclosed in this Letter. We pay but six pence for every twenty shillings or fifty shillings for every hundred pounds received att the Exchequer, besides 7 *lb* 17*s* 6 for a warrant for every two thousand pound. You have no such Warrant, but pay other fees for every summ imprest, an account of wch I presume you have received from your agent. That wch you call a Controlment Certificate from the Exchequer wee call an Imprest Roll & pay year[l]y for it 4 *lb*. And this & the other charges set down in the copy of the Bill wch I send you, are yearly allowed in our accounts. I am

> My Lord

To the Rt Honourable   Your Lordps most humble
The Earl of Lauderdale   and most obedient Servant
at the Mint in   ISAAC NEWTON
     Edinbourgh
   in   Scotland

[*Enclosure*] [2]

Fees & charges paid at the several Offices of the Exchequer
& at the Treasury for the several summs of money issued
out from thence for the use & service of the Mint from
1st January 1722 to the last day of December 1723.

1722		li	s	d
April 8th	Out of Mr Smiths Office £2184. 12*s*. 10½*d*, the six pence per pound	54.	12.	4
	Old fees to the Tellers, Auditor of the Exchequer & Pell Office	5.	10.	0
	Porters & Coach hire	0.	6.	6

		li	s	d			
May 4	New Order for £20000 dated this day.						
	Sign Manual	3.	10.	0			
	Stamps 7*s* 6*d*. Messenger 2*s* 6*d*	0.	10.	0			
	Warrant £1. 10*s*. Order it self 12*s*. 6*d*	2.	2.	6	7.	17.	6
	Clerk for entring & prosecuting the order	1.	1.	0			
	Auditor of the Exchequers Clerk 7*s* Pell Office Ditto	0.	14.	0			
27	Out of Mr Smiths Office £1764. 11*s*. 5½*d* The six pence per pound	44.	02.	3			

	Old fees to the Tellers Auditor of the Exchequer & Pell Office	4.	10.	0
	Porters & Coach hire	0.	6.	0
July 5	Out of Mr Smiths Office £3727. 1. 6½ The six pence per pound	93.	3.	6
	Old fees to the Tellers, Auditor of the Exchequer & Pell Office	9.	7.	6
	Porters & Coach hire	0.	7.	0
Octob. 10	Out of Lord Parkers Office £3725. 14s. ½ the six pence per pound	93.	2.	11
	Old fees to the Tellers, Auditor of the Exchequer & Pell Office	9.	7.	6
	Porters & Coach hire	0.	7.	0
	To the Auditor of the Exchequer for the Imprest Roll	2.	0.	0
	Clerk of the Pells for the same	2.	0.	0
		£327.	0.	0

NOTES

(1) MS. E. 105/64.
(2) This also is in Newton s hand.

1426 PEMBERTON TO NEWTON

?MARCH 1724

From the holograph original in the University Library, Cambridge[1]

Sr.

I make bold to trouble you again with this sheet before it is wrought off; for you to consider farther the first five lines of pag. 96.[2] In the third line the words problematis xiv are rightly blotted out, as that problem has no relation to the present case. But the 15th problem here quoted does not take in enough; for that relates only to the case, when an asymptote and three points are given; whereas, besides this case, with an asymptote two points and a tangent, one point and two tangents, or even three tangents may be given. The first of these last mentioned cases comes under the 16th problem, the second under the 17th, and the third under the 18th. So that it would be more full to put it thus. Concipe tangentis cujusvis punctum contactus abire in infinitum, et tangens vertetur in asymptoton, atque constructiones problematum præcedentium vertentur in constructiones ubi asymptotos datur.[3]

If you approve of this way of putting it, please to insert it in the margent.[4]

I am

Your most humbl.

and most obednt. servt.

H PEMBERTON

To Sr. Is. Newton

NOTES

(1) Add. 3986, no. 8. On the cover Newton has jotted down some calculations relating to Proposition 30 (*Principia*, 1726, p. 105).

(2) Pemberton refers to the third edition, p. 96; this is in sheet N, which Pemberton has not mentioned before in a letter, although he may have indicated his objections on proof sheets sent to Newton.

(3) The wording here proposed was adopted by Newton. Pemberton wrote some further lines at the end of this paragraph, but struck them out.

(4) Another undated, unaddressed paper by Pemberton (Add. 3986, no. 29), in English, sent to Newton, falls in order (though not necessarily in time) between this letter and Pemberton's next. It is a construction of all five species—of which Newton cites only three—of the spiral curve described by a body 'actuated by a centripetal power tending to the center *C*, which is reciprocally as the cube of the distance therefrom'; see *Principia*, Book I, Prop. 41, Coroll. 3. This paper has not affected the text nor is it mentioned in the surviving correspondence.

1427 JOSEPH-NICOLAS DELISLE TO NEWTON

22 MARCH 1724

From the original in King's College Library, Cambridge[1]

For the answer see Letter 1429

Monsieur

Ayant appris par une lettre de mr Taylor a mr l'abbé Conty que la Societé Royale m'avoit fait l'honneur de m'admettre dans son illustre corps; je me suis trouvé forcé par mon devoir dans la liberté que je prens de vous ecrire, car à qui devois je plustot adresser mes trez humbles remerciemens de l'honneur que m'a fait vôtre illustre Societé, qu'a celui qui en est le chef. Permettez moy donc Monsieur de vous protester que ce n'a point eté la vanité qui m'a fait rechercher l'honneur d'etre admis dans vôtre illustre corps: je me rends assez justice pour croire que je ne merite point cet honneur; mais si j'ay j'ay [sic] jamais û quelque ambition, ca eté d'etre en correspondance avec les dignes membres de vôtre Societé sur les matieres d'Astronomie; et cest ce qui m'a fait prendre la liberté d'ecrire à Mr Halley pour lui presenter mon observation de mercure dans le soleil; et lui proposer un commerce reciproque d'observations astronomiques.[2] Vous pouvez etre assuré Monsieur que je me ferai toujours un trez grand honneur et un devoir de me conformer aux vues de vôtre illustre Societe dans les observations que je ferai à l'avenir; et que je ne m'anquerai [*sic*] point d'envoier à Mr Halley ou aux autres astronomes qui voudront bien entrer en correspondance avec moy toutes les observations qui meriteront de leur etre presentées. Je suis, Monsieur, avec tout le respect et la veneration possible vôtre trez humble et trez obeissant serviteur

a Paris ce 2 *avril* 1724 [N.S.]

DE L'ISLE LE CADET

Angleterre

Monsieur
Monsieur Newton President
de la Societé Royale des Sciences
    A Londres

NOTES

(1) Keynes MS. 94(C); microfilm 931.3. Joseph-Nicolas Delisle (1688–1768) was the younger brother of Guillaume Delisle, also a member of the Académie de Sciences, hence he was known as Delisle le Cadet. Having been an *élève*, he became an *adjoint astronome* of the Académie in 1716, and *associé* (in succession to Louville) in 1719. He was Professor of Mathematics at the College de France. Delisle was elected F.R.S. on 12 March 1724.

(2) Delisle, working first at the Paris Royal Observatory and later, after 1725, at the new St Petersburg observatory, carried on a voluminous correspondence with other astronomers, much of which still exists in manuscript in Paris. He also published numerous memoirs on astronomical and geodetical matters (see the article on Delisle by S. L. Chapin in the *Dictionary of Scientific Biography*, IV (New York, 1971)). Thinking that Halley's method of observing the transits of Venus to determine the Sun's parallax could equally well be applied to transits of Mercury, he published in the *Mémoires de l'Académie Royale de Sciences* for 1723, pp. 105–10 and 306–43, two papers concerning the passage of Mercury which took place on 9 November 1723, N.S., the first in preparation for the event, and the second recording his observations. It was presumably about the transit of Mercury that he wished to communicate with Halley.

1428 FATIO TO NEWTON

1 APRIL 1724

From the original in King's College Library, Cambridge[1]

Worcester April the 1. 1724

Honoured Sir,

I take the liberty to send this Letter to You, together with the inclosed project of an Advertisement, which I design to publish, in hopes that it will encrease both the Reputation and the Manufacture of pierced Rubies.[2] For hitherto Watchmakers use but two Rubies in a Watch, and that, for the most part, in Repeating Watches only. The thing being of a publick concern, and what I do much desire, I hope, Sir, you will not disprove that Your Name be mentioned in the Advertisement. I am, with all manner of respect

         Honoured Sir,

Your most humble and most
obedient servant
N. FACIO

To the Honourable Sr Isaa[c] Newton
   at his House
near Leicester Fields
       London

NOTES

(1) Keynes MS. 96(E); microfilm 931.5.

(2) Historians of horology agree that for many years after the invention of jewelled pivots for watches in 1704, jewels were employed only by a very few London makers, and that a century passed before the use of jewels in watches became at all common. Perhaps Fatio meant to print an advertisement in the newspapers. (For Fatio's earlier attempt to take out a patent for jewelled movements see Letter 1243, vol. VI).

1429 NEWTON ₁TO DELISLE

?APRIL 1724

From a holograph draft in the University Library, Cambridge[1]
Reply to Letter 1427

Vir celeberrime

Accepi Literas tuas 2 Apr. 1624[2] datas St. N. Et audiveram antea idque cum voluptate quod observationes tuas Mercurij in Sole nuper visi ad D. Halleium transmiseras. Commercium epistolare inter Astronomos vestros & nostros tendet proculdubio ad scientiam cœlorum promovendam; earum rerum studiosis gratissimum esse debet. Sed et tuo etiam merito effecisti & ut electionem tuam in Societatem nostram promoverem, & ut nos omnes in hac re facile[3] consensum facile præbuerimus. Vale.[3] Eaque de re tibi gratulor. Vale.

Viro Celeberrimo D. De l'Isle, Professori Astronomiæ, apud Parisienses, et Academiæ Scientiarum Socio Is. Newtonus salutem.[3]

Viro celeberrimo D. De L'Isle juniori, Professori Astronomiæ & Regiæ scientiarum Academiæ Socio

Apud Parisienses

Translation

Most famous Sir,

I have received your letter of 2 April 1624,[2] N.S. And I had already learned with great satisfaction that you had sent your observations of Mercury recently seen in the Sun to Mr Halley. A correspondence between your astronomers and ours will undoubtedly tend towards the promotion of knowledge of the heavens and ought to be welcome to all who are studious of such matters. But your own merits also both impelled me to move your election into our Society, and caused us all readily to agree upon it. Farewell.[3] I congratulate you upon this. Farewell.

Isaac Newton greets the celebrated Mr Delisle, Professor of Astronomy at Paris and member of the Academy of Sciences.[3]

To the celebrated Mr Delisle Jnr, Professor of Astronomy and member of the Royal Academy of Sciences at Paris.

NOTES

(1) Add. 3965(13), fo. 368, a very rough draft. The paper also bears a draft relating to *Principia*, Book I, Proposition 56 (1726, p. 147); compare (as regards the time-sequence) Pemberton's Letters 1426 and 1439. It also has some sentences on God and Christ's Kingdom.

(2) *Read:* 1724.

(3) These words are probably meant to be deleted.

1430 JOHN SCROPE TO NEWTON
10 APRIL 1724
From the original in the Mint Papers[1]
For the answer see Letter 1431

Sir

The Lords Comm[it]tee of his Ma[jesty']s most Honoble Privy Councill having had under their consideracon the Representation made to his Ma[jes]ty by the Parliament of Ireland against the Patent granted to Mr: Wood for the Coining of Copper Halfpence and farthings for that Kingdom[2] And having at the desire of the said Mr: Wood been pleased to Order that an Essay should be imediately made of the Fineness value and Weight of the said Coynage and the goodness thereof Compared with the former Coinage of Copper money in the said Kingdome The Lords Comm[issione]rs: of his Ma[jesty']s Treasury desire You in their Lordp's names to depute some person whom you shall know to be skillfull in the Art of Essaying to make the said Essays and Comparison And also to inspect and see that the said Coynage be in all things conformable

272

to the said Patent And the said person is to repair for that purpose with all Convenient Speed to Bristoll where the said Mr. Woods Office for the said Copper Coynage is kept And their Lordp's desire you to furnish him with all such Instruction's as you shall think proper for this Service And to let him know that their Lordp's will gratifye him for his trouble and charges therein in such manner as you shall think reasonable[3] I am

Sir

Your most humble Servant

J SCROPE[4]

Trea[su]ry Chambers
10th April 1724

   Sr: Isaac Newton

NOTES

(1) II, fo. 458; there is a clerical copy at P.R.O., T/27, 23, p. 469.

(2) Compare Letter 1403, note (1), p. 217. The Dublin Parliament protested that the patent had been granted to Wood without its consent and objected to the quality of the coinage. Swift was about to issue his *Drapier's Letters*, and the Privy Council to suspend Wood's mint. Wood defended himself by demanding an assay of his product.

(3) In his reply Newton objected to this proposed procedure and accordingly this instruction was cancelled on 17 April. Instead, Wood and his partners were ordered to send the Pyx containing samples of their coinage to London for assay, waiting upon Newton to receive further detailed instructions about the procedure to be followed (see P.R.O., T/27, 23, p. 471).

(4) John Scrope (d. 1752) had been appointed Secretary of the Treasury on 21 January 1724; he had been called to the Bar in 1692 and was from 1708 one of the Barons of the Exchequer of Scotland. From 1722 onwards he also sat in Parliament. He was a particularly fervent and staunch supporter of Sir Robert Walpole.

1431 NEWTON TO THE TREASURY

13 APRIL 1724

From the holograph original in the Public Record Office[1]
Reply to Letter 1430; for the answer see Letter 1433

To the Rt Honble the Lords Comm[issione]rs of
his Majties Treasury

May it please your Lordps

In obedience to your Lordps Order signified to me by Mr Sc[r]ope in his Letter of 10 Apr. instant, concerning the trial of the Pix of copper moneys coyned by Mr Wood at Bristoll, I humbly represent that the moneys there reserved for a trial are kept lockt up in a box or Pix under the keys of Mr Wood

& the Comptroller of that coynage,[2] or of their Deputies. And that in my most humble opinion the trial thereof may be more authentick & satisfactory & something cheaper if the Box before opening be brought up to London & the moneys be tried in his Majties Mint in the Tower by his Majties Assaymaster before such person or persons as your Lordps or his Ma[jes]ty in Council shall appoint to see the triall performed & report the event to your Lordps, & before the Officers of the said Mint, & before the parties concerned, vizt Mr Wood & the Comptroller of that coynage who are both in town, & their Deputies, & one or two Gentlemen Of Ireland whom your Lordships may give leave to be present if desired.[3] And at the same time any other parcel or parcels of copper moneys old or new may be tried there before the persons above mentioned & the weight & value ascertained & compared with the value of copper moneys now coyned at Bristoll & the trial reported to your Lordps.

But if your Lordps had rather that the copper Pix be tried at Bristoll, I will look out for a man to do it & treat with him & see him instructed & furnished with necessaries with all convenient speed.

All which is most humbly submitted to your Lordships great wisdome

Mint Office
Apr. 13th. 1724

ISAAC NEWTON

NOTES

(1) T/1, 247, no. 39, fo. 224; printed in Shaw, pp. 198–9. There are drafts in Mint Papers, II, fos. 466 and 471.

(2) This was Newton himself but perhaps he meant to indicate Mathew Barton, whom he had deputed to act as Inspector of the Irish coin (see Letter 1431, notes (1) and (3)).

(3) The Lords of the Treasury took Newton's advice, and ordered the Pyx to be brought from Bristol to London; see Letter 1433.

1432 NEWTON TO THE TREASURY
13 APRIL 1724
From the holograph original in the Public Record Office [1]

To the Right Honble the Lords Commissioners
of his Ma[jesty']s Treasury.

May it please your Lordps

The Clerks & under Officers of the Mint whose salaries were not above forty pounds per annum, have lately had their salaries increased by one fourth part by a signe manual, [2] except the Porter of the Mint Mr Robert Low. And I humbly pray that his Salary, wch is but twenty pounds per annum, may in like manner be increased to twenty & five pounds per annum [3]

Which is most humbly submitted to your Lordps great wisdome.

ISAAC NEWTON

Mint Office
April 13th. 1724

NOTES

(1) T/1, 251, no. 29, fo. 73.
(2) In August 1722; see Letter 1393, note (2), p. 203.
(3) Newton had made a representation to the Treasury as long ago as 1722 that Robert Lowe's salary be increased; see Letter 1392, note (1), p. 202. A warrant to Newton to increase Lowe's salary to £25 was signed on 21 April 1724; see P.R.O., T/52, 34, pp. 127–8.

1433 THE TREASURY TO THE MINT
21 APRIL 1724
From the original in the Yale University Library [1]
Reply to Letter 1431; for the answer see Letter 1434

Gent[leme]n

The Lords Committee of His Ma[jesty']s most honble Privy Council having Ordered that an Assay should be imediately made of the Finess Value and Weight of the Copper half Pence and Farthings Coined for the Service of Ireland and the goodness thereof compared with the former Coinage of Copper Money in that Kingdom We on this occasion thought fit by letter dated the 17th. Instant to direct the Patentees and the King's Clerk Comptroller of the said Coinage to cause the Box or Pix containing the Tryal Pieces of the said Coinage to be brought up from Bristoll and lodged in the Office of You the Master Worker of His Ma[jesty']s Mint so as the said Monies might at a time to be agreed on be there taken out and tryed and the said Comparison made before the said Officers of the Mint the said Patentees the Kings Clerk Comptroller and such Other Persons as We should appoint to be present [2] This comes therefore to desire You to fix the time or times for making the said Tryal and Comparison And to be Present Yourselves at the making thereof So as the Report which is to be made to Us in Writing of the said Tryal and Comparison may be Signed not only by Your Selves but also by the Kings Assaymaster and Other the Principal Officers of the Mint the said Patentees and the Kings Clerk Comptroller whom We do Order and direct (upon Notice from You) to attend the said Tryal and Comparison at such time or times as shall be so fixed And so We bid You heartily farewell and remaine

Gent[lemen]
Your very loving Friends

Trea[su]ry Chambers
21st *April* 1724

R WALPOLE
GEO BAILLIE
WILL. YONGE
GEO. DODINGTON

Sr Isaac Newton Mr. Southwell Mr. Scrope[3]

NOTES

(1) MSS. Ph 10018, Southwell 313, Coch. 724, fo. 29.
(2) Thus the Lords adopted Newton's advice in Letter 1431, p. 274.
(3) Edward Southwell was Secretary of the Privy Council of Ireland; John Scrope was Secretary of the Treasury.

1434 NEWTON, SOUTHWELL AND SCROPE TO THE TREASURY

27 APRIL 1724

From the draft in the Mint Papers[1]
Reply to Letter 1433

To the Rt Honble the Lds Comm[ission]ers of his
Majts Trea[su]ry.

May it please your Lordps

According to your Lo[rdshi]ps Order, the Pix of the copper moneys coined at Bristow by Mr Wood for Ireland has been opened & tried at the Mint in the Tower by the kings Assaymaster before us. And by the Comptrollers account[2] (to wch Mr Wood agreed,) there has been coined from Lady day 1723 to March 28th 1724, in half pence fifty & five Tunns, five hundred[weight] & three quarters & twelve ounces, & in farthings three Tunns, seventeen hundred, two Quarters ten pounds weight & eight ounces averdupois. And by the specimens of this coinage which have from time to time been taken from the several parcels coined & sealed up in papers & put into the Pix, we found that sixty half pence weighed fourteen ounces Troy & eighteen penny weight, wch is about a quarter of an ounce above one pound averdupois; & that thirty farthings weighed three ounces & three quarters of an ounce Troy & forty six grains wch is also above the weight required by his Patent; the whole coynage amounting to 59 Tuns, 3 [cwt], 1 Qter. 11 lwt & 4 oz. We found also that both halfpence & farthings when heated red hot spread very thin under the hammer without cracking, as your Lordps may see by the pieces now laid before you. But although the copper was very good & the money one piece with another was full weight, yet the single pieces were not so equally coined in weight as they should have been.

We found also that thirty & two old half pence coined for Ireland in the reigns of Charles the second, king James the second, & king William & Queen Mary, & produced by Mr Wood, weighed six ounces & eighteen penny weight Troy, that is 103½ grains a piece one with another. They were much worn. And if about six or seven grains be allowed to each of them one with another for loss of their weight by wearing (the copper money coined for England in the reign of king William being already as much lightned by wearing,) they might at first weigh about half a pound Averdupois one wth another: whereas thirty of those coyned by Mr Wood, are to be of that weight. They were also made of bad copper. Two of those coined in the reign of king Charles the second wasted much in the fire, & then spread thin under the hammer but not so well without cracking as those of Mr Wood. Two of those coined in the reign of king James the second wasted more in the fire & were not malleable when red hot. Two of those coined in the reign of king William & Queen Mary wasted still more in the fire & turned to an unmalleable substance like a cynder, as your Lordps may see by the pieces now laid before you.

By the Assays we reccon the copper of Mr Woods half pence & farthings to be of about the same goodness & value with the copper of which the copper money is coyned in the Kings mint for England, or worth in the Market about 12 or 13 pence per pound weight Averdupois; & the copper of which the half pence were coined for Ireland in the reigns of king Charles, king James, & king William to be much inferior in value, & almost of no value in the market, the mixture being unknown & not bearing the fire for converting it to any further use untill it be refined.

The half pence & farthing in the Pix coyned by Mr Wood had on one side the head of the king with this inscription GEORGIUS DEI GRATIA REX, & on the reverse a woman sitting with a Harpe by her left side & above her the inscription HIBERNIA with the date. The half pence coined in the reign of King Charles, King James & King William, had on one side the head of king

Charles or king James or King William & Queen Mary, & on the reverse a
Harp crowned.

All which facts we most humbly represent to your Lordps

Apr. 27th 1724

NOTES

(1) II, fo. 467; the draft is in Newton's hand. There are clerical copies at P.R.O., T/1, 248,
fos. 77–8 (signed by Newton, Southwell and Scrope) and in Yale University Library (unsigned).
Further drafts by Newton exist at Mint Papers, II, fos. 468 and 470v. The report was sent by
the Treasury Lords to the Committee of the Privy Council on 6 May 1724 (P.R.O., T/1, 248,
fo. 76). On 24 July this Committee in turn reported to the king (with copies of all the relevant
documents) 'upon the several Resolutions and Addresses of both houses of Parliament of
Ireland...the late Address of your Majesty's Justices and Privy Council of that Kingdom, and
ye Petitions of ye County and City of Dublin' against Wood's patent. The Committee exone-
rated Wood completely from any fault but accepted his offer to reduce the proposed copper
coinage from 360 tons (£100800) to 142 tons (£40000, including the 60 tons already dis-
tributed. But they could reduce the coinage no further. The full Privy Council accepted this
report on 6 August (*ibid.*, fos. 64–75).

(2) See Mint Papers, II, fo. 472.

1435 HAYNES TO NEWTON
29 APRIL 1724
From the original in the Mint Papers[1]

The 29th April 1724

Hond. Sr.

This morning early I went up into ye Mint Office of Receipt, & finding
Mr. Fauqu[ie]r had locked up ye 1 lwt. Averdupois, I was obligd to compare
the 7 lwt. Averdupois wth an exact ballance of Troy wts. The scales & one sett
of the Troy wts. have been lately adjusted, but the other sett of weight[s]
remaining Uncorrected, & there appearing about 6 gr. difference between
those 2 setts of Troy weights, I therefore compard the 7 lwt. Averdupois wth.
both the foremention'd Setts of Troy weights, & also wth. the Troy Stand[ar]d
weights in the Excheqr. & the Acc[oun]t stands thus.

lwt	lwt	oz	dwt	gr	
7. Averdupois	= 8.	6.	0.	18.	Troy at ye Tower—Corrected, &c
7. Do. . .	= 8.	6.	1.	0.	Troy at ye Tower—Uncorrected
7. Do. . .	= 8.	6.	2.	6.	Troy at ye Exchr–Stand[ar]d.

Consequently

lwt	lwt	oz	dwt	gr	
1 Averdupois	= 1.	2.	11.	13.	fere, at ye Towr. Corrected
1. Do. . .	= 1.	2.	11.	11. 5/7.	Troy, At ye towr. Uncorrected
1. Do. . .	= 1.	2.	11.	18	Troy at the Excheqr.[2]

I am Sr

Your most Obedt

& most Humble Servt.

H. HAYNES

Exchequr
Wensday 12 *Morning*

NOTES

(1) I, fo. 134.

(2) One pound avoirdupois should weigh 1 lb 2 oz 11 dwt 16 gr troy.

1436 GUILLAUME CAVELIER TO NEWTON
30 APRIL 1724
From the original in King's College Library, Cambridge [1]

a paris Le 11e *May* 1724

Il m'est tombé en main un petit Manuscrit[2] que lon assure venire de vous, Monsieur, comme votre Nom est tres estimé par toute l'Europe jay voulu le faire imprimer, mais l'on ma assuré quil y avoit des fautes et que cela pouvoit vous faire de la peine de le voir paroitre sous votre illustre Nom ce qui fait que je prens la Liberté de m'adresse[r] a vous et de vous prier, Monsieur, comme le manuscrit que jay est peu de choses, de vouloir bien me marquer, si vous pouviez m'en faire tenir une Copie Corecte *de votre Chronologie*[.] plusieurs personnes qui en ont des Copies defectueuses seront bien aises den' avoir de Correctes et moy Libraire qui ne cherche que de bonnes choses suis persuadé quil ny a rien de meilleur que ce qui part de votre plume j'atens lhonneur de votre reponse et suis avec un profond respec

Monsieur

Votre tres humble et tres obeissant serviteur

G. CAVELIER fils Libraire

rue S. jacques

NOTES

(1) Keynes MS. 138(A); microfilm 1011.22. Printed in More, p. 613. The writer was the younger of the two Guillaume Caveliers, father and son, members of the famous Paris family of booksellers. He became 'Libraire' in 1702 and later rose to a position of importance at the

279

Imprimerie du Roi. (See A. M. Lottin, *Catalogue Chronologique des Libraires et Des Libraires-imprimeurs de Paris*, Part II (Paris, 1789), p. 21.)

(2) Cavelier was eventually to print the manuscript, in French translation, as *Abregé de la Chronologie de M. le Chevalier Isaac Newton, fait par lui-même, & traduit sur le Manuscrit Anglois* (Paris, 1725). It also appeared in volume VII of Humphrey Prideaux, *Histoire des Juifs et des peuples voisins* (Paris, 1725)—another translation from English issued by the same bookseller.

Newton recounts the history of the manuscript in his 'Remarks on the Observations made on a Chronological Index of Sir Isaac Newton, translated into French by the Observator, and published at Paris,' *Phil. Trans.* **33**, no. 389 (July and August 1725), 315–21. (Several drafts of this paper are now in the Yahuda Collection at the Jewish National and University Library, Newton MS. 27.) The Chronological Index was first compiled for 'a particular friend' (Princess Caroline of Wales), some time in 1716, as an abridgement of a longer and as yet incomplete *Chronology*.

Apparently the Princess had asked Newton for a copy of his work on chronology, but since it was not yet finished, Newton drew up his abstract, and transmitted it to Princess Caroline via the Abbé Conti, with instructions that it be kept secret. Conti asked the Princess to intercede with Newton for a copy for himself which she successfully did. Conti made his copy available to a number of friends in France, and it became generally known (see vol. VI, p. 332, note (7)). Nicolas Fréret, a notable French scholar, was responsible for the translation into French and for a number of critical comments which increased Newton's annoyance.

Cavelier, the printer, wrote in the Preface to the *Abrégé* that he had sent three letters to Newton asking permission to print it; the first and third of these letters are Letter 1436, printed here, and Letter 1461, below; the second letter is missing. In the third letter Cavelier said he would take silence for consent, and this he did, obtaining a 'Privelège' for the printing of the *Abrégé* on 21 May 1725. A few days later Newton eventually sent Cavelier a reply (Letter 1469), in which he refused to consent to the publication; but by then printing was under way.

Cavelier sent Newton a copy of the book, which he received on 11 November 1725. Newton's reaction was the printing of his 'Remarks' in the *Philosophical Transactions* (see above), where he claimed that Cavelier must have been aware that he did not want the manuscript made public, for he 'knew that I had not seen the Translation of the Abridgement, and without seeing it could not in reason give my consent to the Impression...He knew that the Translator had written a confutation...under...the Title of *Observations*...'.

For further details of the *Abrégé* and its publication, see Frank E. Manuel, *Isaac Newton Historian* (Cambridge, 1963), pp. 21–36.

1437 SCROPE TO NEWTON
11 MAY 1724
From the original in the Mint Papers [1]
For the answer see Letter 1438

Whitehall Treasury Chambers May 11. 1724

The Right Honble the Lords Com[missione]rs of his Ma[jesty']s Treasury are pleased to Refer this Petition [2] to Sr Isaac Newton Knt:, Master and

Worker of his Ma[jesty']s Mint, who is to consider the same, and Report to their Lordps a true State of the Petitioners case together with his opinion what may be done therein

J SCROPE

NOTES

(1) III, fo. 243v.
(2) The petition, from John Walker, an Edinburgh merchant, is at *ibid.*, fo. 243r. Justus Brandshagen, now dead, had borrowed ten guineas from Walker, who has applied to Brandshagen's daughter for repayment. She has informed him that the Government owed Brandshagen over £100 for his services in respect of the Alva silver mine in Scotland (see vol. VI, p. 317, note (3)), and Walker has therefore been advised to apply to Newton for repayment of the debt.

1438 NEWTON TO THE TREASURY

18 MAY 1724

From the holograph original in the Public Record Office[1]
Reply to Letter 1437

To the Rt Honble the Lords Comm[issione]rs of
his Majties Treasury

May it please your Lordps

In obedience to your Lordps Order of Reference of May 11th instant upon the Petition of Mr John Walker, I humbly represent that Dr Justus Brandshagen received 60 *lb* before his going down into Scotland to survey Sr John Ereskins Mine, & this was for bearing the charges of his journey into Scotland & back again & some other incident charges there, as may appear by your Lordps Warrants to the Earl of Halifax dated 5th Sept. 1716 & 24 July 1617 [*sic*] for issuing out moneys upon this service. And that the said Justus Brandshagen was further paid 129 *lb* for 129 days after the rate of 20*s* per diem during his stay in Scotland upon this service & 25*lb* in discharge of bills for clearing the Mine of rubbish, fitting up a place for making the assays & some other incident charges there. He & Tho. Hamilton left Scotland on Feb. 18th 171$\frac{6}{7}$, & he gave in the Report to your Lordps concerning the Mine 71 days after, vizt on Ap[r]il 29 1717. And when he gave in his Accounts to your Lordps, which was on May 17th following, he demanded 20*s* per diem for those 71 days; but it was not allowed him, the Princes Warrant limiting that allowance to the time of his stay in Scotland upon the service of the Mine. He received therefore only 214*lb*. And when I paid him the last part of this summ (wch was on Aug. 26, 1717) he gave me a receipt in full of all accounts.

The 71*lb* set down by Dr Brandshagen in the first of the three Articles annexed to Mr Walkers Petition, as due to him, was therefore not due to him. And the same is to be understood of the 35*lb* 10*s* set down in the second Article as due to Tho. Hamilton. The 6*lb* 18*s* in the third Article should be 6*lb* 8*s*; & this summ hath been already paid to Dr Brandshagen. And I know of nothing more due to them from the crown then what they have already received & given Receipts for in full of their accounts. [And therefore if Dr Brandshagen's Executor be indebted to Mr Walker, he is to pay the debt himself.][2]

All which is most humbly submitted to your Lordps great wisdome

ISAAC NEWTON

Mint Office.
May 18. 1724.

NOTES

(1) T/1, 247, fo. 277, no. 53. There is an identical holograph draft in Mint Papers, III, fo. 245 and a rougher draft in *ibid.*, fo. 235.

(2) This sentence appears only in the draft, fo. 245.

1439 PEMBERTON TO NEWTON
?MAY 1724
From the holograph original in the University Library, Cambridge [1]

Sr.

In the sheet here sent you in pag. 159. the sixth line from the bottom you will see, that for axem *CO*, I have in the margent directed the expression rectam *PO* to be put:[2] for the axis *CO* is perpendicular to the plane of the ellipsis *PQ*; but the position of this ellipsis in respect of *PO* is so given, that it's lesser axis lies in that line *PO*.

Whereas in the two last lines of this 159th pag. and in the beginning of the next, are these words Inde autem invenietur trajectoriæ vestigium illud *APp* eadem methodo qua curva linea *VIK* in schemate propositionis XLI. (per propositionem illam et ejus corol. 2.) ex similibus datis inventa fuit: it is to be noted, that the curve in that proposition is not found from data like these here given; but only that in corol. 2. the relation is shewn between such data as are here given, and those whereby the curve in that prop. is found. Hence the present expression is liable to objection; might it not rather be put thus.

Inde vero (conferendo prop. XLI. cum corol. suo 2.) ratio determinandi curvam *APp* facile apparet.[3]

I am

Your most humbl.

and most obednt. servt.

H PEMBERTON

To Sr. Is. Newton

NOTES

(1) Add. 3986, no. 9. Newton wrote on the back of this letter: (i) drafts of a scholium intended to follow Book II, Proposition 36 (not used); (ii) notes on the motions of the Moon's aphelion, ending on 29 December [1724].

(2) *Principia*, 1726, p. 159. The reading 'axem *CO*' is not otherwise recorded; all versions read 'rectam *PO*'.

(3) This wording was printed.

1440 MARSIGLI TO NEWTON

?MAY 1724

From the original in King's College, Library, Cambridge[1]

Eruditissimo Ill[ustrissim]oque Præsidi
Reg[iæ] Societatis Anglicanæ
Aloysius Ferdinandus Marsili S.P.D.

Consilia, quæ cum Londini essem a vobis accepi, quibus vestra quoque accedit auctoritas, mihi Danubiale opus meum in memoriam revocarunt, ut ipsum typis traderem ex debito, quod vobiscum conflaram in editione Prodromi huiusce operis 23. abhinc annis exacta. Quare Amstelodamium profectus, ibi de operis eiusdem impressione conveni cum societate quadam Typographorum ad id officium collecta. In patriam inde reversus longævum ab eo pulverem excussi, et tantum in eo laboris impendi, quantum integrum fere annum æquavit. Sic perfectum, quatenus in me fuit, absolutumque Amstelodamium illud misi cum æneis laminis ad id spectantibus, quod, omne cum eo pervenerit, illico typis dabitur. Vos Amstelodamio propinquiores, quam ipse qui paratus sum in Galliam Narbonensem tendere et Marsiliam, ac Theuroentium,[2] et ibi cælo frui mitissimo, ocius atque facilius exemplum huiusce operis accipietis. Ego interim brevi futuram hanc editionem vobis nuncio, primum, ut partibus meis me non defuisse videatis, deinde ut opem vestram quam etiam atque etiam postulo, præstatis mihi, et benignitate, atque humanitate vestra opus publicæ censuræ subjiciendum, prosequamini.

283

Tentamen physicum de Historia naturali maris Regiæ Parisiorum Academiæ inscriptum, prolatum iam puto fuerit in lucem publicam a magna
societate Amstelodamij.[3] Ipse D. Cæsarem Sardi[4] Rogo, ut unum eiusdem
exemplum mittat per D. Boeerhave [sic] ad consocium nostrum D. Sherard,[5]
qui debita veneratione vobis illud meo nomine tradat, quod ipsum stato tempore peragat de opere danubiale.

Apud me sunt adnotationes posteriorum mearum observationum, quas
habui novissimo meo itinere a Leburno Portu ad metropolim vestram et
plurimas circum batava litora, quas ad meum eruditissimum amicum Boeerhave scribo, et addam editionem alteram de eo, quod a me teneram adhuc
ætatem ducente de Bosphoro Tracico observatum fuit, et scriptum.[6]

Si quietem, et incolumem vitam largietur mihi Deus, absolvam Ideam in-
nixam observationibus terræ ac maris, a me peractis disserendi gratia super
probabili organica terræ structura, et ratum est, ut dissertatio[7] ista fronte
ferat nomen Regiæ societatis nostræ, dummodo vos ita eam comitate vestra,
diligentiaque corrigatis, ut tanto honore non omnino indigna videatur.

Specula Instituti huius scientiarum,[8] et Artium absoluta est ea arte, ijs
sumptibus, ea magnificentia, quibus addi nihil potest, at nihil ulterius est
desiderandum. Omnium, quæ in ea fient observationum ad vos collatione
instituentur et procedent. Interim ad vos mitto exemplar observationis[9]
novissimæ solaris eclypsis. Observatorio huic Præfectus est D. Dr. Eustachius
Manfredi; ab eo, quæ potissimum referenda erunt, accipietis.

Epistolas si quas ad me scribere velitis, tradetis D Sherard qui aut Leiden eas
mittet ad D. Boeerhave, aut Amstelodamium ad P. Cæsarem Sardi. Pro in-
columitate tua, Præses eruditissime, nostrorum temporum gloria scientiarum-
que præsidium, eorumque singulorum, qui me istinc proficiscentem tot,
tantisque officijs cumularunt, D.O.M. vota humillima facio, nec ea ullo unquam tempore prætermittam. Ad Ill[ustrissim]um, et Eruditissimum Virum
D. Neuthun [sic] Præsidem dignissimum R. Societ. Anglicanæ
                              Londinum

Translation

Luigi Ferdinando Marsigli sends a grand salute
to the most learned and famous President
of the English Royal Society

The advice which I received from you when I was in London, to which your authority
is also added, has recalled to my mind my work on the Danube, [with the intention]
that I might entrust it to the printers, in accordance with the pledge I made to you in
publishing the *Prodromus* of this work completed twenty-three years ago. Wherefore,

having journeyed to Amsterdam, I there made an agreement about the printing of this same work with a certain company of printers assembled for the purpose. When I had returned thence to my own country I brushed the dust of ages from [the book] and spent as much labour upon it as occupied almost a whole year. After thus perfecting and completing it so far as I was able, I sent it to Amsterdam with the copper plates belonging to it, and when everything has arrived there it will immediately be given to the printers. Being nearer to Amsterdam, you will receive a copy of this work more quickly and easily than I will, since I am about to set out for Provence, and Marseilles, and moreover Theuroentia,[2] where I shall enjoy a most mild climate. Meanwhile I am informing you in a few words of this future edition, first, so that I should not seem to you to be failing in my duties, and also so that you may give me your help, which over and over again I entreat you for, and may bestow your goodwill and kindness upon the work when it comes before public criticism.

A physical essay on the natural history of the sea, addressed to the Royal Academy of Paris, has already been published, I believe, by the great [East Indian] Company of Amsterdam.[3] I am asking Mr Cesare Sardi[4] himself to send one copy of it through Mr Boerhaave to our colleague Mr Sherard,[5] who will deliver it to you with proper respect on my behalf, as at the due time he will do the Danubian work.

At my home are notes on my later observations, which I have made during my most recent journey, from the Port of Livorno to your city, as well as many from around the Dutch coast, which I am writing for my most learned friend Boerhaave, and I may add a second edition of what I have observed and described concerning the Thracian Bosphorus when travelling during my early youth.[6]

If God will bestow upon me a quiet and healthy life, I will complete my idea supported by observations of the land and sea which I have made in order to discourse upon the probable organic structure of the earth, and it is settled that that essay[7] will bear the name of our Royal Society at its head, provided that you will so amend it in your kindly and diligent way, that it may seem not altogether unworthy of so great an honour.

The observatory of this Institute of Sciences and Arts[8] has been completed with such skill, at such costs and with such magnificence that nothing can be added, and nothing further is to be desired. Collections of all the observations which have been made there are being drawn up and sent to you. Meanwhile I send you a copy of the observation[9] of the most recent solar eclipse. Dr Eustachio Manfredi is director of this observatory; you will receive what has been recorded chiefly from him.

Should you wish to write me letters, hand them to Mr Sherard, who will send them either to Leyden to Mr Boerhaave, or to Amsterdam to Father Cesare Sardi. I make a most humble prayer for your safekeeping, most wise President, glory of our times and protector of the sciences, and that of those individuals who heap so many and so great courtesies upon me; nor shall I ever fail to do so.

To the most famous and learned Sir Isaac Newton,
Most worthy President of the English Royal Society, London

NOTES

(1) Keynes MS. 94(F); microfilm 931.3. For Marsigli's last letter, see Letter 1424. The present letter is endorsed by Newton 'received of Mr Innys. Aug. 12. 1724.' It was apparently sent via Hermann Boerhaave and William Sherard. (See G. A. Lindeboom (ed.), *Boerhaave's Correspondence*, Part I (Leyden, 1962), pp. 128–30: Boerhaave to Sherard, 4 August 1724, N.S., where Boerhaave asks Sherard to pass on the letter and adds 'Le bon Compte [Marsigli] a quittè Bologne, & est partie pour Marsigle; je crois, afin d'y finir ses jours en paix, et en tranquillité. que DIEU Luy donne! Il a achevè ses oeuvres du Danube, et son histoire de la mer, qui verront bientost le jour.')

(2) Possibly Marsigli invokes the name, Tauroentium, of an unknown Roman coastal town east of Marseille; in fact he probably meant Cassis, near that city, where he resided more than once. If he went to Provence—an intention also indicated in Boerhaave's letter quoted in note (1) above—he must soon have returned to Italy, settling at Maderno on Lake Garda, upon whose study he embarked. See R. Accademia delle Scienze dell'Istituto di Bologna, *Scritti inediti di Luigi Ferdinando Marsigli* (Bologna, 1930), p. 11.

(3) Marsigli's *Histoire physique de la mer* (Amsterdam, 1725) was published under the direction of Boerhaave, with a Latin preface written by him; see F. W. T. Hunger, 'Boerhaave comme Naturaliste', *Janus*, 23 (1918), 354, and G. A. Lindeboom, *Herman Boerhaave* (London, 1968), p. 170. Publication of the book was paid for by the Dutch East India Company. Marsigli had made extensive sea voyages to England and around the coasts of Western Europe in 1722–3, meeting Boerhaave in Leyden before returning to Bologna.

(4) The reading is doubtful; we have been unable to identify any friend of Marsigli's of this name.

(5) William Sherard (1659–1728) studied law at Oxford, but had a strong amateur interest in botany. He undertook a revised edition of Gaspard Bauhin's Πιναξ *Theatri Botanici* (Basel, 1623), the manuscript for which reached 130 volumes, but which was never published. He corresponded extensively with Boerhaave, and sent him numerous botanical specimens. (See J. Heiniger, 'Some botanical activities of Hermann Boerhaave', *Janus*, 58 (1971), 1–78.)

(6) For these works by Marsigli see firstly A. Righi in *Annuario del R. Liceo Scientifico, 1929–30* (publishing and translating Marsigli's *Epistola continens observationes addendas Tentamini physico naturalis historiæ maris...ad...H. Boerhaave*), and secondly Marsigli's *Osservazioni intorno al Bosphoro tracio ovvere canale di Constantinopoli* (Rome, 1681).

(7) Marsigli does not seem to have completed this planned dissertation.

(8) In 1709 Marsigli had offered the Senate of Bologna a large collection of natural history and geological specimens, of astronomical instruments and experimental apparatus, and of books, and antiquities and in 1712 the Senate formally accepted the gift and finally implemented a plan to use it as a basis for an Istituto di Bologna. This was to be housed in a building constructed for the purpose, which naturally included a tower for making astronomical observations. The observatory was eventually completed in 1725. Eustachio Manfredi was appointed as astronomer in 1711; he himself had earlier founded a Bologna Society known as the Inquieti, which was now officially recognized as the Accademia delle Scienze dell'Istituto di Bologna, one of many academies associated with the Institute.

(9) There is no record of these observations having been sent; certainly they were not published in the *Philosophical Transactions*.

286

1441 NEWTON TO THE TREASURY
18 JUNE 1724
From a holograph draft in the Mint Papers[1]

To the Right Honble the Lords Comm[issione]rs
of his Majties Trea[su]ry

May it please your Lordps

There having been coined almost a million and an half in gold & silver since the last trial of the Pix wch was three years ago, I humbly pray that the Pix of this coinage may be tried this summer[2]

Which is most humbly submitted to your Lordps great wisdome

ISAAC NEWT[ON][3]

Mint Office. June 18*th*
1724.

NOTES

(1) II, fo. 233.
(2) The Treasury ordered that a Trial of the Pyx take place on 3 August 1724 (see P.R.O., T/27, 24, p. 13).
(3) The paper is torn.

1442 BURCHETT TO NEWTON
4 AUGUST 1724
From a copy in the Public Record Office[1]

4 *August* 1724

Sir

The Bearers hereof Messrs: Richard Burridge and Samuel Palmer having represented to my Lord Comm[issione]rs of the Admiralty, that they have something to offer, leading to the discovery of the Longitude;[2] Their Lordships desire you will please to give them an opportunity of communicating their proposal to you, in conjunction with Dr: Halley; and that they may have your opinions thereupon. I am

Sr:

Your most &c

J BURCHETT

Sir Isaac Newton
        The like to
Dr: Halley

(1) Adm/2, 457, p. 244.

(2) Burridge and Palmer (whom we have been unable to identify) attended a meeting of the Admiralty Board on 4 August 1724, and were referred to Newton and Halley (see P.R.O., Adm/3, 35, unpaginated). No reply is minuted in the ensuing months.

1443 PEMBERTON TO NEWTON
?AUGUST 1724
From the holograph original in the University Library, Cambridge[1]

Sr.

I have here inclosed six sheets of your Book,[2] with a few questions upon them. You will observe in the figures I have generally altered them according to the opinion I propose; but with black lead only, so that wherever you do not approve, they are easily restored to what they were before, by rubing of the lead. I just now called in at Mr. Innis's and was very much surprized to hear that the printer complained to you yesterday, that he had no Copy, whereas he hath had four sheets a full week. I suppose you remember that I waited on you tuesday sevennight with the draughts of thirty figures, I had just then received from Mr. Senex.[3] The next day I spent in examining them, and corrected several faults; and on thursday I delivered them with four sheets of the Copy; and the printer has now in his hands all those cuts, except only one I have here inclosed (being the 22d) to have some directions from you concerning it.[4] As soon as I receive this from you with the sheet belonging to it,[5] I shall deliver these likewise to the printer, and the other sheets that come between the sheet I here speak of and the others here inclosed.[6] I am Your most humbl. and most obedient Servt.

H. PEMBERTON

NOTES

(1) Add. 3986, no. 10.

(2) These must have been sheets of the second edition of the *Principia*, since Pemberton means to alter the figures on them and the required new blocks are still to be cut. They must be of the latter part of Book II.

(3) John Senex of Fleet Street, London, best known as a maker of globes; he also made and sold mathematical instruments and engraved plates for mathematical books. He died about 1740.

(4) The twenty-second individual cut in Book II is that on pp. 270 and 272 of the third edition (Prop. 14, sheet Mm). It was a new cut in this edition.

(5) That is, presumably sheet Kk of the second edition equivalent to Mm in the third edition, which could not be reset until the cut just mentioned had been approved.

(6) These 'sheets' must all be second edition, to be reset. The 'other sheets that come between' must, then, begin with Ll and 'the others here enclosed' begin later, perhaps with Nn.

1444 NEWTON TO TOWNSHEND
25 AUGUST 1724
From the printed version in Edleston, *Correspondence*, p. 316 [1]

My Lord

I know nothing of Edmund Metcalf [2] convicted at Derby assizes of counterfeiting the coyne; but since he is very evidently convicted, I am humbly of opinion that its better to let him suffer, than to venture his going on to counterfeit the coin & teach others to do so untill he can be convicted again, For these people very seldom leave off. And its difficult to detect them. I say this with most humble submission to his Majs pleasure & remain

My Lord
your Lordp's most humble & obedient Servant

Mint office Aug. 25, 1724.   Is. NEWTON
Ld. Townshend

NOTES

(1) We have not been able to trace the original. For Charles Townshend (1674–1738) see vol. VI, p. 316, note (2). After Stanhope's death Townshend was reappointed as Secretary of State for the Northern Department (February 1721).

(2) Nothing is known of Edmund Metcalf.

1445 PEMBERTON TO NEWTON
?SEPTEMBER 1724
From the holograph original in the University Library, Cambridge [1]

Sr

According to your order I have here made bold to set down a few thoughts upon the subject I had the honour last to discourse with you upon. [2]

Whereas in corol. 7, 8, 9. of prop. 36 the force of the water upon the circle *PQ* is set down with some latitude, it being only collected to be greater than one third and less than two thirds of the weight of the incumbent column; so that the following proposition, as you were pleased to observe, is left as much at large; I here have presumed to enclose two propositions, [3] which I deduced from considering the method of reasoning on the running of water out of a vessel, which you gave in your first edition. In these propositions, if I have reasoned truly, I have brought out exactly your 37th proposition, that when a cylinder moves in a fluid contained in a vessel of so wide extent as to be in effect infinite in respect of the cylinder; then the resistance of the cylinder moving in the direction of its axis will be neither greater nor less but just equal to the weight of a cylinder of the fluid, whose base shall be equal to the base of the cylinder resisted and altitude half the height whence a body must fall to acquire the velocity of the cylinder. But then you'l see that the case of the second corollary of your proposition comes out by my reasoning a little different; for my proposition makes the resistance to be to the weight of this cylinder in the duplicate ratio of (see your fig. of prop. 37) $2EF^q$ to $EF^q - PQ^q + EF\sqrt{EF^q - PQ^q}$; whereas you put it in the forementioned corollary to be jointly in the duplicate ratio of EF^q to $EF^q - PQ^q$ and in the single ratio of EF^q to $EF^q - \frac{1}{2}PQ^q$: so that my proposition makes the resistance a little less. The difference indeed is not very sensible, unless PQ bear a very great proportion to EF; for if PQ be $= \frac{1}{10}EF$, the resistance of my proposition is to the resistance as in your corollary in the ratio of 100 to 101, and if PQ be $\frac{1}{4}EF$ my proposition diminishes the resistance in the proportion of 1000 to 1067, if PQ be $\frac{1}{2}EF$ I diminish the resistance in the proportion of 100 to 132.

If, Sir you should chance not to discover any paralogism in my propositions, might it not be worth while to consider of making some experiments in globes so large as might make the forementioned difference manifest; provided that you also think the considerations, I am now going to propose, to be of any weight.

I esteem the method you took in the first edition of your book to determine the motion of water running out of a vessel to be just; so far as relates to the quantity of water that will run out in any given space of time; provided the vessel be so large, that the water within it has no sensible motion: The same conclusion you have there given may be readily deduced from the first of my propositions, the demonstration of which scarce differs from the reasoning you there make use of: for if we suppose the water to issue out with that velocity, as to generate the same degree of motion, as the action of gravity would produce in the same time by acting on the incumbent column; it follows from that proposition, that this velocity will be that, which a body would acquire in falling half the height of this incumbent column. I collect from the paper you have given me to be added in pag. 305,[4] that you likewise approve of this proposition in the first edition of your book; by your observing in that paper, that the velocity of the water in the hole it self is what would carry it up to half its height; and your asserting that the increase of velocity in the water arises from the obliquity of its motion.

But if this proposition be true, then must not what is asserted in the beginning of it (1st Edit. pag. 330. paragr. 1.) be also admitted, that every part of the bottom of the vessel sustains the weight of the column of water that is perpendicularly over it? If so a circle fixed within the hole will likewise sustain the weight of the column incumbent over it, when the vessel is so large, that the water with in it has no sensible motion. Whether the difference from this in my experiments arose from the motion of the water within the vessel might be known by making the experiment in a vessel wider.

I have one difficulty further about this 36th proposition, that the method of exhibiting the velocity of the water gives the velocity so much the greater, as the hole bears a greater proportion to the bottom of the vessel; even so that by widening the hole, the velocity here exhibited may be increased to any magnitude whatever short of infinite.

Would it be amiss to examine this proposition by some experiments, where the hole might bear a considerable proportion to the breadth of the vessel?[5] especially since several foreigners have thought fit to controvert this proposition, and are at present disputing about it; so that in all probability such experiments will be made by others.

Sr. I ask your pardon for troubling you with so large a scrawl; but as I shall not allow my self to take up any more of your time upon this head, I was desirous to hint at once all my thoughts. I shall wait on you in a few days for your instructions; and in the mean time shall proceed to prepare for the press some of the following sheets.

> I am
> Your most humbl.
> and most obednt. servt.
> H PEMBERTON

NOTES

(1) Add. 3986, no. 14. The argument for placing this letter at this point in time is as follows: Propositions 36 and 37 of Book II extend through the second edition sheets Pp to Rr, which must have been sent to Newton with Pemberton's last letter. The present letter is a product of his conversation with Newton about those 'six sheets', and cannot therefore have been written long thereafter; moreover, in Pemberton's next letter (according to our ordering) the printed text of the third edition has already advanced to Book II, Proposition 13.

(2) Obviously in question here is the text of the second edition (formerly of so much trouble to Cotes, see vol. v) which Pemberton would prefer to rewrite before reprinting this part of Book II. Newton overruled him and the third edition text remained, with only trifling alterations, as in the second.

(3) We have not found this draft.

(4) The addition is at the end of Proposition 36, case 1 (*Principia*, 1726, pp. 330–1). Newton here asserts that if the diameter of the orifice of outflow be d (water height above it $= h$; acceleration due to gravity $= g$) then the velocity of flow through the orifice is $\sqrt{(gh)}$; if a *vena contracta* is formed by the jet at a distance h' below this orifice, whose diameter is $(21/25)d$, then

the velocity of the water in this *vena contracta* will be about $\sqrt{(2g(h+h'))}$, since $(25/21)^2 \approx \sqrt{2}$. The addition does not alter the sense of the Proposition as it was stated in the second edition.

(5) This was not done.

1446 NEWTON TO THE TREASURY
5 OCTOBER 1724
From the holograph original in the Public Record Office [1]

To the Rt Honble the Lords Comm[issione]rs
of his Ma[jes]ties Trea[su]ry

May it please your Lordships

In obedience to your Lordps Order of Reference of 25 June concerning a new double silver seale made by Mr James Girard for his Ma[jes]ties Bahama Islands in America, we have perused his Ma[jes]ties Warrant for making the same & also the certificate of the delivery pursuant thereto, & find that it was made according to the directions of the said Warrant & sent to the Governour of the said Islands. And by an impression taken from it before it was sent & kept in the Mint, we find that the work was very good. It was weighed in the Mint before it was sent, & there found of the weight expressed in Mr Gerard's Bill, & the price of the silver & work is reasonable in proportion to the prizes heretofore allowed for seales of the like sizes.

All which is most humbly submitted to your Lordps great wisdome

Mint Office   ISAAC NEWTON
Octob.

NOTE

(1) T/1, 248, fo. 154, endorsed '6th octob. 1724' and 'agreed'. Girard's bill, dated 6 April 1724 and amounting to £89. 14s (the seal weighing over 65 ounces) is annexed; on the reverse this sheet (fo. 156) carries Scrope's order of 25 June to the Principal Officers of the Mint to make a report on this bill. There are drafts of Newton's letter in the Mint Papers, III, fo. 471; copies of the bill and order of reference are at Mint 1/7, p. 109.

1447 PEMBERTON TO NEWTON
?OCTOBER 1724
From the holograph original in the Bodleian Library, Oxford [1]

Sr

In this sheet in page 271 [2] l 15, 20, 25, I have changed the expressions in fig cas. 1. prop. XIII, fig. cas. 2. prop XIII and fig. cas 3 prop præced., into in figura prima, in figura secunda, and in figura tertia, because now the figures are drawn in this proposition it self, which they were not before

I am

Your most humble

and most obed. servt.

H. PEMBERTON

To Sr Is Newton

NOTES

(1) Bodleian New College, MS. 361, II, fo. 85; there is a nineteenth-century copy in U.L.C., Add. 3986, no. 12 *bis*, endorsed: 'From original in possession of Revd Jeffery Ekins Sampford Braintree Essex.'

(2) Book II, Prop. 14, the third edition page.

1448 PEMBERTON TO NEWTON

?NOVEMBER 1724

From the holograph original in the University Library, Cambridge [1]

Sr.

In the present sheet [2] the word funependulum is spelt, as I have here written it, with an e between the n and p; but in some of the sheets wrought off it is spelt with an i, thus funipendulum. The only places where it occurs in this sheet are in page 294 and 295. Over-against all of them I have put this mark qr. and if you chuse to have the word spelt every where, in the same manner, be pleased to correct it accordingly. [3]

In page 292. Suppose in line the 10th were added nisi quantum, gravitas specifica mercurii a diverso caloris gradu mutetur; for as the mercury is dilated a little by heat, its specific gravity will be altered. [4]

I am

Your most humbl.

and most obednt. servt.

H PEMBERTON

NOTES

(1) Add. 3986, no. 12.

(2) Pp of the third edition, pp. 289–96, Book II, Section VI.

(3) Newton did not trouble about the inconsistency.

(4) These words were not added.

1449 NEWTON TO HALLEY
3 DECEMBER 1724
From the holograph original in private possession[1]
For the answer see Letter 1455

Dr Halley

I received from you formerly a Table of the motions of the Comet of 1680 in a Elliptic Orb.[2] You there put the Node ascendent in ♑ 2gr. 2'. The Node descendent ♋ 2gr. 2'. The inclination of the plane of the Orb to the plane of the Ecliptic 61gr. 6'. 48″. The Perihelium of the Comet in this plane ♐ 22gr. 44'. 25″. The equated line of the Perihelium Decem. 7. 23h. 9'. The distance of the Perihelium from the ascending Node in the plane of the Ecliptic 9gr. 17'. 35″. The Axis transversus 138,29571. And the Axis conjugatus 1,84812, the mean distance of the earth from the Sun being 1,00000. And in this Orb you computed the places of the Comet on November 3d. 16h. 47'. November 5d. 15h. 37' & Novem. 10d. 16h. 18', as follows

1680 Tempus verum	Long. comp.	Lat. comp.
Novem. 3d. 16h. 47'.	♌ 29°. 51'. 22″	1. 17'. 32″ bor
5. 15 37	♍ 3. 24. 32.	1. 6. 9.
7.[3] 16. 18.	15. 33. 2.	0. 25. 7.

The first of these three places you have inserted into the Table of the motions of this Comet in an Elliptic Orb, wch you have printed in your Astronomical Tables[4] where you treat de motu Cometarum in Orbibus Ellipticis. I beg the favour of you to reexamin the two last of them, vizt those on Novemb. 5d. 15h. 37' and Novem 10d. 16h. 18'.[5]

In the same printed Table, you have calculated the place of this Comet upon March 9d. 8h. 38'. true time. I beg the favour of you to calculate its place in the Parabolic Orb also upon March 9d. 8h. 38' true time, & send me its computed Longitude Latitude & distance from the Sun. For I would add them to the Table of the motion of this Comet in a Parabolic Orb printed in the third book of the Principia Mathematica pag. 459. Ed. II.[6] By its distance from the Sun I mean the distance of its center from the center of the Sun in parts whereof the Radius of the Orbis Magnus is 100000.[7]

I am
Your most humble Servant
ISAAC NEWTON

St Martins Lane
by Leicester fields
Decem. 3. 1724

For Dr Edmund Halley, the Kings
Professor of Astronomy at the
Observatory in Greenwich
in Kent
            These

NOTES

(1) There are in addition two drafts in U.L.C.: Add. 3965(14), fo. 605; Add. 3982, no. 11. The first of these was printed as Letter 538, vol. IV, the second in E. F. MacPike, *Correspondence and Papers of Edmond Halley* (Oxford, 1932), p. 199, where Letters 1455 and 1460 were also first printed. The letter was first printed in Rigaud, *Correspondence*, II, pp. 435–6.

(2) Newton was preparing the new treatment of the comet of 1680–1 which was to be printed on pp. 500–2 of the third edition of the *Principia*. In the first and second editions he had written only of Halley's analysis of a parabolic orbit for this comet; now he was to give to Halley the credit for assigning to the 1680 comet a period of 575 years. Here also he was to print the numbers worked out by Halley for this comet (on the assumption now of an *elliptical* orbit) which are, in fact, all exactly the same (bar one slip) as those quoted by Newton in this letter. We have not found the documentary source of the numbers here quoted (and later printed) by Newton, which do not appear in Halley's extant letters to Newton, although the two men had discussed the hypothesis of an elliptical orbit for the comet of 1680 as early as 1695 (Letters 532, 533, 535, vol. IV). It was, however, Newton himself who first proposed (in Letter 535) that Halley's determination of the places of the nodes be shifted from 1° 53′ in Capricorn and Cancer to 2° 2′ for the elliptical orbit.

When Halley first set his hypothesis of the elliptical orbit for comets before the world ('Astronomiæ Cometicæ Synopsis', *Phil. Trans.*, no. 297 (for March 1705), 1882–99, also published separately in English and Latin), he examined only (and very briefly) the evidence for regarding the comet of 1682 as recurrent—whence he predicted its reappearance in 1758—with a rather doubtful suggestion that the famous comet of 1661 might be identified with one observed by Peter Apian in 1532 (*ibid.*, pp. 1898–99). In fact most of the 'Synopsis' is concerned with the *parabolic* orbit. However, in 1715 the 'Synopsis' was issued again, in English, as a tailpiece to the second volume of David Gregory's *Elements of Astronomy, Physical and Geometrical* (pp. 881–905). In this reprint Halley not only enlarged the discussion of 'Halley's comet' but added a long passage purporting to show that the comet of 1680 was to be identified with ones previously recorded in 44 B.C. and A.D. 531 and 1106. It was this material which Newton incorporated into the third edition of the *Principia*; it was not included in subsequent versions of Halley's 'Synopsis'. Newton did not consider the recurrence of the 1682 comet.

(3) *Read:* 10.

(4) See Letter 1344, note (14), p. 101.

(5) As noted above, Newton was to print all these numbers unaltered.

(6) This table appears on p. 500 of the third edition. It was not extended to give the numbers for 9 March 1681.

(7) The draft in Add. 3982 adds this passage after the valediction: 'In correcting the Parabolic orb of the Comet of 1680 you may assume the same plane of this Orb and that of the Elliptic Orb, and the same Axis in position, and the same time of the Perihelium in both cases: & then by any one good Observation determin the Latus rectum of the Parabola and by another good one examin the determination.'

1450 DELISLE TO NEWTON

10 DECEMBER 1724

From the original in King's College Library, Cambridge [1]
For the answer see Letter 1453

Monsieur

Depuis mon retour en France [2] je n'ay point encore pû trouver le moyen de vous faire tenir les deux derniers volumes de nos memoires de l'Academie, que l'Academie vous doit et qu'elle est bien glorieuse de vous devoir; [3] cest ce qui m'a fait jusquici differer d'ecrire à la societé Royale pour lui envoier mes nouvelles observations; esperant de jour en jour pouvoir envoier le tout ensemble; mais aiant trouvé ici un libraire qui est l'homme de France qui a le plus grand commerce de livres de Mathematiques et qui souhaite d'entrer en correspondance avec Mr Innys; jay crû que sur l'esperance que j'avois de pouvoir vous faire tenir avec son premier envoy les livres de l'Academie je ne devois pas differer plus longtems d'envoyer à la societé ceque j'avois receuilli de nouvelles observations. Je me suis adressé pour cela à Mr Halley avec qui j'entretiendrai le commerce qui j'aurai l'honneur d'avoir avec la societé Royale; [4] mais je n'ay pas voulu le faire cette premiere fois ici sans avoir l'honneur de vous assurer Monsieur de la reconnoissance et du souvenir que je conserverai toute ma vie de l'acceuil favorable que vous m'avez fait lorsque jay û l'honneur de vous voir à Londres.

Le Libraire qui veut entrer en commerce avec Mr Innys se nomme Jombert. il vient d'achever d'imprimer la mechanique [5] de Mr Varignon dont il sur-propose de vous presenter incessament plusieurs exemplaires, pour vous Monsieur et pour vos amis. il imprimera ensuite le commerce litteraire [6] de Mr Varignon et à cette occasion je suis chargé de la part de Mr de Fontenelle secretaire perpetuel de nôtre Academie qui aura la direction de cet ouvrage de vous demander Monsieur la permission d'imprimer vos lettres à Mr Varignon avec celles que Mr Varignon vous a ecrit. vous pouvez etre persuadé Monsieur que Mr de Fontenelle aura soin d'en retrancher tout ce qui s'y trouvera de particulier entre vous et Mr Varignon et qui n'aura pas de rapport à la Geometrie, afin que vous ne puissiez pas vous plaindre que l'on ait divulgué ce que vous aviez ecrit dans le secret à Mr Varignon.

Vous trouverez Monsieur dans le dernier volume imprimé des Memoires de l'Academie (qui est de l'année 1722) l'eloge historique de Mr Varignon. Si ce livre n'est point encore parvenû jusqu'a vous Monsieur voici les principales dates qui se puissent mettre au portrait de Mr Varignon. [7]

Pierre Varignon professeur Royal de Mathematiques et Membre des

296

Academies Royales des Sciences de France, d'Angleterre et de Prusse né à Caen l'an 1654 mort à Paris le 22 dec. 1722 [N.S.].

Jay prié Mr [Halley][8] de vouloir bien me procurer un exemplaire du receuil des observations et tables que Mr Flamsted a fait imprimer à ses depens; comme j'ay offert en revanche à Mr Halley de lui faire part des observations de nôtre Academie quil souhaiteroit dont jay tous les receuils j'espere quil voudra bien s'employer de tout son pouvoir pour me procurer le livre que je lui demande; si vous vouliez bien Monsieur me faire la grace de vous y employer aussi, je pourois compter la chose assurée et je vous en aurois une trez sensible obligation. Je suis Monsieur avec tout le respect et l'attachement possible votre trez humble et trez obeissant serviteur

DE L'ISLE

a Paris ce 21. *dec.* 1724. [N.S.]

NOTES

(1) Keynes MS. 94(D); microfilm 931.3. For previous correspondence see Letters 1427 and 1429.

(2) Delisle had been in London in the autumn of 1724, where he met Newton who presented him with a portrait of himself.

(3) For the arrival of the volumes via Jombert see Letter 1477, p. 333.

(4) See Letter 1427, note (2), p. 270. No observations of Delisle were published in the *Philosophical Transactions*, but it is clear from Halley's article there (33, no. 386 (January and February 1725), 228–37) that Delisle had corresponded with him, and probably sent observations to him.

(5) See Letter 1477, note (3), p. 333.

(6) This was never done.

(7) Clearly Delisle refers back to previous conversations with Newton. The engraved plate was made in London from the portrait that Varignon had presented to Newton in 1720; see Letter 1348, note (2), p. 106.

(8) The ink is smudged; the book is *Historia cœlestis* (1725).

1451 PEMBERTON TO NEWTON
?DECEMBER 1724
From the holograph original in the University Library, Cambridge [1]

Sr.

In this sheet [2] in pag. 323. l. 18, 19, 20. are these words Quare si ad cylindri basem circularem *NAO* erigatur perpendiculum *bHE*, et sit *bE* æqualis radio *AC*, et *bH* æqualis $\frac{BE \text{ quad.}}{CB}$: in which words the line *bE* is spoke of, as if it had not been considered in the proposition before; whereas it has already been mentioned several times. Suppose therefore these words were changed in some such manner as follows. Quare si in bE, quæ perpendicularis est ad cylindri basem circularem NAO et æqualis radio AC, sumatur bH æqualis $\frac{BE \text{ quad.}}{CB}$. [3]

I am
Your most humble
and most obednt. servt.
H PEMBERTON

NOTES

(1) Add. 3986, no. 13.
(2) Tt of the third edition of the *Principia*.
(3) This wording was printed.

1451a PEMBERTON'S QUERIES ON *PRINCIPIA*, 2ND EDITION, pp. 321-60

From the holograph original in the University Library, Cambridge [1]

Queries

Pag. 321. l. 20—28.[2] In this place several causes are enumerated, why the globes fell slower than by the calculation; and among the rest some resistance that might arise from other causes than the density of the fluid: each of which causes promiscuously are implied, to render the experiment not fide dignum. Now whereas in page 318 it is proposed to compare the resistance of bodies, as found by experiment, with the calculation there explained, in order to find out any resistance that may arise from other causes than the density of the fluid; it follows of consequence, that if such a resistance were the only cause of the slow descent of the globes, the experiment would not be any less fide dignum on that account; since you have expressly specified your intention in making these experiments to be to find out how much the resistance of fluids is owing to other causes than their density: so that the credit of the experiment is not lessened because some other kind of resistance may be the occasion of the result; but because it is uncertain whether this or some other of the causes mentioned might effect it. Suppose therefore a distinction were made between these different causes by expressing the paragraph in some such manner as follows[3]

Per theoriam hi globi cadere debuerunt tempore 40″ circiter. Quod tardius ceciderunt, utrum minori proportioni resistentiæ, quæ a vi inertiæ in tardis motibus oritur, ad resistentiam, quæ oritur ab aliis causis, tribuendum sit; an potius bullulis nonnullis globis adhærentibus, vel rarefactioni ceræ ad calorem vel tempestatis vel manus globum demittentis, vel etiam erroribus insensibilibus in ponderandis globis in aqua, incertum esse puto. Ideoque pondus globi in aqua debet esse plurium granorum, ut experimentum certum et fide dignum reddatur

In the 12th line of the paper annext to page 326 after machinam might not the word aliam be properly inserted, as I have done with black lead? because a clock is also a machine, and has been mentioned just before.[4]

Pag. 333. l. 18, 19. &c[5] Is there not some obscurity in the sentence, Et quemadmodum medii partes primæ eundo condensantur, et redeundo relaxantur; sic partes reliquæ quoties eunt condensabuntur, et quoties redeunt sese expandent? May it not be understood to signify, that as long as the parts of the medium move forewards, they are more and more condensed; and begin first to expand themselves again upon their return? But in prop. 47, the line $\epsilon\gamma$ is least of all, when the point ϕ falls on Ω, and this line in preceeding from thence towards ef more and more expands it self till it arrives at eg, where it becomes as much expanded as at EG, whence it began to move: and from hence forth in returning the line eg more and more expands itself, till the point f is returned to the point Ω, where the line ef is most of all dilated; and as it approaches nearer to EG, it again contracts. All this appears from hence, that $\epsilon\gamma$ is put equal to $EG-LN$ in the progressive motion of that part of the medium, and equal to $EG+ln$ in its return. Hence in the foregoing sentence the word condensabuntur means only more dense than before the medium was in motion, and sese expandent means only less dense.

Pag. 336. There is no figure to prop. 46. Should there not be such a one, as I have here inclosed, no. 172.[6]

Pag. 339. l. 18. After acceleratur might not these words in itu, et retardatur in reditu; conveniently be added.[7]

Pag. 349. In l. 6 the words æquator, and in l. 23, 25, 27, the word ecliptica occur both in the same sense: Would it not be better that the word should every where be the same?[8]

Pag. 358. l. 41. After terra might not these words be added, et ab aliis corporibus, in quæ dirigitur?[9]

Pag. 360. l. 15. After demonstratur might there not be added, cum umbræ suorum satellitium ostendunt hos quoque planetas luce a sole mutuato splendere:[10] for if these planets shone with their own proper light, their constantly appearing in full lustre would not prove that their faces obverted to us were also turned towards the sun.

NOTES

(1) Add. 3986, no. 24, fos. [1]–[2]. The pages of the second edition text here cited by Pemberton are reprinted in sheets Yy to Eee (pp. 345–400) of the third. We place these 'Queries' here—being similar in content to Pemberton's letters—since these sheets of the third edition were being printed (as we suppose) about the end of January 1725; see Letter 1457.

(2) Book II, Scholium to Prop. 40.

(3) Essentially, Pemberton rearranges Newton's words in order to distinguish the two sources of discrepancy. This revision was printed on pp. 347–8 of the third edition.

(4) The paper contained the description of Desaguliers' 1719 experiments on the fall of bladders in air ('Exper. 14', *Principia*, 1726, pp. 353–4); the word 'aliam' was added (p. 353, line 23).

(5) Proposition 43; the expression here criticized as ambiguous was left unmodified by Newton; see p. 361, line 35 to p. 362, line 3 of the third edition.

(6) Pemberton's new figure was printed on p. 365 of the third edition.

(7) This was done (*ibid.*, p. 368, line 13).

(8) Newton substituted 'eclipticam' for 'æquatorem' in Cas. 2 (*ibid.*, p. 378, line 32).

(9) This query has been deleted. Pemberton refers to the last of the four short sentences which Newton had added (*Principia*, 1726, p. 389) to the end of Regula III in Book III. He would have had it read: 'In receding from the Earth, [and from the other bodies towards which it is directed] gravity is diminished.' Pemberton's deleted amendment would have implied the existence of bodies without gravity.

(10) Newton added slightly different words (*ibid.*, p. 392, lines 29–30).

1452 NEWTON TO A DUKE
14 DECEMBER 1724
From Messrs Christie's Sale Catalogue[1]

I have discoursed Mr. Crawford about the matter concerning which I had the honour of a letter from your grace and I have made the Deputy of the Warden of the mint acquainted therewith and will acquaint the Warden therewith so soon as he returns to town, and in the meantime will endeavour that the prosecution be carried on in the best manner.

NOTE

(1) This letter, the addressee of which is unknown, was bought by a previous Duke of Newcastle and sold at Christie's on 14 December 1971 (Lot 514). The extract printed here is quoted in the catalogue. We are grateful to the Agent to the Duke of Newcastle, Mr Charles Stableforth, and to Messrs Christie for their courtesy.

1453 NEWTON TO DELISLE
?JANUARY 1725
From the holograph draft in King's College Library, Cambridge[1]
Reply to Letter 1450

Sr

The picture of Mr Varignon is now finished & the inscription wch I had the honour to receive from you, is put upon it. I have desired Mr Innys the bookseller to send two copies of it to his correspondent Mr Jombert the Bookseller at

Paris to shew to the friends of Mr Varignon, [2] that if they desire to have any thing altered, it may be done before I send the plate. The bookseller is to pay nothing for the plate. I make a present of it to the friends of Mr Varignon.

The Letters between me & Mr Varignon relate I think almost wholy to the Edition of my Optiques. [3] But if there be any thing in them wch may deserve the light, I leave it entirely to the discretion of Mr de Fontenelle to publish what he thinks proper out of them & out of Mr Varignons to me. [4] I have spoke to Dr Halley about procuring for you a copie of the tables & observations [5] wch Mr Flamsteed caused to be printed [*ends*]

NOTES

(1) Keynes MS. 142(V); microfilm 931.21.

(2) This was done; see Letter 1477, p. 332.

(3) Had Newton really forgotten his long letters to Varignon about the dispute with Bernoulli?

(4) Fontenelle never published any of Newton's correspondence with Varignon.

(5) Newton means the posthumous 1725 edition of Flamsteed's *Historia Cælestis*.

1454 PEMBERTON TO NEWTON
?JANUARY 1725
From the holograph original in the University Library, Cambridge [1]

Sr.

I beg permission to propose to you the comparing l. 24, 25, 26, 27. of pag. 341 [2] in the sheet here sent you, with the second paragraph of the scholium subjoined to prop. 34. For if what is asserted in the lines before us be universally true without any restriction, how can what is delivered in that paragraph be of any use in the forming of ships?

I am
Your most humbl.
and most obednt. servant
H PEMBERTON

NOTES

(1) Add. 3986, no. 15.

(2) *Principia*, 1726, p. 341. This Scholium, and the Scholium to Proposition 34, both concern the resistance experienced by bodies moving in a fluid. Newton made no changes as a result of Pemberton's comment.

1455 HALLEY TO NEWTON
16 FEBRUARY 1725
From the holograph original in the University Library, Cambridge [1]
Reply to Letter 1449; for the answer see Letter 1460

Honourd Sr

A mistake I committed in considering the scheme of your Comets Orb, which was no less than my taking the Suns motion the contrary way, made me conclude that no other than an Elliptick Orb could suffice to represent the first observations therof with the desired exactness, and you being indisposed out of town, [2] I waited for your return to consult you. Being yesterday at London I guessed by some Symptoms that you take it ill that I have not dispatcht the Calculus I undertook for you, but the aforesd mistake made me despair of pleasing you in it. Being got home last night I was astonisht to find my self capable of such an intollerable blunder, for which I hope it will be easier for you to pardon me, than for me to pardon my self, who hereby run the risk of disobliging the person in the Universe I most esteem. I entreat therefore that you would not think of any other hand for this computus, and that you please to allow me the rest of this week to do it in, [3] being desirous to approve in self [4] in all things

Honrd Sr.
Your most faithfull servt
EDM: HALLEY

Greenwich Feb. 16° 172⅘
To the Honoured
Sir Isaac Newton
In Orbells Buildings
Kensington [5]   By penny post.

NOTES

(1) Add. 3982, no. 7. Obviously Halley had already informed Newton by letter or word of mouth of his willingness to undertake the task laid upon him.

(2) As Letter 1460 also shows, Newton had now left Leicester Fields for the better air of rural Kensington, where he spent his last years. He had suffered a severe respiratory illness in January 1725.

(3) Newton had received the table before 1 March.

(4) *Read:* 'myself'.

(5) According to More, p. 662, footnote 15, these were in the modern Pitt Street, west of Kensington Church Street. The Conduitts had a house in Hanover Square.

1456 NEWTON TO THOMAS MASON
?FEBRUARY 1725
From the holograph original in the Burndy Library [1]

R[evere]nd Sr

A bad state of health makes me averse from minding business. I think I told Mr Robt Newton[2] I would give twelve pounds towards the charge of Erecting a Gallery in your Church, & desired him to pay at next Lady day in part thereof the years rent of a close wch he holds of me in Buckmin[s]ter at 6lb pr ann, the yeare then expiring. And I have desired John Newton[3] of Woollstrope to pay to you towards the same charge the rent wch will be due to me at Lady day next for a close of nine pounds p[e]r ann in the field of Colsterworth next Easton. Taxes must be deducted. And when you have received these two summs pray let me know what they amount unto.[4] I am

Your most humble &
obedient Servant
ISAAC NEWTON

Mr. Tho. Mason

NOTES

(1) Printed in Nichols, IV, p. 60. Since the work was finished before May 1725 (Letter 1464), and the present letter was written before Lady Day in, presumably, the same year, we have placed it arbitrarily here. The recipient, who became rector of Colsterworth, Lincolnshire, in 1720 and died there aged 71 in 1753, is probably the Thomas Mason who was educated at Pembroke College, Cambridge, 1700–7, and was ordained deacon at Lincoln in 1707, and priest in 1709.

(2) Robert Newton (1678–1734), grandson of Newton's uncle Richard Newton, held lands in Colsterworth and Easton. See Appendix II, Newton's Genealogy, and Foster, p. 25. Buckminster was in Leicestershire, about ten miles from Grantham.

(3) Possibly Newton's first cousin once removed, grandson of his uncle Robert Newton. According to Foster, p. 14, this John Newton was a yeoman and carpenter who acted as gamekeeper to Sir Isaac. He was buried on 13 October 1725, aged 60.

(4) See Letter 1465.

1457 PEMBERTON TO NEWTON
?FEBRUARY 1725
From the holograph original in the University Library, Cambridge [1]

Sr.

Mr. Blackborn the correcter of the press having observed to me that he met with some difficulty in the line against which he has put this mark * * in the first page of the sheet whose signature is Zz; [2] I have directed the [om]ission [3] of the word ambientis, and a comma [is] [3] to be put in its stead. [4] I desire the [fav]our [3] of you to cast your eye upon this alteration, and if you approve not of it, to adjust the sentence to your mind. The word ambientis seems to me to burden the sentence and I think the sense of it is very fully expressed by what follows. The word fornica in the second line following I suppose should be fornicata, [5] as Mr. Blackbourn has proposed.

> I am
> Your most humbl.
> and most obednt. servt.
> H PEMBERTON

NOTES

(1) Add. 3986, no. 16.

(2) *Principia*, 1726, p. 353; Pemberton refers once more to the added passage describing Desaguliers' 1719 experiments on the fall of bladders.

(3) The paper is torn.

(4) There is a variation in the printing of copies of the third edition at this point not recorded by Koyré and Cohen. In the copy used by us Newton's original wording of line 12 ('concavæ, ambientis quam...') is preserved, while in the copy used by Cohen and reproduced by him in facsimile the type has been rearranged to permit the reading 'concavæ, quam...' omitting 'ambientis'. Newton seems to have originally meant to say that the surrounding hollow sphere of wood was made wet, and the bladder blown up inside it. Pemberton for some reason wished to omit the word 'surrounding'. It would perhaps have been more satisfactory to write that a moist bladder was inflated inside the hollow sphere: 'formando in orbem sphæricum ope sphæræ ligneæ concavæ ambientis vesicas porcorum, quas madefactas implere cogebantur...'

(5) 'arched'. The word 'fornica' does not exist.

1458 PEMBERTON TO NEWTON
?FEBRUARY 1725
From the holograph original in the University Library, Cambridge [1]

Sr.

I beg leave to propose a couple of queries in relation to the first page of this sheet. [2] The first is concerning the words demittendo ab altitudine pedum 272.

For since it is said that in the preceding experiment the globes were let fall from the top of the church (a culmine) should not some expression here be made use of, that might let the reader understand, that there was in the church a place still higher, that was chose for this experiment? Suppose instead of the words forementioned, some such expression as this was used demittendo ab altiori quam prius, templi loco, ut caderent spatium 272 pedum. [3]

The next query is in relation to the words horologium cum elatere ad singula minuta secunda quater oscillante; which rendred verbatim is a clock with a spring that oscillated four times in a second. Now since the spring of the clock is not the part which oscillates, but the pendulum, suppose elatere were changed into pendulo. [4]

I am sorry I did not think of the first of these queries sooner; because the word culmen does not stand so properly, as one could wish, in the preceding experiment; as the word culmen seems to signify, that the balls were let fall from the very summit of the church; which is inconsistent with what is now said that a place of the church still higher was in this latter experiment made choise of. But this cannot now otherwise be rectified than by reprinting the last leaf of the preceding sheet, and by changing the expression A culmine in the beginning of Exper. 13. into A superiori parte. [5]

I am
Your most humbl.
and most obednt. servnt.
H PEMBERTON

To Sr. Is. Newton

NOTES

(1) Add. 3986, no. 11; presumably written shortly after Letter 1457.

(2) Sheet Zz, p. 353 of the third edition.

(3) The form printed (ll. 14–15) was shorter: 'demittendo ab altiore loco in templo... nempe ab altitudine pedum 272.' These words have been drafted by Newton on the cover of Pemberton's letter.

(4) Pemberton is here obtuse; a clock with a $2\frac{1}{4}$ in. pendulum beating quarter seconds (that is, with a period of $\frac{1}{2}$ second) would hardly be practicable. Newton said, and meant, that a balance-spring watch was used, beating quarter seconds. No alteration was made here, except to substitute *vibrante* for *oscillante* (l. 23).

(5) No such change was made to the lower part of p. 351. For Pemberton's first reading of the account of Desaguliers' experiments see Letter 1451a.

1458a PEMBERTON'S QUERIES ON *PRINCIPIA*, 2nd EDITION, pp. 364-84

From the holograph original in the University Library, Cambridge[1]

Queries

Pag. 364. l. 13, 14, 15, 16 of the additional Paper.[2]

In the last line of this paragraph the reader is referred to a computation, which he has not yet been instructed to make. And indeed the whole paragraph seems to me not to express, what is intended by it, in the fullest manner: your design being to give a reason why you assumed the distance of the moon from the earth a little less than what you shew astronomical observations to make it. Would not this intent be a little more fully expressed after the following manner?

Cum calculus hic fundatur in hypothesi, quod terra quiescit, assumpta est distantia lunæ a terra aliquantulum minor, quam astronomi invenerunt. Si vero habeatur ratio motus terræ circum gravitatis centrum, quod sibi lunæque commune est; distantia hic assumpta augenda est (per prop. LX. lib. I) ut eadem lex gravitatis maneat; et postea (corol. 7. prop. XXXVII. lib. hujus) invenietur circiter $60\frac{1}{2}$ semidiametrorum terræ.[3]

Does not the last paragraph of this additional paper anticipate the following proposition?

Pag. 367. l. 5. Whereas in this line occurrs the expression uti computis quibusdam initis inveni;[4] I beg leave to ask if any other reasoning or calculation is required to prove this, than what is contained in a paper I have here inclosed.[5] If so might it not be convenient to change a little the forementioned expression? Suppose at least the word rationibus were put for computis? for may not the present expression induce the reader to suppose some more intricate process to be pointed at?

Pag. 376. Do not the words in which prop. 14 are [*sic*] expressed seem almost to be contradicted in the demonstration of it? For as in the proposition it is said, that the Aphelia and Nodes remain fixed; in the demonstration it is only shewn, that they would remain so, if they were not moved by certain causes, which, both here and more particularly in the following scholium, are allowed to take effect. Would not the words be less liable to exception, if put into some such form as this?[6]

Prop. XIV. Theor. XIV.

In orbibus planetarum primariorum aphelia et nodi fere quiescunt.

Aphelia quiescunt per prop. XI. lib. I. ut et orbium plana per ejusdem prop. I et quiescentibus planis quiescunt nodi; nisi quantum planetarum, et etiam cometarum actiones in se mutuo motus aliquos vix sensibiles efficiunt.

Corol. 1. Quiescunt stellæ fixæ; propterea quod non aliter mutant situs suos ad planetarum aphelia et nodos, quam pro ratione harum in apheliis nodisque levissimarum mutationum.

Pag. 378. l. 7. of the additional paper.[7]

After parisiensium 57300 might there not be added some such words as these;[8] et mensura senioris Cassini ab his parum differt, quanquam hujus observationibus ea se paulo minus fatendum est, quod per montes transiens eorum altitudines ope barometri non recte æstimavit. Ex mensura autem Picarti &c.

Pag. 384. Should the first paragraph, and the last sentence of the second be crossed out?[9]

NOTES

(1) Add. 3986, no. 24, fo. [3]. We attach this and the following paper at this point because of the order of the pages referred to.

(2) The end of Book III, Prop. 4 (*Principia*, 1726, p. 398), referring throughout to the page numbers of the second edition. The 'additional Paper' contained a rewriting of the final paragraph of Proposition 4 (including the addition of the final sentence: 'Computatio autem iniri potest per prop. LX lib. I') and the text of the following Scholium, new in the third edition.

(3) 'As the computation here is based on the hypothesis that the Earth is at rest, the distance of the Moon from the Earth is taken to be a little less than astronomers have found it. If account is taken of the Earth's motion about the common centre of gravity of the Earth and the Moon, the distance here postulated must be increased (by Prop. 60, Book I) so that the law of gravity may remain the same; and afterwards (corol. 7, Prop. 37 of this book) it may be found to be about 60½ terrestrial radii.'

This is clearly intended as a substitute for the final paragraph of Proposition 4, but Newton did not approve such a change.

(4) The second edition reads here: 'calculis quibusdam initis' and the third finally reads 'calculo quodam inito' (p. 401).

(5) See Number 1458*b*, following.

(6) This proposed change was rejected by Newton. See also Letter 1462.

(7) This contained the new version of the first three paragraphs of Book III, Prop. 19, as printed in the third edition.

(8) See *Principia*, 1726, p. 413, line 4. The addition was not made.

(9) This was indeed Newton's intention—doubtless the omissions were indicated by square brackets round the passages—and accordingly these passages do not appear in the third edition (pp. 418–19).

1458*b* THE ENCLOSED PAPER

From the holograph original in the University Library, Cambridge[1]

The reason of the assertion in pag. 367. l. 1–5 Princip. is this[2] that the motion of the Satellite cannot be lasting, except it be attracted just as much less than the primary planet for one half of its revolution round the primary, as it is attracted more during the other half of it's revolution. And this ballance in the sun's attraction will be effected by the situation of the center of the orbit now proposed, and no[t] otherwise.

If the orbit of the Satellite be eccentric to Jupiter, the Satellite will describe about Jupiter an Ellipsis, Jupiter being in one of the Foci, except so far as the action of the Sun disturbs the Satellite's motion. Let therefore $ABCD$.[3] represent the eccentric orbit of a Satellite, I the focus, in which Jupiter is placed, the Sun being in S. Suppose E to be the center, and to be placed at the distance from the Sun wherein the Satellite would be as much attracted by the sun as Jupiter is in I. Draw SE, and likewise BD through E perpendicular to ES. Then the Satellite, all the time in between the line BD, and the Sun, will be attracted more strongly than Jupiter; but all the time it is on the other side BD it will be attracted less. and the difference between the attraction of the Satellite and Jupiter by the sun will be about the same, for the same distance of the Satellite from the line BD, so that at any distance from this line it shall be attracted as much more than Jupiter, when it is between the sun and that line as it shall be attracted less than Jupiter, when it shall be at the same distance beyond that line. In the Satellites passage therefore from D through A to B the action of the sun upon the satellite shall fall as much short of it's action upon Jupiter, as the action of the sun upon the satellite shall exceed it's action upon Jupiter, while the Satellite is passing from B through C to D. Or rather the defect of the sun's action upon the satellite shall a little surmount the excess, because the Satellite takes up a little longer time in passing from D to B than in returning from B to D through the other half of its orbit. Whence if the center E be placed but a little nearer to the sun than has been supposed, the total action of the Sun upon the Satellite in the compass of the Satellites entire revolution, shall equal it's total action upon Jupiter in the same time; and consequently the motion of the Satellite shall be as regular as it can be with the sun's action upon it. But if the center of the Satellite's orbit were either nearer or further off than tis required hereto, the action of the sun upon the satellite shall in the first case exceed, and in the latter case fall short of the sun's action upon Jupiter, and the Satellites motion be rendered very irregular. So that from the great regularity found in the motion of all the Satellites it is manifest, that if they are not attracted as much as Jupiter at equal distance from the sun, the centers of their orbits must be placed as has here been said. And whereas the force wherewith the sun acts on the satellite at the distance *SE* is to the force wherewith it will act on the same in the distance *SI* in the duplicate ratio of *SI* to *SE*; it is evident that *SI* must be to *SE* in the subduplicate ratio of the force wherewith the sun acts on Jupiter in *I* to the force wherewith it will act on the satellite at the same distance.

NOTES

(1) Add. 3986, no. 31, fos. 1–3. This is clearly the paper mentioned in Number 1458*a* referring to the revised second edition sheets of Book III, Prop. 6.

(2) In Book III, Prop. 6, Newton argues that at a given distance from the Sun a planet and its satellite experience gravitational forces towards the Sun in proportion to their masses, so that the average action of the Sun upon the satellite is equal to its action upon the primary planet; otherwise the satellite's orbit would be markedly eccentric, and such eccentricities are not observed. Pemberton's explanation does not seem to make Newton's argument clearer, and certainly does not account for Newton's saying that the satellite's eccentricity would be *nearly* proportional to the square root of the ratio between the gravitational forces of satellite matter and planet matter (supposing these to be different). Pemberton demonstrates the eccentricity to be exactly as the square root of this ratio.

(3) The figure may be reconstructed as shown below.

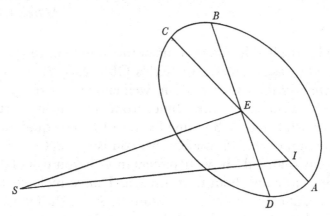

1459 NEWTON TO JOHN ARMSTRONG

?EARLY 1725

From a holograph draft in the Bodleian Library, Oxford[1]

To Collonel Armstrong surveyor of the Ordnance
at his house in the Tower of London

Sr

The other day, I signed a Letter to you without duly considering it being sick at Kensington.[2] I hope in a few days to be well enough to come abroad &

as soon as I am able I intend to wait upon you at your house & explain the Letter with the business it concerns. I am

NOTES

(1) Bodleian New College MS. 361, II, fo. 8. John Armstrong (1674–1742), major-general and colonel of the Royal Regiment of Foot in Ireland, was elected F.R.S. on 2 May 1723. He had been appointed 'Chief Engineer of England' in 1714. Later he became Surveyor-General of the Ordnance. The present letter probably concerned disputes between Mint and Ordnance over the occupation of buildings in the Tower, although no other letters relating to such matters are extant for this period.

(2) See Letters 1455 and 1456. We tentatively date the present letter shortly after the date of Newton's removal to Kensington.

1460 NEWTON TO HALLEY
1 MARCH 1725
From the holograph draft in the University Library, Cambridge[1]
Reply to Letter 1455[2]

Orbells buildings in Kensington
March 1st 172⅘

Dr Halley

I thank you for the Table you sent me of the motion of the Comet of 1680 in a Parabolic Orb so as to answer to Kirk's Observations as well as to Flamsteed's. It answers all their Observations well enough for my purpose. But you have omitted the distances of the Comet from the Sun in parts of the mean distance of the earth from the Sun divided into 100000 equal parts: such parts as the Latus rectum of this Parabolic Orb consists of 2508.[3] These distances you have computed already in your papers in wch you calculated this Table, & you need only to copy them from thence. I have inclosed a copy of your Table with a vacant column for these distances, & beg the favour of you to fill it up by inserting these distances out of those your loos papers in wch you made your calculations of this Table. The distances are inserted in your Table published in the second edition of my Principles pag 459.[4] I intend still to keep that Table & add this new one to it if you please to fill up the column of distances in the same manner that the two Tables may be like one another.[5] And by the help of this new Table I shall be able to make the schemes of the motion of this Comet more perfect. I am

Your humble servant
ISAAC NEWTON

To Dr Halley

NOTES

(1) Add. 3982, no. 8. No. 9 is an antecedent draft.

(2) Or perhaps rather to a later letter from Halley, Letter 1455 being the last extant.

(3) In the printed text of the third edition of the *Principia* (p. 500, line 7) as in the second (p. 459, line 2) Newton retained the figure of 2430 parts for the latus rectum (one astronomical unit = 100 000 parts).

(4) The table on p. 459 of the second edition appears without important changes on p. 500 of the third, with a misprint in column 1 (Dec. 29). The discrepancies between observed and computed positions for the comet are also different, because in the second edition Newton had printed Flamsteed's observed positions of the 1680 comet without alteration (p. 455: 'Hujus motum a *Flamstedio* observatum Tabula...') whereas in the third (p. 496) he gives the observations of Flamsteed as corrected by Halley. The net result appears to be some reduction in the discrepancies ('errores') as printed in the third edition. Newton also added (p. 502) the new table in which Halley computed the positions of the 1680 comet according to the *elliptical* hypothesis (not present in the earlier editions). This table does not list the distances of the comet from the sun.

(5) It is a little hard to know precisely what Newton means; he modified the old table of parabolic motion (*Principia*, 1713, p. 459; 1726, p. 500) and added a new table of elliptical motion (p. 502); he did not add a second table on parabolic motion. As just noted, the comet's distances from the sun are the same in the tables of both editions (apart from the misprint in the third edition, noted in its corrigenda). It is probable that Halley also was puzzled to know what more Newton wanted from him; perhaps to have the solar distances corrected, as the comet's places had already been corrected by Halley from Flamsteed's observations.

1461 CAVELIER TO NEWTON
9 MARCH 1725
From the original in King's College Library, Cambridge[1]
For the answer see Letter 1469

Monsieur

il y a environ six mois[2] que jus lhonneur de vous marquer quil m'etoit tombé en mains une Copie de votre Chronologie, je vous priois de me marquer si vous n'aviez point d'additions et Corrections a y faire par ce quil y avoit quelques fautes de la part de celui qui en avoit fait la traduction. comme les scavans souhaitent avec empressement tout ce qui part d'un aussi habil que vous, Monsieur jay lhonneur de vous ecrire cette seconde[3] Lettre pour vous prier instament de me marquer si vous ny voulez rien changer que si je ne recois point de vos Nouvelles je prenderay votre silence pour un Consentement qu'elle paroisse tel qu'elle est et je la donneray au public avec des Remarques[.] je suis avec respec

Monsieur
a paris ce 20e
Mars 1725 [*N.S.*]

Votre tres humble et tres obeissa[nt] serviteur   G. CAVELIER fils
Libraire rue S. jacques

NOTES

(1) Keynes MS. 138(B); microfilm 1011.22. For the background to the letter see Letter 1436, note (2), p. 280.

(2) Hence Cavelier's second letter to Newton must have been written about October 1724.

(3) Presumably the second letter in which Cavelier says he will print the *Abrégé* if he does not hear from Newton, and his third letter in all.

1462   PEMBERTON TO NEWTON

?APRIL 1725

From the holograph original in the University Library, Cambridge[1]

Sr.

In page 410[2] is not the first corollary of prop. XIV. expressed a little to[o] bare? for if the fixed stars did preserve always the very same position in respect of the Aphelion of any one of the planets, they would not then be at rest. Might not this corollary rather be put in some such manner as this, Quiescunt stellæ fixæ, propterea quod positiones suas ad planetarum aphelia nodosque non aliter mutant, quam pro ratione motus horum lentissimorum jam memoratorum.[3]

In page 411. the last sentence in the scholium containing these words Et hi motus, ob parvitatem, negliguntur in hac propositione; seems to me as if it might better be omitted; because these motions are not really neglected in the proposition but expressly taken notice of.[4]

In pag. 412. l. 21. for ubi might not ubicunque[5] rather be put as I have done in the margent; for the force of the argument consists in this phænomenon's taking place in every situation of the earth in respect of jupiter.

> I am
> Your most humbl. and most obednt. servt.
> H. PEMBERTON

NOTES

(1) Add. 3986, no. 17. This cannot have been written very long before Pemberton's next dated letter of 17 May 1725 (Letter 1468), when the printed text of the third edition of the *Principia* had reached p. 451.

(2) The text of the third edition. Pemberton had previously commented on this very point in Number 1458a (p. 307).

(3) 'The fixed stars remain at rest, because they in no way change their positions with respect to the aphelia and nodes of the planets than as a result of the very slow motion of these last, just mentioned.' Pemberton was logically correct but Newton made no alteration.

(4) Nevertheless, they were printed.

(5) Perhaps the 'ubicunque' in line 16, p. 412, is meant.

1462a PEMBERTON'S QUERIES ON *PRINCIPIA*, 2nd EDITION, pp. 389–426

From the holograph original in the University Library, Cambridge[1]

Queries

In pag. 389. l. 3. The expression nondum observatæ was more suitable to the first edition of your book, than it will be to the present, now the world has been so long acquainted with your discoveries. Might not the expression rather be changed into a prioribus astronomis non observatæ, or the like?[2]

Ibid. l. 33. Would it be amiss to give some brief hint at the principle, whence the precept contained in this line was deduced.[3]

Pag. 405. l. 9. Should not the letters *Lmp* be *lmP*.[4]

Pag. 433. Whereas the semicircle *BAD* has many of its letters the same with the letters of the other figure; would it not be better to put other letters to this semicircle? Suppose those I have writ upon the draught, here sent, in black lead. If you so approve I will take care to alter the text accordingly.[5]

Pag. 390. l. 11. Here besides the 19th. corol. of prop. XVI. lib. I. should also be cited the additional paragraph you have added to the 20th corol. of that proposition.[6]

I have long debated with my self, whether I ought to give you any farther trouble upon this place; but your favourable reception of my endeavours hitherto has at length emboldened me to desire your patience, while I trouble you with what follows.

I have a small suspicion that one consideration of some moment is omitted in the paragraph now mentioned. According to what is there observed,[7] when the moon is in the meridian of any place, it will elevate the water with the greatest force; and because as it declines from thence this power of the moon gradually diminishes so as about three hours after wholly to cease; it is there concluded that with the ceasing of this action the waters should cease to rise, unless so far as they may for some short time be carried on by a fluctuating motion. Now my suspicion is that the moon will cease to elevate the waters sooner than this, for the following reason. The water that is raised by the moon will surmount in gravity the other parts of the water; so that the moon can continue to elevate the water no longer than while the elevating power of the moon exceeds this additional weight of the water raised.

313

Again if we suppose even that the fluctuating motion of the water may continue the rising of it so much longer than the moon has any power to raise it, as to cause the greatest height to fall out at the time here said: I endeavoured to demonstrate in the paper I formerly sent you upon this subject (and which I have now by me)[8] that the water will subside again in the short space of 42'. 40". though this be repugnant to experience. I apprehend your design in proving the water to rise till three hours after the moons coming to the meridian to be that the time of the tides, as collected from the theory, may correspond with what is found in those coasts that lye most open to the ocean. But it may not only be doubted, whether there be any port, before which the water is not much shallower than in the free ocean, through which shallowness the water must take some time in flowing: but in the philosophical transactions Vol. XVI. No. 185. pag. 220. I find some ports where the tide is made sooner than three hours after the moons appulse to the meridian; and upon consulting a map I find these ports to be on a more prominent part of the coast, than the adjacent ones where the tide comes in later.

Pag. 426. In the proposition of this page[9] the effect of the sun upon the water is computed by the greater hight the sun will cause the water to have under itself, and in the opposite point of the earth, more than in the parts 90 degrees distant: and in page 429. this computation is applied to the moon;[10] and there considered as the whole measure of the height, to which the sun and moon elevate the waters. But if the greatest height of the waters be three hours after the appulse of the luminaries to the meridian, and the least height three hours before; it is evident that the hight, when the sun or moon is on the meridian, will be but a very small part of the height, to which the water will rise: if indeed the hight in this case will sensibly differ from the height when the luminaries are rising or setting; the times of the luminaries coming to the meridian and the times of their rising and setting all falling in the middle between two tides.

NOTES

(1) Add. 3986, no. 31, fos. 6–7. Again, these queries relate to pages of the second edition of the *Principia* containing matter subsequent to that of the third edition pages discussed in Letter 1462, which were already partly reprinted by the time of Letter 1467.

(2) Book III, Prop. 22; Newton wrote that there were lunar inequalities not yet observed; Pemberton proposes (and Newton will agree) to read: 'inequalities not observed by earlier astronomers'.

(3) Book III, Prop. 23 (*Principia*, 1726, p. 423, line 29). Newton asserts that the motion of the apse-lines of the satellites must be diminished from that which he has just defined in the ratio of 5 to 9 or 1 to 2 'for a reason which I have not time to explain'.

(4) This alteration was made (*ibid.*, p. 439, line 6).

(5) This was done (*ibid.*, p. 472).

(6) '& 20' was added (*ibid.*, p. 424, line 8).

(7) Probably Pemberton refers to the first form of a sentence added to the beginning of Proposition 24 in the third edition (p. 424, lines 20–5): 'Sed vis...sit vadosum'. The sentence is a vague qualitative explanation of why the highest tide should occur when the moon is some three hours past the meridian of the place. See Pemberton's Letter 1467.

(8) We have not found this, presumably because it was retained by Pemberton.

(9) Prop. 36.

(10) Prop. 37, Corol. 1. Newton seems to have ignored Pemberton's comments.

1463 J. T. DESAGULIERS TO NEWTON
29 APRIL 1725
From a copy in the University Library, Cambridge[1]

April the 29th. 1725

Sir,

According to the order of the last Council of the Royal Society,[2] I herewith send the Figures of the Apparatus us'd in such Experiments as I made before the Society, since I gave in the Figures (representing the Machines made use of during the foregoing year) a few meetings before Saint Andrew's day was Twelvemonth.[3]

When I had complied with the former order, I desir'd some Members of the Council to tell me whether it was expected of me that I shou'd make Experiments at every Meeting or every other Meeting of the Society, tho' there shou'd be nothing new to be explain'd and prov'd by Experiment, or none of the Experiments formerly made by some of our Members call'd in question so as to render it necessary to make them again. I told them that I wou'd not be wanting in anything that was my Duty; and that if repeating Experiments Which had been made before, by way of Entertainment, was thought agreeable, I cou'd easily do something every Thursday; having by me a very large Apparatus, which I us'd at my Courses of Philosophy.

But I apprehended that unless I had made some new Discovery, or had Models of new and useful Engines to offer, I shou'd only offend by taking up the Society's time in making a shew to no End and Purpose. That whenever I had made some Experiments not entirely new, I had observ'd some Members were displeas'd: and that I imagin'd it was my Business to take my Directions from the President, and also to put in Execution any Experiment recommended to me at a Meeting of any Member of the Society, if the President did not give orders to the contrary, That I shou'd always attend to take Directions, and never had neglected or wou'd neglect what shou'd be committed to my care.

I was confirm'd in my notions by the Gentleman I spoke to, and accordingly the last year attended constantly, except when I went out of Town to observe the great Solar Eclipse[4] at Bath (where it was central) and tho' I was disappointed by the cloudy weather, I made the same Preparations to observe it as if I had been sure of what I intended. I look'd upon this as doing a service to the Society; and, if I remember right, I ask'd leave before I went.

Now I find by the last order, that I was mistaken; and shou'd have made more Experiments; which I had certainly done if I had thought it my duty: and therefore now I beg the Favour of the Council to give me my Directions in Writing, if I am wrong in my Notion of what is requir'd of me. If not I shall be very ready to take Directions from such Gentlemen as think that I do not make Experiments often enough, and shall never think much of the Pains and Expence of preparing a new Apparatus for performing any Experiments that they shall contrive.[5]

Not to multiply words I have only sent the Figures[6] here of the few, but new machines I have made use of this Year; because large Descriptions of 'em have already been given in: But as I attend here I am ready to explain them if desir'd.

I am, Sir
Your most oblig'd
and most humble Servant
J. T. DESAGULIERS

P.S. I don't know whether I shou'd call part of my last Year's service the Directions which I gave the Smith and Bricklayer for making the contrivance to convey Heat for warming and keeping dry the Repository. But it was done by order.

To The President

NOTES

(1) Add. 4007, fo. 669; a nineteenth-century copy. We have been unable to locate the original. John Theophilus Desaguliers (1683–1744) was educated at Christ Church College, Oxford, and became a lecturer in experimental philosophy at Hart Hall in succession to Keill. In 1712 he moved to London and began giving private courses of demonstration-lectures, which were extremely popular. In 1714 he was elected Fellow of the Royal Society, and was frequently called upon to perform experiments at their meetings. He was eventually officially appointed and paid as Curator of Experiments, a post which he retained until his death. His textbooks on experimental philosophy, in which he describes many of the experiments he performed in his lectures, were very successful.

(2) On 15 April 1725 Desaguliers was asked to 'lay before the Council a List of such Experiments as he has performed before the Society since the last payment made to him.' There is no record of any such experiments in the Journal Book of the Royal Society for that

period; nevertheless at the next Council Meeting (29 April 1725) he was ordered £30 for the year ending on 30 November 1724.

(3) At a Council Meeting on 24 October 1723 Desaguliers had in the same way been ordered to produce a list of experiments performed before the Society during the previous year. At the next meeting (14 November) he was awarded a gratuity of £40 for his services, although the Journal Book for that year records only one group of experiments, performed by Desaguliers at a meeting on 6 December 1722.

(4) Of 11 May 1724.

(5) At the meeting of the Society preceding the Council Meeting on 29 April, Desaguliers demonstrated some experiments on cohesion, presumably in a last minute attempt to make up for his failure to produce any experiments in the previous year. But in the ensuing months there is no record in the Journal Book of any further experiments performed by him.

(6) These are now missing. It seems possible that Desaguliers made machines (that is, apparatus for use in experiments) under the auspices of, and at the expense of, the Royal Society, and then used them in his private lecture demonstrations. There seems to have been a slight lull in Desaguliers' activities at the Royal Society during the period 1723–5; previously, and in subsequent years, he seems to have taken his duties as Curator of Experiments much more seriously.

1464 NEWTON TO PERCIVAL

12 MAY 1725

From the printed version in Nichols, Vol. IV, p. 60

London, May 12, 1725.

Sir,

I desire you to acquaint John Groves, and the rest of the neighbours in the parish of Colsterworth and Wolsthorpe, that I agree to the design proposed to me, of bringing their commons to a rule; suppose, by allowing eighty sheep-commons[1] to a farm, and ten to an ancient cottage, and settling the beast commons according to ancient right, to be set down in a list of them; and where any dispute arises, the commons may be proportioned to the annual value of the farm or cottage. And I should be glad to see the settlement finished. There are one hundred and twenty sheep commons due to me by ancient right, on account of the royalty. I am, &c.

ISAAC NEWTON.

NOTE

(1) One Thomas Percival is mentioned as one of Newton's tenants in Letter 950, vol. v (p. 347). The matter of sheep commons is also raised there.

1465 NEWTON TO MASON
12 MAY 1725
From the printed version in Nichols, IV, p. 61

London, May 12, 1725

Sir

I am very glad to understand that the gallery in your church is finished to your mind.[1] And as for the 1*l.* 14*s.* 4*d* which remains in your hands over and above the 12*l.* which I gave towards it, you may apply it to the use of the young people of the parish that are learning to sing Psalms, as you desire. I have herein sent you an acquittance, which I desire you deliver to my cousin Robert Newton, for his year's rent. I am your most humble and your most obedient servant,

ISAAC NEWTON

NOTE

(1) Compare Letter 1456.

1466 GEORGE NEEDHAM TO NEWTON
16 MAY 1725
From the original in the Public Record Office[1]

Chester May ye 16*th*: 1725

Sir Isack Newton,

Your great, wise, and good genious, so much known to mankinde, leads me, (tho unknown) to trouble you with the Melancholly acct: of an unhappy youth, tho a most hopefull young Gentleman, & son of the best of Mothers, who has been always Esteemed an Extraordinary Wife to her husband, & looked on to be most valuable in the deportment of her minde which I hope altogether may deserve your advice, & consideration, toward this young Gentleman, & doubt not, but if you will please to give him leave to make him-selfe known to you, that you will finde him deserveing of yr instructions and favor because, I beleive him, to be a branch of yr Family, as I judge may appear, by the following acct. vizt:

This young Gentleman, Robert Newton[2] now upwards of 15 years old is the heir of Heightly[3] in the County of Salop, which Estate is well Wooded, & upwards of £800 per annum: This sd Robert Newton, is Lineally descended from Sir Richard Newton who was Lord Cheife Justice of the Common Pleas

318

in Reigns of 3 of the King Henerys, & allso descended Linially from Sr Peter Newton, mentioned in Bakers Chronicles &ca.[4]

This sd Robt: Newton, Minor, is kept out of his Estate by two Daughters of his Fathers Eldest Brother, & tho there was a good Settlement made att the Marriage of his Grandmother, on the Male heirs, yet his uncles two Daughters pretend there has been fines past, and that their Father & GrandMother by Lease & Release has created it fee simple but this sd Robt: Newtons Father who was next in the intail was not taken in, He was Minister of St Peters Church in this Citty of Chester, When he dyed, he left 5 young Children, the Eldest, but 12 years old, and this present youth, then 4 months old, their Mother, through her great care and industry, gave them all, a good Education the Eldest son, Chosen fellow, of Brazon Noze Oxen, and took his Master of Art's degree, he afterward, was Reader of St Stephens Walbrook London.[5] In the year 1720 he preferred a Bill in Chancery against the two Daughters of his Unckle & continued in suit near two years, but finding his Mothers substance Exhausted, & not able to proceed dismissed his bill, and soon after dyed, and ever since his death, the two daughters are cutting down the Wood & makeing great devastation in the Estate, as appears by the Oaths of Creddible Persons

Sr Isack my years & Eyes will not permitt me to write, what may be said on this terrible case, but as Justice & convenience, are the motives of troubling you, I hope may Excuse me as I am with great Respect,

Sir Isack,
Your most humble servant
GEO: NEEDHAM

The Widdow Newtons house, is in the Abby yeard here I will wait on you when I return to London—where I hope I may be able, to show, that this Family & youth does merit yr Protection & Patronage

Sir Isack Newton att his house[6]
near Leisterfeilds
London

NOTES

(1) Mint/19, 5, fo. 11. Nothing is known of the writer of the letter.

(2) Robert Newton is here claimed to be a descendant of Sir Richard Cradock (?1370–?1448) of Newtown in Montgomeryshire, who took the name Newton. Richard Newton was appointed a justice of the common bench in 1438 and knighted in 1439. One of his descendants, John Newton of Barr's Court, Gloucestershire, received a baronetcy in 1660, with remainder, default his own male issue, to Sir Isaac Newton's relative, John Newton of Gonerby, born 1626. This latter John Newton, according to the generally accepted genealogy which Newton himself constructed, was descended from the Newtons of Westby, and there is no positive

indication that the two Newton families were in any way connected. However it seems unlikely that the baronetcy should revert to a totally unrelated family.

For details of Newton's family tree, see Foster, especially p. 9. The question of Newton's relationship to the Newtons of Newtown is also discussed in *Notes and Queries*, 3rd Series, **1** (1862), 158–90 (it is there also suggested that Richard Newton derived his name from Newtown in Glamorgan, not Montgomeryshire).

Perhaps we should take this opportunity to correct some minor errors in earlier volumes of this *Correspondence* concerning this branch of Newton's family. Letter 248, vol. II, is addressed to the second, and not (as note (1) to that Letter implies), the third, Baronet. It is curious that the letter should be addressed to Barr's Court; if the Letter is correctly dated, Sir John was still at this time a burgess in Parliament for Grantham. Note (1) of Letter 583 (vol. IV, p. 265) is in error in suggesting that Sir John Newton was *first* of Barr's Court, and *later* of Thorpe, Lincolnshire. The Newtons were a Lincolnshire family; the Barr's Court property was presumably inherited with the Baronetcy. Note (2) of Letter 703 (vol. IV, p. 461) implies that Isaac's great great grandfather, John Newton, was brother to Sir John's greatgrandfather, William Newton, in fact he was his father. (Reference to Appendix II, Newton's Genealogy, will clarify this.) The same errors are made in note (1) to Letter 719 (vol. IV, p. 489).

(3) Possibly Highley, in Shropshire near Bridgnorth.

(4) Richard Baker, *A Chronicle of the Kings of England From the Time of the Romans' Government unto the Death of King James* (first published in London in 1643, with numerous later editions). We have not traced the reference to Sir Peter Newton.

(5) According to Joseph Foster, *Alumni Oxonienses...1715–1886* (London, 1887; Oxford, 1891), a Peter Newton matriculated at Brasenose College, Oxford, 24 March 1716, aged 18, became B.A. on 20 February 1720, and M.A. in 1722. His father was another Peter Newton, clergyman at St Werburgh, Chester. The biographical details given in the letter make it likely that these two Newton's may be identified with Robert Newtons eldest brother and father.

(6) The writer clearly did not know of Newton's recent move to Kensington.

1467 PEMBERTON TO NEWTON
?MAY 1725
From the holograph original in the University Library, Cambridge[1]

Sr.

In pag. 424. l. 25,[2] After the word profundum, and in the room of the words vel horarum plurium si sit vadosum; might not some such sentence as follows be a little more convenient? Sed sæpius ad littora spatio horarum trium circiter, vel etiam plurium si mare sit vadosum.

These words will imply the reason why you use the third hour so much afterwards, without making any further mention of these earlier hours here taken notice of.

I am
Your most humble
and most obednt. servt.
H PEMBERTON

NOTES

(1) Add. 3986, no. 18.

(2) This is now the third edition text, in some version not recorded; Koyré and Cohen, *Principia*, p. 613, give no variant for this sentence, nor does the word 'profundum' appear at this point. The words proposed by Pemberton were added to the text.

1468 PEMBERTON TO NEWTON
17 MAY 1725
From the holograph original in the University Library, Cambridge[1]

Sr.

I here send you a sheet inclosed. In the Scholium contained in page 451[2] you'l see I have put into the margent some alterations. After Mr Machin's name I have put his title of Professor of Astronomy at Gresham College;[3] and have put out the letter D before his name, judging it useless. In the second line, after the word invenerunt, I have added a short sentence alluding in the slightest manner I could to my having published something of this method in the epistle upon Mr. Cotes's book:[4] for as it does not seem to me improper to hint at that particular; so to the best of my remembrance you took some Notice of it in the Scholium, which you once drew up; and which as far as I can recollect it, I have made that pattern of this here written.

But before you return the sheet, you'l please to adjust the whole of this Scholium as shall be most agreeable to your mind.

What is written at the bottom of this 451 page is for the direction of the Printer.

I am

Your most humble

and most obednt. Servt.

H PEMBERTON

May 17. 1725

NOTES

(1) Add. 3986, no. 19. About this time Pemberton must have dealt with the sheets containing Book III, Prop. 39 of the *Principia*, dealing with the Precession of the Equinoxes, in which no change was made. Probably shortly after considering this Proposition Pemberton received a letter (now owned by the Philadelphia Historical Society) from Brook Taylor dated 27 May 1725 in which Taylor criticized Newton's treatment of this problem, and argued that the value of the precession dynamically computed should be $57\frac{1}{2}''$ rather than $50''$ (thus differring markedly from observation). Taylor concludes: 'I submit these reflexions to your judgment and shall be glad to know whether you think any thing in the new edition of the Principia deserves to be alter'd upon account of them.' Pemberton took no action.

(2) Of the third-edition text; the Scholium (to Prop. 33) contains Machin's two propositions on the motion of the Moon's nodes. Newton, introducing it in three short sentences, writes that this matter has been mentioned elsewhere; that he received two papers, each containing a couple of propositions, one from Machin and the other from Pemberton, agreeing with each other; that he prints the former, which he received first. It is not clear that any of these sentences came from Pemberton.

(3) He was appointed in May 1713; see Letter 997, vol. v (p. 408).

(4) Newton gives no specific reference. Pemberton refers to his publication *Epistola ad amicum de Cotesii Inventis, Curvarum Ratione, quæ cum Circulo & Hyperbola Comparationem Admittunt* (London, 1722), addressed to James Wilson. There Pemberton discusses the content of Cotes' *Harmonia Mensurarum*, published posthumously in 1722. On pp. 5–6 of the *Epistola* he shows that

$$\int \frac{z^{\frac{1}{2}\eta-1}\sqrt{e+fz^\eta}}{g+hz^\eta}\,.dz = \int \frac{2vx^{-1}}{\eta h}\sqrt{\frac{eh-fg+fx^{2v}}{-g+x^{2v}}}\,.dx$$

if we make the substitution $x^v = \sqrt{(g+hz^\eta)}$. The integral is thus, as Pemberton points out, equivalent to Form 11 in Newton's *De Quadratura Curvarum*. This integral, Pemberton then explains, may be used in the computation of the Moon's nodes, but he gives no details of how; these may be found in the manuscript he had originally submitted to Newton (see note (2) above) in U.L.C., Add. 3966(11), fos. 92–4 (see also Cohen, *Introduction*, p. 263).

1469 NEWTON TO CAVELIER

27 MAY 1725

From the printed version in the *Philosophical Transactions*[1]
Reply to Letter 1461

I remember that I wrote a Chronological Index for a particular friend, on condition that it should not be communicated. As I have not seen the manuscript which you have under my name, I know not whether it be the same. That which I wrote was not at all done with design to publish it. I intend not to meddle with that which hath been given you under my name, nor to give any consent to the publishing of it. I am, your very humble servant,

Is. Newton

London, May 27, 1725. St. Vet.

NOTE

(1) The Letter is printed in Newton's 'Remarks on the Observations made on a Chronological Index of Sir Isaac Newton, translated into French by the Observator, and published at Paris', *Phil. Trans.* for 1725, **33**, no. 389, 315. For the background to the letter see Letter 1436, note (2), p. 280.

1470 PEMBERTON TO NEWTON
31 MAY 1725
From the holograph original in the University Library, Cambridge[1]

Sr.

After I parted from you on thursday morning I met with a disappointment in my intention to sollicite my lord chief Justice King;[2] which I shall more particularly inform you of, when I have the honour of seeing you. And my expectation being thus diminished, I thought my self not at liberty to put you to trouble, when the success might be doubtful. But though your most kind intention has happened not to be put in execution, I shall always make the same grateful acknowledgement, as if I had received all the advantage I hoped for from it.

I understand you did not see Dr. Halley at the Royal Society; I intend therefore to morrow or next day to take a turn down to Greenwich to see him.

I am

Your most humbl. and most obedt. servt.

Monday   H PEMBERTON

May 31. 1725

NOTES

(1) Add. 3986, no. 20.

(2) Peter King (1669–1734), Chief Justice of the Common Pleas, was about to be raised to the peerage and created Lord Chancellor. What appointment Pemberton sought from him is not known.

1470a PEMBERTON'S QUERIES ON *PRINCIPIA*, 2ND EDITION, pp. 464-74
From the holograph original in the University Library, Cambridge[1]

Questions.

Pag. 464. l. 23. Should not the word fere be omitted for according to the present alteration in this place the comet moved through nine signs compleat?[2]

Page 467. l. 5, 6. These words, quod duratio caloris, ob causas latentes, augeatur in minore ratione quam ea diametri, I have compared with the 11th question annexed to your book of Optics.[3] And in that question you seem to have a different sentiment from what is expressed here; for there you suppose that the sun by reason of its great magnitude may retain its heat for ever without any diminution. If so the time that a globe will retain its heat must increase in a greater proportion than its diameter increases.

323

Pag. 471. l. 27, 28.[4] It is here said, that the vapour which constitutes the tail of a comet, by participating [in] the motion of the comet itself, so ascends from it as to accompany the comet; and the heavens are concluded to be without resistance by reason that these thin vapours can move so freely. Again in l. 23, &c of pag. 469[5] the tails of comets are compared to the smoke arising from bodies in our air. and the incurvated figure of those tails is likened to the same figure observed in the smoke that ascends from a body moving in the air. But if it be here after said, that the smoke participates of the motion of the body itself (as without doubt it does) it will follow that the smoke in ascending should always continue perpendicularly over the body, if the resistance of the air did not obstruct its progressive motion: and therefore by this comparison the heavens should rather be concluded to have some resistance though small. Though I think it is not necessary to suppose any such resistance to account for this deviation of comet's tails from a direct opposition to the sun. But from what is observed in pag. 472 l. 20,[6] &c, that the vapour of these tails will proceed to move in conic sections about the sun along with the comet itself; this property of the tail may very well be explained, for the motion of the vapour in ascent being compounded with the progressive motion of the comet will cause each part of the tail to describe an orbit of a different species and in a different position from the orbit of the comet, though in the same plane.

Hence I presume to ask whether the comparison drawn between the appearances in the tails of comets and in the smoke of fuming bodies in our air, as it is set down in the foresaid page 469 might not be omitted, or some way modified?

Likewise in consideration of what has been last said, might not the word ellipsibus in pag. 472. l. 22. be conveniently changed into the more general expression sectionibus conicis?

Pag. 472. in the beginning. Concerning the cause that carries up these vapours from the comet I beg leave to observe, whether it be not done by the same means as fumes are carried up from heated bodies in vacuo, where fumes ascend with a less heat than in the open air. Since the heavens are void of any matter that can give a sensible resistance to the progressive motion of this vapour, what is that aura ætherea[7] which by its motion in ascent can carry this vapour along with it?

Pag. 472. l. 18, 19. I here observe that comets which never descend very near the sun do emit tails; even that (I suppose) which you treat on in page 477, that did not descend quite so near the sun as the earths aphelion distance.

Pag. 474. l. 17. Upon the word Portugallia I observe that Lusitania is the latin name of Portugal.[8]

NOTES

(1) Add. 3986, no. 31, fos. 4–5. We place these questions here on the supposition that their writing must have preceded that of Letter 1472, when the printing of the third edition of the *Principia* had reached this same point. The page numbers in these questions are those of the second edition.

(2) *fere* was omitted (*Principia*, 1726, p. 506, line 6).

(3) The sentence remained (*ibid.*, p. 509, lines 12–14). Pemberton's comment lacks acuity. In this passage of Book III, Prop. 41, Newton is concerned with the cooling of externally heated bodies, such as comets. In *Opticks*, Query 11, Newton considers a different case, where stars have been made (by God?) so intensely hot as to emit light very copiously; this light Newton further supposes so to interact with the matter of the star as to create more heat, to compensate for that which is radiated. There is thus no exact equivalence between Newton's view of the radiative cooling of comets and his hypothesis of the 'astrophysical' generation of heat.

(4) *Principia*, 1726, p. 514, lines 4–6 (unchanged).

(5) *Ibid.*, p. 511, line 33ff. (unchanged).

(6) *Ibid.*, p. 514, line 36ff. The word 'ellipsibus' (p. 515, line 1) is retained.

(7) The 'auram ætheream' remains (*ibid.*, p. 514, line 26).

(8) The change to 'Lusitania' was made (*ibid.*, p. 517, line 1).

1471 RICHARD PINDAR TO NEWTON

9 JUNE 1725

From the original in the Bodleian Library, Oxford[1]

Most Hon'red Sir

Haveing come thus fare in order to pay your Hon'er a visit I thought it proper to present my case before your Honour by a few Lines (with great Submision Humbly asking pardon for my boldness) Honour'd Sir I have by the assistance of almighty God been diligent in my trade and discharged my duty both to God and my family upward of 20 years and getting but little more then to support my self and family and I find that to keep a little stock togather with my trade will be to duble advantage but can doe but little myself I take boldness to address myself to your Honour to assist me by Lending me 15 or 20 pounds for the space of 2 or 3 years in which time I hope to retorn it for I have a fair prospect of advantage being seated very conveniant for that porpass and now if your Honour please to oblige your unworthy Relation in so great a favir you will ingage me in my retorns at the thrown of God's grace to implore for a Blessing in the world to come that your Honour may receive a heavenly reward which is all I can Retorn Who is your honour's unworthy sarvent

RICHARD. PINDAR

June 9 1725

For Sir Isaac Newton

(1) Bodleian New College MS. 361, II, fo. 58. A note added in Newton's hand states that 'Richard Pindar is a weaver & lives at Gosberton in Lincolnshire near Boston'. He was presumably the grandson of Newton's Uncle Richard (see Appendix II, Newton's Genealogy).

1472 PEMBERTON TO NEWTON
22 JUNE 1725
From the holograph original in the University Library, Cambridge[1]

Sr

I have taken a fresh into consideration the figure of the place of the comet; in order to find the reason why these places did not fall in a right line as they ought to have done. For this purpose I found the several places of the comet by the longitudes and latitudes set down in page 496 of this edition of your book.[2] which page I have here inclosed for your fuller satisfaction. The places of the comet thus laid down agree well enough with the former in p, P, R, S: but in Q and V they differ a little, in T and in X and Y they differ more. The places which I have now found from their Longitudes and Latitudes, I have drawn with black lead without erasing the former places, that you might see the difference.[3] In your observations it is not mentioned on which side of the line $\gamma\delta$ the comet in X lay, and I was misguided to place it on the wrong side of that line from the erroneous place I had given to the comet in T.[4] But as for this place of the comet in T, I beg leave to observe that the longitude which you have ascribed here to the comet and according to which I have now put down this place T with black lead, is not consistent with your observations on the comet in this place as set down in page 498 of this proof sheet now sent you: so that I suspect there may be some error in this place. You have ascribed a greater longitude to the comet in T than what the star M has; the longitude of the star M being $\aries$ 29°. 18'. 54", and the longitude of the comet $\aries$ 29°. 20'. 51": but by the observations the comet in T should have a less longitude than the star M.[5] You observe that the line drawn through M and T fell four times nearer to the star B than to the star F: therefore this line must decline towards B from the circle of latitude; which passes through M; for M differs in longitude from B more than three times what B differs from F in longitude.

As soon as you return me this scheme, with your directions; I will take care, that it shall be altered, to your mind.

Be pleased to compare the numbers, by which you give the several distances of the stars from each other in the beginning of page 497, with your original paper, wherein you set them down, if you have it by you: because I suspect errors in some of them. For instance AI is $27\frac{7}{12}$, BI $52\frac{1}{6}$ and the sum of these is $79\frac{3}{4}$, less than AB which is $80\frac{7}{12}$; whereas the sum of these two ought to be greater than AB.[6]

I am
Your most humble
and most obedient servt.
H PEMBERTON

Tuesday Morning
Jun. 22. 1725

NOTES

(1) Add. 3986, no. 21.

(2) *Principia*, Book III, Prop. 41; consideration of the comet of 1680. In the third edition Newton altered the tabulated 'corrected' positions of the comet by up to one minute (usually less). The figure showing these positions (*ibid.*, p. 497) is redrawn to include additional positions; in this figure (as in that of the second edition) the successive positions of the comet do not lie precisely on a straight line though the approximation is fairly close.

(3) Presumably Pemberton is marking the figure as printed in the second edition.

(4) The star positions γ, δ, and the places of the comet in T and X were new in the third edition (though a place T was also given in the second).

(5) The longitude of the star M as printed by Newton (and here quoted by Pemberton) was determined by Pound. The longitude of the comet on 5 March 1680 at 11.30 p.m. was always previously stated by Newton to be as Pemberton here quotes it; but the rather vague description of the comet at this moment, in relation to star M (*Principia*, 1726, p. 498) and the figure on the previous page, certainly give the comet a smaller longitude. Accordingly, in the third edition Newton arbitrarily redefined the comet's longitude on 5 March as 29° 18′ 0″. The fact is that Newton did not have a *measured* position for the comet on 5 March.

(6) The three stars A, I and B are not precisely in a straight line; therefore $AI+BI > AB$. Newton (wisely) did not further fudge his numbers, but let them stand.

1473 PEMBERTON TO NEWTON
17 JULY 1725
From the holograph original in the University Library, Cambridge[1]

Sr

What is contained in pag. 509. l. 12, 13, 14. in the sheet which comes to you with this note is an expression, which I think is inconsistent with what you have said at the end of your Optics; viz. in Qu. 11.[2] Here in this sheet you declare it as your opinion, that the time which globes retain heat increases in a less proportion than their diameters increase; whereas the sentiment expressed in your Optics implies the contrary, viz. that the time which globes will retain their heat must increase faster than their diameters; because you suppose that if a globe amounts to a certain magnitude like that of the sun or fixed stars it shall for ever retain its heat. [3] Therefore the time, which the sun or a fixed star will retain its heat will bear a greater proportion to the time that a lesser globe will continue hot, than the diameter of the sun or star bears to the diameter of the lesser globe; for the time which the sun or star will naturally retain its heat, is supposed by you to be infinite.

The blank space left in pag. 506 is for a copper cut of the comet.

> I am
> Your humble
> and most obednt. servt.
> H PEMBERTON

Jul. 17. 1725
To Sr. Is. Newton
to be sent with the sheet

NOTES

(1) Add. 3986, no. 22.

(2) Pemberton repeats with respect to the proof sheet (Ttt, pp. 505–12) of the 1726 *Principia* the objection he had already raised in Number 1470*a* with respect to the second-edition copy. His doing so supports the notion of a considerable lapse of time between reviewing the copy and correction of the proof.

(3) Pemberton has missed the point of the difference; see Letter 1470*a*, note (3), p. 325.

1474 BURCHETT TO NEWTON

12 AUGUST 1725

From a copy in the Public Record Office[1]
For the answer see Letter 1476

12 *Augt.* 1725

Sr.

Mr Jacob Rowe[2] having by his Petition to my Lords Comm[ission]rs of the Admi[ral]ty represented, [3] that he hath lately made a further Improvement on the Latitude in proposing more frequent opportunities for observing the same and a new Invented Instrum[en]t of the greatest exactness and usefullness in taking Altitudes at sea, [4] as also that he proposes several Rules[5] whereby Our Method of keeping an Acct. of a ships Distance sailed by the Log will be much more exact than the Common sand Glasses, which Invention on Tryal at Sea he doth believe will appear to be of service in discovering the Longitude; I am commanded by their Lord[shi]ps to desire that you will be at the trouble of examining into what Mr Rowe hath proposed and give their Lordps. your Opinion of the same. I am &c

J Burchett

Sr Isaac Newton

NOTES

(1) Adm/2, 458, p. 46. Newton's reply shows him at last losing patience with the continual demands of the Admiralty that he should assess untenable methods of improving longitude determinations.

(2) Rowe had been engaged for many years in devising instruments for use at sea, and the Admiralty seems to have looked favourably on a number of his inventions (see P.R.O.: ADM/2, 453, pp. 442–68, and ADM/2, 454, pp. 33 and 498).

(3) Rowe's petition was read to the Admiralty on 12 August 1725 (see P.R.O., Adm/3, 35, unpaginated).

(4) Rowe had submitted an instrument for this purpose to the Admiralty in October 1719, on which they had given a favourable report (see P.R.O., ADM/2, 453, pp. 442 and 468) and for which Rowe had been suitably rewarded (P.R.O., ADM/1, 4102, fo. 118).

(5) Newton gives some details of Rowe's methods in his reply.

1475 NEWTON TO COLIN MACLAURIN

21 AUGUST 1725

From the holograph original in the Stanford University Libraries, California[1]
For the answer see Letter 1481

Sr

I am very glad to hear that you have a prospect of being joyned with Mr James Gregory[2] in the Professorship of Mathematicks at Edinburgh, not only because you are my friend, but principally because of your abilities, you being acquainted as well with the new improvemts of Mathematicks as with the former state of those sciences. I heartily wish you good success, & shall be very glad to heare of your being elected. I am with all sincerity

Your faithfull Friend &
most humble Servant

London Aug. 21
1725.   Isaac Newton
For Mr Mac Laurin

NOTES

(1) Printed in Brewster, *Memoirs*, II, p. 385 and elsewhere. Colin Maclaurin (1698–1746) became Professor of Mathematics at Marischal College, Aberdeen, in 1717. In 1719 he visited London, where he met Newton and was elected Fellow of the Royal Society; the following year he published his *Geometrica Organica, sive Descriptio Linearum Curvarum Universalis* (London).

329

From 1722 to 1724 he travelled in France, and there wrote his 'Demonstration des loix du choc', which won him the 1724 prize of the Académie Royale des Sciences. In 1724 he returned briefly to Aberdeen, but in November 1725 became joint Professor of Mathematics with James Gregory at Edinburgh (for further details see Letter 1482). In 1742 Maclaurin published his *Treatise of Fluxions* (Edinburgh), and on his death left in manuscript *An Account of Isaac Newton's Philosophical Discoveries*, which was published in London in 1748 with Patrick Murdoch's biographical preface, 'An Account of the Life and Writings of the Author'.

(2) James Gregory (1666–1742), younger brother of Newton's friend David Gregory and nephew of the great mathematician James Gregory (vol. I, p. 15, note (1)), was educated at Marischal College, Aberdeen, and proceeded M.A. at Edinburgh University in 1685. Three years later he became Professor of Mathematics at St Andrews, and then in 1692 succeeded his brother as Professor of Mathematics at Edinburgh. He was forced to retire on account of ill-health in 1725.

The younger James Gregory is not comparable in stature to his brother or his uncle, but he was one of the earliest lecturers on Newton's *Principia*. His sole printed work, a thesis dated 1692, is a compendium of the *Principia*. (For the Gregory family see P. D. Lawrence, Ph.D. thesis, Aberdeen University, 1971.)

1476 NEWTON TO THE ADMIRALTY

26 AUGUST 1725

From the holograph original in Christ Church College Library, Oxford[1]
Reply to Letter 1474

To the R[ight] Honble the Lords Commissioners
of his Majties Admiralty.

May it please your Lordships

The Longitude will scarce be found at sea without pursuing those methods by which it may be found at land.[2] And those methods are hitherto only two: one by the motion of the Moon, the other by that of the innermost Satellit of Jupiter. Sublunary motions are too unconstant. The first method hath been long practised by Geographers, & Geography hath been setled thereby: but the Theory of the Moon is not yet exact enough for the sea. It hath lately been made exact enough for finding the longitude at sea without erring above three degrees, & if it were exact enough for finding it without erring above one degree it would be very usefull, if without erring above forty minutes it would be more usefull, & if without erring above half a degree it would scarce be further improveable, as I told a Committee of Parliament in writing when this matter was referred to them.[3] And thereupon the Parliament passed the Act to reward him or them who should find it to a degree or to forty minutes or to half a degree. But nothing hath been done since for making it more exact then it was at that time. Dr Halley hath been observing the Moon the three last years, [4] & finds her Theory as exact as I affirmed: but to make it exact enough for sea affairs is a work of time. At present its errors sometimes amount to six minutes. When it shall be made a little exacter so as never to err above three or four minutes, it may be time to begin to apply it to sea affairs.

The other method of finding the Longitude is by observing the eclipses of the innermost Satellit of Jupiter. This is the easier & exacter method at land, & hath much corrected Geography: but Telescopes of a sufficient length for seeing those eclipses are not manageable at sea. And what may be done by short reflecting Telescopes wch take in much light & magnify but little, hath not yet been tried.

A good jewel-watch kept from the air in a proper case, & examined every fair morning & fair evening by the rising & setting Sun, & kept in an eaven heat, may be sufficient for knowing the time of an observation at sea till better methods can be found out. Or else such a pendulum clock [may be used] as the Quaker tryed in going from hence to Portugall. Or such a clock as Mr Case Billingsley proposed to be tryed. But these clocks will be affected by the variation of gravity in varying the Latitude: & the quantity of that variation is not yet sufficiently known.

The hour glasses of Mr Rowe [5] made with sand of Tin, are a very good piece of art, the sands being weighty globular & small [& of an equal size]. [6] These glasses run very eavenly, & in sailing by the log may with advantage be used instead of the vulgar hour glasses made with common sand, if those that are well made can be had at a moderate price, & the vulgar glasses be not exact enough for this purpose. For upon examining one of the vulgar glasses I found that its greatest errors in running too fast or too slow did scarce exceed a minute in an hour.

Also the new methods of Mr Rowe for finding the motion of a ship with respect to the sea water at the depth of some fathoms below the surface of the sea rather then at the surface may add to the improvement of sailing by the Log, provided these new methods will not be too troublesom to be chearfully used by the seamen. But this deserves to be well considered. For the motions of the sea arising from winds are most uncertain at the surface, & reach not very deep unless the wind be lasting: but those of lasting currents seem to go deep, & so do those arising from the cause of tydes.

Mr Rowe's instruments for taking altitudes at sea more exactly & frequently then heretofore, are well contrived, & deserve to be tryed at sea for finding the Latitude. They may be tried by taking the altitude of the Sun or of any star in the meridian several times to see if the observations agree with one another better then those made with the instruments hitherto in use. They should also be examined before they are bought to see how well they agree with a good large Quadrant in taking altitudes at land, because they are not equally divided into degrees.

All which is most humbly submitted to your Lordps great wisdome

ISAAC NEWTON

Mint Office Aug. 26
1725

NOTES

(1) MS. 350/4; the letter is endorsed 'Recd. from Sr. Isaac Newton the 26. of August 1725, & read.' A draft of the letter is in U.L.C., Add. 3972, fo. 30.

(2) Newton uses the opportunity of Rowe's petition to inveigh against the Admiralty once again for sending him futile suggestions concerning the determination of the Longitude; compare Letter 1377 and notes.

(3) See Number 1093*a*, vol. VI.

(4) See Letter 1344, note (14), p. 101.

(5) For Jacob Rowe's work on marine instruments, see Number 1380, note (2), p. 177.

(6) The brackets are Newton's.

1477 CLAUDE JOMBERT TO NEWTON
1 SEPTEMBER 1725
From the holograph original in King's College Library, Cambridge[1]

De Paris ce 12e 7bre 1725: [N.S.]

Monsieur

J'ay receue avec la plus vive recognoissance la planche gravé [*sic*] du Portrait de Mr. de Varignon que vous avez eû la bonté de m'envoyer par le S. Innys;[2] mais le public vous est icy encor plus redevable d'avoir bien voulu ajouter cette ornement à l'Edition que je vien de faire de sa Mecanique.[3] Cette atention de Vostre part fait seule l'Eloge de ce grand homme que vous avez honnoré de Vostre Confiance et de Vostre Estime; Je prend la Liberté Monsieur de joindre au rémerciment que J'ay l'honneur de Vous faire huit Exemplaires Reliés de la ditte Mecanique dont vous disposerez s'il vous plaist en faveur de vos amis. Je m'estimerois fort heureux Monsieur si je pouvois imprimer aussy quelqu'un de Vos ouvrages qui vous attire l'admiration de toutte l'Europe; Je vous suplie tres humblement en cas que vous soyé disposée à en faire paroistre encore quelqu'un en france de Vouloir bien Jetter les yeux sur moy il n'est rien que je ne fisse pour maquitter de mon devoir d'une maniere à Vous Satisfaire; Je suis avec un tres profond respect

Monsieur

Vostre tres humble et tres obeissant serviteur

JOMBERT

J'ay chargé Monsieur Le Sr. de Varenne de la caisse dans la quelle vous trouveray deux Memoires de l'Academie que Mr. de lisle ma remis pour vous. [4] J'ay aussy charge le d[ite] Sr. de Varenne Libraire a Londre par ordre Mr. de fontenelle de quatre Exemplaires en blanc de la ditte Mecanique; dont un pour vous.

NOTES

(1) Keynes MS. 142(W); microfilm 1011.26. A second version also holograph, and identical except for orthography, is at Keynes MS. 142(X). Possibly Jombert sent both copies to Newton. Claude Jombert, one of the famous Jombert family of book-publishers in Paris, became 'Libraire' on 21 December 1700; he died in 1733. (See A. M. Lottin, *Catalogue chronologique des libraires et des libraires-imprimeurs de Paris*, Part II (Paris, 1789), p. 93).

(2) See Letter 1453.

(3) Pierre Varignon, *Nouvelle mecanique ou statique, dont le projet fut donné en M.DC.LXXXVII* (Paris, 1725). Chez Claude Jombert, rue S. Jacques, au coin de la rue des Mathurins, a l'Image de Notre-Dame. No portrait of Varignon is bound into either of the two volumes of the work. In the Preface is mentioned the projected *Commerce de Lettres* which Jombert planned, but never published; perhaps he intended to include Varignon's portrait in this.

(4) For Delisle's part in this, see Letters 1450 and 1453.

1478 SCROPE TO THE MINT

23 SEPTEMBER 1725

From a clerical copy in the Public Record Office [1]
For the answer see Letter 1483

Gent[lemen]

Mr. Kidder Assay Master in Ireland having by direction of the Lord Lieutenant there made an Assay of several new Portugal Gold Coins specified in his Report which comes inclosed. The Lords Comm[issione]rs: of his Ma[jesty']s Trea[su]ry are pleased to direct you to cause the weight and Assay of the like Coins to be taken here and to report the same to their Lordps with your Opinion at what value they may reasonably go in Ireland in case his Ma[jes]ty: by his Proclamation shall think fit to make them current in that Kingdom [2] I am &c 23 Sept: 1725

J SCROPE

May it Please your Lordps

In obedience to your Lordps Orders I have exactly weighed the several new Portugal Gold Coins recd: from your Lordps and have made a return of them underneath and have also assayed them severally and have reported them as they answered by Assay.

333

		Dwt	Gra	M		Dwt	Gra	M
No: 1 One Port Peice	Wt:	1 „	2 „	16	Better than standard	— „	$\frac{3}{4}$.	—
No: 2 One Do.	Wt	2 „	6 „	15	Better . . .	— „	$\frac{3}{4}$.	—
No: 3 One Do.	Wt	4 „	14 „	15	Better . . .	— „	$\frac{3}{4}$.	—
No: 4 One Do.	Wt	9 „	5 „	—	Better . . .	— „	$1\frac{1}{4}$.	—
No: 5 One Do.	Wt	18 „	8 „	15	Better. . .	— „	1.	—

All which is humbly submitted to your Lordps
this 6th: day of April 1725.
By your Lordps
most humble & obedient servt:
VIN: KIDDER[3]

NOTES

(1) T/27, 24, p. 110; Scrope sent a reminder to the Mint concerning the matter on 5 November 1725 (see P.R.O., T/27, 24, p. 116).

(2) On 31 July 1725 Holles Newcastle had sent to the Treasury Lords a letter he had received from the Lord Lieutenant of Ireland asking for the new Portuguese gold coinage to be made current in Ireland; he wrote again on 21 September 1725, enclosing a petition from the merchants of Dublin and Cork; see P.R.O., T/1, 253, fo. 217.

(3) The writer was Assaymaster of the Dublin Mint; compare vol. VI, pp. 112–13 and 147.

1479 NEWTON TO THE TREASURY
4 OCTOBER 1725
From the holograph draft in the Mint Papers[1]

To the Rt Honble the Lords Comm[ission]ers
of his Ma[jesty]s Treasury

May it please your Lordps

The Pix of the copper money made since the last trial thereof, remains to be tried whenever your Lordps please to appoint a day for the trial & a person to see the same & report it to your Lordps.[2] Which is most humbly submitted to your Lordps great wisdom

Mint Office
Octob. 4th. 1725   ISAAC NEWTON

NOTES

(1) III, fo. 316.

(2) According to a report by Henry Kelsall (P.R.O., T/1, 251, fo. 235) the Pyx was opened and tried on 7 October 1725. Random samples of copper coin were weighed, and some coins were tested by hammering when red hot. These tended to split at the edges, although a few beat thin without splitting.

1480 CHAPMAN TO NEWTON
23 OCTOBER 1725
From the original in the Bodleian Library, Oxford[1]

Honourable Sir

I owe my most dutifull and Gratefull Acknowledgement for the Candid Countenance and Reception which you afforded my humble suit by the Revd Mr Cranor I addressed you since by the Revd and truly worthy Dr Clarke, and I flattered my self you would Condescend to Honour mee (Tho I own I am undeserving) with some small notice. Tho I came to town on Purpose to pay my duty to you to Render an Account of my Conduct and Circumstances, I would not Intrude my self till I waited on Doctor Clarke yesterday (Just after your Honour Called there) who advised me to apprise you of it, and request an Audience to be Ingennuous in every thing, and submit it to your wisdom and goo[d]ness. My affairs (I thanke god and your Honours Bounty) are not so perplext and Cumbersome as of late they have been. I have made the Burden lighter and supported my family (my wife and Infant son) comfortably and Reputably by an Indefatigable Incessant Industry not presuming to Attend your Honour, till I had Reduced things to an Easie narrow Compass. I have Commands from my Mother to your Honour; and beg your good pleasure, and answer to my humble request of a desired Interview, and I question not, but I shall Accquitt my self to your Honours satisfaction and approbation, as becomes, Honble Sir,

> Yours most obliged, and
> most obedient servt,
> and Kinsman
> NEWTON CHAPMAN

oct ye 23 1725

> To
> The Honourable Sr
> Isaac Newton

NOTE

(1) Bodleian New College MS. 361, II, fos. 61–2. For the Chapmans and their possible relationship to Newton see Letter 1314, note (2), p. 33.

1481 COLIN MACLAURIN TO NEWTON
25 OCTOBER 1725
From the holograph original in Hale Observatories Library, Pasadena, California[1]
Reply to Letter 1475

Honoured Sir

I am much obliged to you for your kind letter that Mr Hadly[2] transmitted to me. It has been of use to me.[3] However the Provost (or Mayor) of the toun has thought fit to consult yourself directly on that subject, because I made some scruples to make your letter that was addressed to me publick. I flatter myself you will as soon as your Convenience can allow give an answer to his letter, that the want of it may not obstruct the Affair.[4]

I have lately had a dispute with a Gentleman here[5] who attack'd your Prop. 36, Lib. 2. of the Principles, & is supposed to be a pretty good Mathematician. 'tis about the pressure[6] on the Circellus PQ (Prop. 36. Lib. 2. Cor. 7, 8, 9, 10)[7] He finds by a calcul of his that the pressure on PQ is to the Cylinder on the base PQ of the height $\frac{1}{2}GH$ as $2AB^2$ to $AB^2+EF^2-PQ^2$,[8] whereas you make that proportion as $2EF^2$ to $2EF^2-PQ^2$ in Cor. 10.[9]

I can demonstrate (and he allows it) that when CD & EF are equall the pressure on PQ is to the Cylinder on PQ of the height $\frac{1}{2}GH$ as $2EF^2$ to $2EF^2-PQ^2$. But he Objects that tho' the proportion must be allowed in that case, yet it cannot be generall, and that it ought to vary with AB tho' AB does not enter into your proportion Prop. 36, Cor. 10.

I have answered this and have showed that when AB is very great the pressure on PQ should be the weight of the whole Cylinder above PQ according to him; because the ratio of $2AB^2$ to $AB^2+EF^2-PQ^2$ in that case is a ratio of $2AB^2$ to AB^2 or of 2 to 1. And this I think absurd since by the very idea of the Cataract PQ cannot bear the whole Cylinder above it—

But I trouble you no further, I am more and more satisfied that your Book will triumph over all that oppose it. and that as it has met with resistance from the prejudices & Humours of men it will prevail the longer. I am with much gratitude and the greatest Respect

Honoured Sir
Your most oblidged
Most Humble Servant
COLIN MC LAURIN[10]

Edinburgh
Octo[be]r. 25. 1725

336

If at any time you have any Commande for me a letter directed to Mr Turn-bull's shop will find me.

NOTES

(1) Printed in Brewster, II, pp. 385–6.

(2) Possibly a mistake for Edmond Halley, or else presumably John Hadley (1682–1744; Fellow of the Royal Society, 1716), later inventor of the reflecting quadrant, a friend of Francis Hawksbee Junior, Clerk-Housekeeper to the Royal Society.

(3) Compare Letter 1482, note (1), p. 339.

(4) The reply to the Provost's letter drafted by Newton was probably never sent; see Letter 1482, note (1), p. 338.

(5) Possibly George Campbell (F.R.S. 1730; one of this name proceeded M.A. at Edinburgh in 1721) whom Maclaurin tried to assist, as he had deprived him of the hope of succeeding James Gregory (see Charles Tweedie, *James Stirling* (Oxford, 1922), pp. 15, 57–72, 93, 193–4; the third of these references contains Campbell's undemonstrated expression for the quantity of water evacuated through a hole radius r under a head a, in a letter to Stirling of 1728). Mr J. H. Chillington III, who has kindly furnished much of the information in our notes concerning Maclaurin, states that correspondence between him and Campbell on Newton, *Principia*, Book II, Prop. 36, is in Aberdeen University Library, MS. 206.

(6) That is, weight.

(7) In Corollaries 7–10 to Prop. 36, Book II of the *Principia* (1726) Newton considered the problem of water flowing out of an annular aperture at the centre of the base of a cylindrical vessel (see Koyré and Cohen, *Principia*, I, pp. 485–6).

(8) We may guess how the 'gentleman' arrived at this result. Suppose, following Newton, we consider the system to be equivalent to a cataract of water falling through the annular aperture defined by the ice surfaces AE, HP, HQ and BF. The particles of water are considered to suffer no resistance either from each other or from the surfaces of the ice. In the general case, where $CD \neq EF$ we may use the principle of continuity of flow together with the assumption that the water falls under gravity through the height HG without resistance to its vertical motion to show that:

$$\frac{\text{total volume of}}{\text{solid ice}} \Big/ \frac{\text{volume of cylinder,}}{\text{diameter } AB, \text{ height } HG} = \frac{AB^2 - EF^2 + PQ^2}{AB^2 + EF^2 - PQ^2}.$$

In the special case $AB (= CD) = EF$ clearly the whole of the solid ice rests on the little circle PQ, so that we have:

$$\frac{\text{weight of ice on } PQ}{\substack{\text{weight of cylinder of ice,}\\ \text{diameter } PQ, \text{ height } \frac{1}{2}HG}} = \frac{2EF^2}{2EF^2 - PQ^2}$$

which is the result Newton gives.

Returning to the more general case, if we make the unwarranted assumption that the volume of ice is divided between the outer part of the funnel resting on the annulus CE, and the inner core resting on the little circle PQ, in proportion to the areas of their respective bases, that is as $AB^2 - EF^2$ to PQ^2, then in general:

$$\frac{\text{weight of ice over } PQ}{\substack{\text{weight of cylinder of ice,}\\ \text{diameter } PQ, \text{ height } \frac{1}{2}HG}} = \frac{2AB^2}{AB^2 + EF^2 - PQ^2}.$$

(9) Newton had arrived at this expression by interpolating between two special cases, a procedure which he himself clearly regarded as only approximate.

(10) Maclaurin spelt his name variously; up to 1726 he used the form McLaurin and then (like Newton) he adopted the form MacLaurin. We have used the modern spelling (Maclaurin) throughout the editorial matter in this volume.

1482 NEWTON TO JOHN CAMPBELL
c. NOVEMBER 1725
From the holograph draft in Hale Observatories Library, Pasadena, California[1]

Mr Ld

I received the honour of your Letter[2] & am glad to understand that Mr Mac Laurin is in good repute amongst you for his skill in Mathematicks. For I think he deserves it very well.[3] And to satisfy you that I do not flatter him, & also to encourage him to accept of the place of assisting Mr Gregory in order to succeed him, I am ready (if you please to give me leave) to contribute twenty pounds per annum towards a provision for him till Mr Gregories place becomes void if I live so long,[4] & I will pay it to his order in London.[5] When your Letter arrived at London[6] I was absent from thence wch made it the later before I received it. Otherwise I might have returned an answer a little sooner. I am

> My Lord
> Your Lordps most humble
> & most obedient servant
> ISAAC NEWTON

To his Lordp the
Provost of Edinburgh
    in Scotland

I reccon him well skilled in Arithmetic, Algebra, & Astronomy & Opticks, wch are the Mathematical sciences proper for an University, & abundantly sufficient for a Professor.[7]

NOTES

(1) Printed in Brewster, *Memoirs*, II, p. 387. Another, rougher draft precedes it on the same folio. Compare Letters 1475 and 1481.

There is no evidence that the letter was ever sent; Mr Chillington informs us that it is not entered in the Edinburgh Town Council Minutes or Letter Books, and that Maclaurin himself did not know of it until much later when John Conduitt discovered the drafts amongst Newton's papers.

Nonetheless, it is clear from the Edinburgh Town Council Minutes that the Council knew of Newton's support for Maclaurin's candidacy. For at the meeting on 3 November 1725, when Maclaurin was elected and nominated, the committee report (see MS. Edinburgh Town Council Minutes, LV, pp. 15–19) states (p. 16) 'That Mr. Colin McLaurin...had made surprising appearances in that part of Learning and these so very well known to all the Learned that tho he had a very favourable character bestowed on him by very Great Men and even by Sir Isaac Newton himself, he did not seem to need any of those to convince us.' Maclaurin officially accepted the position on 10 November 1725 (*ibid.*, p. 24). Perhaps Maclaurin had shown the Provost Newton's earlier letter (Letter 1475) (he says in Letter 1481 that the letter had 'been of use' to him) but had hesitated to employ it as an official recommendation.

John Campbell had been replaced by George Drummond as Lord Provost of Edinburgh on 25 October 1725, but Newton presumably addressed his letter to Campbell. According to Conduitt's account of Maclaurin's election (see Robert Wodrow, *Analecta: or materials for a History of Remarkable Providences; mostly relating to Scotch Ministers and Christians*, IV (Edinburgh, 1842), pp. 215–16) Campbell was in the chair, but himself supported the election of a namesake (seemingly George Campbell, see Letter 1481, note (5)), and raised difficulties over the payment of salaries to Gregory and Maclaurin.

Much later, in 1731, Maclaurin sent a letter to Martin Foulkes, President of the Royal Society, reporting that Conduitt had just informed him of Newton's offer of monetary help, complaining of the nature of his duties as Professor, and giving further details of his election. He wrote, 'I have been drudging here these six years in an intolerable way. My whole Time almost has been employed in teaching the Elements of a Science...I have taught six hours in a day I am so fatigued that I can read nothing if it is not of a much easier Nature...The salary that was settled on me at my admission was 50 *lb st*[*erling*] yearly, but I was obliged to pay Mr. Gregory a considerable sum for consenting to my settlement which I was obliged to take of my own funds, and indeed it cost me several years to make it up for he got from me in all betwixt 260 & 270 *lib. st.*...' (Royal Society, MS. FO. 4, no. 2.)

(2) This letter, now lost, is referred to in Letter 1481.

(3) The rougher draft adds here, 'He is well settled both in the new geometry and the old.'

(4) Deletions in the rougher draft show that Newton had difficulty in deciding for how long the annual payments should continue. He wrote, and deleted, 'five', 'six' and 'seven years'.

(5) The rougher draft has here the deleted passage, 'For I have a kindness also for Mr Gregory upon his brothers account, & should be glad to have a hand in helping to a coadjutor.'

(6) Newton moved to Kensington (then considered as 'out of London') earlier in the year for his health's sake (see Letter 1455, note (2)).

(7) This passage, which Newton probably intended to insert somewhere in the letter, is on the back of the sheet.

1483 NEWTON TO THE TREASURY
10 NOVEMBER 1725
From the holograph draft in the Mint Papers[1]
Reply to Letter 1478

To the right honble the Lords Commissioners
of his Majties Treasury

May it please your Lordps

In obedience to your Lordps Order of Reference of Sept. 23, 1725, that we should cause the weight & assay of five sorts of new Portugal gold coins to be taken, & report the same to your Lordps, with our opinion at what value they may reasonably go in Ireland in case hisMa[jes]ty should think fit to make them current there by Proclamations. We have caused five pieces of the said gold coynes, one of each sort, new out of the Portugal Mint, to be weighed & assayed. And they proved as follows.

Pieces	Weights	Assays	[Value in England][2]		
	oz. dwt gr		£	s	d
1	0. 1. 3	Sta. weak	[0	4	4]
2	0. 2. 8	Sta. strong	[0	9	1]
3	0. 4. 15	Sta. weak	[0	18	0]
4	0. 9. 5	Sta. strong	[1	15	10]
5	0. 18. 9	Standard	[3	11	6½]

By these assays the coyn is standard, & by the weights of the pieces the five species are in the progression 1, 2, 4, 8, 16 in weight. The pieces of the biggest species weighing 18 dwt 9 gr are worth 3*lb*. 11*s*. 8*d* each as they come fresh out of the mint supposing a Guinea to be worth 1*lb*. 1*s*. And according to this valuation, the next in bigness are worth 1*lb*. 15*s*. 10*d* each; the middlemost are worth 17*s* 11*d* each; the least but one are worth nine shillings each abating an halfpenny; & the least are worth 4*s* 6*d* each abating a farthing. These are their values in England one with another when they come fresh out of the Portugal Mint. But the Merchants will be apt to pick out the lightest pieces for Ireland, & send the heaviest pieces to the melting pot. And those pieces wch are full weight when they come fresh out of the Mint, will soon grow lighter by wearing in Portugall before they come into Ireland. And the smallest pieces will weare fastest in proportion to their weight & value. And some abatement in the value ought to be made for this lightness. If therefore this Portugal money is to be made current in Ireland, we are humbly of opinion that the biggest pieces should not be current for more then the value of three Guineas & the third part of a Guinea in Ireland or not for above 4 pence more. The next in bigness may be current for half the value of the biggest: the middlemost for a quarter of that value: the least but one for the eighth part of that value & the least for a sixteenth part thereof, if these be not too little to be made current.

All wch is most humbly submitted to your Lordps great wisdome

NOTES

(1) II, fo. 229; there is another draft at P.R.O. Mint/19, 5, fos. 12 and 11 v. The letter itself, dated 10 November 1725, and signed by W. Cary and Is. Newton, is at P.R.O., T/1, 253, no. 48, fo. 237; this document is printed in Shaw, pp. 203–4.

The Treasury reported their findings to the King on 5 January 1726; a copy of the report is at P.R.O., T/54, 29, p. 464. Whilst claiming to adopt Newton's suggestions, they advised that one Portugal piece be valued at 5 Irish shillings, with the larger coins in proportion. Even allowing for the fact that an English shilling was worth 13 Irish pence, this still gives one Portugal piece as about 4s. 7d in English money, considerably greater than the value of no more than 4s. 4½d recommended by Newton.

(2) This column was added in the final document.

1484 RAWSON TO NEWTON

31 DECEMBER 1725

From the original in the Bodleian Library, Oxford[1]

Hond Sr

I make bold to present to you a new almanack wishing your Hon[ou]r a happy new Year with my humble service. I remain your humble servant

CHA RAWSON
Stationer to ye Mint

Decemb 31th
    1725
To
  The Honble Sr Isaac
Newton Master Worker
of his Ma[jes]ties Mint within
the Tower of London
    Pr[es]ent

NOTE

(1) Bodleian New College MS. 361, II, fo. 81. Compare Letter 1417.

1485 JOHANN PETER BIESTER TO NEWTON
1 FEBRUARY 1726
From the original in the Jewish National and University Library, Jerusalem[1]

Illustrissime Domine!

Veniam dabis, quod Epistolam meam de removendis difficultatibus, praxin Methodi inveniendi longitudinem impedientibus et ad te, et per publicam impressionem dederim.[2] Ad te illa videbatur dirigenda, cum ipse, solida ratione, difficultates movisses; per impressionem vero, quo et aliis, quibus eædem dubium forsan moverant, remotio illarum innotescere possit. Nolui tibi esse molestus, quamdiu illæ remoram dabant meæ Methodo, potius habens, Londini tam diu degere, licet domi[3] negligendo, hic consumendo propria, usque dum tam illi difficultati de calculo remedium invenirem, quam illi de observatione iam tum remedium scirem; Crimen enim censeo, asseverare, aliquid esse solidum, cuius soliditatem non vera suffulciat ratio. Jam vero, invento remedio utrique nullus dubito de tuo favore erga meam Methodum; cum non solum, dum mihi concedebatur honor te visendi, ipse me doceres, quod Acta Illustrissimi Parliamenti usque ad 2000£ Sterl.[4] (quæ hic non requiruntur) concesserint, ut iisdem, si aliqua propositio de invenienda longitudine reperiatur rationalis, rei succurratur; sed et benevole promitteres, si haberi possit calculus Lunæ præcisior, successum rei tibi habere commendatum. Non male vertas, Illustrissime Domine, quod in te primariam fiduciam huius rei collocem; duo sunt mihi argumenta, quare id? nempe, quod in te mihi sint cognitæ, et facultas dicendi sententiam, et Authoritas. Adiuves itaque me ex benevolo promisso, exque animo propenso promovere scientiarum incrementum, simul cum reliquis Dominis, quibus ex Actis Illustrissimi Parliamenti negotium de invenienda longitudine est commissum, ut sicut propria hactenus inveniendæ longitudinis gratia fiderim, ita, quod restat eandem vere ducere in actum, absque ulterioribus meis sumptibus fieri possit, cum Acta Illustrissimi Parliamenti, ex tua relatione, non desiderent profusionem propriorum sed necessaria succurrendæ rei destinaverint, Tibique cum reliquis Illustrissimis ac Doctissimis D[omi]nis, in Actis denominatis, plenariam huius rei potestatem concesserint. Si otium id tibi permittet, Illustrissime Domine, humillime sistam et supputationes reliquarum observationum, a Dno Halley mihi communicatarum, collectas ex iisdem principiis, et procedendo præcise eodem modo, quo, quod nulla in iisdem sit differentia nisi in minutis 2^{dis},

appareat;[5] imo paratus sum, si id placebit, demonstrationem trigonometri-
cam cuiusvis, ex mea Theoria Lunæ, addere.

Qui de Cætero sum et ero
Illustrissime Domine
Observantissimus Cultor
J P BIESTER

Londini die 1 *Febr.*
172⅚

Translation

Most illustrious Sir,

   You will forgive me for having presented my letter about removing the difficulties
hindering the application of the method for finding out the longitude both to yourself
and to the public press.[2] It seemed proper to address it to you, since you yourself, with
sound reasoning, have put forward the difficulties, and proper also to do this by print,
so that others in whom perhaps the same may have occasioned doubts, may learn of their
removal. I did not wish to trouble you as long as these [difficulties] remained a hindrance
to my method, preferring, even at the cost of spending my wealth here and neglecting
my home,[3] to remain in London until I had found a solution for the difficulty concerning calculation, just as I already knew a solution for that concerning observation.
For I consider it a crime to assert anything as sound, the soundness of which is not supported by true reasoning. Now, indeed, having found a solution to both [difficulties],
I have no doubts concerning your approbation of my method; since you not only
informed me, when I was granted the honour of calling upon you, that the Acts of the
Most High Parliament have granted up to £2000 sterling[4] (which in this case are not
needed), so that if any proposal for finding the longitude be judged rational, there will
be funds for it from this source; but you also kindly answered [me] that the success of the
[proposal] would seem the more likely to you if a more precise calculation of the Moon
could be obtained [from it]. Do not turn against me, most illustrious Sir, because I
place my chief trust in this matter in yourself; for so doing I have two reasons, namely
that I recognize in you both the ability to put forward an opinion, and the authority
[to maintain it]. Give me your help, therefore, because of your kind promise and because of your disposition to promote the growth of science (together with the rest of the
gentlemen to whom the business of finding out the longitude is entrusted by the Acts of
your Most High Parliament). Just as I have so far invested my wealth for the sake of finding the longitude now what remains to put [the method] into practice, may be done
without further expense on my part, since the Acts of the Most High Parliament (by
your account) do not seek the expenditure of private wealth but supply the funds
necessary for the business, and grant you (with the rest of the famous and learned men
named in the Acts) a full authority in the matter. If leisure permits you, most illustrious
Sir, I will humbly remain and [work out] the computations of the rest of the observations to be furnished me by Dr Halley, collected from the same principles and proceeding
in precisely the same way, by which it may appear that no difference is evident between
them save in the seconds of arc;[5] what is more I am ready (if you please) to add any

343

geometrical demonstration whatever from my theory of the Moon; who for the rest am, and will be,

Most illustrious Sir,
Your most devoted admirer
J. P. BIESTER

London, 1 *February* 1725/6

NOTES

(1) Yahuda Collection, Newton MS. 7(4). Little is known of the writer of the letter, Johann Peter Biester. He was apparently a pupil of Georg Wolffgang Wedel, Professor of Medicine at Jena University from 1673 to 1721, and published medical dissertations in the 1690s.

He also published two short tracts on the determination of the longitude, both in 1726. One of these, *Methodus inveniendi Longitudinem Meridianorum tam in Mari, quam in Terra ex Astronomicis* (London, 1726), is a discussion of how the longitude may be determined from lunar measurements, if these are accurate enough. In the second, *Observatio Satellitum Jovis per Telescopium ope Machinulæ quæ et Observatorem et Telescopium portat in Maris Possibilis Possibilis in Mari* [sic] (London, 1726), the longitude was to be determined from observations of the eclipses of Jupiter's satellites, without relying on a clock. This involved the use of a machine Biester had invented called the 'Portatoris'; Biester gives a vague description of the device, which seems to include a crude form of gimbal (already proposed by Huygens for his marine chronometer in about 1659).

(2) It is not clear to which of the two pamphlets mentioned in note (1) above Biester refers here. Possibly they were bound together.

(3) Probably Jena.

(4) The Act (see vol. VI, p. 160, note (2)) offered £2000 towards the cost of experiments and instruments for finding the longitude.

(5) Some omission seems to have occurred here; the sense seemingly is that he will work out from his theory lunar positions for stated times when the Moon was observed by Halley; the discrepancy between his calculated and Halley's observed positions, he asserts, will only be some seconds of arc.

1486 PEMBERTON TO NEWTON
9 FEBRUARY 1726
From the holograph original in the University Library, Cambridge[1]

Sr.

I suppose you remember what I formerly observed to you upon the 23d and 24th prop. of the first book of your Principia,[2] that the line XY will not always cut the section; which occasioned your adding a short paragraph at the end of prop. 24, in pag. 87.[3] But as this paragraph only acknowledges the defect of the demonstrations, but does not shew how to supply it, I beg leave to propose the cancelling the leaf, and instead of the present paragraph to insert something like this which follows.[4]

344

In hac propositione et casu secundo superioris constructiones eædem sunt sive recta XY trajectoriam secet sive non, si vero non secet demonstrationes perfici possunt sumendo in tangente AI punctum tam prope puncto contactus, ut duæ rectæ duci possint trajectoriæ occurrentes; quarum altera in hac propositione parellela sit rectæ DC, altera tangenti PL; et in casu secundo propositionis præcedentis harum rectarum altera parallela sit rectæ BD, et altera rectæ PC.

This paragraph will stand nearly in the same room as the present, and will perhaps answer your purpose somewhat more fully. That this paragraph here proposed is true, will appear as follows. Having drawn in the latter figure of pag. 85 the line QTV parallel to DB, and QRS parallel to PC, which shall meet the section in T,V,R,S, I prove as follows, that the ratio of HAq to AIq is compounded of the ratio of $CG \times GP$ to $DG \times GB$ and of the ratio of $BH \times HD$ to $PI \times IC$...[5]

Again the demonstration of prop. 24 turns upon proving that $CI \times ID$ is to $CL \times LD$ as SIq to SLq;...[5]

In the last paragraph of page 47 there occurr these words circulum concentrice tangit, aut concentrice secat; and this double expression is here made use of, because the circle here spoke of does more frequently cut the curve than touch it. Notwithstanding this afterwards the word contactus twice occurrs without any other expression joined with it to include the more frequent case, when the circle and curve cut. This impropriety may be avoided by cancelling this leaf and writing this 3d Corollary after some such manner as this[6]

Coroll. 3. Si orbis vel circulus sit, vel circulum concentrice tangat aut concentrice secet, id est, eandem habeat curvaturam eundemque radium curvaturæ ad punctum P cum circulo per istud punctum descripto; et si PV chorda sit &c.

If you approve of these corrections; these two leaves together with two others which are likewise to be cancelled will make one compleat sheet to be reprinted.[7]

I am

Your most humble and most obedient servant

H PEMBERTON

Feb. 9. 1725–6

NOTES

(1) Add. 3986, no. 23.

(2) The former letter (if such it was) is now lost; it would presumably have been written between Letters 1422 and 1426 (compare Koyré and Cohen, II, p. 833).

(3) Lines 1–4; Newton remarks (in effect) that demonstration in the cases where XY does not intersect is obvious from the cases where intersection occurs.

(4) No such change was made; see Cohen, *Introduction*, pp. 272–3.

(5) We have omitted the long and tedious proof.

(6) This was not done. Compare Letter 1416 above; and Cohen, *Introduction*, pp. 273–4, for further details. Briefly, Newton had already modified Book 1, Prop. 6, Coroll. 3 from its form in the second edition; presumably as a result of this suggestion from Pemberton Newton again revised this Corollary (though less elaborately than Pemberton proposed) which necessitated the reprinting of p. 47 which was then inserted as a 'cancel'.

(7) Pp. 47–8 and 65–6 are cancels since Newton did not agree to the reprinting of pp. 87–8 or pp. 351–2 (Letter 1458), hence only two leaves (four pages) were reprinted.

1487   DU QUET TO NEWTON
28 MARCH 1726
From the holograph original in the Royal Society of London[1]

Monsieur

je crois ne pouvoir pas mieux faire que de me adresser a vous qui estes un des plus sçavants hommes de l'europe, pour vous prier tres humblement de m'honorer d'une reponce, a fin de me faire sçavoir de quelle maniere je m'y doits prendre pour decouvrir sans risquer de perdre le fruit de mes aplications, au suiet de la longitude par mer, que je crois avoir trouvee; au cas que je m'y sois trompé, ce sera la premiere fois que je le serai, ainsi que vous la prendrez par le memoire cy jointe[2] qui contient mes productions precedentes;[3] outre la longitude par mer dont la proposition[4] est separee du memoire, vous y trouverez des choses, Concernant la navigation qui doivent aussi interresser la nation Angloise[5] cest pourquoy Monsieur jespere que vous aurez la bonté, de ne me pas refuser vos bons avis. j'ai lhonneur destre avec toute la veneration düe a vostre Elevation et lEstendue de Vostre genie

    Monsieur

                          Vostre tres humble
               et tres obeyssant serviteur du Quet
                          ingenieur rüe de larbre sec[6]
                            visavis le petit paradis
                                 a Paris

Monsieur le chevalier Newton

NOTES

(1) MS. Early Letters, Q–R, 1, no. 7. Du Quet sent this letter, with additional papers enclosed (see below) via Monsieur de Voulouze, and a second, similarly worded letter, dated 8 April 1726, N.S. (Bodleian New College MS. 361, II, fo. 59) direct. We assume that the same date applies to both versions of the letter. For Du Quet, see Letter 1379.

(2) This memoir (Royal Society, MS. Early Letters, Q–R, 1, no. 8; not in Du Quet's hand)

enumerates Du Quet's inventions, mostly concerning carriages and ships, the earliest of which is dated 1691. Curiously, the paddlewheel method of propelling a ship (see Letter 1379) is not specifically mentioned. The memoir has been wrongly recorded in the Letter Book of the Royal Society, xv, as dated 2 November 1722; there is no record of its ever having been read before a meeting of the Society.

(3) The version in the Bodleian Library adds: 'que j'ai mis entre les mains de Monsieur de Voulouze en le priant de vous le faire rendre avec une pareille lettre a celle cy.'

(4) On a separate sheet (no. 7 *bis*) in Du Quet's hand (in which he refers to himself in the third person as a 'French Engineer'). He reveals no details of his method stressing only its simplicity, but demands assurances from the Admiralty officials that his right to the invention will be recognized before he will disclose it.

(5) The version in the Bodleian Library adds: 'beaucoup plus attentive que la nostre aux bonnes productions ainsi qu'il paroist par les recompenses que les actes du parlement promettent;'

(6) On the right bank of the Seine, near the Louvre.

1488 NEWTON TO MASON

10 MAY 1726

From the holograph original in the Babson College[1]

Reverend Sr

I am indebted to you for your trouble in getting the floor of your church to be repaired, & have sent you a note drawn upon my Cousin Robert Newton[2] to pay you three pounds out of such money as he hath of mine in his hands, & to take your receipt in discharge for the same. I hope you have finished the floor according to your mind, & remain

Your most humble
& most obedient servant
ISAAC NEWTON

London, May ye 10*th*
1726
For Mr Tho. Mason Rector
of Colsterworth neare
Post Witham in
          Lincolnshire

NOTES

(1) Sir Isaac Newton Collection, MS. 428; printed in Nicholas, p. 61. Compare Letters 1456 and 1465.

(2) See Letter 1456, note (2), p. 303.

1489 HENRY JACKSON TO NEWTON
8 JUNE 1726
From the original in the Bodleian Library, Oxford[1]

Sr

When I was Last to Wait upon you at your house, You was pleas'd to favour me with the Priviledge of acquainting you of some New Mathematicall Instruments of my Invention, and Particularly of my Instrumentall System of the Sun, Moon, Earth, and Stars, which is An Everlasting Table of the Moons True Place and its appulses to the Fixed Stars and Planets.

And you was Likewise so kind and Generous, as in your great Condescention and Goodness, to advise me to Print my Book,[2] Concerning the uses of my Instruments.

Accordingly I Immediately Put it in the Press, But a Long & Severe sickness has so Retarded my Progress, that it is but of Late, that I could Effectually Pursue that work; But now it being Allmost Printed, And my first Publique Notice of it being Lately given, in the London Journall; I have thought myself Oblig'd now to wait upon you again, to give some further account of my Inventions.

And therefore I humbly Pray, you will be Pleas'd, (If now at Leisure,) to favour me with an Oppertunity of Speaking to you upon this subject.

I am with all due Respects

Sr:

Your most Obedient Humble Servant

J. JACKSON

June 8th: 1726

To

The Honble

Sr: Isaac Newton

These

NOTES

(1) Bodleian New College MS. 361, II, fo. 60. What little is known of the author is derived from the meagre autobiographical statements in his book (see note (2) below). He describes himself as a citizen of London 'for above thirty years', but as having spent most of this time on trading voyages to all parts of the world.

(2) *Longitude & Latitude in Conjunction, by Sea and Land, by Day and Night, Found out and Demonstrated* (London, 1727); the last two sections of the book had been printed in 1725. Of the several dedications prefacing the book, one is to the Royal Society. Jackson says that 'last summer' he shewed his '*New Invented Sea Quadrants*, both *Single* and *Double*, to your *Great* and

Judicious President, and the *Numerous Assembly* then present,' and that the instruments met with complete approbation. (There is no record of this meeting in the Journal Book of the Royal Society.) In the body of the book, Jackson claims the invention of a number of modifications to astronomical instruments for use at sea, but does not describe them in detail. The only method of determining longitude which he deems reliable is the astronomical method involving the determination of the Moon's position relative to the fixed stars.

1490 NEWTON TO JOHN CONDUITT
25 JUNE 1726
Source unknown[1]

Mr Conduit.

Sr

There are now come into the Mint, 134 journey of gold and it will all be coined into money by Wednesday next. But Wednesday next is a holiday being S. Peter & S. Pauls Day and therefore the delivery may be made upon Thursday June 30. If you please to come to London upon Monday or Tuesday next and let the Cashier of the Bank or the Officers of the Mint have notice to attend on Thursday morning at the Mint I hope all will prove right. I hope your Lady[2] and Kitty[3] are well. My service to them.

Your most obliged humble servant
ISAAC NEWTON

Kensington June 25 1726

NOTES

(1) A typed copy of this letter was found amongst the papers of Dr J. F. Scott (editor of vol. IV of this *Correspondence*). We have been unable to locate its source.

(2) Newton's niece, Catherine Conduitt, née Barton; see Index, vol. VI.

(3) Catherine, the only child of John and Catherine Conduitt.

1491 NEWTON TO FONTENELLE
?June 1726.   For the answer see Letter 1492

According to the *Sotheby Catalogue* (p. 23, Lot 136) this is the autograph draft of a Latin letter sending Fontenelle a copy of the third edition of the *Principia*. The draft is now in the possession of M. Emmanuel Fabius.

1492 FONTENELLE TO NEWTON
3 July 1726.   Reply to Letter 1491

According to the *Sotheby Catalogue* (p. 23, Lot 136) this is an autograph letter dated 14 July 1726 N.S. thanking Newton for his gift. This letter also is now in the possession of M. Emmanuel Fabius.

1493 NEWTON TO SCROPE
AUGUST 1726
From the holograph draft in the Bodleian Library, Oxford[1]
For the answer see Letter 1494

Sr

Mr Bull[2] the second ingraver in the Mint is dead & upon inquiring after another who upon the death of the first engraver Mr Croker, may be fit to succeed him, I can hear of no better artist then Mr Rollos[3] the kings engraver of Seales. The place in the mint is 8*lb*[4] per annum wth a house & the prospect of succeeding the principal graver when he dyes. And Mr Rollos is willing to accept of it to live & work in the Mint. And as he is the fittest person that I can hear of I take the liberty to recommend him to your consideration. I am

Your

NOTES

(1) Bodleian New College MS. 361, II, fo. 10v.

(2) See Letter 1398.

(3) John Rollos was appointed Assistant Engraver, at £80 per annum, by a warrant dated 29 September 1726 (see P.R.O., Mint/3, 168, fo. 3 and T/52, 33, p. 267). He was apparently too busy with other affairs to perform his duties adequately, and was relegated to his original post of Chief Engraver of seals, without pay. He remained in this position until 1745.

(4) *Read:* £80.

1494 SCROPE TO THE MINT
16 AUGUST 1726
From a copy in the Public Record Office[1]
Reply to Letter 1493

[To the] Warden, Ma[ste]r & Worker & Comptr[oller]
of ye Mint abt: Mr: Jno: Rollos

Gent[lemen]

The Lords Com[missione]rs of his Ma[jesty']s: Trea[su]ry being informed that the Place of Assistant to the Chief Graver of his Ma[jesty']s: Mint is Vacant by the death of the Mr: Sam[ue]l Bull; And observing that, by a Clause in the Letters Patent constituting the said Chief and Assistant Gravers, it is directed that in case the said Assistant's place shall be void it may be filled up by such probationer or Apprentice as shall be presented to his Ma[jes]ty: by the Warden, Master & Worker, and Comptr[oller] of the Mint, by Warrant under his Ma[jesty']s Royal Sign Manual, Their Lordps are pleased to recommend Mr. John Rollos (to whom for his Skill and Abilities in Engrav[in]g his Ma[jes]ty has been lately pleased to Grant the Office of Engraver of his Signets, Seals, Stamps, & Arms,) to be presented to You for the said assistant Engraver's Office.

I am &ca: 16th: Augst: 1726

J SCROPE

NOTE

(1) T/27, 24, p. 175. Presumably this constitutes a formal request to Newton to recommend Rollos for the post of Assistant Engraver, Newton himself having put the suggestion to the Treasury.

1495 NEWTON TO THE KING

AUTUMN 1726

From a holograph draft in the Mint Papers[1]

To the Kings most excellent Majesty

May it please your Majesty

The Office of Assistant to the Chief Graver of Your Majties Mint being vacant by the death of Mr Samuel Bull, & by a clause in the Letters Patent constituting the said Chief & Assistant Gravers it being directed that in case the said Assistants place shall become void it may be filled up by such probationer or apprentice as shall be presented to your Majesty by the Warden, Master & Worker & Comptroller of the Mint, by warrant under your Majesties Royal signe Manual: We the said Officers of your Majesties Mint do most humbly present to your most sacred Majesty Mr John Rollos (to whom for his skill & abilities in engraving your Majesty hath been lately pleased to grant the Office of Engraver of your signets seales stamps & arms) p[r]aying that the said Office of Assistant Graver may be filled up by the said Mr John Rollos by warrant under your Majesties Royal signe Manual

All wch is most humbly submitted to your majesties good Will & pleasure

NOTE

(1) I, fo. 177. The letter was presumably composed in consequence of Letter 1494.

1496 HAYNES TO NEWTON

6 SEPTEMBER 1726

From the original in the Mint Papers[1]

Sr

The Southsea Comp[an]y have lately recd. from their Agents in the West Ind[ie]s 6 or 7 parcels of Gold, wch they intend to have Coyn'd: and accordingly have imported 2 parcels, & have part of a 3d. parcel now in ye Assay office, ready to be weigh'd.

Upon a view of ye af[oresaid] parcel I inspected the Gold & when all ye Ingots were assayed I causd 4 to be retd to ye Comp[an]y to be refin'd. In a 2d. parcel I found 7 wch I suspected to be of a bad mixture; & therefore Orderd Mr. Whitaire to acquaint Mr. Utbert, the Comps. Agent, & Mr. Blatchford, their Refiner, that those 7 suspectd. Ingots shoud be taken back & refind, unles they wou'd indemnify ye Office They both gave Mr. Whitaire strong assurances that there was no bad mixture in them, upon which they were all taken in, before I had heard their answer. Whereupon I caution'd Mr. Vanderesh not to sett them out for melting, Intend[in]g to Apprize You of ye Affaire, that ye sd Suspected Ingots might be further exam'd before their delivery to ye Melter. A timely caution being necessary to make us safe, for the future.

I have taken all the Care I cou'd in examining what has already been assayd, at ye Cutting, hammering, sheering, appearance on the Copples, & after nealing: but because ys mayn't be sufficient, I woud further propose that all the suspected Ingotts be broke in Yr presence, the Comps. Refiner & Agent attend[in]g, & also Yr Melter, when We may be better assur'd of the Nature of the Mass by a careful inspect[io]n of the breaches, than we can be, by an Assay taken off at one Corner.

The Comp[any]s Agent & Refiner prest the admittance of ye Gold now in ye assay office: but I conceive it altogether reasonable that the office in gen[era]l & the Matter in particular shoud be safe, & indemnifyd. However the whole is submitted to Your considerat[io]n that this affair may be settled upon a good foot to secure Us for the future. If you please to Appoint to morrow being Wensday, or Thursday morning, I will attend in my place

I am
Hond. Sr.
Yr. most Obedt. H. Servt
H. HAYNES

Tuesday 6. Sept[em]ber. 1726

(1) 1, fo. 108.

1497 J. F. BACHSTROHM TO NEWTON
21 OCTOBER 1726
From the original in the Royal Society[1]

Vir Perillustris atque Excellentissime
Præses Academiæ Scientiarum
Anglicanæ Dignissime

Pervenit et ad nos a cultioribus literis nimium quantum remotos obscura quædam Fama, constitutum esse ab Inclyta Academia Scientiarum Ei pretium, qui Causam Gravitatis præ cœteris eruere valeret.[2] Res equidem ardua, sed tentare cuivis licet. Non equidem de Pretio obtinendo adeo sum sollicitus quandoquidem Tuum Vir Illustriss: Judicium mihi longe majus omni pretio erit. Hoc unicum est Vir Illustriss: quod a Te in Oris hisce incultioribus, adeoque Erroribus magis obnoxiis summo cum studio expeto atque expecto, Præcipue cum plurima[3] ad huc sint, quæ Tecum Vir Illustriss: *De Motu Fluidorum* imprimis de *Æstu Maris*, de *Magnete*, de *Motibus Moralibus*[4] minus vulgaria communicanda habeo, ubi primum de favore Tuo in suscipiendis literis meis me certiorem reddideris. Si dignaberis ad hæc respondere, Epistolam una cum literis publicis Comitis de Finos Legati in hoc Regno transmittere licebit, quoniam ejus Gratia atque Humanitate præprimis utor, nullus enim certior atque opportunior huc mittendi modus. Non est, quod addam, nisi quod me Favori atque Benignitati commendem atque maneam Tibi

Præses Perillustriss:
Excellentissime ac
Doctissime
ad omnia Obsequia paratissim[um]
Cliens atque Servus
Johannes Fridericus Bachstrohm
Med. D.

Grodnæ[5] *Cal. Nov.*
Ao MDCCXXVI. [*N.S.*]

Translation

To the most famous and excellent Gentleman, the most worthy
President of the English Academy of Sciences

An obscure report has reached us, so very far as we are from civilized shores, that a prize has been established by [your] renowned Academy of Sciences for whoever manages to bring to light the cause of gravity before anyone else.[2] The thing is, of course, difficult, but anyone may try it. Indeed, I am not so very anxious to obtain this prize, because your opinion, most illustrious Sir, will be far more important to me than all prizes. It is this judgement only, most illustrious Sir, which I desire and expect from you with great eagerness, here in these uncivilized parts, which are accordingly more susceptible of error. Particularly since there are still many out-of-the-ordinary things[3] which I have [by me] for communicating to you, most illustrious Sir—*On the motion of fluids*, and especially *On the waves of the Sea*, *On the Magnet* and *On the movement of human beings*[?][4]—whenever you give me greater certainty of your goodwill in receiving my letters. If you consider it worthwhile to reply to this, you will be at liberty to convey a letter together with the official letters of the Count of Finos, Ambassador in this kingdom, since I enjoy his particular friendship and kindness, for no method is more to be depended upon and more convenient that this for sending [them].

I would add to this only that I commend myself to your favour and kindness and remain, most famous, excellent and learned President, your follower and servant most ready to submit in all things,

JOHANN FRIEDRICH BACHSTROHM
Doctor of Medicine

Grodno[5]
1 *November* 1726 [*N.S.*]

NOTES

(1) MS. B2, 120. According to J. C. Adelung (ed.), *Jöchers Allgemeines Gelehrten-Lexikon*, I, Ergänzungsband A und B (Leipzig, 1784; reprinted Georg Olms, Hildesheim, 1960), pp. 1323–4, Johann Friedrich Bachstrohm was born in Silesia and studied theology at Halle. In 1717 he became a professor at the Gymnasium at Thorn. From 1720–8 he was chaplain of a regiment stationed at Warsaw. During this time he studied medicine; according to Adelung, he also became a member of the Royal Society but we have found no record of this. In 1729 he founded a press at Constantinople, and undertook a translation of the Bible into Turkish.

He published books on a variety of subjects, including scientific ones. He also wrote theological works under the pseudonym Christianus Democritus Redivivus.

(2) The Royal Society did not offer prizes of this nature. Possibly Bachstrohm refers to a prize offered by the Académie Royale des Sciences. The prize for 1728 was to find the physical cause of universal gravity (see G. B. Bülffinger's winning entry in *Recueil des Pièces qui ont remporté les prix de l'Académie Royale des Sciences*, depuis l'année 1727 jusqu'en 1737, article XI (Paris, 1728), pp. 1–40). Presumably the subject of the competition would be announced considerably in advance.

Bachstrohm published a tract *Exercitatio sive Specimen de Causa Gravitatis cui Adjecta sunt*

354

Nonnulla de Originibus Rerum tanquam Fundamenta Physices Novæ Antatheisticæ (Dresden, 1728); the discussion in this work is for the most part theological. On p. [2] Bachstrohm again implies that it was the Royal Society which put forward, 'not long ago', the question of universal gravity.

(3) Presumably he refers to works as yet unpublished. In 1734 his *Nova æstus Marini Theoria ex Principiis Physico-mathematicis Detecta & Dilucidata. Cui accedit, Examen Acus Magneticæ Spiralis, quæ a Declinatione & Inclinatione Libera esse creditur* appeared.

(4) The reading is difficult; possibly *Motibus Mortalibus* is intended. No book of this title by Bachstrohm is recorded.

(5) A town now in the U.S.S.R. near the Polish border, almost 150 miles north-east of Warsaw.

1498 NEWTON TO MASON[1]
4 FEBRUARY 1727
From the printed version in Nichols, p. 61

London, Feb. 4, 1726–7

Sir,

I have procured some assays to be made of the pieces of ore which your friend at Wolsthorpe left with me when you were last at London; and they hold no metal, but run into a black brittle substance, without one grain of malleable metal therein. You may please to acquaint the owner of those pieces of ore with the success of the assays which I have procured to be made.

I am, Sir, your most humble and most obedient servant,

ISAAC NEWTON

NOTE

(1) The letter is possibly related to Letter 1512.

1499 MASON TO JOHN CONDUITT
23 MARCH 1727
From the original in King's College Library, Cambridge[1]

Good Sir

This morning I rec'd from you ye Melancholy News of yt truly great and good Gentleman's Death Sr. Is: N: And I have according to your desire made Sr Isaac's Heir & Representative, who is ye Bearer of ys, acquainted wth it, but God knows a poor Representative of so great a man, but ys is a case yt often happens.—There are two Families of ye Newton's in ys Parish, both descended from ye 2d & 3d Brothers of Sr. Is. Newton's Father—the 2d Brother was

Robert Newton from w[ho]m ye Bearer of ys John Newton is descended, ye 3d was Richard from w[ho]m descends Robert Newton now living in ys Parish, so yt wthout dispute John Newton ye Bearer is heir to ye estate not devised by will.

· · ·

Worthy Sir
Your most obedient Humble Servant
THO. MASON.

P.S. Sr Isaac Newton in ye Days of his Health & Prosperity used to talk pretty much ab't founding & Endowing a School in Woolsthorpe for ye use of ye Parish as ye Neighbours & his relations inform me. I myself never knew him but in his declining years, having been but six years Rector, but He used to talk wth me pretty much upon yt subject, tho his dying wthout a will leaves no room for any such hopes.

Colsterworth March 23th 172$\frac{6}{7}$

NOTE

(1) Keynes MS. 134; microfilm 931. For the writer see Letter 1456, note (1), p. 303. Newton died on 20 March 1727; he left no will. In the Library of King's College, Cambridge (Keynes MS. 127(A)) are a number of documents concerning the winding up of his estates.

Newton having died intestate, there was considerable discussion over who should administer his personal estate. John Conduitt wrote (Keynes MS. 129):

He died worth about £32,000 personal estate wch is divided between his 4 nephews & 4 nieces of the half blood the land wch he had from his father & mother went to his heir of the whole blood John Newton whose great Grandfather [was] Sr Isaac's uncle[.] a little before he died he gave away an estate in Berkshire to the sons & daughter of my wife's brother [*Colonel Robert Barton*] who by their father's dying before Sr Isaac had no share of the personal estate, & an estate he bought at Kensington of about the same value to my daughter [*Catherine Conduitt*].

For those relatives of Newton who survived him, see Appendix II, Newton's Genealogy.

UNDATED CORRESPONDENCE

Note on arrangement

Where possible, each letter has been assigned a tentative date and included in the main chronological sequence. The remaining undated letters are arranged below in two sections. The first section comprises letters from Newton arranged in alphabetical order of addressee. Those letters for which the addressee is not known are grouped together at the end of the section according to the MS. collections from which they are derived. The second section consists of letters addressed to Newton, again arranged alphabetically. We have not considered it worthwhile to print in full all the begging letters and other trivia addressed to Newton; these are merely calendared in their place in the alphabetical sequence.

There are a number of letters addressed to Newton which we have been unable to trace, and the date of which we do not know. These were all sold as Lot 129 at Sotheby's sale in 1936 (see *Sotheby Catalogue*, p. 21). The writers are listed there as including P. Allix, J. A. Arland, J. P. Biester, F. Bianchini, T. J. De Lagny, G. P. Domcke, G. Grandi, G. J. 'sGravesande, J. Hermann, B. Laurentius, J. G. Liebknuss, L. A. Muratorius, P. v. Musschenbroek, L. a. Ripa, J. J. Scheuchzer, N. Tron, J. J. M. Ursius, J. Valletta, L. Vincent, D. de Volder, D. Waeywel and H. Wetstein, père. They are described as letters 'mostly of compliment from foreign Scholars (in Latin)'.

1500 NEWTON TO PIERRE ALLIX

n.d.

From a holograph draft in the Jewish National and University Library, Jerusalem[1]

Reverendissimo Viro D. P. Allix S.T.D.
Is. Newton S.P.

Vir dignissime,

Quamvis linguæ Latinæ multo minus assuetus quam vernacula, tamen ut Responsum tibi magis gratum sit, rescribam in hac lingua licet stylo mediocri[.] In suo anno 37 Captivitatis Jehojachin (2 Reg. 25. 27) anno Nabonass[aris] 166 [*read* 186] completo et anno 187 currente juxta Canonem.[2] Aufer annos 37 & captivitas Jehojachim incidet in annum Nabonass[aris] 150. Adde regnum Zediciæ annorum 11 et interitus Hierosolymorum incidet in an. Nabonass. 161. Ann[us] autem quartus Darij primi incidit in An. Nabonass. 230 justa Canonem. Et hic annus est septuagesimus ab interitu Hierosolymorum inclusive...

Translation

Isaac Newton presents a grand salute to the
very reverend Mr P. Allix D.D.

Most worthy Sir,

Although the Latin tongue is much less familiar to me than the common speech, to
render my reply more acceptable to you I reply in that language, though in a mediocre
style. In the thirty-seventh year of his captivity Jehoiachin [was redeemed by Evil-
merodach King of Babylon] (II Kings, 25, v. 27) [hence this year corresponds to] the
complete year of Nabonassar 186 and the current year 187, according to [Ptolemy's]
Canon. Subtract 37 years and the captivity of Jehoiachin began in the year 150 of
Nabonassar. Add the reign of Zedekiah of eleven years and the destruction of Jerusalem
falls in the year of Nabonassar 161. However, the fourth year of Darius the First falls in
the year 230 of Nabonassar according to the Canon. And this year is the seventieth inclusive from the destruction of Jerusalem...

NOTES

(1) Yahuda Collection, Newton MS. 24(3). This is an exceedingly rough draft with numerous deletions and interlineations; we have constructed as much text as we could from the
microfilm in the University Library, Cambridge. The text is unfinished, and has been completely struck out. The only clue to its date is the appearance on the same folio of a draft of
Letter 1021, vol. VI (7 November 1713).

The recipient, Pierre Allix (1631–1717), a Protestant theologian, migrated from France to
England after the revocation of the Edict of Nantes. He founded the French Anglican church
in England and became a protégé of Gilbert Burnet, Bishop of Salisbury. He wrote a number
of theological works, and in 1707 published *An Answer to Mr. Whiston's late Treatise on the Book
of Revelations*, where he accused Whiston of errors he had copied from Jewish writers about the
chronology of Jerusalem's destruction and the Jewish captivity under Nebuchadnezzar,
followed by their liberation by Cyrus 51 years later in Jehoiachin's third year (see *ibid.*
pp. 35–9). Possibly Allix consulted Newton on this chronology before publishing. (Our thanks
are due to Professor E. R. Briggs for supplying this information.)

1501 NEWTON TO CONRADE DE GOLS

n.d.

From the holograph original in Trinity College Library, Cambridge[1]

Sr

Please to place to the Acco[un]t of John Read all such Stock as is due to me
for £1000 paid in upon my Name in the 4th. Subscription for sale of South Sea
Stock

ISAAC NEWTON

To Mr Conrade de Gols
Trea[sure]r to the South Sea
Company

NOTE

(1) R. 16.38*b*, no. 435. The fourth subscription of South Sea Stock, sold at 1000 per cent, took place at the beginning of June 1720. Presumably the present letter dates from this time or later.

1502 NEWTON TO GARDINER

n.d.

From the holograph original in Trinity College Library, Cambridge[1]

Sr

I am greatly sollicited by my Cousin Tonstall[2] (grand-daughter to my Unkle Ayscough) to entreat you to use your endeavour to procure for her husband Mr Tonstal, one of the two places wch are now voyd among the poor knights of Windsor.[3] He has served about 32 years in the third regiment of Horseguards.[4] General Compton (his officer) formerly gave me a very good character of him: & you may have his character from any of his Officers. And what kindness you can do for him, I shall take as done to my self. I am

Your most humble
and most obedient Servant
Is. NEWTON

Coll. Gardiner

NOTES

(1) R. 16.38 A⁹. Colonel Gardiner may probably be identified with Robert Gardner Katherine Greenwood's second husband. Robert Gardner is referred to as 'Colonel' in a manuscript relating to the estate which Newton gave Katherine Greenwood's children by her marriage to Robert Barton (see P.R.O., PROB/3/26/66, and Appendix II, Newton's Genealogy).

(2) See Appendix II, Newton's Genealogy, p. 486 (note (14)).

(3) In 1348 Edward III founded the Order of the Garter and, associated with it, an ecclesiastic 'College' centred on St George's Chapel at Windsor Castle. The latter was to support twenty-six *milites pauperes*, 'English warriors reduced to great poverty, who in accordance with the prescribed number shall perpetually worship God' (College Statutes, 1352). The *milites pauperes* became known as the 'Poor Knights of Windsor' in 1429, and later, in 1883, changed their name to the 'Military Knights of Windsor.' Tonstall is not listed in Edmund H. Fellowes, *The Military Knights of Windsor, 1352–1944* (Windsor, 1944) as one of the Poor Knights. Since the present letter was written on one of the relatively rare occasions when two places were void simultaneously, if we assume that the letter was written when Tonstall was in his fifties or, more probably, sixties, and hence very late in Newton's own life, then its date must be either 1716 or 1720.

(4) This was founded as the Duke of York's Troop, raised in 1660; in 1670 it became known as Her Majesty's Third Troop of Horse Guards, or the Third Troop of Life Guards. It was disbanded in 1746. Hatton Compton became Lieutenant of the regiment in 1691/2, Brigadier General in 1702, and Lieutenant General in 1707.

1503 NEWTON TO ?ROBERT GAYER

n.d.

From the holograph draft in the Jewish National and University Library, Jerusalem[1]

Sr

Since I had your letter I had some thoughts of coming by Windsor to Stoak & talking with you about what relates to the R[oyal] S[ociety] but now having no prospect of such a long journey I can only return my thanks to you by Letter for your care of my Niece & your interceding with the Prince in favour of the R.S. & acquaint you that it being the long vac[a]tion in wch the R. Society do not meet & several of their principal members are in the country I know not how to lay the matter of ye letter before them at [present], & therefore beg the favour that you would be pleased to let it rest till they begin to meet again

NOTE

(1) Yahuda Collection, Newton MS. 7(3), fo. 31. It is probably addressed to Robert Gayer, son of Sir Robert Gayer of Stoke Poges (see vol. IV, p. 476, note (2), and Letter 1540), a close friend of Newton, with whose family Newton's niece, Catherine Barton, may well have stayed. Possibly the letter dates from the summer of 1705, when Prince George of Denmark became involved in the vexed problem of the publication of Flamsteed's observations.

1504 NEWTON TO JANSSEN

n.d.

From a holograph draft in the Bodleian Library, Oxford[1]

Sr Theodore

I humbly begg the favour of you to get leave that Mr Cha Gregory may be admitted to subscribe 500*lb* or 1000*lb* in the next subscription in the south sea [*draft ends here*]

NOTE

(1) Bodleian New College MS. 361, II, fo. 45. For Sir Theodore Janssen, see vol. V, p. 125, note (1).

1505 NEWTON TO JOHN LACY

n.d.

From a holograph draft in the Jewish National and University Library, Jerusalem[1]

Cl[arissime] D[omi]ne

Etsi ego non is sum cujus authoritati in dijudicandis Mathematicorum controversijs tantum deferas ⟨humanitati tamen, tuæ injurius essem⟩ exigit tamen humanitas tua ut postulatis tuis reservando uti faciam exhibeam tibi. De lite autem vestra non est quod multa dicam[2] siquidem æque notum ⟨constat⟩ est planam superficiem, esse minimam omnium inter suos terminos ac lineam rectam inter suos brevissimam. Hoc vel Mathesis ignaris constare potest, nam quemadmodum filum distensum (puta chorda Cytharæ) si a rectitudine distrahatur, plus in longum dilatatur donec tandem per nimiam tensionem disrumpitur; sic planum distensum (puta pellis Tympani) si a planitie similiter distrahatur, dilatabitur in longum [et] latum donec tandem per nimiam tensionem dilaceretur. E tuis etiam ratiocinijs res luce clarior est siquidem omnes norunt superficiem pyramidis vel coni vel segmenti sphæræ aut Sphæroidis aut Conoidis cujuscunque majorem esse plano baseos. E.g. sit

ABCD mons pyramid $\overset{\text{icus}}{\underset{\langle\text{idalis}\rangle}{}}$ æquilaterus cujus vertex *D*, et erunt triangula

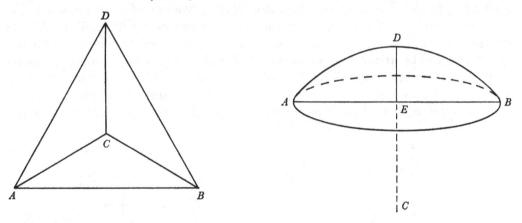

ABD, DBC, DCA, ABC æqualia adeoque superficies montis (quæ ex triangulis *ABD, ABC, DCA* conflatur) triplo major base *ABC* sit. Insuper sit *ADB* mons sphericus cujus centrum *C*, Basis *AB* vertex *D* ac perp *DE*. Superficies montis duobus circ. æquabitur centro *E* radijs *ED* descriptis,[3] hoc est superabit basem alio circulo cujus semidiameter sit altitudo montis: Id quod et Archimedes alijque, demonstraverunt.[4] Mensurantur autem loca montosa, ut nosti, distribuendo distinguendo in pluras quorumlibet solidorum segmenta quorum superficies Geometris notæ sunt. Sed hæc methodus non nisi a Geometriæ peritissimis idque cum labore maximo adhibere potest, & veritatem satis appropinquabit si montis superficies in plura momenta quamproxime plana resolvatur nisi ubi prior methodus in aliqua parte montis adversarij [*illegible phrase*] commodem obveneret [*illegible word*] methodus, ubi spero post [*illegible word*] æquationes præstiti non posteriorum ubi error omnis quamvis est etiamnum in defectu versitur.[5]

Vale & ama

Tui obsequientissimus

Is. NEWTON

These

For Mr Jo Lacy, one of his Majesties surveyors. To be left at Mr Meemis a Coffeman neare ye Kings Playhouse [*deleted*] in London

Translation

Most famous Sir,

Although I am not the person to whose authority in settling mathematical controversies you defer so courteously yet ⟨I should be unjust to you⟩ your courtesy demands that I offer you [an answer] to your demands. However, there is not much I can say as to your dispute,[2] since it is the case ⟨is well known⟩ both that a plane surface is the smallest of all between its boundaries and a straight line is the shortest [distance] between its [ends]. This can be made evident to those ignorant of mathematics; for just as a taut thread (say the string of a guitar) is stretched longer if it is pulled out of the straight line until finally it is broken by excess tension, so a stretched plane (say the skin of a drum) will be stretched in length [and] breadth if it is likewise pressed out of the plane until in the end it is ruptured by excess tension. From your own arguments too the thing is as clear as daylight since everyone knows that the surface of a pyramid or of a cone or of the segment of a sphere, spheroid or conoid of any kind whatever is greater

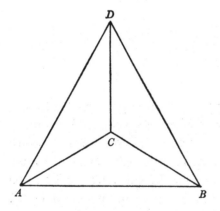
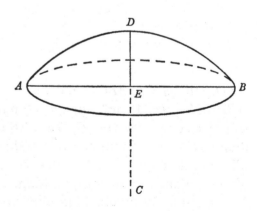

than the plane of its base. For example, let *ABCD* be an equilateral pyramidal mountain whose vertex is *D*; the triangles *ABD*, *DBC*, *DCA*, *ABC* will be equal and so the surface of the mountain (which is comprised of the triangles *ABD*, *DBC*, and *DCA*) will be three times as great as the base *ABC*; further, let *ADB* be a spherical mountain whose centre is at *C*, *AB* the base, the vertex *D* and *DE* perpendicular. The surface of the mountain will be equal to two circles described with centre *E*, radii *ED*,[3] that is it will exceed the base by another circle the radius of which is the height of the mountain. Which is demonstrated by Archimedes and others.[4] Mountainous regions are measured, however, as you know, by distributing and distinguishing [them] into numerous segments of solids of all kinds, the surfaces of which are known to geometers. But this method can only be applied by those most skilled in geometry with great effort, and it will approximate closely enough to the truth if the mountainous surface can be resolved into many almost plane small sections... [5]

Farewell and love

Your most devoted

Is. NEWTON

NOTES

(1) Yahuda Collection, Newton MS. 31, fo. 1. It is very hard to decipher and is heavily corrected. We have used angle brackets ⟨ ⟩ to indicate those phrases which Newton probably meant to delete. John Lacy is presumably not the comic actor of the same name, but may have been the translator of Tacquet's *Military Architecture* (London, 1672). Otherwise we have discovered nothing about him.

(2) It seems that the dispute turned on the question, is the area of a mountainous region, derived by taking horizontal measurements only, the same as that of a plane region with the same measurements? Obviously, as Newton goes on to show, it is not.

(3) Something is amiss here—apparently Newton miswrote his draft, omitting the second radius *EA*.

(4) Archimedes, *On the Sphere and Cylinder*, I, Prop. 42: since the area of the convex surface of the segment is proved in this proposition to be πAD^2, it is also equal to πAE^2 plus πDE^2.

(5) We have omitted the partially read lines from the translation. Newton seems to hope that 'the prior method' will be preferred to 'the second' where every error stands out.

1506 NEWTON TO ?SIR JOHN NEWTON

n.d.

From the holograph draft in the University Library, Cambridge[1]

Sr John

I am desired to beg the favour of you to read the inclosed Letter, I believe the matter is truly represented & has nothing more in it then a piece of indiscretion committed by the Gentlewoman who writes it.

363

(1) Add. 3984, fo. 13v. On the back is Newton's draft of Letter 937. For Sir John Newton see Letter 1466, note (2), p. 320.

1507 NEWTON TO JOHN NEWTON
n.d.

I understand that Tho. Hubart is not willing to stand to his bargain by wch I was to allow him tenn pounds for his manure and he agreed with you to leave his Farm at Lady Day.

(Untraced. See *Sotheby Catalogue*, Lot 170, where this excerpt is printed. John Newton was probably either grandson or greatgrandson of Isaac Newton's uncle Robert. See Appendix II, Newton's Genealogy.)

1508 NEWTON TO ?OLDENBURG
n.d.
From a holograph draft in the University Library, Cambridge[1]

Sr

I r[eceive]d yours & understand by it my buisines has not yet done troubling you.[2] I thank you yt you were pleasd to deposit yt Guiney[3] & have ordered John Stiles[4] to repay it at ye delivery of this letter. So Sr wth all my hearty thanks for yr former favour as well as this I rest

NOTES
(1) Add. 3970(3), fo. 532v. It seems likely that this draft was intended for Henry Oldenburg, from the allusion to Newton's payment of (almost) a half-year's subscription to the Royal Society. In which case it was written before 1675 (compare Letter X. 153), possibly in 1673 or 1674.

(2) The controversies following the publication of Newton's first optical paper in 1672.

(3) The reading is doubtful.

(4) The Cambridge carrier.

1509 NEWTON TO PROCTOR

n.d.

From a holograph draft in the Sir Isaac Newton Collection, Babson College[1]

For Mr Proctor an Atto[r]ney at [?Grantham]

Sr

The last summer your Brother Mr Parkins[2] of Coulsterworth spake to me yt you might succeed your Father in the management of my concerns.[3] I thank you that you will be pleased to do me that kindness. I left wth your Br a bond of Mr Tongue's for 80*lb*, & ordered Mr Tongue to pay in ye money this Christmas. If the money be ready I desire you would receive it wth ye interest & deliver in ye bond & send me twenty pounds of the money by the Cambridge Carrier. There are two yeares interest already paid, since ye date of ye bond. The rest of ye interest I desire you would receive till Christmas & no longer if ye money be ready. There is other money due to me of wch I will give you an acc[oun]t if you please to let me know that you have received this letter.

NOTES

(1) Babson 421. Proctor is mentioned in Letter 950, vol. v. On the reverse is the beginning of another, presumably related, draft letter, 'Sr[,] Your Brother obliged me ye other day by calling upon me in his way to Scotland, & giving me hopes of having your company here ere long'. On both sides of the leaf there are notes on chemistry.

The letter was clearly written while Newton was still at Cambridge, that is, before 1696. It is possibly related to Letters 457 and 465, vol. III, and to Letter 1520.

(2) Probably William Parkins, rector of Colsterworth 1690–1720 (see Turnor, p. 155).

(3) 'about Grantham' deleted.

1510 NEWTON TO ——

n.d.

From a holograph draft in the Bodleian Library, Oxford[1]

Sr

I received your kind present of a collar of very good brawn & return my hearty thanks for it. I hope you have your health well & wish you a happy new year.

NOTE

(1) Bodleian New College MSS. 361, II, fo. 16. A similar, undated draft is in the Jewish National and University Library, Yahuda Collection, Newton MS. 7, packet 3.

1511 NEWTON TO ——

n.d.

From the draft in the Bodleian Library, Oxford[1]

[May] it please your Lordp

I do not know of any book of Letters in wch Robt Stevens[2] saith what Manuscripts had & what had not the Epistles. I am
My Lord

NOTES

(1) Bodleian New College MS. 361, II, fo. 17v.

(2) Possibly Robert Stephens (1665–1732), founder member of the Society of Antiquaries (1717), and collector of the letters and memoirs of Francis Bacon. He was Chief Solicitor of the customs until 1726, when he succeeded Thomas Madox as Historiographer Royal. He was a relative of Oxford, to whom this letter may be addressed. No book of letters of the type Newton describes here seems to have been written by him.

1512 NEWTON TO ——

n.d.

From a holograph draft in the Bodleian Library, Oxford[1]

May it please your Lordp

I have procured an assay to be made of the Oar wch your Lp sent to me & send you inclosed the Report of the Assayer. He found neither silver nor Lead nor any other metal in the Oar: But in assaying it some part of it evaporated in a sulphurous fume & the rest became a cinder without yeilding any metall. He tells me that if he had had a sufficient quantity of Oar he would have made two or more assays. For a single assay is scarce sufficient to ground a report upon by reason of unforeseen accidents & the different natures of Oares. He tells me also that the Oare wch I gave him was scarce sufficient to make a single assay & that to enable him to make a Report with assurance, there should have been a pound of the oar or at least half a pound, & that if I can help him to any more of the Oar he will repeat the Assay. I am My Lord

Your Lo[rdshi]ps most humble & most obedient servt

(1) Bodleian New College MS. 361, ɪɪ, fo. 18v. This letter may help to dispose of the unlikely legend that Newton himself carried out assays and other chemical or metallurgical work in the Mint. The Letter is possibly related to Letter 1498.

1513 NEWTON TO ——

n.d.

From a holograph draft in the Bodleian Library, Oxford[1]

Sr

The passage in ye ℞[2] is to be thus intended.—Et hujus sublimati partes tres abstrahantur primum a duabus Vitrioli deinde a tribus vel quatuor cerussæ. Postea de cerussa illa cum aqua pluviali (addito si opus est aceto destillato q.s. sed præstat aceto non uti.) extrahatur saccharum.[3] I have seen Mr. Craigs new piece[4] but had not time to read it. If your friend should go into Flanders or any thing else should fall out so yt you cannot go to work this winter, what if you should spend the winter here. About a fortnight since I was taken ill of a distemper wch has been here very common, but am now pretty well again.

(1) Bodleian New College MS. 361, ɪɪɪ, fo. 34. It is impossible to date the Letter with any certainty. The style and the hand indicate that it is an early letter; clearly it is addressed to a close friend. 'Craigs new piece' may be his 1693 publication (see note (4) below). Possibly the chemical procedure discussed relates to the work of John Francis Vigani (?1650–1712), an Italian who emigrated to England in 1682, the same year that he published his *Medulla Chymiae* (Danzig, 1682). After spending a number of years at Cambridge as a tutor and private lecturer in chemistry and pharmacy, in 1692 he was invited to write a book on chemistry, presumably by the University. This was never completed. In 1703 he was granted the honorary title of Professor of Chemistry by the Senate of the University. According to Humphrey Newton (see Brewster, *Memoirs*, ɪɪ, p. 92), Vigani was a friend of Newton's. (See also L. J. M. Coleby, 'John Francis Vigani', *Annals of Science*, 8 (1952), 46–60.) The present letter may be addressed to Vigani, but is more probably the letter Newton addressed to Fatio de Duillier on 2 May 1963 (see vol. vɪ, p. 365).

(2) 'Recipe'.

(3) 'And three parts of this sublimate are drawn off, first by two of vitriol, next by three or four of white lead, and then the sugar is extracted from this white lead with rain water (with distilled vinegar added as necessary, but it is better not to use vinegar).' We have been unable to trace the source of this recipe; it may derive from Vigani or Robert Boyle.

(4) This book by John Craige (see vol. v, p. 118, note (3)), was presumably his *Tractatus Mathematicus de Figurarum Curvilinearum Quadraturis...* (London, 1693).

1514 NEWTON TO ——
n.d.
From a holograph draft in the Bodleian Library, Oxford[1]

Sr

I received your Letter by wch I understand that you want a few more Bibles to be disposed of to poor people, & I have therefore delivered thirty to Mr Auditor Foleys[2] Clerk who will send them to you. And I pray you to accept of them & dispose of them to poore people as you shall find occasion. I am glad to heare of your good health, & wish it may long continue,

I remain

NOTES

(1) Bodleian New College MS. 361, II, fo. 39.

(2) Thomas Foley was appointed Auditor of Imprests on 22 January 1713, and retained the post for many years. Hence this letter must be of 1713 or later.

1515 NEWTON TO ——
n.d.
From a holograph draft in the possession of Lord Lymington[1]

I have received both your Letters & thank you for the general accounts sent me therein. The roads begin now to be good, & the weather after this month of March is over, will also be mild & warm, & if you please then to take a journey to London with the particular accounts, your charges shall be born. Mr Tompson is upon marrying my Neice Mrs Smith,[2] & I desire to have her accounts stated in order to a marriage settlement. For wch end I should be glad to see you here in Easter week or the week after at the furthest if your occasions will permitt. I am

NOTES

(1) This rough draft is scrawled on the back of a letter-cover postmarked 19 March addressed to Newton in St Martins Lane, so the letter must be later than October 1710. Newton has jotted on the cover 'Fatio 30. Neice Smith 25. Landlady, 21. Cook. 8. Tankard 12½. Guineas 27. Rastal 2. Pocket 30. Linnen.' Hence we assume he decided on a settlement of £25.

(2) 'Mrs Smith' must be one of the three children of Newton's half-brother, Benjamin Smith; she married Carrier Tompson. See Appendix II, Newton's Genealogy.

368

1516 NEWTON TO A NOBLEMAN
n.d.
From the holograph draft in the University Library, Cambridge[1]

May it please your Grace

An Italian[2] as I heare has lately published a book[3] about things being infinitely infinite, & if that Author should say that the number of points in the line *AB* is infinitely infinite, & by consequence that an infinite number of points will take up but an infinitely small part of the line *AB*, I am not able to confute him.[4] On the contrary others may dispute that the number of points in a line infinitely long is no more then infinite & therefore ye points *A* & *B* must be at an infinite distance to be the extremes of an infinite number of points. And others may dispute whether questions of this kind are Mathematical or Metaphysical. For my part I must acknowledge them far above my understanding & remain

Your Grace's most h[umble] & most
ob[edien]t S[ervant]
I.N.

NOTES

(1) Add. 3965(13), fo. 543v. This must be an answer to a previous letter.

(2) 'Pere Grandi' deleted; presumably Newton had in mind Guido Grandi, *De Infinitis Infinitorum, et Infinite Parvorum Ordinibus Disquisitio Geometrica* (Pisa, 1710).

(3) 'in Italy' deleted.

(4) 'it would be requisite to see his Demonstration before I give my opinion in the matter' is deleted; evidently Newton knew the title of the book only.

1516A NEWTON TO ——
n.d.
From a holograph draft in the University Library, Cambridge[1]

Sr

Tis now a month since your last, & I have not yet heard from your Brother. I question not but you have seen him by this time & therefore since I cannot yet heare from him I beg ye favour of a line or two from you to let me know how things stand now & when I may expect to have some money & ye security you told me of. The least troublesome way for me will be to take bond, but ye bondsmen I desire to know, & in what readiness things are.

NOTE

(1) Add. 3964(8), fo. 22. Possibly the letter relates to Proctor and his 'brother' Parkins; compare Letters 1509 and 1520.

1517 NEWTON TO ——

n.d.

From a holograph draft in the University Library, Cambridge[1]

Sr

These are to certify that the bearer hereof Edward Carter[2] is Waterman & Servant to the Mint & one of that corporation & on that account is exempted from all Parrochial Services by the Charters of the Mint that he may attend her Majesties service therein. And as these privileges are of ancient standing & have ever been allowed & were granted in our Charters for enabling the Officers & Ministers of her Majts Mint to perform their duties without interruption & to carry on the coinage wth dispatch for the publick service, so their services are not wanted in their respective Parishes, there being great choise of other people to perform parish duties: so tis hoped that her Majesties Officers in all other stations will have that regard to her Majesties government & to the rights of the Crown & services under it as not to give her Office of ye Mint any trouble in this matter. For if the constitution of the Mint be disturbed in this case, the people imployed in the coinage will soon become liable to be taken from service at such times as we cannot spare them without putting a stop to the coinage till we can get others capable of serving in their room. For if the coinage be stopt in any one part thereof it must stand still in every part to the dammage of the Merchant & the disabling of ye Master & Worker to perform his contract wth her Majesty. And hereof all her Majties Justices of the peace & all other her Officers herein concerned are desired to take notice

Is. NEWTON Master & Worker
of her Majties Mint.

NOTES

(1) Add. 3968(30), fo. 439v.

(2) One Edward Carter, waterman, died before September 1711 and was succeeded in the post by his son, also an Edward, when he became old enough, in July 1714. (See Mint/1, 7, pp. 72–3 and *Cal. Treas. Books* XXVIII (Part II), 1714, p. 337.) The manuscript context is of the summer of 1714.

1518 NEWTON TO ——

n.d.

From the holograph draft in the University Library, Cambridge[1]

Sr

My Tenants acquaint me that some of ye Commissioners for ye present Tax, considering that ye clause in ye this Act[2] for excusing ye Universit[i]es is otherwise worded then it was in ye last Act, & does not excuse Colleges, designe now to tax ye lands in Thurleigh & Risely belonging to my Lectureship:[3] This made me think it proper to acquaint you that my Lands are not in the nature of College lands but of a salary or stipend belonging to my Lectureship, & that tho the words of ye Act run so that they do not excuse Colleges from being taxed for their lands yet it excuses all particular members of Colleges from being taxed severally for the salaries & profits belonging to their several places whether they receive those salaries or profits from ye King or from ye Universities or from any of ye Colleges or out of any other estate. Colleges are excused for the scites of their Colleges but not for their revenues, but Readers are excused for all their profits whatsoever. For ye words of yt Act run thus. *Provided that nothing in this Act be extended to charge any...Reader Officer or Minister of ye said Universities &c for or in respect of any stipend wages or profits what soever arising or growing due to them in respect of their several Places &c.* These words I think include me plainly as one of the Readers and by consequence excuse me expresly from paying for any of the profits of my Professorship. Sr now I have represented ye case to you I beg ye favour yt you would let the rest of ye Comm[ission]ers know it at your next meeting & then I leave it their Justice. Sr. I remain

NOTES

(1) Add. 3970, fo. 645. The letter is clearly addressed to one of the Commissioners of Taxes for the University and Town of Cambridge, probably the Vice-chancellor of the University. It was probably written in the early 1670s.

(2) Presumably Newton meant to delete the article 'ye'. It is impossible to identify exactly the Act to which Newton here refers. The clause from which he quotes appears in all the Acts for granting aid to the crown by general taxation throughout the period of Newton's lectureship as Lucasian Professor of Mathematics (1669–1701). The wording of the clause does indeed vary from Act to Act, but the phrases Newton here quotes remain virtually identical.

It seems unlikely that the letter dates from the period (1688–1695) when Newton was himself a Commissioner of Taxes for the University and Town of Cambridge, and was also resident in Cambridge; during this time he presumably had ample opportunity for personal contact with the other Commissioners and in any case probably attended the regular meetings of the Commission. Two possible explanations present themselves. Newton may have confused a general Taxation Act with one of the frequent polls for raising money, which normally included both a tax *per capita* and an additional estate tax for the better-off (see for example Statutes of the Realm, Charles II, An. 29/30, Cap. I, [1677/8]). These latter Acts did not always include the exemption clauses Newton discusses here. Alternatively, in 1695/6, after Newton had left Cambridge for London, an additional clause was added to the taxation Acts (see William III, An. 7, 8, Cap. V, and subsequent taxation Acts),

> Provided always that nothing contained in this Act shall bee construed or taken to discharge any Tenant of any Houses or Lands belonging to the said Colleges...or any of them who by their Leases or other Contracts are and does stand obliged to pay and discharge all Rates Taxes and Impositions whatsoever but that they and every one of them shall be rated and pay all such Rates Taxes and Impositions Any thing in this Act contained to the contrary not withstanding.

Clearly Newton was in a somewhat ambiguous position *vis-à-vis* this clause.

(3) The executors of Henry Lucas' estate purchased lands in Bedfordshire on his behalf, bringing in a rent of £100 p.a. to provide a stipend for a professor of mathematics (the Lucasian Professorship). This endowment was made in 1663. (See John Willis Clark, *Endowments of the University of Cambridge* (Cambridge, 1904), pp. 165–6 and 632.) These estates included Thurleigh (also spelt 'Thirleigh') and Riseley (see the estate papers in Cambridge University Archives, D.XI. 1–18; this information was kindly supplied to us by Miss E. S. Leedham-Green, Assistant to the Keeper of the Archives).

1519 NEWTON TO ——

n.d.

From a holograph draft in the Jewish National and University Library, Jerusalem[1]

Sr

I send you the book I promised & hope you will please to accept of it tho I beleive amidst your business of consequence you have no time to spend on these matters. Considering your great learning I presumed last wednesday to propose you to the Council of the Royal Society for their assent that you may be proposed to the Society at a full meeting to be elected one of their members, & the Council gave their assent unanimously. I hope you will pardon me for doing this without your leave it being the effect of my good will to your self & the cause you are solliciting. I am

> Your most humble and
> most obedient servant
> Is. NEWTON

NOTE

(1) Yahuda Collection, Newton MS. 7(4). It is impossible to be certain to which future member of the Royal Society Newton here refers.

1520 NEWTON TO ——

n.d.

From a holograph draft in the Jewish National and University Library[1]

Sr

I wrote to you about 4 or 5 months since, but you were then from home & not having heard from you since, I have sent this to beg a line or two from you to let me know how my affairs stand & when I may find you & your brother at home.[2] I desire you would put your brother in mind to have his arrears ready against ye time you appoint, & I will wait upon you. In ye mean time I rest

NOTES

(1) Yahuda Collection, Newton MS. 14, fo. 90. Clearly the letter relates to Newton's affairs in Lincolnshire.

(2) Possibly Proctor and his 'brother' Parkins; compare Letter 1509 and also Letter 465, vol. III.

1521 NEWTON TO ——

n.d.

From a holograph draft in the Jewish National and University Library, Jerusalem[1]

Sr

When you were last here you told me it would be necessary for me to go over into Bedfordsh. to take some order about Mr Day's tenure before our Lady day. If you please to let me know by a line or two ye next opportunity what time I may find you at [*draft ends here*].

NOTE

(1) Yahuda Collection, Newton MS. 14, fo. 118. Possibly the Letter dates from shortly before June 1672, when Newton was in Bedfordshire. See Letter 69, vol. I, p. 194.

1522   JOHN ARNOLD TO NEWTON

20 March.   Original

Begs continuance of his bounty. (Bodleian Library, New College MS. 361, II, fo. 95v.)

1523   ARNOLD TO NEWTON

6 November.   Original

Has lived four or five years at Islington and contracted small debts. Desires 40s and will not trouble Newton again. (Bodleian Library, New College MS. 361, II, fo. 96.)

1524   ARNOLD TO NEWTON

14 April.   Original

Requests money to make up cloth which he has into a suit of clothes. Newton has often very generously supplied his wants before. (Mint Papers, II, fo. 462.)

1525   ARNOLD TO NEWTON

22 May.   Original

Has raised money to put out his son as apprentice to a painter; requests help to furnish him with clothes. Thanks for 'constant kindness for so many years past'. (U.L.C., Add. 3965(18), fo. 679.)

1526   ARNOLD TO NEWTON

5 November.   Original

Reluctant after many and great favours, but because of sickness in the family begs a guinea. (U.L.C., Add. 3968(14), fo. 252r.) The year seems to be 1721.

1527   ISAAC BANASTRE TO NEWTON

n.d.   Original

Regrets misunderstanding of his former Latin letter. Is in difficulties with bills from the 'Upholder' [i.e. the undertaker] and Apothecary after his wife's death, amounting to £10. 'Dr Bentley, Dr Knipe [1638–1711], Dr Freind, the ArchBp of York Bp of Ely & others have done wonders for me'. Is to go as tutor to a country gentleman's son. Seeks a loan only from Newton. (Jewish National and University Library, Jerusalem; Yahuda Collection, Newton MS. 7(4).)

1528   CATHERINE BARTON TO NEWTON

n.d.

Letter concerning the 'French Ladys friend'. (Untraced. See *Sotheby Catalogue*, Lot 176.)

1529   A COMPOSITOR TO NEWTON

n.d.   Original

Could not rightly understand the boy, but thinks he has done 'it' correctly; would welcome a tip. (Bodleian Library, New College MS. 361, II., fo. 896.)

1530   A COMPOSITOR TO NEWTON

n.d.   Original

Sends last revise; hopes to drink the author's health. (Bodleian Library, New College MS. 361, II, fo. 87.)

1531   A COMPOSITOR TO NEWTON

n.d.   Original

Had hoped for a tip when the work was finished. Fears the boy has wronged him. (U.L.C., Add. 3965, fo. 651 r.)

1532   ROBERT CORBEY TO NEWTON

n.d.   Original

Is settling in business in the country; asks permission to thank Newton personally for favours received. (Bodleian Library, New College MS. 361, II, fo. 79.)

1533   FRANCIS CRESSENER TO NEWTON

n.d.

From the original in the Bodleian Library, Oxford[1]

Sr

I just received a letter from my son who is at Frederickshall in Norway it requires an answer by to morrow Post I would Gladly Sr see you this evening or to morrow to advise wth you about ye buisness he hath wrot & am

your most humble servant

FRA: CRESSENER

12th

NOTE

(1) Bodleian New College MS. 361, II, fo. 91 v. The date is partly torn away. The unusual name 'Cressner' is associated with East Anglia; it was borne by several Cambridge graduates during Newton's lifetime, but none was named Francis.

1534 CRESSENER TO NEWTON

n.d.

From the original in the Jewish National and University Library, Jerusalem[1]

Sr

I thank you for yr kind inquiry after me & much rejoyce at yr recovery of yr health indeed Sr I am greatly ashamed to ask pardon for not performing my promis to you in ye feathers. they be bought but no shipps happened to goe to yt port soe late I doubt not of giting ym over in time this Summer & hope in God you will use ym many winters & am

Sr
Your very Humble
Servant
FRA: CRESSENER

Jan ye last
To
Sir Isaac Newton

NOTE

(1) Yahuda Collection, Newton MS. 7, packet 4.

1535 CRESSENER TO NEWTON

n.d.

From the original in the University Library, Cambridge[1]

Sept 22d

Sr

I much want to advise further with you in my little affaire if Convenient to you [I] should be Glad to see you sum time to day & am wth all respect

Sr
your most Humble Servt
FRA CRESSENER

To
Sr Isaac Newton
in St Martins Street
by Lester Fields

NOTE

(1) Add. 3965, fo. 376v. Newton has used the back of the letter for notes on observations of a comet.

1536 ISAAC DAVIES TO NEWTON
n.d.
From the original in the Mint Papers[1]

Sr

I have waited on y Gentleman abt ye Italian Coins, & I have obtain'd their perusal for 3 or 4 d[ay]s, So yn you please to vouchsafe ye Honr of calling before I return 'em, shall be glad may pass yo[u]r observacon, & then if appears amongst 'em any Thing Material where several of a Sort, believe may purchase 'em

I'me scarce ever from ye Counting House except betwixt 2 & 3 Nor ever expect to see any Gentlemn there for whose Charactr shall entertain a more sensible Veneration

Sr

Foster Lane   Yo[ur] most humble & obe[dient][2]
Frydy. 19 *June* 9 *a Clock*   Servt,
ISAAC DAVIES

To
> Isaac Newton Esq.[3]
> At Her Majestys Mint in ye
>> Tower

NOTES

(1) II, fo. 11v. It is impossible to date this letter precisely. On the reverse Newton has drafted a list of values of foreign coins, mostly Italian.

(2) The edge of the paper is torn.

(3) The form of address indicates that the letter was written before Newton was knighted; that is before April 1705.

1537 FELIX DE ALVARADO TO NEWTON
n.d.
From the original in the University Library, Cambridge[1]

Sr

Your Honrs known unbounded goodness, a singular Patron & favorer to all Schollars & Learned men, is my sole motive for enterprizeing this small present to your Honr. a Rude & Unpollisht Booke, Begotten in Spaine, & brought forth in England. I shall be extremly obliged to your Honrs. favor in countenanceing ye same, haveing only a sett No. assigned me for ye performance,

377

wch. I hope to dispose off, to those worthy Gentn. my Benefactors, who did me ye favor, to accept my Translation,[2] ye Booke of Common Prayer into Spanish, my Indigent circumstances did not permit me to Undertake it on my owne account, I am in all Duty to Command

Sr. Yr. most Obedt. servant

FELIX DE ALVARADO

Sr. Be pleased to Cast an Eye thereon, there being therein, Uncommon Varietie. I shall not trouble your Honr. any more on any acco[un]t.

To the Honble Sir Isaac Newton
     These

NOTES

(1) Add. 3968(41), fo. 126; on the back are fragments relating to Des Maizeaux's *Recueil*.

(2) Felix Antonio de Alvarado was translator of *La Liturgia Inglesa, o el libro de Oracion Commun* (London, 1707, 1715). It is the second edition which is probably referred to here, as De Villamil records a 'Common Prayer Book, in Spanish, calf gilt leaves, 8vo. 1715' as being amongst the volumes in Newton's Library. Hence the 'Unpollisht Booke' to which De Alvarado refers is presumably a still later publication, possibly his *Dialogos Ingleses y Espanoles* (London, 1719).

1538   P[ETER] GARDNER TO NEWTON

        n.d. Original

Reduced by sickness, the writer, 'qualified for french, English, Highgerman and Danisch', craves compassion. (Bodleian Library, New College MS. 361, II, fo. 49.)

1539   PETER GARTNER TO NEWTON

        n.d. Original.

Is sorry to apply again, because of illness; sends 'translation of the English [book] against Popery.' (U.L.C., Add. 3965(12), fo. 421v.) Presumably the writer is the same as that of Letter 1538.

1540   R[OBERT] GAYER TO NEWTON

        5 April. Original

Thanks Newton for present of wine and sends Lady Betty's service. (Private possession.) The writer was Robert Gayer of Stoke Poges, born in 1673; his wife ('Lady Betty') was Elizabeth Annesley; see vol. IV, p. 476, note (2).

1541 BARTHOLOMEW GOODDAY TO NEWTON

n.d.

From the original in the Bodleian Library, Oxford[1]

Worthy Sr

The Improvement of Navigation has been the study of several Ingenious Mathematicians of late years, and although some of their Inventions have fall'n short of the end proposed, the Authors of them have been rewarded for their good Endeavours; which together with the hopes that I have Discovered a real Improvement, makes me bold to trouble your Worship.

I have taken great pains, and have been at some charge in this Art; the Pamphlet Intitled Mercator Improved is part of my small Performances, the Instrument there in Described is not yet truly made, for I find it difficult to get a new device done according to directions.

I have likewise finished the Description of a new Instrument to take the Altitude of the sun without a Horizon, as mentioned at the end of my book; the which description if your Worship pleas to appoint some person to Inspect, we will get a compleat Instrument made to lay before your Worship, and the Lords of the Admiralty; and that it may answer the end proposed, is not only the desire, but sure and certain hope; of him who desires nothing more than worthily to subscribe himself;

Sr

your most humble, and
most obedient servant.
BARTH: GOODDAY

February: 25

PS

I have desired a friend of mine
who is acquainted with Mr Edw. Jones.
to beg the favour of his approbation.

For
Sr Isaac Newton
These

NOTE

(1) Bodleian New College MS. 361, II, fo. 93v. The writer was a teacher of mathematics giving an address in the Strand. His pamphlet, *Mercator Improved, or the Description of a New Instrument* [*for Navigation*] (London, 1725), is summarized in E. G. R. Taylor, *The Mathematical Practitioners of Hanoverian England* (Cambridge, 1966), p. 159.

1542   DANIEL HARRISON TO NEWTON

n.d.   Original

Craves compassion; was formerly chorister at Trinity College, Cambridge, and his father was 'Combination [Room] Keeper'. (Bodleian Library, New College MS. 361, II, fo. 80v.)

1543   ?HICKSTAN TO NEWTON

n.d.   Original (Latin)

Recommends himself to Newton. (Bodleian New College MS. 361, II, fo. 92; there is a facsimile in Frank E. Manuel, *Isaac Newton Historian* (Cambridge, 1963), facing p. 10.) The signature is difficult to read, but may be 'Niam Hickstan'. The writer presumably hoped for an interview.

1544   WILLIAM HOWARD TO NEWTON

n.d.   Original

'The Author of the Poem on the Resurrection to You humbly Dedicated now humbly waits in hopes of Your favourable acceptance of it.' (U.L.C., Add. 3965, fo. 489.) Perhaps the writer was author of *The Resurrection. A Poem*, of which the catalogue of the British Library lists the fourth edition [1750?].

1545   J. LEON TO NEWTON

n.d.   Original (Latin)

Apprised by Conti of Newton's integrity, desires permission to converse with him. (U.L.C., Add. 3965(10), fo. 120.)

1546 EDMUND LONGBRIDGE TO NEWTON

n.d.

From the original in King's College Library, Cambridge[1]

Most honoured Sir,

Suffer me to seek ye Continuance of yt favour wch. you was pleased to vouchsafe me in speaking to Mr Hall[2] for me & forgive me the liberty wch: in this experience of your goodness I have taken to represent wth: all due Submission some of those reasons wch: would seem to make it an Act of ease ye Comptr[oller] & of his Justice to me that I be restored: There is a necessity of having one experienced Clerk in Comptr[oller']s Office, otherwise it is of no manner of service to ye Crown; & Complaints will of Consequence be made both by ye Comm[issione]rs & Auditors of ye Imprests. w:ch Complaints would ye more affect ye Comptr: who is supposed of Ability to master cheque, & by no means to retard beyond ye Limits of ye Statute ye proper Acc[ounts] That Youth for whom I was sent a Starving, could not but give occasion for these, who was unable or unwilling to [bear] ye Weight of that buisness. Not to be always repeating to yr Hon:r ye Injustice as well as discouragem:t it is to Men of liberal Education who have chearfully submitted to long labour & poverty in lieu of encou[ra]gem:t to receive Wants & Miseries; while Novices & Fo[ot]men are advanced on their Ruins; to reap not only ye fruit, but ye Credit too of their Industrys: permit me therefore once more most earnestly to beseech you to [be]come my Advocate with ye Compt[rolle]r as without whose favour [my] Family & my Wife who now lies languishing on a Bed of Sicknes must in humane probability perish: I have no need to mention to yr Hon:r ye Christianity of Charity, whose Good Heart is filled wth every Virtue of that Religion.

Almighty God, by whose Power and Bounty all things are upheld, vouchsafe yr Hon:r a length of days remarkable, & a Prosperity equal to your Wisdom & Knowledge.

I humbly beg leave to Subscribe myself

Most Hon:d Sir,

Your afflicted

most obedient, faithful

Servant

EDM: LONGBRIDGE[3]

To
Sr Isaac Newton
att his House in St Martins
Street Near Leisester feilds
London

NOTES

(1) Keynes MS. 148(C), 3. A fold in the letter has caused some obliterations.

(2) Thomas Hall, Comptroller of the Duties on Salt and Rock Salt.

(3) On 27 January 1716 Thomas Hall signed a recommendation, presumably made at Newton's instigation, for the appointment of Edmund Longbridge, then a clerk in his office, to the post of Deputy Controller. A Warrant was made for his appointment on 14 September 1716 (see P.R.O., T/53, 24, p. 549). Presumably at some later time he was dismissed, and in the present letter requests reinstatement.

1547 JOSEPH MORLAND TO NEWTON
n.d.

From the original in King's College Library, Cambridge[1]

Sr

I have done and will do my best while I live to follow your advice to repent and believe I pray often as I am able that god would make me sincere & change my heart. Pray write me your opinion whether upon the whole I may dye with comfort. This can do you no harm written without your name. God knows I am very low & uneasie & have but little strength

Your most humble servt
JOS. MORLAND

Pray favour me with one line because when I parted I had not your last words to me you being in hast

Direct For Dr Morland in Epsom Surrey

For Sr Isaac Newton
in St Martins Street
in Leicester Feilds
London[2]

NOTES

(1) Keynes MS. 127; microfilm 1011.19. Also printed in Brewster, *Memoirs*, II, p. 359.

Presumably the writer is Joseph Morland (?1671–1716), a relative of Sir Samuel Morland (probably his brother). He proceeded M.D. at Leyden University on 18 June 1699 at the age of 28; he was elected Fellow of the Royal Society on 30 November 1703. He published at least two medical treatises, and was editor of Samuel Morland's *Hydrostaticks, or Instructions concerning Waterworks...* (London, 1697).

(2) The address indicates that the letter must have been written between September 1710, when Newton moved to St Martin's Street, and 1716, when Morland died.

1548 THOMAS SAMPSON TO NEWTON
n.d. Original

Sends a book which has borne several impressions, hoping it will amuse Newton during his 'Retirement', which he hopes will soon be over. (U.L.C., Add. 3965(17), fo. 659.)

1549   HANNAH TONSTALL TO NEWTON

   n.d.   Original (highly illiterate)

Begs assistance in state of debt; the bearer her cousin will explain. (U.L.C., Add. 3965(12), fo. 230v.) The writer was a granddaughter of William Ayscough, Newton's maternal uncle. Compare Letter 1502.

1550   JOSEPH TREVOR TO NEWTON

   n.d.   Original

Begging letter written from a coffee-house because 'I cant gett a Letter In at Yr: Own House.' (U.L.C., Add. 3965(18), fo. 707.)

1551   MARGARET WARNER TO NEWTON

   19 January.   Original (torn)

Apparently, thanks for a present of wine. (U.L.C., Add. 3965, fo. 638.) The writer was Catherine Barton's sister.

1552 GEORGE WATSON TO NEWTON

n.d.

From the original in the Sir Isaac Newton Collection, Babson College[1]

Sr

Mrs Barton[2] failing me off ye payment off ye five hundred pounds, if you wod be pleased to pay ye money & take ye assignement off ye Morgage it wod be a great convenience to me; & you know yt ye title is undoubtedly very good; I will waite on you any time to day, if you will name ye time. Sr

                                                        I am your humble

March ye 4th   Servant

                                                        GEO: WATSON

NOTES

(1) Babson MS. 438. Nothing is known of the writer of the letter.
(2) Catherine Barton.

1553   —— TO NEWTON

   n.d.   Original

Concerning the time of a meeting between Newton, Halley and William Lowndes. (Bodleian Library, New College MS. 361, II, fo. 88v.) The writer was probably Lowndes' clerk, and the meeting was possibly about the longitude (?1714).

APPENDIX I. ADDITIONS AND CORRECTIONS TO EARLIER VOLUMES

Introduction

It is inevitable in a work of this size and nature that only as it nears completion should certain omissions and errors become apparent. Documents which were previously undatable begin to fit into the chronological jigsaw; reappraisal of manuscripts produces answers to old puzzles; and, not least, friends and colleagues supply information and material that necessitate changes in opinions already printed. But whilst we feel we need give no apology for the inclusion of an appendix to this final volume, perhaps we require some excuse for its size. The vast majority of the added material concerns Newton's work at the Mint, and is derived from the mass of records of the Mint's activities now preserved at the Public Record Office in London. Much of this material is uncalendared and lacks detailed indices; hence there were large groups of manuscripts which were never discovered by the editor of volume IV of this *Correspondence*, and which we have since unearthed either during our own routine searches or as a result of hints from other historians.

So great is the quantity of this material, covering the years 1696–1709, that it casts quite a new light on Newton's activities at the Mint. His post—first as Warden, then as Master—was clearly no sinecure. One of his earliest tasks was the setting up of the Country Mints in 1696 to receive and recoin the old hammered money, and thus to remove from circulation badly clipped and devalued coin. The Mint at Chester, under the charge of Edmond Halley, was the most troublesome (see Letter X.566), and this is touched upon in volume IV; but none of the five Mints ran really smoothly, and although the recoinage was complete by 1698, accounts were not finally settled until 1703 (see Letters X.597 and X.663.2). Another early concern was with the prosecution of counterfeiters (see Letter X.593); a number of letters are extant in P.R.O., Mint/15, 17, many addressed to Newton, giving evidence against coiners, and the volume also contains a number of depositions witnessed personally by Newton.* Much of this material is difficult to date precisely, but all falls within the period October 1698 to June 1699. Even after Newton was relieved of the task of prosecuting coiners, the prevention of counterfeiting was very much his concern, and we find him making a report (Letter X.630.2) on the treatment of counterfeit money by the official receivers, many of whom simply returned it, instead of cutting it as they should; and later Newton proffered further recommendations for preventive measures against coiners (Letter X.643.1).

Financial stringency was a perpetual bugbear. Many of the Mint's buildings were sadly in need of repair, and friction arose with the Ordnance over their occupation and rebuilding (see Letters X.610.2, X.612, X.616, X.626.1). The moneyers constantly nagged for better wages (Letters X.610.3, X.633.6 and X.709.1). Theirs was a two-fold

---

* See also Sir John Craig, 'Isaac Newton and the Counterfeiters', *Notes and Records of the Royal Society of London*, **18** (1963), 136–45.

problem; first an earlier Provost of the Moneyers had had special terms attached to his appointment whereby his salary continued for life even after he had resigned his post; and second, the Moneyers were being paid piece rates, and since, with the coinage nearly at a standstill, there was very little work for them, they were receiving nothing at all by way of payment. Newton, on the other hand, with only £3000 to cover costs of both maintenance of the Mint and salaries of Mint employees, could scarcely meet their demands, and eventually the Treasury had to find him an extra £500 p.a.

But by far the greatest headache for Newton must have been the recoinage at the Edinburgh Mint in the period 1707–9; a number of letters relating to this matter have already been printed in volumes IV and V, but the additional ones printed here considerably clarify the story (see Letter X.715.2 and Letters following). The overall impression we receive is of bungling inefficiency on the part of both the Mint officials in London and their counterparts in Edinburgh. The root of the problem lay in the conflict between the Act of Union—which stated that administrative and practical procedures in England and Scotland must be identical—and the intrinsic differences between the two Mints which made an achievement of this identity impossible (see for example Letters X.729.1 and X.729.2). The situation was aggravated by the failure to make proper arrangements for the payment of personnel (including Newton himself) until some time after the recoinage had begun (see Letters X.735, X.736 and X.738), and by the death of Stuart (the Collector of Bullion at Edinburgh) in April 1708 (see Letter X.738.2) which effectively froze a large part of the financial resources of the Mint there for a considerable time.

Of the remaining letters in the Appendix not many deserve special notice. Fatio de Duillier has been hard done by in previous volumes—a letter to Newton of 1690 (Letter X.353) was totally omitted; two more were printed out of order (Letters 463 and 464)—not unintentionally, but we mention them here because the reader, looking for them in their correct chronological position, might miss them; and three further letters were wrongly dated (Letters 395, 396 and 397). Three somewhat cryptic letters dating from around April 1705 concern Newton's candidacy for the University of Cambridge seat in the Parliamentary elections, and show him calculating his chances of success. Only a few items in the Appendix concern Newton's scientific work. An interesting draft of a letter to John Wallis (Letter X.398.2) shows Newton already in doubt over the chronology of his early papers on the calculus. There are two letters concerning chemistry, one written by Newton about the publication of some of Boyle's manuscripts (Letter X.398.1, c. 1692), the other (Letter X.704, c. 1705) to Newton from William Yarworth, whom Dr Karin Figala has suggested had considerable connections with Newton, and whom she identifies with the alchemical writer, Cleidophorus Mystagogus. We include too (Number X.826) the draft of a paper related to Cotes' correspondence with Newton over the second edition of the *Principia*. The paper, which Newton sent to Cotes on 24 March 1710/11, relates to Proposition 37 of Book II, which dealt with the resistance experienced by a cylinder moving through a fluid medium; Cotes, in his reply to Newton (see vol. V), pointed out that Newton had made a mistake in his consideration of the relative motions of the system, and Newton subsequently rewrote the Proposition.

The reader may already have noticed that Letter 1 in this *Correspondence* is not a true letter but an exercise copy; however, we believe that forgeries of letters to and from Newton do not appear in these volumes. The most famous of these are the 622 'Newton' letters sold by Vrain-Denis Lucas to Michel Chasles, a number of which were printed in the *Comptes Rendus Hebdomadaire de l'Académie des Sciences*, **65** (Paris, 1867). These include letters to Rohault, Des Maizeaux, Malebranche, Cassini, Fontenelle and others; the most notorious amongst them are the Newton–Pascal letters, the earliest of which is dated 1659! Lucas's forgeries were quickly detected by everyone except Chasles himself. Another spurious letter, from Newton to Dr Law (presumably William Law, the Quaker mystic) was printed in *Nature*, May 12 (1881), p. 39, and *The Popular Science Monthly*, August 1881, p. 572; we have not traced its source, but its style is clearly not Newton's, nor even contemporary.

Note on arrangement and numbering

The Appendix comprises all our additions and corrections to the main sequence of letters, arranged in continuous chronological order. We felt it expedient to assign numbers to the additional letters we have printed to indicate where they should have appeared in the original numerical series. Hence a letter omitted from volume IV, which should have been printed between Letters 574 and 575, is here assigned the number X.574. Where more than one letter needs to be interpolated at a single point in the sequence, we have added suffixes 1, 2, 3 etc. to the letter number. Hence Letters X.398.1 and X.398.2 fall chronologically between Letters 398 and 399 of volume IV.

Not all the additional letters have been printed in full, although all have been assigned numbers with the prefix X. We have given in full all letters of which holograph drafts by Newton exist, or of which there are originals or copies bearing Newton's autograph, even if the text of the letter is in the hand of a copyist. Letters to Newton, and letters from the Mint where there is no particular evidence of Newton's direct concern in them, we have either printed in full or merely calendared, depending on our estimate of their importance.

Corrections and additions to editorial comment are listed under the number of the letter to which they apply. Hence the entry preceded by the number 557 is a correction or addition to Letter 557, vol. IV. Where the dating of a Letter has been found incorrect, it has been listed under its old number and date, with a cross-reference at its new date. For example Number 663, a Royal Warrant originally given the approximate date 1702/3, can now be assigned the exact date 10 February 1702/3, so that its proper place is between Letters 658 and 659. In our Appendix this additional information is given under the heading 663, but there is a cross-reference, with the number of the letter given in angled brackets ⟨663⟩, at the correct chronological point in the sequence. We have given similar unnumbered cross-references to those few Letters added out of sequence by former editors at the ends of volumes III and IV.

We have not reprinted or noted here the editorial corrections previously printed at the end of volume III.

110   NEWTON TO COLLINS

20 May 1673

Note (3) (vol. I, p. 282): The book referred to in the letter is not by Hendrick van Heuraet, but is Grégoire Huret's *Optique de portraiture et peinture* (Paris, 1670 and 1672). See Hall and Hall, *Oldenburg*, IX, p. 554, note (13).

X.132 NEWTON TO OLDENBURG

Late JANUARY 1674/5

From the holograph draft in the Jewish National and University Library[1]

Sr

The kind profer wch you were pleased formerly to make to me about omitting ye payments required by my being a member of your Society,[2] my state suggests to me now to embrace, For ye time draws near yt I am to part wth my Fellowship,[3] & as my incomes contract, I find it will be convenient that I contract my expenses. This gives me occasion Sr to trouble wth ye payment for ye last half year wch I have sent you by John Stiles[4] ye bearer hereof. The note wch you sent in your last for Dr Castle[5] was delivered to him, & sometime after I took occasion to speake wth him about his answer, but I know not whether he ever sent one

Your humble servant
I. NEWTON

NOTES

(1) Yahuda Collection, Newton MS. 14, fo. 42; printed in Hall & Hall, *Oldenburg*, XI, Letter 2597. The letter is dated from internal evidence, see notes below.

(2) See Letters 101 and 102, vol. I; Newton in March 1673 tried to resign from the Royal Society, and Oldenburg then offered to waive Newton's membership fees if he withdrew his resignation. Newton must have declined the offer, but now wishes to take advantage of it. Oldenburg brought the matter up at a Council Meeting on 28 January 1675 (see vol. I, p. 149, note (7)), and hence we assume Newton wrote this letter shortly before that date.

(3) Newton was to have resigned his Fellowship at Trinity College in the spring of 1675, in accordance with a statute which required Fellows to take Orders after a certain lapse of time; in the event he was granted a Royal Patent on 27 April 1675, exempting the Lucasian Professor from this rule; see Edleston, *Correspondence*, p. xxv.

(4) The Cambridge carrier.

(5) Edmund Castell (1608–85) had been elected Fellow of the Royal Society on 16 April 1674, but was 'admitted' (i.e. attended a meeting and signed the roll) only on 28 January 1675; it could well be in connection with this that he had been in correspondence with Oldenburg.

166   COLLINS TO OLDENBURG FOR LEIBNIZ

14 June 1676

Note (1), line 7 (vol. II, p. 49): For 'June' read 'July'. (A correction kindly supplied by the late Professor L. Rosenfeld.) The mistake arose in Newton's lifetime: See J. E. Hofmann, *Leibniz in Paris* 1672–76 (Cambridge, 1974), p 231.

X.205 NEWTON TO THE MASTER AND FELLOWS OF ST CATHARINE'S COLLEGE, CAMBRIDGE

?1677

From the holograph original in the archives of Queens' College, Cambridge[1]

Supposing ye ridge of Cath. Hall building terminate ye sight of ye skies to an eye placed in ye midst of ye bottom of Queens Coll. Chappel window & yt ye ridge is 22 foot higher then that eye, & yt if yt ridge were continued it would butt upon Queen's Coll. new building 34 foot from yt eye northward if ye building extend 10 foot into Qu. Coll. ground or 44 foot if it do not extend into Q. Coll. ground:[2] then in ye first case ye side continued infinitely from ye Chappel would hide from yt eye a spherical triangle of ye skies of 32 deg. 54′ supposing ye whole skies otherwise visible to yt eye to be exprest by 180 deg. But in ye second case a triangle of only 26 deg. 34′ would be hidden. The difference of these is 6 deg. 20′. That is in ye first case $\frac{2}{11}$ parts of ye skies, in ye second $\frac{1}{7}$ would be hidden: the difference of wch is about $\frac{1}{28}$,[3] wch is ye loss of skye-light by ye 10 foot advance on Q Coll. ground.

The length of ye little barn subtends an arch of about 48^d to ye said eye considered as center.[4] The height of its ridge above yt eye is 13^d. The product of these is 624. The content of ye whole skies in parts of the same denomination is rad. $\times$ 180^d that is 57$^d \times$ 180^d or 10260 whereof 624 is about ye 17th part. About thus much of ye skies is therefore hidden from yt eye by that building. But if ye eye were at ye top of ye window, ye barn (which is not so high) would hide none of ye skies from it; if it were in ye middle between ye top & bottom ye barn would hide about half as much skye: & so proportionally of intermediate places. That is considering ye whole window ye barn hides about ye 34th or 36th part of ye skye light.

In like manner some allowance is to be made in ye other building for ye eye's being placed in severall parts of ye window, but not so much because that building is higher & further off. If about ye 34th part of sky-light be recconed intercepted in both cases the error I think will not be considerable.

388

Yet all this light is not lost: For ye buildings themselves reflect some light perhaps $\frac{1}{3}$ or $\frac{1}{6}$ of what ye intercepted skies would reflect. And so we may reccon about ye 40th part of ye whole light lost by each building, & rather less then more.

The chief reason why this account falls so much short of wt I [supposed] reconing before wth Mr James by guess, is that we yn reconned what light would be intercepted by ye side of ye building continued on Queens Coll: ground without deducting what would be intercepted by that side not extended on their ground.

Had there been a loss of ye 8th part of ye whole light, (that is if all ye ground on wch ye building is to stand had belonged to Queens College [and] was to be reconned[)] I guess it might have caused a fortnight or threeweeks candle light in Spring & Autumn that is about a week or 10 days in spring & as much in Autumn in Chappel time.   I. NEWTON

NOTES

(1) MS. QC 16/1/i. The manuscript was borrowed by Isaac Milner, President of Queens', on 15 February 1799 and never returned to St Catharine's. We are grateful to the Master and Fellows of Queens' College for permission to print it, and to Mrs Dorothy Owen and Dr Whiteside for bringing it to our attention. For the date see below.

(2) It appears that Newton had been asked to give an opinion on the degree of impairment of the light within Queens' College Chapel caused by the new buildings of St Catharine's rising on the other side of Queens' Lane. In 1673 virtually all the existing buildings on the St Catharine's site had been pulled down, leaving only Bull Court standing; the rebuilding was completed in 1704. In January 1676 Queens' College gave permission for St Catharine's to build on a strip 36 yards long and 10 ft wide belonging to the former College but forming a part of the St Catharine's site. Probably Newton's investigations relate in part to buildings erected on this strip. It is known that the Hall and Buttery of the new College were built by Whitsuntide 1675, and the Combination Room and Library by February 1676/7. The new Master's Lodge, extending on to the Queens' College strip, was probably built about this time (W. H. S. Jones, *The Story of St. Catharine's College, Cambridge* (Cambridge, 1951), pp. 55–61).

(3) So Newton appears to have written, but $\frac{1}{26}$ is nearer.

(4) The barn was presumably one of the stables and outbuildings on the site, some of which had been once used by the Cambridge carrier, Thomas Hobson (1544–1631). The sense of the paragraph is that this existing barn already blocks out about as much light from Queens' College Chapel as the new structure would do.

X.240 NEWTON TO SIR ROBERT MARKHAM
?JUNE 1680
From the holograph draft in the Jewish National and University Library, Jerusalem[1]

Sir Robt.

Your son[2] is well arrived hither, by whom I received 5 guineys, An acquittance for it you will receive inclosed in his letter. He shall have money of me as he has occasion for it, & his accounts you shal receive quarterly as the custome is. I suppose it may be enough to to [*sic*] keep his french if a French Mr come to him once a week. If you think more necessary, be pleased to let me know your mind. Mr. Battely[3] has alread [*draft ends here*]

NOTES

(1) Yahuda Collection, Newton MS. 2. Sir Robert Markham, Bart., d. 11 August 1690, was one of the Markhams of Sedgebrook, near Grantham, Lincolnshire.

(2) George Markham was admitted Fellow-commoner at Trinity as Newton's pupil on 26 June 1680; see Edleston, *Correspondence*, p. xlv. He became a Fellow of the Royal Society in 1708 and died in 1736.

(3) We are not able to identify this person.

⟨463⟩   FATIO TO NEWTON

24 February 1689/90

Printed out of chronological order as Letter 463 (vol. III, pp. 390–1).

X.353 FATIO TO NEWTON
17 APRIL 1690
From the holograph original in the Brotherton Library, Leeds[1]

Sir

If you have not yet met with a servant[2] that may serve your turn I think I may help you to one; he is recommended from many good hands: he has been bred up in Paul's School, and was just fit for the University. But his relations dying and leaving him poor he will be now very glad to serve you, and he was inquiring for a Master. He seems to be very modest and good natured, and he promiseth to do any thing that you bid him, both to coppy out your papers, and to serve you in your experiments and in all your other concerns. He shaveth but very indifferently, but that may soon be learned. I do think that his age of about eighteen only or sixteen, his little experience of ye world, his having been kept strictly at ye School, his understanding of Latin, and his having no other hopes left but upon ye kindness of his Master will make him a very good servant to you, chiefly he beeing prepared to receive any sorts of commands from you and having allready served in another place. I do really think Sir that tho' you may be by this time provided with another, yet this is ye very man yt you wished for, and that it will prove in time more convenient to you to have parted with another and accepted of this. I do expect Sir your answer about him. In ye mean while I will inquire farther in what relates to ye young man and his abilities. Mr Hugens his brother[3] hath caused me to guess from his discourses and his answers to some questions of mine that Mr Hugens is not about ye making of one of your telescopes, but onely some very large object glasses, that is very broad ones, which may, tho' but short, bear a vast aperture for to discover more easily by them ye satellits allready known and their Eclipses, the fixed starrs and perhapps new Planets &c. For if that only be intended what occasion is there for an image accurately defined. I find that such Eclipses may be best observed with an oblong aperture A whose length seen from ye ocular glass may seem to cut at right angles the line drawn from ye center of $2\!\!\downarrow$ to ye Satellit.[4] I did write lately to Mr Hugens.[5] Mr Hampden presents his services to you. I am with all my heart

Sir

Your most humble and most
obedient servant
N. Fatio de Duillier.

London Aprill ye 17th. 1690
For Isaac Newton Esqr
at Trinity Colledge at
                    Cambridge

    present

NOTES

(1) MS. 248. For Fatio de Duillier see vol. II, p. 477, note (1). Also compare Letter 355, vol. III.

(2) Humphrey Newton had been Newton's amanuensis from 1685 to 1690; presumably Fatio knew Newton was looking for someone to replace him.

(3) Constantyn Huygens, then living in London. Presumably Fatio, who had considerable correspondence with Christiaan Huygens, was a frequent visitor to Constantyn. In a letter to his brother, dated 25 April 1690, N.S. (see Huygens, *Œuvres Complètes*, IX, pp. 412–13), Constantyn reports on discussions with Fatio concerning Christiaan's latest telescope. The telescope itself is described in Christiaan's letter to Constantyn of 21 March 1690, N.S. (see *ibid.*, p. 396); it was 12 ft long, but with a very large diameter, and hence an aperture far greater than that of an ordinary 12-foot telescope. No other details were given.

(4) Presumably this arrangement was intended to give greater illumination without loss of resolution in the direction of the line joining planet and satellite.

(5) See Huygens, *Œuvres Complètes*, IX, pp. 407–12; the letter was dated 11 April 1690, and sent via Constantyn Huygens.

392   NEWTON TO WALLIS

27 August 1692

D. T. Whiteside suggests that the letter is of a later date, being drafted in the winter of 1692/3, in reply to Wallis' request for comments on the proofs of pp. 390–6 of Wallis' *Opera Mathematica*, II (Oxford, 1693). See also Letter X.398.2.

395   FATIO TO NEWTON

17 September 1692

The original is dated '9ber ye 17th 1692'; that is, 17 November 1692, not 17 September 1692 as stated in vol. III.

396   NEWTON TO FATIO

21 September 1692

The original is clearly dated 21 November 1692, not 21 September 1692 as stated in vol. III.

397   FATIO TO NEWTON

22 September 1692

The original is dated '9ber the 22th 1692'; that is, 22 November 1692, not 22 September 1692 as stated in vol. III.

⟨395⟩   FATIO TO NEWTON

17 November 1692

See correction to Letter 395 above.

⟨396⟩   NEWTON TO FATIO

21 November 1692

See correction to Letter 396 above.

⟨397⟩   FATIO TO NEWTON

22 November 1692

See correction to Letter 397 above.

X.398.1 NEWTON TO ——

?1692

From the holograph drafts in the University Library, Cambridge[1]

Sr

There being left with me by Mr Boyle two experiments to be published after his death: I beg ye favour that you would print them & add them to ye end of one of his books. I make this request to you because divers of his books have been printed for you. The Expts are are [*sic*] as follows.

Exper 1.

Exper. 2.[2]

I have tried neither of these experimts & so cannot affirm the success of them on my own knowledge but must leave ye trial to ye Reader.

Exper. 1. butter of Antimony.[3] Let it run per deliquium in a cellar, Pour of the clear liquor. You may distill ye feculent residue & let ye butter wch destills, run again per deliquium. When you have as much butter run per deliquium in a clear liquor as you desire: digest it in a convenient heat in a glass well stopt & it will putrefy & grow black & afterwards become clear again & then be a menstruum for resolving bodies like ye Alkahest, but not so potent. The putrefaction is caused by ye moisture wch ye butter in running per deliquium imbibes out of the air. The just degree of heat for ye digestion I do not know but conceive it must equal that of blood or of a balneum not hotter then the hand can endure.

The menstruum prepared by the first of these two experimts was proposed by Mr Boyle as a thing wch might be of good use in medicine for analysing & subtiliating bodies. I have tried neither of these experiments my self so cannot affirm ye success of them on my own knowledge, but must leave ye trial to ye Reader. For I cannot tell whether Mr Boyle tried ye first or had it only by communication, tho I believe he had tried ye second himself. [If you please to do Mr Boyle yt right as to print this paper I shall take occasion hereafter to acknowledge ye obliga][4]

NOTES

(1) Add. 3965, fo. 469r and v. The letter was perhaps intended for John Locke, who in 1691–92 was engaged in preparing for the press Boyle's *General History of the Air* and *Medicinal Experiments*, both of which books were printed after Boyle's death (see K. Dewhurst, *John Locke, Physician and Philosopher* (London, 1963), pp. 283–4 and vol. III, Letters 389 and 391). The draft does not correspond to any of Newton's known letters to Locke, however, nor have we encountered these additional experiments.

The sheet bears drafts for the *Principia* and a note of naval vessels launched or rebuilt in 1692; this may have been added later, of course.

(2) This is not described by Newton.

(3) Antimony chloride. To 'run *per deliquium*' is to deliquesce.

(4) Editorial brackets; Newton has deleted this sentence.

X.398.2 NEWTON TO WALLIS

LATE 1692

From a holograph draft in the University Library, Cambridge[1]

Sr

I meet wth no other amendments wch need be made beside these wch follow.[2]

pag. 368 lin. ult. pro 1665 lege 1666.

pag. 392 lin ult. pro $-2\dot{x}oo\dot{y}\dot{y}$[3]$+xoo\dot{y}\dot{y}+\dot{x}o^3\dot{y}\dot{y}$ lege $-2\dot{x}oo\dot{y}\dot{y}-xoo\dot{y}\dot{y}-\dot{x}o^3\dot{y}\dot{y}$

pag 393. 1.2. pro $+xo\dot{y}\dot{y}+\dot{x}o\dot{y}\dot{y}$ lege $-xo\dot{y}\dot{y}-\dot{x}oo\dot{y}\dot{y}$.

The plague was in Cambridge in both ye years 1665 & 1666 but it was in 1666 yt I was absent from Cambridge & therefore I have set down an amendmt of ye year.[4] I wrote to you lately[5] that I found ye method of converging series in the winter between ye years 1665 & 1666. For that was ye earliest mention of it I could find then amongst my papers. But meeting since wth the notes[6] wch in ye year 1664 upon my first reading of Vieta's works Schooten's Miscelanies & your Arithmetica Infinitorum, I took out of those books & finding among these notes my deduction of the series for the circle out of yours in your Arithmetica Infinitorum: I collect yt it was in ye year 1664 that I deduced these series out of yours.[7] There is also among these notes Mercators series for squaring the Hyperbola found by ye same method wth some others.[8] But I cannot find yt I understood ye invention of these series by division & extraction of roots or made any further progress in this business before the winter wch was between ye years 1665 & 1666. But in yt winter & ye spring following by ye use of Division & extraction of roots I brought ye method to be general,[9] & then the plague made me leave Cambridge. But I do not think it requisite that you should make a particular mention of these things. I beleive you have said enough in ye beginning of your 91th Chapter.

In your third Chapter you have given us a collation of the Arabic cyphers with ours both old & modern.[10] The other day looking into Taverniere's travells into India[11] lib. 1 pag 23 I met wth ye cyphers now used all over India. They are these.

Our old figures 2, 4, 5, & 8 (thus marked $\mathcal{Z}$ $\mathcal{R}$ $\mathcal{Y}$ $\mathcal{S}$) seem to be borrowed from these.

I communicated the Postcript of your letter[12] to our Vicechancellor[13] together wth such other Papers as your Messenger shewed me. His answer was that he had been so much pressed by the heads of Colleges upon several occasions to admit nothing of this kind without full evidence yt the prisoner was a true object of charity, that he feared some of ye Heads would be displeased if he should not give leave without better evidence then he had in this case least he should meet wth a new check from ye Heads. He exprest a great respect to your letter & said that he lived within five miles of Sandwich[14] & knew the hands of most of ye gentry there & had the case come recommended by some of them whome he could have confided in he would have admitted it, but for ye seale of ye corporation he represented that he knew they sometimes granted it upon too slight occasions to get rid of troublesome people & therefore he could lay no great stress upon that. Mr Deeds he said he knew, but seemed to lay no stress upon his testimony. I beleive he mistook the man, because I since heare there are more of that name. He desired that you should be well satisfied about the business & so telling him that I would represent it to you I took my leave & giving your Messenger ten shillings was fain to dismis him with this answer. Last summer when ye Vicechancellor stood to be our Burgess I voted against him & so can challenge no interest in him tho he is very fair & civil to me.[15] My hearty thanks for the sheets of your book. I am very glad it is so neare finishing. You have done our nation a great honour in putting so useful a work into Latin for ye perusal of forreigners.

NOTES

(1) Add. 3977(7). This letter (which was brought to our notice by D. T. Whiteside) and the closely related Letter 392 (vol. III, pp. 219.20) both date from the winter of 1692/3, and represent Newton's reactions to proof-sheets Wallis had sent him of his *Opera Mathematica*, II, which was finally published on 28 August 1693. The part of the text Newton is concerned with here is the Latin translation, with amendments and additions, of Wallis's English *Algebra* of 1685.

(2) None of these changes was made in the printed text, nor included in the table of errata at the end of the volume. Hence it is probable that Newton never sent the letter to Wallis.

(3) Wallis in fact prints the correct term, $-2\dot{x}oo\dot{y}$. See Whiteside, *Mathematical Papers*, VII, xvii, note (37).

(4) Wallis had written, as part of the introduction to Chapter 91 (*Opera Mathematica*, II, pp. 368–9), 'In [Epistola Posteriore] ait se ob Pestem *Cantabrigiæ* grassantem (quod accidit anno 1665) a *Cantabrigia* excedentem, hanc speculationem interrupisse.' The sentence originally appeared in English in Wallis's *Algebra* (Oxford, 1685), p. 330, as 'In [the Epistola Posterior] he says, that by the Plague (which happened in the Year 1665,) he was driven from *Cambridge*; and gave over the prosecution of it for divers years.' Compare vol. II, p. 152, note (16) and p. 166, note (1). Newton was apparently in Cambridge from September 1664 to June 1665,

from 20 March 1666 to June of that year, and then again from April 1667 onwards (see White-side, *Mathematical Papers*, I, p. 8, note (20)).

(5) Newton possibly refers here to one of his two letters to Wallis, now lost, dated 27 August and 19 September 1692 (see vol. III, p. 220, note (6), and Wallis, *Opera Mathematica*, II, p. 391).

(6) See Whiteside, *Mathematical Papers*, I, pp. 19–24 for bibliographical details of the works mentioned; for Newton's annotations of Viète, Schooten and Wallis, see *ibid.*, pp. 25–142. Newton only rarely placed the date at the head of his mathematical papers; his own accounts of the chronology of his researches are not always clear or entirely consistent with one another. Whiteside, basing his dating in part on Newton's autobiographical statements (*ibid.*, pp. 7–8) and partly on the handwritings, suggests that these annotations were made between 1664 and Autumn 1665.

(7) Elsewhere Newton states that he made these annotations 'in winter between the years 1664 & 1665' (Whiteside, *loc. cit.*, pp. 7–8 and p. 97, note (1)). For the annotations he refers to, see *ibid.*, pp. 96–111.

(8) *Ibid.*, pp. 112–13. Nicolaus Mercator stated the area under the hyperbola $y = 1/(1+a)$ as $\left(a - \dfrac{a^2}{2} + \dfrac{a^3}{3} - \dfrac{a^4}{4}\right) \dots$ in his *Logarithmotechnia* in 1668.

(9) It is not clear what papers Newton refers to here. Some work on reduction to series by division and extraction of roots is included in his early annotations of Wallis, and perhaps Newton intended his generalization of the binomial expansion. The next clear statement appears in the October 1666 Tract (see Whiteside, *Mathematical Papers*, I, pp. 122–34, 403, 410).

(10) In his *Algebra* (Oxford, 1685), p. 8, Wallis had printed both contemporary Arabic and Greek numerals, and had compared them with those older ones found in a manuscript of Sacrobosco, who claimed they came originally from Indian sources. Wallis, if he ever received Newton's comment, added nothing to his text as a result of it; see *Opera Mathematica*, II (1693), p. 10.

(11) John Baptista Tavernier, *The Six Travels of John Baptista Tavernier, Baron of Aubonne through Turky and Persia to the Indies During the space of Forty years* (London, 1684). Pages 15–214 of Part II are entitled 'Travels in India', and it is to p. 23 of this section which Newton refers. He has copied with extreme care the symbols there printed.

(12) Now lost.

(13) George Oxenden (1651–1703) became Regius Professor of civil law at Cambridge in 1684. He was Master of Trinity Hall from 1689 until his death; in 1692 he was Vice-chancellor of the University and from 1695 to 1698 he represented it in Parliament.

(14) Oxenden's father, Sir Henry Oxenden, lived near Wingham, about six or seven miles from Sandwich, and was M.P. for Sandwich.

(15) The death of Sir Robert Sawyer, who, together with Edward Finch, had been elected Burgess for the University of Cambridge in February 1690, necessitated a by-election in November 1692. Henry Boyle was returned on 21 November 1692. According to Henry Cooper's *Annals of Cambridge*, IV (Cambridge, 1842–1908), p. 19, Newton voted for John Brookbanks, and Oxenden was detained in London by illness, and was not admitted. At the next election, on 25 October 1695, Boyle and Oxenden were returned.

It is curious that Newton mentions the Vice-chancellor as having stood 'last summer', since the election was in November, and Oxenden apparently unable to stand because of illness.

⟨392⟩   NEWTON TO WALLIS

Winter 1692/3

This is differently dated in vol. III; see the correction noted above.

⟨464⟩   FATIO TO NEWTON

11 April 1693

Printed out of chronological order as Letter 464 (vol. III, pp. 391–2).

417   MEMORANDA BY GREGORY

30 June 1693

Note (10) (vol. III, p. 275): Hofwijck, Huygens' villa, was at Voorburg, then a village, now a suburb of the Hague. The house mentioned in the note was his winter residence, at no. 6 or no. 10 Noordeinde, the Hague. (A correction kindly supplied by the late Professor L. Rosenfeld.)

427   NEWTON TO LEIBNIZ

16 October 1693

On p. 286, line 1 (vol. III) the [s] in subtili[s] is redundant, and the correct translation (compare p. 287, line 13) of the sentence containing it is, 'But he seems [to me] to fill the heavens too much with some [kind of] subtle matter.' (A correction kindly supplied by the late Professor L. Rosenfeld.)

X.436 NEWTON TO ——

c. 1693

From the holograph draft in the University Library, Cambridge[1]

Mr Oughtred Clavis[2] being one of ye best as well as one of ye first Essays for reviving ye Art of Geometrical Resolution & Composition I agree with ye Oxford Professors that a correct Edition thereof to make it more usefull & bring it into more hands will be both for ye honour of our nation & advantage of Mathematicks.

Is. N.

NOTES

(1) Add. 3965(6), fo. 35v. The draft appears on a page of drawings of scientific instruments, reproduced as Plate IV, facing p. 176, in Hall & Hall, *Unpublished Scientific Papers*.

(2) William Oughtred's *Clavis Mathematicæ* was first published in Oxford in 1652; by 1693 it had reached its fifth Latin edition. But probably Edmond Halley's English translation of 1694 is referred to here. (See Whiteside, *Mathematical Papers*, I, p. 16, note (7), where the whole draft is also printed.)

457   NEWTON TO BUSWELL

?June 1694

Newton's draft, and Buswell's original letter, were sold again at Sotheby's on 29 October 1975 (Lot 142; see *Sotheby Catalogue* for the sale, p. 62).

532   HALLEY TO NEWTON

28 September 1695

Note (6) (vol. IV, p. 172): In the Letter Halley refers to the 'great comet' of 1680/1, while it is 'Halley's Comet' of 1682 which has a period of 75 years. (We are grateful to Mr Patri J. Pugliese for this emendation.)

538   NEWTON TO HALLEY

late October 1695

This letter is wrongly dated; it is a draft of Newton's Letter to Halley of 3 December 1724, printed as Letter 1449.

⟨559⟩   NEWTON TO THE TREASURY

25 May 1696

See correction to Letter 559 below.

550   THE MINT TO THE TREASURY

8 June 1696

This Memorandum was presented in person by Newton, who, together with other Mint Officers, was present at the Treasury on this date (see P.R.O., T/29, 8, p. 319). The business of the recoinage and the setting up of the Country Mints had been the subject of a number of earlier discussions at the Mint, at some of which Newton was present (*ibid.*, pp. 294–319). It appears that initially there were to be Mints at Warwick and Hereford; it is not clear when the five towns listed in note (2) were finally selected.

553   NEWTON TO THE TREASURY

July/August 1696

George Macey is mentioned as Newton's clerk in dealing with counterfeiters in a Treasury Minute dated 18 August 1696 (see P.R.O., T/29, 8, pp. 369–70, and compare vol. IV, p. 210, note (2).) Possibly Letter 553 is a reply to the Treasury's comments.

557   NEWTON AND NEALE TO THE TREASURY

10 December 1696

A draft by Newton, dated 1 December 1696, is in Mint Papers, I, fo. 194.

559   NEWTON TO THE TREASURY

? 1696

This Memorandum was laid before the Treasury by Newton on 25 May 1696 (see P.R.O., T/29, 8, p. 306), and was probably written on the same day.

X.560   THOMAS NEALE TO THE TREASURY

15 January 1696/7. Holograph original

Neale points out that although not much poor quality hammered coin is received for recoining, if inspection of the coin were to cease, bad coin would be brought in. The inspection and first melting of the coin is slowing down the recoinage, both at the Tower Mint and at the Country Mints; moreover the cost of the waste from this operation should be borne by the Crown. In London the weighing and inspecting of coin received could be done at the Exchequer, and Newton is willing that this should be so. In the country special personnel should be employed for the task. (P.R.O., T/1, 43, no. 14, fos. 61–2.) Newton has added a signed postscript, 'I have seen this Proposal & believe that if good bargains be made for ye Country Mints it will be for his Ma[jes]ties service. Is. Newton.'

X.566 NEWTON TO HALLEY

21 JUNE 1697

From a clerical copy in the Public Record Office[1]

To Mr Edmond Halley &c.

Sr.

Mr. Molineux Acquainting me that he had rec[eiv]ed a Letter from you importing [sic] a strong suspicion of some foule Play either among the Tellers or in the Melting house or both whereby the Money comes out worse than heretofore, We have sent downe a Teller[2] & desire you to employ him not only in telling but in inspecting the other Teller's & to Turne out those Teller's who either thrô unskillfullness or Corrupcon receive bad Money wch may easily be discovered by letting him examine some of the Money rec[eiv]ed by them

ISAAC NEWTON

London 21°. *June* 1697

NOTES

(1) Mint/10, 2, fo. 42v. This is one of a number of copies of letters from Newton concerning the setting up and operation of the Chester Mint which are to be found in P.R.O., Mint/10, 2. The Chester Mint went into production in October 1696, and closed down in the summer of 1698; it was one of several Country Mints established in order to remove clipped coin from circulation and replace it by new coin (see Letter 550, vol. IV).

In July 1696 officers for the Chester Mint were appointed: Robert Weddell as Warden, Thomas Clarke as Master and Worker, Edmond Halley as Comptroller, Peter Pemberton as Assaymaster, Francis Fosbrook as Surveyor, John Colins as Clerk and Edward Webb as Porter, all acting as deputies for their counterparts in the Mint in the Tower. (See Mint/10. 2, fos. 1–6; the first of these, appointing Robert Weddell, is signed by Newton.)

An extensive correspondence followed between the officers of the Chester Mint and those of the Mint in the Tower, some of which is already printed in vol. IV; see Letters 555, 562, 563, 570, 578, 582. Further letters concern troubles at the Chester Mint, and the difficulties experienced by the Mint in the Tower in extracting adequate accounts from Chester. A few of these letters are signed by Newton alone; these we print here and below (Letters X.574, X.577, X.584). The majority are signed by a number of Mint Officers, and these we have omitted except where Newton seems to have been particularly concerned in them (see Letter X.571).

At Mint/10, 2, fo. 72v is a copy of the letter (Letter 586, vol. IV) which Newton sent to all Country Mints concerning their closure. Two postscripts are added intended for the Chester Mint in particular, one by Thomas Neale, saying 'I approve this, I would have it punctually complied with', and one by Newton, saying 'I desire Mr. Halley to take particular care of this matter'. In June 1698 the Mint Officers at London sent the Chester Mint a request to wind up its affairs, and to send all its equipment up to London (Mint/10, 2, fo. 78r).

(2) The present letter is followed by one from Thomas Molyneux to Halley, naming the Teller sent to Chester as William Greenall. There had already been considerable feuding between officers of the Chester Mint (see vol. IV, pp. 320–1).

Greenall was eventually appointed permanently in place of Edward Lewis, the Comptroller's Clerk, who was found unable to render adequate accounts. But although Lewis was officially suspended and replaced by Greenall on 22 July 1697 (see Mint/10, 2 fol. 50r, 48v), it is not clear that he in fact then gave up his office (compare Letter 578, vol. IV, and Letter X.574).

X.571 THE MINT TO THE CHESTER MINT

?AUGUST 1697

From a clerical copy in the Public Record Office[1]

To the Officers of [the] Chester Mint

Gent[le]men

Wee are much concern'd to hear of ye continu'd quarrells amongst you at the Mint & the storys relating thereto are so directly opposite one to the other that all Wee can say to ye matter, till Wee hear both sides together, is, that Wee believe both sides much in the wrong, & resolve to come & hear it Our Selves

Butt that the publick business may not stopp in the mean time, Wee do hereby strictly charge & command you, that the melting & coynage goe on wthout any interuption, yt money to ye severall Importers be duly issued in course, or soone as pix'd and assay'd, & that ye same be actually paid ('till the Dep: Master returne) by Mr Lewis, & Mr Robinson wth whome the Masters Key will be trusted & You the Warden & Comptroller are to see the same be done accordingly.

Till Wee come let there be no further quarrelling, but let the publick business be peaceably carry'd-on, as it ought to be: for the Mint will not allow of the drawing of Swords, & assaulting any, nor ought such Language, Wee hear has been, be used any more amongst You

We are

Your Lov[in]g Fr[ien]ds

Is. NEWTON

THO NEALE

THO: MOLYNEUX

1697

NOTE

(1) Mint/10, 2, fo. 53v. Compare Halley's Letter to Newton of 2 August 1697 (Letter 570, vol. IV, p. 246) to which this may be a reply. Even after this strong tone had been adopted by the Mint at the Tower, the Chester Mint continued to be tardy with its accounts, and needed constant chivying from London. Another similar letter, demanding that quarrelling be stopped, was scr.t to Chester a little later (Mint/10, 2, fo. 63v).

X.574 NEWTON TO THE CHESTER MINT
?SEPTEMBER 1697
From a clerical copy in the Public Record Office[1]

To the Warden, Master, & Comptroller of the Mint
at Chester

Gentm

You are required to suspend the payment of all Moneys due or wch shall become due from your Mint to Mr Bowles & Mr Lewis, untill such time as the accounts of Mr Lewis shall be ballanced, & those betweene Mr Bowles and Mr Weddell Examin'd. I am

Your Loving Friend   Is NEWTON

Memorandum[2]

On Satturday 25. of Septr in presence of Mr Neale Mr Clark called for and commanded the Money I had stop't upon Bowles's acco.t and then pd it to Lewis's in Contempt of this Order

Test. ROBT WEDDELL

NOTES

(1) Mint/10, 2, fo. 60r. Compare Letter X.566, note (2), p. 400. In October 1697, the London Mint issued an ultimatum demanding that the Clerks come to London to render their accounts, on pain of suspension (see Mint/10, 2, fo. 63r). Lewis remained unable to render his accounts. Despite this, and an earlier attempt to suspend his salary in July 1697 (see Mint/10, 2, fo. 50r), Lewis seems to have continued in office until at least December 1697 (see Letter 578, vol. IV, p. 254). A little later he disappeared, and was dismissed on 2 February 1698 (see *Cal. Treas. Books*, XIII, p. 59).

(2) The Memorandum, in a different hand, has clearly been added later.

X.577 NEWTON TO HALLEY
25 DECEMBER 1697
From a clerical copy in the Public Record Office[1]

Edm. Halley
Lond. Decem[be]r 25th 1697

Sr

Since your Clerk[2] has finished his Day Book, I believe he may best be spared to come wth Mr Whitleys[3] money unless you had rather come with it your selfe, Take your choise. The Master & Worker appointed one the last time[4] & tis but reasonable that I should appoint one now. I am &c.

yours IS NEWTON[5]

NOTES

(1) Mint/10, 2, fo. 67r.

(2) Richard Morley; there was some confusion over who should bring the money, a task coveted for its financial recompense. Thomas Clarke, Master and Worker, wished to come with the money himself, but was out of town when Morley set out with it; Clarke hoped to overtake him on the way to London, to make it appear that he himself had accompanied the money (Mint/10, 2, fo. 67v).

(3) Morgan Whitley, Receiver for Cheshire and South Wales. The sum of money to be delivered to London was £8886. 15s. 00d.

(4) Thomas Williams, the Master and Worker's Clerk, had been appointed to the task on the previous occasion.

(5) A postscript, now illegible, follows.

X.584 NEWTON TO WEDDELL

10 MARCH 1697/8

From a clerical copy in the Public Record Office[1]

To Robt Weddell Warden of ye Mint at Chester

Sr

I understand that Mr Clark has order'd the Smiths fourge to be built under the Governours Lodgings wch must needs prove inconvenient & Offensive to him. Pray take care that it be remov'd from thence & built in some other more convenient place I am

Your loving Friend

Is NEWTON

NOTE

(1) Mint/10, 2, fo. 70v. The rebuilding of the smithy at Chester was also the concern of an earlier letter; see Mint/10, 2, fo. 69r.

⟨761⟩ NEWTON TO THE GOVERNOR OF CHESTER CASTLE

16 April 1698

Printed out of chronological order, as Letter 761 (vol. IV, pp. 543–4).

X.591.1 THE MINT TO THE TREASURY

18 August 1698. Clerical copy

Report on the possibility of erecting a Mint in Ireland. This is opposed on the grounds that it might 'draw Bullion and Coynage' from England, and thus prejudice trade and manufacture. 'Like the other plantations, [Ireland] is and ought to be inferior to this Kingdom and subservient to its interests.' (Mint Papers, II, fo. 216.) The signature has been torn out. The possibility of setting up a Mint in Ireland was discussed again in 1701; see Letter 643, vol. IV.

X.591.2 THE MINT TO THE TREASURY

20 August 1698. Clerical copy

Report on a bill from Robert Yates for his expenses at the Bristol Mint, referred to the Mint Officers on 6 May 1698. The Mint Officers consider his bill reasonable. (P.R.O., Mint/1, 6, fo. 22.) A copy of Yates' bill, amounting to £229, is at fo. 21v. A copy of the Treasury Warrant authorizing payment is at fo. 22.

X.593 JOHN BLACKWELL TO ?NEWTON

8 October 1698. Clerical copy

Blackwell, the Mayor of Bristol, reports on attempts to apprehend William Hulbert, James Field and Thomas Smith, all counterfeiters, who have been informed upon by John Bale. (P.R.O., Mint/15, 17, no. 102).

Mint/15, 17 is a volume formerly in the Royal Mint, entitled *Mint Depositions*, which covers the period 1697–1706. A few of the documents and letters in the volume are printed in vol. IV, but the extent of Newton's business in the prosecution of counterfeiters is not sufficiently emphasized there. (See Letter 553 and notes, vol. IV, pp. 209–10.) The majority of documents in Mint/15, 17, are depositions of information against counterfeiters, and a great number of these have been witnessed by Newton. The volume also contains a number of letters falling into three categories: letters from officials in provincial towns discussing the prosecution of coiners there; letters giving information against counterfeiters; and letters from counterfeiters asking for pardon, usually on the grounds of having informed against other counterfeiters. The letters are calendared in this Appendix; see Letters X.595, X.596.1–4, X.606.1–11, X.608.2–3 and X.613.

Bale's reports precede this letter at nos. 99 and 100. For Bale, see also Letter X.595.

X.595 BLACKWELL TO ?NEWTON

24 October 1698. Clerical copy

Encloses further information from John Bale on the counterfeiting activities of Thomas Bodily, William Trowman, John and William Harvey, Adam Clempson and William Johnson. (P.R.O., Mint/15, 17, no. 106.) The enclosures are at no. 105. For Bale, see also Letters X.593 and X.596.1; for Bodily, see also Letter X.608.1.

X.596.1 BLACKWELL TO NEWTON

12 November 1698. Clerical copy

Reports that further information from John Bale concerning Thomas Bodily has resulted in Bodily being arrested at Bath and committed to Chester Gaol. Adam Clempson has also been committed to gaol. (P.R.O., Mint/15, 17, no. 109.) Bale's information is at no. 108. Compare also Letter X.595.

X.596.2 WILLIAM NABB TO ?NEWTON

Late 1698. Clerical copy

Letter concerning information (enclosed) against counterfeiters, given by Francis Batho. (P.R.O., Mint/15, 17, no. 178.)

X.596.3   BLACKWELL TO ?NEWTON

28 November 1698.   Clerical copy

Reports information from Humfries and Henry Prewet that Field, a counterfeiter, has been seen in London. (P.R.O., Mint/15, 17, no. 109, [2].)

X.596.4   ISAAC WALLINGTON TO NEWTON

3 December 1698.   Clerical copy

Reports from Cambridge that one Sadler and his son-in-law are in Ely Gaol convicted of counterfeiting, but that 'the cheef of their Gangue one Thornton' is still at large. Wallington has sent Mr Gail (bearer of this letter) and Mr Woodham (Keeper of Ely Gaol) up to London to see if they can apprehend him. (P.R.O., Mint/15, 17, no. 110.) Compare Letters X.608.3 and X.613.

X.597   REPRESENTATION FROM THE MINT TO THE TREA-SURY

December 1698.   Clerical copy

The Mint Officers claim to have paid the officers of the Country Mints their expenses and salaries, and, due to the pressure of business in the Country Mints, they are now out of pocket. They are ready to render their account, but begg 'yt Wee may be admitted to pass the several yearly Acc[oun]ts in one, by reason they are so entangled one year with another, yt it is not possible to distinguish them'. (Mint Papers, II, fo. 475.) There is another clerical copy at fos. 495–6. The accounts of the Country Mints were not finally settled until 1703. Compare Letters X.633.2 and X.663.2.

X.606.1   THOMAS CARTER TO NEWTON

?Late January 1698/9.   Clerical copy

Carter and Chaloner have both been detained at the Fleet prison, but now Chaloner has been released after trial. Carter hints that he has information against Chaloner, which he will divulge if released. (P.R.O., Mint/15, 17, no. 130, [1].) Chaloner had been detained in prison several times on counterfeiting charges; it is not clear which particular release Carter refers to. The letter may be of somewhat earlier date. Compare vol. IV, pp. 232, note (3), pp. 259–60 and pp. 305–9.

X.606.2   CARTER TO NEWTON

?Late January 1698/9.   Clerical copy

Gives further information against Chaloner. (P.R.O., Mint/15, 17, no. 130, [2].)

X.606.3   JOHN IGNATIUS LAWSON TO NEWTON

?Late January 1698/9.   Clerical copy

Lawson writes from Newgate to beg for his liberty. He hints that he has evidence implicating Chaloner. (P.R.O., Mint/15, 17, no. 131.)

X.606.4   JOHN WHITFIELD TO NEWTON

25 January 1698/9.   Clerical copy

Whitfield, a counterfeiter, begs an interview with Newton. Clearly he hopes to win his freedom by informing on others. (P.R.O., Mint/15, 17, no. 129.) Presumably Whitfield was already in Newgate Prison; compare Letter X.606.5. A number of depositions and counter-depositions by Whitfield and others, informing against John Ignatius Lawson, Thomas Carter, William Chaloner, and other counterfeiters detained at Newgate, are at P.R.O., Mint/15, 17, nos. 115–26.

X.606.5   WHITFIELD TO NEWTON

9 February 1698/9.   Clerical copy

Begs Newton to come to the Dogg Tavern [at Newgate]. He writes 'wee are now at a stand by shamming lunaticks and the more for want of the honour of your presence for one $\frac{1}{2}$ hour'. (P.R.O., Mint/15, 17, no. 134.)

X.606.6   LAWSON TO NEWTON

9 February 1698/9.   Clerical copy

He has evidence to divulge, and requests to be examined. (P.R.O., Mint/15, 17, no. 199 [2].)

X.606.7   LAWSON TO NEWTON

13 February 1698/9.   Clerical copy

Encloses a paper giving evidence against a number of people who have been illegally re-edging clipped money, and also counterfeiting new money. (P.R.O., Mint/15, 17, no. 199, [3].)

X.606.8   LAWSON TO NEWTON

?February 1698/9.   Clerical copy

Writes that Chaloner is pretending sickness at Newgate Prison, but to no avail. He claims that the jurors at Chaloner's trials have been subverted. (P.R.O., Mint/15, 17, no. 132.)

X.606.9   CHALONER TO NEWTON

?February 1698/9.   Clerical copy

Claims he cannot engrave well enough to counterfeit successfully. (P.R.O., Mint/15, 17, no. 133, [3].) Compare Letter 606, vol. IV, pp. 305–6.

X.606.10   CHALONER TO NEWTON

?February 1698/9.   Clerical copy

Claims he 'never could work at the Goldsmith's Trade in [his] life'. He states that Patrick Coffee [against whom earlier depositions had been made] was responsible for the engraving. (P.R.O., Mint/15, 17, no. 174.)

X.606.11   CARTER TO NEWTON

?March 1698/9.   Clerical copy

Chaloner has been hanged, and his acquaintances are now trying to get Carter hung as well. He begs for liberty. (P.R.O., Mint/15, 17, no. 130, [2].)

607   THE MINT TO THE TREASURY

January/March 1698/9

The Treasury referred Crichlow's petition to the Mint on 17 January 1699, hence the present letter must have been sent in January. (See *Cal. Treas. Books*, XIV (1698–9), p. 243.)

X.608.1   ROKEBY TO NEWTON

21 MARCH 1699
From a clerical copy in the Public Record Office[1]

Judge Rokebys Letter to Isaac Newton
Esqr Warden of his majties Mint

Taunton 21 March [16]9⅘

Sr.

At my coming to this place last night I found yours of 14 instant wth ye Informacon therein inclosed & the letter from the Major of Bristoll[2] to you & this afternoon the major of Bath came to me & brought wth him an Informacon sent to him by the Major of Bristoll upon wch ye Mayor of Bath apprehended Boddily[3] & sent him to the Gaol of this County[4] where I found upon my Gaol Callender that he was Committed by the Mayor of Bath on suspicon of counterfeiting the Coyn of this Realme but the Major of Bath tells me that this Jo Bates[5] who gave the Informacon is removed from Bristol to Stafford he supposes to give evidence there for the King I have no evidence at all ag[ains]t Boddily in this County but it appears by Bates's Informacon that he can give evidence ag[ains]t him in Worcestershire. Therefore I send all the papers concernd [in] this matter inclosed herein to you and I purpose to leave Boddily in the Gaol here to be sent by Hab[eas] Corpus to the next assizes for Worcestershire that so you may cause him to be prosecuted there if you think fitt and if you have Bates in your power to give evidence ag[ains]t Boddily in Worcestershire and this is all I can do by Law to prevent Boddily escaping Justice I am

Sir

Your very humble servt

THO ROKEBY

NOTES

(1) Mint/15, 17, no. 160.

(2) John Blackwell; see also his earlier letters, possibly to Newton (Letters X.593, X.595 and X.596.1).

(3) Thomas Bodily (see Letters X.595 and X.596.1).

(4) At Chester.

(5) Possibly John Bale (see Letters X.595 and X.596.1).

X.608.2   LAWSON TO NEWTON

3 April 1699.   Clerical copy

Writes 'I am sorry you are incensed ag[ain]st me' and blames this on the false reports of others, which should be examined. Begs mercy. (P.R.O., Mint/15, 17, no. 199, [5].)

X.608.3   JOHN PYKE TO NEWTON

3 April 1699.   Clerical copy

Written from Cambridge. Thomas Sadler and his wife, committed to prison for counterfeiting (to which charge they have pleaded guilty) have given evidence against others at a court in Bury.

This evidence was judged entirely false, hence they cannot claim the King's mercy for divulging it. The original evidence against the Sadlers must be made available at Ely Assizes, where they are shortly to be tried. (P.R.O., Mint/15, 17, no. 289.) Compare Letter X.596.4.

X.610.1 THE MINT TO THE TREASURY
8 APRIL 1699
From the original in the Mint Papers[1]

To the R[ig]ht Honble the Lds Commiss[ione]rs of His Majtys Treasury

May it please yr L[ordshi]ps

Wee have considered the Petitions[2] annexed of Mr. Wallis & Mr. Bovey, & belive that their Presses may be used in Coynage and tho' one or two Presses might perhaps be safely licensed, yet they may draw on more Petitions, and after the precedents of licensing these presses, it may be more difficult to refuse the rest: and it may be of ill-consequence to license too many. But Wee humbly submitt the Whole to your Lordshipps Wisdome

Tower Mint Office
April 1699

Is NEWTON
THO NEALE
THO: MOLYNEUX

Mint Office Apr 8. 1699[3]   1699

NOTES
(1) I, fo. 458. The letter is in a clerical hand, but has been signed by Newton.

(2) The petitions precede at fos. 454–7. Gerard Bovey required a press for making tin buttons; Jacob Wallis for making 'Hooks and Chains for watches'. Compare Letter 607, vol. IV, pp. 306–7.

(3) The date has been added by Newton.

X.610.2 THOMAS FOWLE TO THE MINT
27 April 1699. Copy

Reports on the pulling down of the Smith's house at the Mint. Fowle has been at the Mint almost 27 years, and remembers conversations between Mint Officers, now deceased, that indicate that although the Smith to the Ordnance sometimes worked in the Smith to the Mint's house, the house nonetheless belonged to the Mint, not the Ordnance. (Mint Papers, III, fo. 415.) Compare Letter X.612, note (3), p. 411.

X.610.3 THE MINT TO THE TREASURY
May 1699. Clerical copy

Report on a petition from the Moneyers, referred to the Mint by the Treasury on 11 May 1699. The Moneyers ask for additional payment of £587. 18s. 5d for expenses incurred since March 1696. The Mint Officers suggest they be paid £400. (P.R.O., Mint/1, 6, fo. 63v.) The Moneyers' petition is copied at Mint/1, 6, fo. 63r, but without their account. Compare Letter 610, vol. IV.

X.612 NEWTON AND NEALE TO THE TREASURY
16 JUNE 1699
From a clerical copy in the Public Record Office[1]

To the Rt. Honoble the Lords Com[missione]rs
of his Ma[jesty']s Trea[su]ry

May it please your Lordps.

The Mint being the Treasury of ye Nation, that it might be well guarded on all sides, was placed in the Tower between ye Lines, wth a gate at each end, & a porter in that Gate of ye Tower wch answers to the Street of ye Mint. But the Guarrison & Office of Ordnance[2] have of late years prest into it, & by degrees possess'd themselves of severall places in it, & mingled with us, and being too strong for us, continue still to crowde us out, whereby ye custody of the gold & silver is rendred unsafe, to our hazard & ye discouragem[en]t of Importers and ye Mint is brought into disorder & ye Officers thereof grow dayly more & more weary of liveing in it, & by degrees desert it, to the neglect of ye King's businesse, and begin to want houses; And particularly the Office of Ordnance haveing for some yeares employed our Smith, & built themselves conveniencies at his shopp, & perhaps done some repaires to his house, they have now without our leave pull'd downe the whole house,[3] being a large place 90 foot long, in the very middle of ye Mint, & are laying ye Foundations of a new house there, as if the place were their own, Our Smith, (as wee suspect) trecherously prompting them to it for his own ends; If these things be allowed (besides that it is a Nusance) it will be impossible for us to guard our gates, or guard ourselves from ye Guarrison or reduce the Mint any more to good order, the want of wch has dammaged the King & the Moniers more thousands of pounds in the late recoynage of ye hammer'd monies than all ye houses taken from us are worth, & caused ye moneys to be wors coyned than it would otherwise have been. These things wee most humbly lay before yor Lo[rdshi]ps to be regulated, as shall appeare most reasonable & most for ye Service of ye king & Governm[en]t praying that ye Officers of ye Ordnance least they mispend the King's money, may have speedy notice to desist from building in ye Mint, till matters be adjusted, wch wee desire may be speedily done, & hopeing wee may know our bounds & shutt our gates, the Mint being

placed in a Guarrison not to be broke open & invaded, but better guarded.
All wch is most humbly submitted to your Lordps.

Is NEWTON
THO. NEALE

Mint Office, June ye 16th, 1699.

NOTES

(1) Works/3, 1, fos. 12v–13r; printed in *The Wren Society,* **18** (1941), 151–2. The Treasury referred Newton's letter to the Officers of the Works (see the Treasury reference, dated 21 June 1699, at Works/3, 1, fo. 13r). The Officers of the Works then arranged a meeting between themselves, the Mint Officers and the Ordnance Officers, which took place on 12 September 1699. We have found no record of the outcome.

At Mint Papers, III, fos. 416–17, is a draft, not in Newton's hand, of a Memorial dated 17 June 1699 from the Mint Officers concerning their privileges, and the abuse of them by the Ordnance. It is similar in essence to the letter we print here, but in greater detail.

(2) For earlier trouble with the Ordnance Officers, see Letter 569, vol. IV.

(3) For Thomas Fowle's letter to the Mint reporting the pulling down of the Smith's house see Letter X.610.2. Newton's notes on the employment of a Smith by the Mint since 1643 are at Mint Papers, III, fo. 413. (Compare also Letter X.617.) Copies of documents concerning the use by the Mint of buildings and space in the Tower are at Mint Papers, III, fos. 418–23.

In defence of the Ordnance, it must be said that the Mint buildings at this time seem to have been in a state of disrepair (see Letter X.626.1). Since repair of buildings and payment of Mint salaries all had to come from the same fund of £3000 p.a., attempts were made to cut down on salaries in order to pay for building costs. (See for example Letters X.633.6 and X.643.2; compare also Letter 726, vol. IV.)

X.613   ROGER JENNINGS TO NEWTON

30 June 1699.   Clerical copy

Written from Ely. 'I rec'd yours of the 24th with Mr. Mountagues inclosed.' Encloses information from Thomas Sadler against William Thornton and Henry Holder. (P.R.O., Mint/15, 17, no. 290.) Compare Letter X.596.4.

X.616   FOWLE TO NEWTON

?1699.   Clerical copy

A petition. Thomas Fowle entered employment at the Mint in July 1672. He reports on earlier Wardens: Sir Anthony St Leger, Thomas Wharton, Sir Phillip Floyd, the Earl of Denby, Dr Owen Wynn, and Benjamin Overton. All of these treated the post more or less as a sinecure; Newton differs from them in this. He reports on precedents for the enforcement of the Mint's privileges in regard to the arrest of Mint Officers. Fowle himself has now been arrested. (Mint Papers, I, fos. 21–3.) For Fowle, see also Letter X.610.2.

⟨762⟩   NEWTON TO THE GOVERNOR OF CHESTER CASTLE

23 November 1699

Printed out of chronological order as Letter 762 (vol. IV, pp. 544-5).

X.617   OFFICERS OF THE ORDNANCE TO ?NEWTON

2 DECEMBER 1699

From a copy in Newton's hand in the Mint Papers[1]

Sr

By direction of the Rt Honble Henry Earle of Romney Master Gen[era]l of his Ma[jesty']s Ordnance This is to acquaint you that his Lordp will give an answer to your Report concerning ye House belonging to ye Smith of this Office in the Mint when it shall be laid before the Treasury and sent by their Lordps to him,[2] and for ye Title he has so clear a one, that there can be no dispute thereof the possession wch can be fully proved having bin time out of mind in the Office
We are

Your humble servants
C MUSGRAVE

JON CHARLTON

WM BOULTER

Office of Ordnance
2d *Decr* 1699

NOTES

(1) III, fo. 412. Compare Letter X.612, note (3), p. 411.

(2) It seems that Newton had sent a report direct to the Ordnance Officers, who felt it should have come through the intermediary of the Treasury. Possibly this was the detailed memorandum referred to in Letter X.612, note (1), p. 411. Presumably the report was sent, as an extract of a letter dated 13 February 1699/1700, from the Ordnance to the Treasury, is at Mint Papers III, fos. 428-30, where the Ordnance Officers reiterate their claim to the Smith's house, and add that the Mint Officers in any case have more room than they need in the Tower.

X.619.1   NEWTON TO ——

?1699

From a holograph draft in the Bodleian Library, Oxford[1]

Sr

I have lookt into De Omerique's Analysis Geometrica[2] & find it a judicious & valuable piece answering to ye Title. For therin is laid a foundation for restoring the Analysis of the Ancients wch is more simple more ingenious & more fit for a Geometer then the Algebra of the Moderns. For it leads him more easily & readily to the composition of Problems & the Composition wch it leads him to is usually more simple & elegant then that wch is forct from Algebra.

NOTES

(1) Bodleian New College MSS. 361, II, fo. 14v.

(2) Antonius Hugo de Omerique, *Analysis Geometrica, sive Nova et Vera Methodus Resolvendi tam Problemata Geometrica quam Arithmeticas Quæstiones. Pars Prima de Planis* (Cadiz, 1698). The book is listed in the Musgrave catalogue; see De Villamil, p. 89.

X.619.2 NEWTON TO THE TREASURY
?EARLY 1700
From a holograph draft in the Mint Papers[1]

In obedience to your Lordps order of reference of ye 28th of Nov. 1699[2] we have considered ye annexed Petition of the Governour & Company of ye East Indian Merchants for procuring by your Lordps leave two Mills & two Presses wth ye Utensils there unto belonging to be sent to Fort St George for facili[t]ating their coynage there: and upon enquiry we find that the said Merchants by their Charter?[3] have a Mint at ye said Fort for coyning Pagodes Rupees & other East Indian money by the Hammer & it is represented to us that their money so coyned is not so fair & specious as the like money coyned by the Dutch & for that reason not so much coveted by the natives, for wch reason & for making dispatch in their coynage the said Merchants desire Mills & Presses as they represent.[4] We find also that by a late Act of Parliamt mentioned in ye said Petition, it is made High Treason for any person to make, procure or knowingly to have in his custody any Press for coynage or Cutter for cutting out ye Blanks by force of a skrew unless by your Lordps License but ye making & having of Mills is not forbidden nor requires your Lordps Licence. We do not find any barr to the erecting of a mint by ye mill & Press at Fort St George by ye said Merchants excepting ye said Act of Parliament wch barr your Lordps have power to remove by giving them leave to procure two Presses & Cutting Engins. But we humbly conceive that ye opinion of ye Council of Trade should be first had in this matter least the erecting of such a Mint should promote the exportation of money out of England or hinder ye importation of Gold & Silver into his Majties Mint in this kingdom & also yt the opinion of ye Attorney Gen[era]l should be had concerning the legalness of their Mint & the form of a licence, & lastly that sufficient security be given to your Lordps that the instruments licensed by your L[ordship]s be not put to any other use then ye coyning of East Indian money at Fort St George & for yt end be shipt away to ye said Fort upon their delivery from ye Tower

NOTES

(1) I, fo. 509. This is apparently the later of two drafts concerning the Mint set up by the East India Company at Fort St George, Madras. Compare Letter 618, vol. IV; we assume this letter was written shortly after its receipt. The earlier draft (Mint Papers, I, fo. 507) gives more detail about the legal prohibitions against the operation of such a Mint, but only in the draft we print here are to be found the closing sentences concerning the flow of coin and bullion to and from England.

(2) See *Cal. Treas. Books*, xv, 1699–1700, p. 219.

(3) Newton has inserted a question mark over the word 'Charter'.

(4) On 23 January 1700 the Treasury issued a Warrant to the Mint to 'sell to the East India Company two mills, two presses and two cutting engines to be sent to India for the service of their coining of pagodes, rupees and other species of money of that country.' (See *Cal. Treas. Books*, xv, 1699–1700, p. 258.)

X.626.1   THE MINT TO THE TREASURY

11 May 1700.   Clerical copy

The Press house and part of the Provost of Moneyers' house need rebuilding, at an estimated cost of £580. The Comptroller's house is in need of repairs estimated at £335. 10s. (P.R.O., Mint/1, 7, p. 6). The Tresaury referred this Mint report to the Officers of His Majesty's Works on 18 May, and on 25 May it was ordered that the repairs be carried out; see Mint/1, 7, pp. 7–8. A number of similar discussions concerning repairs required to Mint buildings are minuted at Mint/1, 7, pp. 6–20. Compare also Letter 726, vol. IV, and notes.

X.626.2   THE MINT TO THE TREASURY

25 May 1700.   Clerical copy

Report on Thomas Weddell's account of his expenses incurred in prosecuting Joseph Horton, a counterfeiter. (P.R.O., Mint/1, 7, p. 8.) For Weddell see vol. IV, note (3), p. 386, and compare Letter X.645.2 below; for Horton see Letter 762, vol. IV, pp. 544–5.

X.630.1   THE MINT TO THE TREASURY

c. January 1700/1.   Clerical copy

Report on George Bannister's petition. Bannister has asked for £110. 6s for his expenses as Deputy Warden at York Mint. The Mint Officers do not agree that he should be granted the whole amount, but leave it to the Treasury to make a decision. (P.R.O. Mint/1, 7, p. 24.) Bannister's petition is at Mint/1, 7, p. 14. Compare also Letter X.633.2.

X.630.2   NEWTON AND STANLEY TO THE TREASURY

c. January 1700/1.   Clerical copy

The Treasury has asked if the public receivers are cutting all counterfeit coin received, as they are bound to do by Act of Parliament. The Mint Officers reply that this *is* done at the Exchequer, the Custom House, and the Excise and Stamp Offices, but at the Bank of England the receivers merely return counterfeit coin. Also in the country receivers do not cut counterfeit coin as they should, which may explain the prevalence of counterfeiting there. (P.R.O., Mint/1, 7, p. 25.)

X.631.1   NEWTON AND STANLEY TO THE TREASURY

22 JANUARY 1700/1

From a clerical copy in the Public Record Office[1]

To the Right Honble: the Lords Comm[issione]rs
of His Majestys Treasury

May it Please Your Lordships

Pursuant to your Lordships Order of Reference of the 16 December we have examined the Peticion[2] and Bill of Mr. Thos. Birdikin Deputy Warden of His Mats. Mint at Exon[3] wch is hereunto annexed and as to the first Article of the Bill wherein he prays the allowance of £22:10 for Lodgings in the Mints so far as there was Room for them but where there wanted Room the Officers provided Lodgings for themselves in other places without being hitherto allowed anything by the king for the Same.

The Second Article where he prays allowance of £30 for the Quarters Salary from Midsummer to Mich[ael]mas 1698. we humbly Conceive to be reasonable because He and the Kings Clerk after the other Officers were dismissed staid at Exeter with the Deputy Master to assist him in Coining some Moneys after Mid Summer & to take Care of makeing up some of the Kings Sweep of the first melting & principally to Quiet the Importers some of which were not paid off till Michaelmass.

As to the third Article where he prays £46 allowance for his pains in attending above 131 days in the melting house above 20 hours each melting & for his Expences in Encouraging the Workmen & Charges in Letters about the Mint Business: We find that when the Melter of that Mint demanded an allowance for the Wast in the first melting of the Hammered Money which the Petitioner believed too much, the Petitioner undertook to oversee the melting of about 114000 Cwt of Silver and took the Sweep into his Owne Custody for

4¹5

the King & took care to see it made up giving a Constant attendance so that Tricks might not be playd to Increase the Waste of the Meltings or diminish the Profitt of the Sweep. And this Attendance he gave at the Meltings twice a Weeke & sometimes thrice a Weeke about 20 hours a day for above a year And by this his care and Diligence he Saved 220£ to the King out of the allowance wch. would otherwise have been made to the Melter for the same. For this service therefore considering it was above his Duty & for his Charge in Encouraging the Workmen during this service & for the Charge of Letters during the whole Business of that Mint, we humbly Conceive this Article of the Peticion to be reasonable.

All wch. is most humbly submitted to your Lordps: great Wisdom

J: STANLEY
IS: NEWTON

Mint Office
Jan: 22: 17$\frac{00}{01}$.

NOTES

(1) Mint/1, 7, p. 27.

(2) A copy of Birdikin's petition is at Mint/1, 7, p. 26. He asks for a total of £98. 10*s*; that is, £22. 10*s* for lodgings, £30 for salary, and £46 for attendance at the Melting House. A Warrant to Newton, dated 7 February 1700[/1], authorizing payment of £76 only, is in the British Museum, Add. 5751, no. 101, fo. 339.

(3) The Exeter Mint was one of five Country Mints which began operation in the late summer of 1696 (see Letter 550, vol. IV). All were closed down about two years later, but the settlement of their accounts was not complete until April 1703 (see Letter X.663.2). When the accounts came to be rendered, there was considerable evidence of embezzlement and corruption amongst the Mint Officers, but Exeter was the only one of the mints that closed with its balance in credit.

X.631.2 NEWTON TO THE TREASURY

?LATE JANUARY 1700/1

From a holograph draft in the Mint Papers[1]

To the Rt Honble the Lords Comm[ission]ers of his Majties Trea[su]ry

May it please your Lordps

In obedience to your Lordps commands that we should give our opinion whether Pistoles ought to be valued in England by comparing them with our new coynd monies of due weight & fineness or by comparing them with the current cash of the nation wch by wearing is grown lighter then standard we have considered the matter & humbly represent that the Par of Exchange

416

between the monies of two several nations is to be adjusted by considering the monies of both as Bullion & comparing them by the weight and assay without any regard to the stamp of the monies of either nation. But in valuing monies in the Markets of one & the same nation (wch is the present case) the money of that nation having there a currency by it[s] stamp is not to be considered as bullion but valued at what it was coyned for[,] altho by wearing it be grown lighter then standard. And the monies of other nations wch have not there a currency by their stamp are to be considereed as bullion & valued only by their weight & fineness. Pistoles therefore in ye markets of England are to be considered as bullion & valued by comparing them with English money of due weight & allay wthout making any allowance or abatement for ye lightness of ye coin by wearing

In our last memorial we valued Pistoles one wth another at 17s. 1d or thereabouts. This value we gathered from their weight & assays finding them one wth another to weigh 4 dwt $7\frac{1}{4}$ gr & to be worse then standard $\frac{1}{3}$ of a grain or thereabouts. But by many assays made ye last week & this we find more exact that Pistoles one wth another are $\frac{1}{2}$ a grain worse then standard & weigh only 4 dwt $7\frac{1}{8}$ gr as nearly as we can determin. According to wch recconing they are one wth another worth 17s & $\frac{7}{15}$ parts of a penny.

All wch we offer to your Lordps consideration wth most humble submission to your great Wisdome.

NOTE

(1) II, fo. 142; there is a copy at P.R.O., Mint/1, 7, p. 28. There is another, rougher draft at fo. 148, probably for the same letter, but with considerably different wording and stating that the value of the pistoles is 'between two limits 17s & 17s. $0\frac{1}{4}d$' and that this 'is their true value as nearly as I can reccon'. These drafts were probably written shortly after Letter 631 (vol. IV, pp. 352–3), but before 31 January 1700/1, when Newton wrote a paper entitled 'The value of French Pistoles stated' (Mint Papers, II, fo. 152) where he suggested, after much crossing out, a value of 17s or, if the worst coins be included, 16s. $11\frac{3}{4}d$. Compare Letter 631, note (3) (vol. IV, p. 353).

X.633.1   THE MINT TO THE TREASURY

2 April 1701.   Clerical copy

Report on a memorial from Thomas Weddell, requesting his salary of £30 for quarter ending Michaelmas 1698. (P.R.O., Mint/1, 7, p. 28.) For Weddell see vol. IV, note (3), p. 386.

X.633.2 NEWTON AND STANLEY TO THE TREASURY
19 APRIL 1701
From the original in the Public Record Office[1]

To the Rt Honble the Lords Comm[ission]ers of his Ma[jesty]s Trea[su]ry

May it please your Lordps

The Petition hereunto annexed of several Officers and Clerks of the five late Mints in the Country in behalf of themselves & others being referred to us by your Lordps Order of 8 Sept 1699, we deferred to make a Report upon the same untill ye Accompts of the said Mints were stated least in making up those Accompts we should want their assistance after we had fully dismist them by our Report. But those Accompts being now in the Auditors hands we most humbly lay before your Lordps that we have considered the said Petition & find that the said Petitioners have not been allowed any thing for the charge of their journeys in going to the several Mints & returning back again, & that they continued there two years or something above without any other allowance or recompence then their bare salaries except Mr Burdikin[2] Mr Banister[3] & Mr Weddell[4] & that their business proved far greater then was expected & that they were paid their salaries only till Midsummer 1698 but staid in the said Mints in his Majts service two or three months longer & then attended at the Tower about making up the Accompts of the Country Mints. For wch services Mr Burdikin Mr Banister & Mr Weddell have lately upon their several Petitions been allowed their Salaries & we are humbly of opinion that the rest do equally deserve the like allowance, vizt

	£	s	d
Mr Annesley[5] for the Qter after Midsummer for himself	30.	00.	00
& for his clerk John Rowe	10		
Mr Walter[6] for the same Qter	12.	10.	00
Mr Spicer[7] for ye same	12.	10.	00
Mr Reynel for the same	12.	10.	00
Mr Tewley[8] for the same	12.	10.	00
Mr Banister for his Clerk Henry Marshal for the same Quarter	10		
Mr Morley[9] for the same	10.	00.	00
Mr Combes for the same	10		
	120.	00.	00

We also lay before your Lordps that Mr Reynel Deputy Kings Clerk attending at Exeter Mint after all the other Clerks had quitted their Posts made up the

Books wch were left deficient by the other Clerks & so soon as the business of that Mint was over, a debate arising concerning a deficiency of about 500 *li*[10] in the Accompts of Mr Israel Hayes Comptroller of that Mint, the said Mr Reynel by order of the Officers of his Majts Mint in the Tower came from his house neare Exeter to London to assist the said Officers in stating the said Accompts of Mr Hayes & by reason of the delatoriness of Mr Hayes he staid in London attending on that business till about ye 19th of December, for which services & the charges of his journey hither & back again to Exeter we are humbly of opinion that the said Mr Reynel may deserve the further summ of 20*li*

And we further lay before your Lordps that the coynage Duty is so limited by Act of Parliament as not to extend to these services, but there is money imprest to Mr Neale & now remaining in the hands of his agent Mr Fauquier out of which these services may be rewarded by your Lordps Warrant to ye Auditors to allow the same now in Mr Neales Acc[oun]ts now in the Auditors hands

All which is most humbly submitted
to your Lordps great wisdome
J STANLEY
Is NEWTON

Mint Office 19th Aprill 1701

[11]Mr Woo[d]noth Comptroller of Bristoll Mint came from thence at Midsummer to visit his sick Father but left his Clerk there to act for him & attended many days at ye Tower about the Accts & other business of Bristoll Mint for wch service he may deserve for himself half ye Michaelmas Quarter's salary vizt 12*li* 10*s* & for his Clerk Mr Calverly 10*li*.

NOTES

(1) Mint/10, 18. The letter is in Newton's hand. On the same page is a copy of the petition referred to, dated 2 April 1699. Not all the signatories of the petition are included in the list of salaries to be awarded. The payments Newton suggests were made (see the record of payment on 29 April 1701, at P.R.O., Mint/1, 6, p. 68 and also *Cal. Treas. Books*, XVI (Part II), 1700–1, p. 253). Compare also Letters X.597 and X.663.2.

(2) Thomas Birdikin, Deputy Warden at the Exeter Mint; for his earlier petition for payment see Letter X.631.1.

(3) George Bannister, Deputy Warden at York Mint. For the Mint report on his petition see Letter X.630.1.

(4) Thomas Weddell, Deputy Warden at Chester Mint; for the Mint report on his petition see Letter X.633.1.

(5) Archibald Annesly, Deputy Warden at the Bristol Mint.

(6) Charles Walter, King's Clerk.

(7) Edward Spicer, Weigher and Teller.

(8) Michael Tewly, King's Clerk.

(9) Richard Morley, Weighing Clerk.

(10) Compare Letter X.643.3.

(11) By a mark in the margin, Newton has indicated that this paragraph is to be inserted before the final paragraph of the letter.

X.633.3 THE MINT TO THE TREASURY

c. April 1701.   Clerical copy

Report on Benjamin Overton's memorial for his own and his clerk's expenses in prosecuting counterfeiters. (P.R.O., Mint/1, 7, p. 30.) For Overton, see vol. IV, note (2), p. 196.

X.633.4 THE MINT TO THE TREASURY

30 April 1701.   Clerical copy

Report on a petition from the Mine Adventurers. The Mint Officers recommend that they be allowed to engrave the Arms of Wales on money coined from silver taken from Welsh lead mines, since there is a good precedent for this kind of distinctive marking. (P.R.O., Mint/1, 7, p. 31.) The Warrant, dated 19 May 1701, granting this request, is at Mint/7, 84.

X.633.5 NEWTON TO THE TREASURY

30 APRIL 1701

From a copy in the Public Record Office[1]

To the Right Honble the Lords Comm[issione]rs of His Majestys Treasury

May it Please Your Lordships

At the foot of my Acc[oun]ts as Warden for prosecuting Clippers and Coyners, wch: are now fully past and determined there remains due to me from his Majesty the sum of 169:10:07¾ which sum I most humbly Pray Your Lordships to allow me out of the Moneys Imprest to Mr. Neale and now remaining in the hands of Mr. Fauquier, by issuing Your Lordships Warrant to the Auditors to allow the same in Mr. Neales Acco[un]ts now in their hands

All which is most humbly submitted

Is: NEWTON

30*th*: *Aprill* 1701

NOTE
(1) Mint/1, 7, p. 32. Compare Letter 617, vol. IV, and notes.

X.633.6   NEWTON AND STANLEY TO THE TREASURY

30 April 1701.   Clerical copy

Report on a petition from John Braint, Provost of the Moneyers. Braint complains that
Newton continues to pay to Thomas Doyley, Braint's predecessor as Provost, the £100
p.a. due to the Moneyers. This is the result of a Royal Patent allowing Doyley, in
recognition of his services to the crown, to retain the salary even after the Provostship
had been transferred to Braint. Braint demands that he himself and the Moneyers
should also be granted a salary of £100 p.a. to be divided amongst them. The Mint
Officers report that they cannot afford to pay two salaries of £100 p.a. for effectively
the same appointment, and suspend both salaries awaiting further instruction from the
Treasury. (P.R.O., Mint/1, 7, pp. 33–4, including a copy of Braint's petition, which
had been referred to the Mint on 27 March 1701; see *Cal. Treas. Books*, XVI (Part II),
1700–1, p. 235.)

The Treasury replied on 15 May 1701 that Doyley's salary should be cancelled, but
made no mention of a salary for Braint and the Moneyers (Mint/1, 7, p. 35). A Warrant
awarding a salary of £100 to the Moneyers, backdated to their last salary payment, was
eventually issued by Godolphin and sent to Newton on 21 July 1702 (see the copy at
P.R.O., Mint/1, 5, fo. 34). For John Braint, see vol. IV, p. 331, note (1).

X.633.7   THE MINT TO THE TREASURY

7 May 1701.   Clerical copy

Report on a petition from Edward Courtney, Muster Master of the workmen in the
Exchequer. Courtney petitions for payments which the Mint Officers do not consider
his due. (P.R.O., Mint/1, 7, p. 32.)

⟨643⟩   NEWTON TO THE TREASURY

May or June 1701

Printed as Letter 643. See correction to Letter 643 below.

X.639   NEWTON TO ?GODOLPHIN

?JULY 1701

From a holograph draft in the Mint Papers[1]

May it please your Lordp

I herewith send your Lordp a Paper[2] of the steps by wch the Triall of the
Pix proceeds, together with an Extract of so much of the Indenture of the Mint

as relates to this Triall. And in anything in wch I can serve your Lordp I
should be glad to receive your Lordps commands, being

My Lord
Your Lordps most humble
& most obedient servant
Is NEWTON

NOTES

(1) I, fo. 236. The letter is clearly related to Number 639, vol. IV, dated ?July 1701, hence
we assign it a similar date, but the dating is by no means certain. It might almost equally well
concern the discussions of 1710 about a Trial of the Pyx for the Edinburgh Mint (see Letter
759, vol. V), or perhaps was sent to Oxford before his first experience, as Lord High Treasurer,
of a Trial of the Pyx.

(2) Printed as Number 639, vol. IV, from Mint Papers, I, fos. 228–9. There is another draft
at fo. 232.

641   NEWTON TO THE TREASURY

September 1701

The holograph draft at Mint Papers, I, fo. 121–2 bears the date 5 September 1701.
Other drafts are at Mint Papers, I, fos. 111 and 117–25. Copies of the petitions from
Haynes, Philip Shales, Charles Brattell, George Foord and Thomas Edwards for the
post of Weigher and Teller are at fos. 112–17.

643   NEWTON TO THE TREASURY

Late 1701

Lowndes referred the Memorial and Report (discussed in the letter) to Newton on
27 May 1701 (see P.R.O., T/27, 16, p. 349). Hence the letter was probably written soon
after this date, in fact it is probably the report mentioned in the Treasury Minutes (see
Cal. Treas. Books, XVI (Part II), 1700–1, p. 72) as having been read at a meeting of the
Lords on 6 June. Compare vol. IV, p. 382, note (1). Further holograph drafts, differing
considerably from the version printed, are to be found at Mint Papers, II, fo. 222 and in
the Bodleian Library, New College MS. 361, II, fo. 20v.

X.643.1   THE MINT TO THE TREASURY

December 1701. Clerical copy

The Mint Officers have been ordered to consider methods of preventing the clipping and
counterfeiting of coin. They recommend the continuation of Acts of Parliament already
passed concerning the matter, but with a number of amendments: (i) The death penalty
for passing counterfeit coin is too severe, and discourages juries from making convictions.

(ii) The £40 reward for convicting a counterfeiter should be paid by the Receiver General instead of by the County Sheriffs, who are frightened of being out of pocket. (iii) It should be made illegal to possess moulds for casting metal. (iv) All counterfeit money received should be cut and recoined, but if on cutting it prove good, its exchange value should be given. (P.R.O., Mint/1, 7, p. 37). Compare Letter 553, vol. IV, Newton's earlier letter concerning the conviction of counterfeiters.

X.643.2   THE MINT TO THE TREASURY

c. End of 1701.   Clerical copy

The three Roettiers, Engravers at the Mint, are paid £450 p.a. for engraving seals and £325 p.a. for engraving coin. All this has to come from the £300 p.a. allowed the Mint for salaries and repairs. The Mint cannot afford this, and since the £450 p.a. is nothing to do with the coinage it should not come from this source. Further, John Roettiers is the only one of the three left claiming the salary, and he has not served the present king as Engraver of either seals or coin. (P.R.O., Mint/1, 7, p. 35.) For the Roettiers family, see vol. IV, p. 241, note (2). The letter (undated) appears in a series of letters from 1701, but possibly it is itself of earlier date, since it seems curious that the Roettiers should still be receiving payment as engravers of coin, when they were discharged from this work in 1697.

X.643.3   THE MINT TO THE TREASURY

10 December 1701.   Clerical copy

Report on the petition of Thomas Tipping, who is trying to settle the accounts of the Exeter Mint. Israel Hayes, late Deputy Comptroller of that Mint, died indebted to the crown £1100. Tipping points out that this debt must be settled before the accounts can be closed. The Mint Officers suggest that in view of Hayes' services to the crown, the debt be discharged from public funds. (P.R.O., Mint/1, 7, p. 36.)

X.645.1   THE MINT TO THE TREASURY

18 March 1701/2.   Clerical copy

Report on petitions from Ralph Locke (Melter at the Bristol Mint) and Samuel Welford (Melter at Exeter) who ask for £277. 18s. 6d and £277. 7s. 0d, respectively, to cover their expenses. The Mint Officers recommend a somewhat reduced sum. (P.R.O., Mint/1, 8, p. 23.) A copy of Godolphin's Warrant, authorizing the payments recommended by the Mint, is at Mint/1, 8, p. 24.

X.645.2   THE MINT TO THE TREASURY

26 March 1702.   Clerical copy

The Mint Officers report that Weddell still has not received the £60. 13s. 6d due to him, and has petitioned again for the money. (P.R.O., Mint/1, 7, p. 9.) Compare Letter X.626.2.

X.645.3   JOHN BATCHELOR TO NEWTON

4 April 1702.   Original

Summons to a meeting of the Trustees of Archbishop Tenison's Chapel 'on Tuesday in Easter weeke at 9. in the morning precisely' to elect officers. (Mint Papers, II, fo. 73v.) Compare Letter 642, vol. IV, Letter 1235, vol. VI, and Letter 1384.

647   NEWTON TO GODOLPHIN

15 April 1702

Undated holograph drafts are at Mint Papers, III, fos. 340 and 341. Orders for medals were received up to June 1702 (see P.R.O., Mint/1, 8, pp. 16–20; one of these orders is printed as Number 648, vol. IV) hence the date of the letter must be later than June 1702.

X.648   SAMUEL GARTH TO NEWTON

?April 1702.   Original

Letter suggesting a Latin inscription for Queen Anne's coronation medal, 1702. (U.L.C., Add 3968(41), fo. 41v.) The letter is difficult to read, and the suggested inscription almost illegible. Compare Mint Papers, III, fos. 289 and 310.

⟨760⟩   NEWTON TO HALLEY

2 June 1702

Printed as Letter 760 (vol. IV, p. 543), but headed NEWTON TO BURMAN. See the correction to Letter 760 below.

X.649.1 NEWTON TO THE TREASURY

?JUNE 1702

From the holograph draft in the Mint Papers[1]

To the Rt Honble Sidney Lord Godolphin
Lord High Treasurer of England

May it please your Lordp

By Order of Council and of the late Comm[ission]ers of her Majties Treasury I coyned 300 Medals of Gold & 1200 medals of Silver for her Majesties Coronation the charge of wch amounts unto 959*li*. 19*s*. 4½*d*. I coyned also by your Lordps Warrants 518 Medals of Gold for the Honble House of Commons and 40 Medals of Gold (most of them double ones) for forreign Ministers & Persons of quality the charge of wch amounts unto 1525 *li.* 18*s.* 11*d.* The whole charge is 2485 *li.* 18*s.* 3½*d* the particulars of wch are set down in the Bill hereunto annexed.[2] And for the discharge of this Accompt I humbly pray your Lordps Order.

All which is most humbly submitted to
Your Lordps great wisdome
[ISAAC NEWTON][3]

Mint Office

NOTES

(1) III, fo. 336. Another draft is at fo. 338. The last batch of medals was ordered on 2 June 1702, so the letter must have been written shortly after this. Compare Letter 647 (which is wrongly dated; see the correction noted above).

(2) Not now there.

(3) The signature has been torn out in both drafts.

X.649.2   THE MINT TO GODOLPHIN

30 June 1702.   Clerical copy

Report on John Roettiers' petition. The Mint Officers recommend he be paid arrears of salary, both because of his straitened circumstances and in view of his past services to the Mint, but they cannot do this without further Warrant. (P.R.O., Mint/1, 7, p. 39.) For John Roettiers, see vol. IV, p. 241, note (2). Compare also Letter X.643.2.

⟨647⟩   NEWTON TO GODOLPHIN

?Summer 1702

Printed as Letter 647 (vol. IV, pp. 385–6), but wrongly dated. See above.

650   MINT TO GODOLPHIN

7 July 1702

There are a number of related, but considerably different, drafts in the Mint Papers. In particular a signed, holograph paper at Mint Papers, II, fos. 104–5, is dated 1 July 1702, and bears the title 'The value of Gold in proportion to silver in several parts of Europe.' Further drafts and notes are at fos. 63–4 and fo. 74.

X.650.1   NEWTON TO GODOLPHIN

JULY 1702
From a holograph draft in the Bodleian Library, Oxford[1]

To the Rt Honble Sidney Ld Godolphin Ld High Treasurer of England

May it please your Lordp

The Petition of Mr Cha Fryth for an allowance in his Accts now depending of 370. 8. 9 upon two Tickets of the M[aste]r & W[orke]r of Chester Mint wth ye Report of the Comm[ission]ers of Excise upon it we received 30 March 1702 & in obedience to the Order of ye then Lds Commers of ye Treasury upon them we have examined ye matter & humbly conceive the true state of it to be as follows

Mr Fryth imported into Chester Mint several parcels of hammered money before Lady day 1697 for all wch he received back 5s pr oz in new monies & endorsed all the Tickets before ye end of July following. In May he imported two other parcells for wch at 5s pr oz he was to receive back 807 *li* 10s & in part thereof received in August of Williams (Mr Neales Clerk) 400 *li* & afterwards of Lewis another Clerk 300 *li* more & then recconing wth Lewis concealed the 400 *li* pd by Williams & deducted only the 300 *li* & took Lewises Note for the remaining 507 *li*. 10s & endorsed the Tickets, whereas he should have deducted also the 400 *li* & taken a Note only for 107 *li*. 10s the true deficiency. And this misrecconing is ye grownd of ye Petition.

After these importations there were two others in July & August for wch at 5s pr oz Mr Fryth was to receive back 370 *li*. 8s. 9d. And in October November & December following he did receive of Lewis at 4 payments 500 *li*, wch makes up the aforesaid deficiency of 107 *li* 10s & pays off ye 370. 8. 9. due upon these last two Tickets for wch an allowance is now petitioned & leaves 22 *li*. 1s. 3d in his hands in part of the 8d pr oz wch he was further to be allowed. For the 500 *li* was paid out of the Treasury of ye Mint in satisfaction for silver imported by Mr Fryth & therefore ought to be deducted in his Accts from ye allowance of 5s 8d pr oz according to ye words of ye Act of Parliamt wch run thus. 'And all & every such Receivers General & Collectors in their respective Acc[oun]ts to his Majty shall be allowed the deficiencies occasioned by the recoyning of the said hammered money that is to say the difference between the summ of the hamm[ere]d money brought into ye Mint computed at 5s 8d an ounce & the summ in tail of the new money wch he or they do receive back from ye Mint for the same.'

Therefore instead of granting the Petition we are humbly of opinion that

Mr Fryth be further charged to her Ma[jes]ty wth 22. 1. 3 in his Acc[oun]t now depending. All wch is most humbly submitted to your Lordps great wisdome

NOTE

(1) Bodleian New College MSS. 361, II, fo. 21; a rough holograph draft of the case against Frith at fo. 30 gives further details. Charles Frith, a collector of excise in Cheshire, submitted a petition in June 1701 that he had delivered 1481 ounces of hammered silver money to the Chester Mint in 1697, for which he had had no reimbursement, and so could not clear his accounts (*Cal. Treas. Books*, XVI (Part II), 1700–1, p. 297). On 4 July 1701 Lowndes wrote to the Mint asking for their views on this petition (T/27, 16, p. 359) and clearly Newton's draft was in reply to this request. It was read at the Treasury on 4 August 1702 (*Cal. Treas. Books*, XVII (Part I), 1702, p. 68). We may presume Newton's draft, therefore, to have been written in July 1702, after Godolphin became Lord High Treasurer.

There is another draft at Mint Papers, II, fos. 476–7, differing considerably from that we print here.

Frith was apparently suspended from office on account of his debts, but in June 1703 the Treasury requested his reinstatement as the debts had by then been discharged (see *Cal. Treas. Books*, XVIII (Part II), 1703, p. 309).

X.650.2 THE MINT TO GODOLPHIN
5 AUGUST 1702
From a holograph draft in the Mint Papers[1]

To the Rt Honble Sidney Lord Godolphin
Lord High Treasurer of England

May it please your Lordp

In obedience to your Lordps Order of Reference of July 3 upon the annexed Petition of Mr Anthony Redhead[2] late Master & Worker of the Mint at Norwich, we have considered the matter and humbly represent that ye Petitioner for a debt of above 3300 *li* due to the Treasury of that Mint had an Extent taken out against his body & goods by Mr Neale about Michaelmas 1699 & was thereupon committed to Prison in Ludgate (as he represents) where he hath continued ever since. That we do not know what private engagements there were between Mr Neale & the Petitioner upon the setting up of that Mint nor how poor the Petitioner is But considering that he hath lain long in prison & that his estate if he hath any doth not appeare to us we beleive it reasonable that he should be set at liberty so soon as it can be done by a due method. For if her Majty sets him at liberty she discharges Mr Neale & his security from the debt. The Petitioner was Mr Neale's Prisoner & to set him at liberty belongs to Mr Neales Administrators & security & yet if they

427

set him at liberty they discharge him of the debt. In order to his liberty we are humbly of opinion that the Accts of the Country Mints wch lye in the Auditors hands be dispatcht with all convenient speed[3] & that Mr Neales security do attend the passing of them And when it appears how recconings stand between her Ma[jes]ty & them, we shall be ready to advise with the Queens Counsel about the next step for bringing things to an issue

All wch is most humbly submitted &c

NOTES

(1) II, fo. 485. A clerical copy in P.R.O., Mint/1, 7, p. 40 is dated 5 August 1702. No further action seems to have been taken until 1703; see Letters 657, 658 and 661, vol. IV. Redhead again petitioned for release in July 1704 (see P.R.O., T/1, 91, no. 52).

(2) For Redhead see Letter 657, note (3), (vol. IV, p. 397). His petition claimed that he was indebted to the Crown because Neale had sent him unjust and unskilful servants. See *Cal. Treas. Books*, XVII (Part I), 1702, p. 279.

(3) The accounts were not completed until April 1703; see Letter X.663.2.

⟨652⟩   NEWTON TO GODOLPHIN

20 August 1702

See below, addition to Letter 652.

X.651   THE MINT TO THE TREASURY

?October 1702.   Clerical copy

Thomas Weddell and others have petitioned for a reward (advertised in the *London Gazette* for 3 December 1700) of £100 for apprehending 'Carter alias Browne alias Williams', a counterfeiter (P.R.O., Mint/1, 7, p. 9). For Weddell, see vol. IV, p. 386, note (3).

652   NEWTON TO GODOLPHIN

16 October 1702

A holograph draft of this letter, with slightly different wording, and dated 20 August 1702, is at Mint Papers, I, fo. 65.

654   NEWTON TO ?LOWNDES

Late 1702

Sir John Craig suggested, in private correspondence with Dr J. F. Scott, that this letter should be dated 1717. He gave a number of reasons for this, the chief of which are as follows:

(1) The draft includes the phrase, 'if His Majesty please to give order', hence it must have been written during the reign of either William III (who died on 8 March 1701/2) or George I.

(2) The tin stock in William III's reign did not exceed the capacity of the 'Irish Mint'. In fact, the earliest discussions concerning the storage of tin at the Mint in the Tower seem to have taken place at the Treasury on 22 October 1703 (see *Cal. Treas. Books*, XVIII (Part II), 1703, p. 78 and compare Letter 667, vol. IV).

(3) Drafts on the back of the letter concerning the value of the guinea relate to one of two reports, respectively of 7 July 1702 and 21 September 1717. Hence the letter draft is probably of shortly before one of these dates.

Hence 1717 is the probable date. Compare Letter 1251, vol. VI.

656   NEWTON TO GODOLPHIN

1702

The letter was probably written shortly before 21 October 1702, on which date a Warrant was issued authorizing payment of £60 p.a. to Bull, together with piece rates for medals made. See *Cal. Treas. Books*, XVII (Part I), 1702, p. 370.

657   THE MINT TO GODOLPHIN

January 1702/3

See correction to Letter 661, below.

X.657   THE MINT TO GODOLPHIN

13 January 1702/3.   Clerical copy

Report on Dorothy Wynne's petitions for £49. 18s. 6d due to her husband Owen Wynn, late Warden of the Mint. The Mint Officers recommend that the sum be paid. (P.R.O., Mint/1, 7, pp. 40–1.)

658   NEWTON TO THE TREASURY

15 January 1702/3

See correction to Letter 661, below.

X.658   THE MINT TO GODOLPHIN

20 January 1702/3.   Clerical copy

Report on Anne Morris' petition for payment of her late husband's expenses for prosecuting counterfeiters. The Mint Officers recommend that £40 be paid. (P.R.O., Mint/1, 7, pp. 41–2.) The recommendation was accepted on 16 March 1703; see P.R.O.,

T/27, 17, p. 145, but Lowndes *then* asked Newton from what fund the money should be drawn. Newton's reply is Letter 660, vol. IV.

⟨663⟩   **ROYAL WARRANT**

10 February 1702/3

See correction to Number 663, below.

661   **THE TREASURY TO NEWTON AND OTHERS**

22 March 1702/3

The 'enclosed papers' referred to include Letters 657 and 658. Clerical copies of all these appear in P.R.O., Mint/1, 8, pp. 33–8; the originals are at P.R.O., T/1, 84, no. 29. Letter 657 is dated 15 January 1702/[3], and bears Newton's name as the sole signatory. Note (5) (vol. IV, p. 402): the drafts mentioned in the note are at Mint Papers, II, fos. 487 and 489.

663   **ROYAL WARRANT TO NEWTON**

1702/3

A clerical copy of the Warrant, dated 10 February 1702/3, is at P.R.O., Mint/1, 8, p. 31.

X.663.1   **COMMISSIONERS FOR PRIZES TO NEWTON**

21 April 1703.   Original

The Commissioners have received an order from Godolphin to pay Sir Cloudesly Shovel £1000 in Spanish coin. They now order Newton to make this money available. (P.R.O., Mint/18, 38.) Newton was able to provide the money out of the Vigo Booty. See Letter 665, vol. IV.

X.663.2   **THE MINT TO THE TREASURY**

24 April 1703.   Clerical copy

A long Mint report about the accounts of the Country Mints. At their closure there was a net deficit of £5289. 7s. 7½d outstanding. All the Mints were in debt except the Exeter Mint. (P.R.O., Mint/1, 8, pp. 39–45). The Country Mints had ceased operation in 1698; see Letter 586, vol. IV. The business of settling their accounts had been long drawn-out; compare Letter X.633.2.

X.664 DAVID GREGORY TO NEWTON

16 MAY 1703

From the original in the Bodleian Library, Oxford[1]

Much Honoured Sir

According to your desire I searched the publick Library here for Papias.[2] There is nothing of him to be found here save 6 or 7 lines *De quatuor Marijs* in Latin. This and all the other fragments that remain of him, are put together in Grabij *Spicilegium Patrum.* Vol. 11. lately printed here, from page 30 to page 35.[3] I am with all respect

Much honoured Sir
Your most humble and
most oblidged servant
D GREGORY

Oxon. 16 *May.*
    1703.

NOTES

(1) Bodleian New College MS. 361, ii, fo. 51v. The letter to which this is the reply is now lost.

(2) Papias of Hierapolis (probably second century A.D.) was one of the 'Apostolic Fathers' and an important secondary authority on the writing of the Gospels. His *Exposition of the Lord's Oracles* is quoted by Eusebius of Caesarea, but no work of his survives.

(3) Johann Ernst Grabe, *Spicilegium S.S. Patrum* (Oxford, 1698–9). The fragment Gregory refers to is given in full by Grabe (pp. 34–5) and recorded as 'ex Cod. 2397. Bibliothecæ Bodleianæ Anno 1302 & 1303 in Abbatia Osneyensi prope Oxonium scripto fol. 286. pag. 2.' The 'four Maries' referred to are Mary Mother of God, Mary wife of Clopas, Mary Salome wife of Zebedee, and Mary Magdalene.

665   NEWTON TO ?LOWNDES

16 June 1703

A Warrant from Godolphin to Newton, dated 22 March 1702/3, ordered that the Vigo Bay treasure (see vol. iv, p. 404, note (2)) be melted down in the presence of the Commissioners for Prizes. A further Warrant, dated 23 June 1703, authorized the selling of all the Vigo Bay plate remaining at the Tower. (See copies of both Warrants at P.R.O., Mint/1, 8, p. 32.) Records of public auctions of some of the remaining treasure are to be found at P.R.O., Mint/18, 39 and Mint/18, 40.

Gold, silver and copper medals were struck to commemorate the Vigo Bay capture (see vol. vi, Plate v and Mint Papers, iii, fos. 305–7). A number of medals were also struck from brass ordnance which was part of the Booty (see Letter X.672, below).

X.666   THE MINT TO GODOLPHIN

1 September 1703.   Clerical copy

Report on William Bond's petition for payment for the prosecution of counterfeiters. The Mint Officers recommend he be paid, but without naming a sum. (P.R.O., Mint/ 1, 7, pp. 42–3). See also Letter X.667.1. Bond's petition had been referred to the Mint on 12 August 1703 (see *Cal. Treas. Books*, xviii (Part ii), 1703, p. 368). He claimed to have detected £500–£600's worth of clipped coin.

667   NEWTON TO GODOLPHIN

30 October 1703

Further holograph drafts (dated 28 October) are in the Mint Papers, iii, fos. 506 and 508, and (dated 30 October) at fo. 473. Newton's notes concerning the letter are at fo. 494.

X.667.1   THE MINT TO GODOLPHIN

4 November 1703.   Clerical copy

In reply to a request from the Treasury to name a sum to be paid to Bond, the Mint Officers recommend £60. (P.R.O., Mint/1, 7, p. 43.) Compare Letter X.666, above. The Warrant authorizing payment, dated 30 November 1703, and the receipt, are in the British Museum, Add. 5751, no. 100, fo. 338.

X.667.2   THE MINT TO GODOLPHIN

9 December 1703.   Clerical copy

In reply to a petition from the Lords Commissioners for Trade and Plantations, the Mint Officers append a list (not now with the copy) of exchange rates for foreign coin in Her Majesty's Plantations in America. (P.R.O., Mint/1, 7, p. 45.) The petition is dated 23 November 1703 (copy at Mint/1, 7, p. 44). It was referred to the Mint on 2 December 1703; see P.R.O., T/27, 17, p. 86. See also Letter X.673.1, below. The table of rates is printed in *Cal. Treas. Books*, xix (Part ii), 1704–5, p. 224.

X.669 NEWTON TO GODOLPHIN

EARLY 1703/4

From a holograph draft in the Mint Papers[1]

May it please your Lordship

The Officers of the Mint being appointed by your Lordship to sell her Majties Tin & cause order the price thereof to be paid to a Receiver whom your

Lordp shall appoint & who shall account annually for the same: if your Lordship, for lessening the number of Officers & making the business more expedite & less chargeable, shall think it for her Majties service that the said Receiver be an Officer of the Mint, I am willing, as I am your Lordps Undertreasurer or Receiver for the business of the coynage, so by ye assistance of my Deputy & Clerk to undertake the charge of receiving & accounting for the price of the Tin. And as your Lordshp understands best what is fittest for her Majties service so I make this offer with absolute submission to your Lordps great Wisdom, being

My Lord
                                 Your

NOTE

(1) III, fo. 475. Compare Letter 669, vol. IV. Presumably the present letter is slightly later in date. By April 1704 the Mint and the Treasury were clearly ready to begin the sale of tin, and Lowndes ordered Ellis to insert a public notice in the *London Gazette* advertising the sale of tin at the rate of £3. 16s. 0d per cwt, merchant weight.

X.672   THE MINT TO GODOLPHIN

19 April 1704.   Clerical copy

Report on some 'unserviceable brass ordnance' remaining from the Vigo Booty, and so far unsaleable. Samuel Proctor has offered to sell this off for the Mint by coining it into small medals bearing the word 'Vigo' on one side and the Queen's effigy on the other. He asks for twenty-five per cent of the gross profit in excess of £70 per ton. (P.R.O., Mint/1, 7, p. 47.) A copy of the Treasury reference to the Mint, dated 28 March 1704, together with Proctor's Memorandum, dated 22 March 1704, is at P.R.O., T/27, 17, p. 348. At Mint Papers, II, fo. 344 is a note by Newton that Mr Eyres, a refiner of copper at Southwark, who believes that the brass ordnance will not sell for more than £60 per ton, will guarantee to deliver seven tons of fine copper worth £100 per ton for every 10 tons of ordnance. Clearly Proctor's offer is marginally more attractive.

673   NEWTON TO GODOLPHIN

15 May 1704

A further holograph draft of this letter, dated 8 May 1704, is at Mint Papers, III, fo. 448. Samuel White countered Newton's report by writing to Lowndes that Newton was unfit to judge of matters of this sort. See P.R.O., T/1, 91, no. 52. Discussions between White and Newton had taken place on 16 February 1704, and they met again on 4 July 1704 (see *Cal. Treas. Books*, XIX (Part II), 1704–5, pp. 10 and 40).

X.673.1   LOWNDES TO THE MINT

20 May 1704.   Clerical copy

Orders the Mint Officers to prepare a draft proclamation giving the rates of exchange of foreign coin in the Plantations. (P.R.O., Mint/1, 7, p. 45.) Compare Letter X.667.2.

X.673.2   NEWTON TO GUIDO GRANDI

26 MAY 1704

From the holograph draft in the Bodleian Library, Oxford[1]

Viro Clarissime D. Guidoni Grando Isaacus Newton salutem.

Accepi Librum[2] D. Viviani de Locis solidis ut et libros tuos in quibus[3] Geometrice demonstras Problemata Viviani[4] et Hugenij,[5] et pro tanto munere gratias ago quam plurimas. Geometriam Veterum adhuc florere et vestris eximijs inventis ac demonstrationibus auctam esse valde gaudeo. Hyeme præterita Librum[6] de rebus Opticis et origine colorum olim scriptum in lucem edidi cujus exemplar ad te mitto. Anglice scriptus est, et sub finem invenies libellum unum et alterum de rebus Mathematicis idiomate Latino, quorum gratia totum mitto, Utinam tanto Judici non displiceant. Vale

Londini. VII *Kal. Jun.*
         MDCCIV

Translation

Isaac Newton greets
the most famous Mr Guido Grandi.

I have received Mr Viviani's book about solid loci[2] as also your books[3] in which you demonstrate geometrically the problems of Viviani[4] and Huygens,[5] and for such a great gift I owe you the best possible thanks. I am extremely glad that the geometry of the ancients still thrives, and that it is improved by your skilful inventions and demonstrations. Last winter I published a book[6] written long ago about optics and the origin of colours, of which I send you a copy. It is written in English, but at the end you will find a couple of little treatises concerned with mathematics, which are in Latin, on account of which I send the whole book to you. May they not be found wanting by so great a judge! Farewell.

London, 26 May 1704

NOTES

(1) Bodleian New College MS. 361, III, fo. 235v; there is another draft in the Newton MSS. (Container 3, Folder 30) of the Frederick E. Brasch Collection at Stanford University.

Guido Grandi (1671–1742) taught mathematics in Florence from 1694 until 1700, when he became Professor of Philosophy at Pisa. In 1709 he became a Fellow of the Royal Society, and in 1714 Professor of Mathematics at Pisa. He defined, discussed and re-named a number of geometrical curves—the *versiera*, the *rodonea* and the *clelia*. His most important work was *Quadratura Circuli et Hyperbolæ* (Pisa, 1703).

(2) Vincenzio Viviani, *De Locis Solidis Secunda Divinatio Geometrica* (Florence, 1701).

(3) Guido Grandi, *Geometrica Divinatio Vivianeorum Problematum* (Florence, 1699) and *Geometrica Demonstratio Theorematum Hugenianorum circa Logisticam seu Logarithmicam* (Florence, 1701).

(4) Viviani's problem was to construct four equal-sized windows in a hemispherical dome so that the remaining area of the dome is quadrable. Viviani gave his solution to the problem in *Formatione e misura di tutti i cieli* (Florence, 1692) but without a demonstration.

(5) For Huygens' problems concerning the logarithmic curve see *Discours de la Pesanteur* (Leyden, 1690), pp. 176–80.

(6) Newton refers here to the first English edition of the *Opticks*, published in 1704, to which were appended his *Enumeratio Linearum Tertii Ordinis* and his *Tractatus de Quadratura Curvarum*. These were omitted from some later editions.

X.673.3   LOWNDES TO THE MINT

8 August 1704.   Original

This is the reference mentioned in Letter 674, vol. IV. (Mint Papers, I, fo. 140.) Petitions from John Fowler, William Parsons and John Croker are at *ibid.*, fos. 141–5.

674   THE MINT TO GODOLPHIN

23 August 1704

At P.R.O., Mint/1, 7, p. 48, is a clerical copy, dated 12 October 1704, apparently the final version of the letter. Newton's amendments in the draft of 23 August are not implemented, and the final section of the letter is slightly different. It includes the sentence, 'Mr. Le Clerc is a quick and skillfull graver butt we humbly desire more time to consider of filling the third gravers place.' Compare Letter X.678 and Letter X.684.

X.674 A PROPOSAL BY NEWTON CONCERNING ENGRAVING

8 SEPTEMBER 1704

From the holograph original in the Mint Papers[1]

A Proposal for regulating the Gravers Office[2]

That the Gravers do work from time to time in such tasks as the Master & Worker for ye time being shall appoint for making dispatch in the coynage of

money & Medalls according to her Maj[es]ties Orders. That wthout her
Majts Orders or your Lordps they shall dispose of no medals with her Majts
Effigies nor convey out of the Mint any Dyes or Puncheons for making such
Medals, nor sell such Medals at a higher rate then shall be allowed by the
Warden Master & Comptroller of her Majts Mint.

Sept. 8. 1704

Is. [NEWTON][3]

NOTES

(1) I, fo. 147.
(2) This is presumably the 'Question' referred to in Letter 676, vol. IV, p. 419.
(3) The signature has been partly torn out.

⟨674⟩   THE MINT TO GODOLPHIN
12 October 1704

See correction to Number 674 above.

X.678   THE MINT TO GODOLPHIN
15 November 1704.   Clerical copy

Report on the Engravers' arguments over their accommodation in the Mint. The Mint
Officers submit that this matter is always to remain under the jurisdiction of the
principal Mint Officers. (P.R.O., Mint/1, 7, p. 48.) Compare Letter 674, vol. IV, *ad
finem*.

X.684   THE MINT TO GODOLPHIN
10 January 1704/5. Clerical copy

Le Clerk is to be appointed Third Engraver, as he has obtained a discharge from the
Duke of Zell [later George I]. His salary is to be £80 p.a. (P.R.O., Mint/1, 7, p. 49.)
There was considerable trouble over the appointment of Le Clerk as an engraver; see
Letter 674, vol. IV, Letter X.740.1 below, and vol. V, p. 308, note (4). A copy of the
Warrant for his appointment at a salary of £80 p.a. is at Mint/1, 5, fo. 35v.

X.691.1   NEWTON TO ——
c. APRIL 1705
From the holograph draft in the Bodleian Library, Oxford[1]

Sr

[I] received yours & thank you heartily for the favour...particularly for
the trouble you gave your self of asking the opinion [of se]veral friends about

the matter of wch I wrote to y[ou]...your advice about what you think proper for me to do.[2] [I] shall endeavour to make the best use of every thing....can be adjusted wch I hope will be in a little...to wait upon you, & in the mean time I continue to beg the f[avour] of my friends that they would reserve a vote for me. I am w[ith] all respect & afection

NOTES

(1) Bodleian New College MS. 361, II, fo. 28, badly damaged by fire and imperfectly legible.

(2) This draft probably relates to Newton's rather doubtful Parliamentary candidature at Cambridge in 1705; compare vol. IV, Letters 688, 689, 693. Parliament was dissolved on 5 April. Compare Letter X.691.2 following.

X.691.2 NEWTON TO ?HALIFAX

?APRIL 1705

From a holograph draft in the Bodleian Library, Oxford[1]

May it please your Lordp

.   .   .[2]

I have inclosed an account of the votes for Burgesses of ye University as I stated it for my self when last at Cambridge.[3] Since ye stating of this Acc[oun]t two or 3 votes are gone off from me to A, so that A is now about 26 or 28 votes above me, but his interest depends more upon out-lyers then mine.

If W[4] should desist I should gain about 17 votes in Trin. Coll. & 4 or 5 in other Colleges & A might lose 4 or 5 votes or above in Trin. College & gain 2 or 3 in other Colleges. And Mr G would gain about 24 votes in Trin. College & about 12 or 15 or perhaps 20 in other Colleges & so would be able to spare 10 or 15 votes wch would secure me. I should scarce have wanted this last assistance had W desisted before ye rising of the Parliament. For the opposition of W & the vogue against me of late have discouraged my friends & checkt & diminished my interest & inclined indifferent persons against me.

I do not expect that W will desist, he declares he will not, & they reccon at Cambridge that he is under very firm obligations to A. But I have stated this matter to your Lordp., being better prepared to do it now I have looked over my papers then on Wednesday when your Lo[rdshi]p last askt me about it. I am

Your

NOTES

(1) Bodleian New College MS. 361, II, fo. 32; a very rough draft. Compare Letter X.691.1 above and Letters 688 and 693, vol. IV.

(2) The first part of the draft is apparently a rough preliminary version of what follows; it is written in the third person. Clearly it was not intended as part of the final letter, so we have omitted it.

(3) The Cambridge Parliamentary candidates were (besides Newton): (i) Hon. Arthur Annesley ('A'; see vol. IV, p. 440, note (10)), (ii) Hon. Dixie Windsor ('W'; see below) and (iii) Hon. Francis Godolphin ('G'; see vol. IV, p. 440, note (8)). The election took place on 17 May; Annesley and Windsor were elected. Presumably the present letter was drafted—but perhaps never sent—in late April or May. It may have been a reply to Letter 693.

(4) Hon. Dixie Windsor, ?1673–1743, second son of Thomas Windsor, first Earl of Plymouth. Windsor entered Trinity College, Cambridge in 1691, and became a Fellow in 1697. He represented Cambridge as a moderate Tory from 1705 to 1725. (See Romney Sedgwick, *The House of Commons 1715–1754*, II (London, 1970), p. 549.)

X.691.3 NEWTON TO ?FRANCIS GODOLPHIN[1]

?APRIL 1705

From the holograph draft in the possession of Lord Lymington

Sr

That there may be no misunderstanding enterteined between us, I would acquaint you that at your motion I have joyned wth you so far as ye frame & temper of ye University permits, & how far that was I told you the other day at your chamber. I may now say further that our College has been more hearty & zealous for you then your's has been for me & yt I have done more for you then you have done for either me or your self.

NOTE

(1) Compare Letters X.691.1 and X.691.2 above, and vol. IV, Letters 688 and 693. Possibly the letter was intended for Francis Godolphin (see vol. IV, p. 440, note (8)).

X.693.1 NEWTON TO JANSSEN

MAY 1705

From a holograph draft in the Bodleian Library, Oxford[1]

Sr Theodore

My Lord Treasurer has referred your Proposal to the Officers of the Mint & we humbly beg the favour of you to meet us on Thursday morning at ten a clock at Sr John Stanleys Office in the Cockpit[2] to discourse the business in order to our making a Report. I am

Your most humble Servant
Is. NEWTON

NOTES

(1) Bodleian New College MS. 361, II, fo. 37 v. The date is established from the fact that Sir John Stanley was Warden of the Mint (1699–1708), and the reference of Sir Theodore Janssen's proposal to the Treasury for the export of 250 tons of tin to Leghorn to the Officers of the Mint on 14 May 1705 (*Cal. Treas. Books*, xx (Part II), 1705–6, p. 253).

(2) Sir John Stanley was Secretary to the Lord Chamberlain in 1705, and in this capacity he presumably had an Office in the Cockpit, Whitehall.

X.693.2 NEWTON TO GODOLPHIN

?1705

From a holograph draft in the Mint Papers[1]

To the Rt Honble the Earle of Godolphin
Lord High Treasurer of England

May it please your Lordp

It being your Lordps pleasure to send Tin in Barrs to storehouses in Lisbon Genoa & Leghorn for supplying Portugal Spain Italy & Turkey with Tin, as Mr Taylor has signified to us, & for this end that we should send it in her Majties Transports to Portsmouth to be there shipt off into her Majties Men of Warr:[2] we humbly lay before your Lordp that there being but few opportunities of Transports from hence to Portsmouth it will be more expeditious & also cheaper to send it from Cornwall to Portsmouth in the Tin-ships. And therefore we humbly propose that the Commissioners for the Tin affair in Cornwall may be ordered to run into barrs & ship off annually to Portsmouth three or four hundred Tunns (or such quantity as your Lordp shall think fit) of the Tin at Truro, taking the blocks together as they come to hand that the Pewterers here may have no occasion of complaining that the Tin sent to the Tower has been culled; & packing up the barrs in deale boxes, it being found by experience that barrells are too weak for this service. By this means the charge of carryage from Portsmouth hither & of unlading here & loding again into Transports will be saved, & the Transports will be freed from the trouble of this carriage, & Portsmouth may be always without any difficulty supplied with such a stock of Tin as shall be sufficient for Lisbon Genoa & Leghorn

All which is most humbly submitted to your Lordps great wisdome

Mint Office

NOTES

(1) III, fo. 514. Possibly the letter is the outcome of the discussions at the Treasury mentioned in Letter X.693.1.

(2) For difficulties concerning the transport of the tin in men-of-war see Letter X.707.

X.694.1   PETITION FROM THE MINT TO THE QUEEN

25 July 1705.   Clerical copy

The Mint Officers complain that they are not always obtaining their allowed exemption from public duties (such as jury service, appointment as Sheriff, etc.), and from arrest and imprisonment without prior permission from the Warden of the Mint. (P.R.O., Mint/1, 7, pp. 50–1.) Compare Letter 560, vol. IV.

X.694.2   STANLEY TO NEWTON

AUGUST/SEPTEMBER 1705

From the original in the Bodleian Library, Oxford[1]

Cockpit wednesday

Sr

I send you inclosed Mr Williams explanation of his former proposall about tin, wch he desires we may consider before next mint day, I have therefore appointed him saturday next at four in the afternoon at my office at ye Cockpit & desire you & Mr Ellis will meet at that time if it be not inconvenient.

I also just now receivd another proposall about ye tin referred to us by Ld Treasurer[.] Mr Tindall ye proposer will meet us this day seven night at ye mint at ten a clock to consider of it. I am

Sr

Yr most humble servt

J STANLEY

NOTE

(1) Bodleian New College MS. 361, II, fo. 84. The date is indicated by the reference of John Williams' proposal concerning tin to the Officers of the Mint on 26 July 1705, by the similar reference of William Tindal's scheme for tin in August, and the reading of the Officers' reply to both on 31 October (*Cal. Treas. Books*, XX (Part II), 1705–6, pp. 149, 155 and 171).

X.703   PETER LE NEVE TO NEWTON

c. November 1705

'I have herewith returned the coppys of the two affidavits which when settled if you please to reconvey to me I will engross them on stampt paper and send them back for Sr John [Newton] and yourself to be sworn to...there will be a necessity for a crest to be assigned to you for that of Sir John Newton you can have no right to...' (Untraced; see *Sotheby Catalogue*, Lot 177(5) where this extract is printed.) The letter concerns Newton's pedigree, and the choice of a suitable coat of arms subsequent to his knighthood in April 1705. Compare Letter 703, vol. IV.

X.704 WILLIAM YARWORTH TO NEWTON
?1705
From the original in the Bodleian Library, Oxford[1]

Most Honoured Friend

I have Presum'd to send to thee for the wanted Alowance because to day is Market-day, and ye Queen's Tax, Window Lights, Poor and Skavenger &c. are all upon me, soe that I have been these two daies very uneasy, and ye more because my Land-lady hath been with me: I know thy Wisdome is such as to Conceive aright of ye Reasons of my Pressure, for without that, wch Necessity forces, I shou'd wholy omitt it: I have sent thee a Book[2] of my weak Labours, wch I hope upon thy Judicious Consideration will satisfy thee from Acetum to Elixer, and how long that may be justly said to Reign, sc. even to ye Production of Azoth: I humbly desire that ye Book may pass thy most Nice and Curious Examination and be Pleas'd to give me thy Sentiments thereon, wch shall be to me as a Golden Touch-stone to know its value by; for I am well satisfyed that no Person Living is more Capable than thyself, therefore a Smile to this Request will be as Acceptable to me as ye. Sun in its Meridian Altitude to the benum'd and Bewinter'd Nature: thou wast Pleased to tell me, that this Age was not worthy of my General therefore thy sentiments on ye whole shall be my Law and Dictate; ye. Number Printed is but small, & so done that I can stop it as I please, this being ye first Book wch hath pass'd out of my hand, wch I present as my Mite into Minerva's Treasure, so wth Prayers to God for thy long life and Prosperity every way I Rest thy true friend to serve in & at all Comands

W. Y.

NOTES

(1) Bodleian New College MS. 361, II, fo. 89v.

(2) William Yarworth (somewhat transparently disguised in print as Y—worth) was the author of numerous books—or versions of the same book—dealing with wine-making, pharmacy and spagyrical chemistry. The date of his letter is suggested by the following considerations: the Window Tax was introduced in 1697, while the reference to the Queen's Tax suggests a date later than 1702. In fact it is possible that the volume presented to Newton by the author was *The Complete Distiller: Or the whole Art of Distillation practically stated... To which is added, Pharmacopœia Spagyrica nova: or an Helmontian Course; being a Description of the Philosophical Sal-Armoniack, Volatile Salt of Tartar, and Circulatum Minus, &c. The Second Edition...* (London, 1705). The first edition, of which Newton probably possessed a copy, had appeared in 1692. We are grateful to Dr Karin Figala for correspondence about this letter; in an unpublished article she discusses the connection between Newton and Yarworth, and identifies the latter with Cleidophorus Mystagogus, pseudonymous author of a number of alchemical tracts.

X.705   THE MINT TO GODOLPHIN

2 January 1705/6.   Clerical copy

Report on a petition (dated 21 December 1705) from the Company for smelting down
lead with pit-coal and sea-coal. The Mint Officers recommend that the Company be
allowed to stamp a distinguishing mark on coin produced from their smeltings. (P.R.O.,
Mint/1, 7, p. 50.) The petition is at Mint/1, 7, p. 49. Godolphin's Warrant, dated
20 April 1706, granting permission, is at Mint/1, 8, p. 60.

X.707   THE MINT TO GODOLPHIN

? 1705/6

From the original in the Public Record Office[1]

To the Rt. Honble the Lord High Treasurer of Great Britain

May it please your Lordp

The time of the Year approaching for exporting Tin, into the streights and
an hundred Tunns being the last summer melted into Barrs for that purpose.
We humbly pray Your Lordships directions whither yt Tin and so much more
as shall be thought convenient shall be put on Board her Ma[jes]ties men of
Warr[2] in the next Convoy into the Mediterranean, to be carried to Leghorn
and Genoa or the Trade in the Mediterranean be left to the Merchants
management

All which is most humbly submitted to your Lordps Great Wisdom

C. PEYTON
Is. NEWTON
JN ELLIS

NOTES

(1) T/1, 111, no. 33. The letter is in a clerical hand, but the signatures are autograph.

(2) Tin was loaded onto the men-of-war, but as a result of a discussion at the Treasury on
14 June 1706, it was relanded, as there was no certainty as to the destinations of the men-of-
war. (See *Cal. Treas. Books*, xx (Part II), 1705–6, p. 84.) Clearly Newton's letter must antedate
this discussion.

X.708.1 NEWTON TO GODOLPHIN

5 MARCH 1705/6

From a clerical copy in the Public Record Office[1]

To the Rt. Hon. the Lord High Treasurer of England

May it please your Lordp.

In pursuance of your Lordps. Verball Order[2] for Coyning her Ma[jes]ties [gold] which came from Portsmouth, the same has been opened, Weighed, melted into Ingots and Coyned. It was an Ounce and a quarter above Weight, but the dust Gold was very foul & being Examined wth a Loadstone was found full of Iron filings and therefore in the melting lost something more then two pounds in Weight, and remained very Brittle. The gross weight by the Invoyce[3] was 65 Lwt: 8 oz. 3 dwt. The standard Weight after melting and toughning was 65 Lwt: 10 oz: 11 dwt: 5 gr. and being Coyned it made 2944 Guineas and 22 grains over wch. after the rate of 21*s*: 6*d* the Guinea and 2*d* the grain, amounts to 3164£: 15 shillings and 8 pence[4] as in the Weighers and Tellers Account annexed, with the Warden and Comptrollers Certificate.

In the Invoyce the Gold is recconed at £5: 5*s* the ounce supposing it perfectly fine and recconing a Crown piece at 6 Shilling, wch is the rate in ye Plantations and by this recconing the Gold is valued in the Invoyce at 4137£: 19*s*: 9*d*[5] by abating one shilling in six, and allowing for the waste in melting and Toughning and for the want of perfect fineness the value comes down to 3164£: 19*s* and 8*d* as above.

The charges of sending an officer of the Mint to Portsmouth and bringing the Gold from thence & melting the same into Ingots were £9: 9*s*: 6*d*;[6] as in the bill Annexed[7] without allowing the Officer anything for his Trouble who in modesty would make no demand on account. I humbly pray your Lordships Order[8] for paying the surplus above the Charges into the Exchecquer

All which is most humbly submitted to Your
Lordships great Wisdom

Is: NEWTON

Mint Office
5th March 1705/6

I believe 12 Guineas may be a reasonable reward to ye officer above mentioned, if your Ldp. shall so think fitt.

Is: NEWTON

NOTES

(1) Mint/1, 8, p. 57. There is a draft by Newton, dated 4 March 1705/6, at Mint Papers, II, fo. 569.

(2) A copy of Godolphin's order to the Mint, dated 3 January 1705/6, is at Mint/1, 8, p. 55. The gold had been seized from Captain John Quelch who had been convicted of felony and piracy in New England. It had been brought to Portsmouth on the *Guernsey* by Captain John Huntingdon.

(3) A copy of the invoice is at Mint/1, 8, pp. 55–6.

(4) A clerical error for £3164. 19s. 8d.

(5) The invoice states £4137. 15s. 9d, apparently the correct value.

(6) The draft reads £9. 19s. 6d.

(7) A copy of the bill is at Mint/1, 8, p. 58.

(8) A copy of Godolphin's Warrant is at Mint/1, 8, p. 59. It included an allowance of £12. 18s (i.e. 12 guineas reckoned at 21s. 6d per guinea) for G. Foord, who had carried the gold from Portsmouth.

X.708.2   THE MINT TO GODOLPHIN

10 May 1706.   Clerical copy

Report on the Pyx of 3 August 1701, and a statement of gold and silver standards current. (P.R.O., Mint/1, 7, p. 53.) The order of reference requesting this report is minuted at P.R.O., T/27, 18, p. 209.

X.708.3   NEWTON TO ?SAMUEL GARTH

?6 JUNE 1706

From a holograph draft in the Jewish National and University Library, Jerusalem[1]

I was this morning at your lodgings to acquaint you that yesterday there was a full meeting of the Roy. Society & I did myself the honour to procure you & the Marquiss de Guiscard to be elected members of the Society. And if you please upon any wednesday to see one of their meetings I will wait upon you thither. I am

NOTE

(1) Yahuda Collection, Newton MS. 7(4). The draft is on the same page as the related draft of Letter X.708.4, hence we assign it the same date. Samuel Garth M.D. (1661–1719), successful physician and versifier, was the only Englishman elected on 5 June 1706, hence we assume that the letter is addressed to him. Garth was certainly known to Newton (compare Letter X.648), but he appears never to have been admitted as a Fellow.

X.708.4 NEWTON TO THE MARQUIS DE GUISCARD
6 JUNE 1706
From a holograph draft in the Jewish National and University Library, Jerusalem[1]

Honoratissimo Dno Marchesio de Guiscard
Is. Newton salutem

Hon D[omi]ne

Cum iudicissem ex sermonibus tuis te non minus artibus et scientijs quam disciplinis bellicis pollere hesterna die occasionem nactus id obtinui a Societate Regia ut in numerum sociorum societatis elegerit. Et quamvis claritate nominis tui ex hac electione nihil accedat volui tamen hoc facto animum meum in te benevolum contestari, Et spero quod in bonam partem accipies. Vale

Dabam 6to die Junij
1706

Translation

Isaac Newton greets the most honoured Marquis de Guiscard

Honoured Sir,

As I judged from your conversation that you excel as greatly in the arts and sciences as in military affairs, yesterday when opportunity presented itself I persuaded the Royal Society to elect you as one of its Fellows. And although this election adds nothing to your reputation I meant by this act to make my goodwill towards you evident, and I hope you will take it in good part. Farewell.

6 *June* 1706

NOTE

(1) Yahuda Collection, Newton MS. 7(4). The draft is on the same page as Letter X.708.3. Antoine, Marquis de Guiscard, Abbé de la Bourlie (1658–1711) began his career as the Abbé of a noble French family, but later became embroiled in political activities. He travelled to Vienna, where he assumed the title of Marquis de Guiscard, and then to the Hague, where in 1706 he stirred up feeling against Louis XIV. There, too, he became acquainted with Marlborough and St John, and presented plans for an English attack on the French coast, suggesting that the French people were ready to rise against the Crown. In July 1706 the plan was put into operation, but with total lack of success. Guiscard was now in receipt of £500 p.a. pension from Queen Anne, but in 1711 Harley saw fit to reduce this to £400 p.a., with no guarantee for life. This angered Guiscard, who revealed to the French the past plans for the invasion of France. The English discovered his treachery, and he was examined before a committee of the Privy Council on 8 March 1711. During the trial he attempted to knife Robert Harley, and

subsequently died in Newgate as a result of wounds he himself received during the fracas. (See I. S. Leadam, *The Political History of England*, IX (London, 1921), p. 91 *passim*.)

We have found no evidence of Guiscard's interest or ability in scientific matters.

X.708.5   THE MINT TO GODOLPHIN

20 June 1706.   Clerical copy

Report on a petition, dated 8 May 1706, from the Engravers asking to be allowed to mint medals. The Mint Officers recommend that permission be granted to mint medals bearing 'plain historical designs and inscriptions in Memory of Great actions'. (P.R.O., Mint/1, 7, pp. 53–4.) A Warrant from Godolphin, granting permission, and dated 2 November 1706, is at Mint/3, 18. Compare Letter 676, vol. IV, and see also vol. VI, Plate V, facing p. 233.

X.709.1   THE MINT TO GODOLPHIN

5 September 1706.   Clerical copy. Continued in Letter X.709.2

Report on a petition from the Moneyers. The Moneyers are short of work, and since they are paid piece rates, not salaries, they are now impoverished. The Mint Officers suggest that £400 p.a. be granted to split between the 21 Moneyers and that their number be gradually reduced to 16 by taking no new apprentices. This will give the remaining Moneyers £25 p.a. each. (P.R.O., Mint/1, 8, pp. 62–3.) The request was granted by Godolphin in a Warrant (copy at Mint/1, 8, p. 64) dated 19 November 1706.

X.709.2   THE MINT TO GODOLPHIN

25 September 1706.   Clerical copy. Continuation of Letter X.709.1

The Moneyers have petitioned the House of Commons, which has granted an extra £500 p.a. to the Mint over and above the £3000 already received annually for salaries and maintenance. The Mint Officers suggest that, following their former recommendations, £400 of this additional income be passed to the Moneyers. (Mint Papers, I, fo. 206.) Godolphin's warrant to Newton, authorizing payment, is at Mint Papers, I, fo. 206v.

X.714   GEORGE YEO TO NEWTON

?February 1706/7.   Printed notice

Summons to a Court at Christ's Hospital on 4 March 1707, to consider the admission of children at Easter, the election of a Beadle, and other matters. (Bodleian Library, New College MS. 361, II, fo. 72.) For Newton's involvement with the Mathematical School at Christ's Hospital, see Index to vol. IV, and Letter X.749.1. He had been appointed a governor of the Hospital in 1697 (E. H. Pearce, *Annals of Christ's Hospital* (London, 1908), p. 124).

X.715.1 SAMUEL NEWTON TO NEWTON
18 MARCH 1706/7
From the original in the Bodleian Library, Oxford[1]

Hond Sr.

This Gentleman the Bearer, was the sole Inventor of hanging Coaches: or Calashes so that they cannot possibly overturne: he is very ambitious that You should see it, (before he makes it publick), least any unforseen fault in the contrivance, should appear to You, wch. when found out, I dare affirme he will be able to rectify. He knows nothing of Mathematics, but I take him to be one of the best Mechanicians in England. & if You will please to appoint the time & place when & where to view this contrivance, he will esteem it the height of his happines to have the Honr. of wayting upon You. I am

> Most honrd. Sr.
> Your most obedient
> humble servt.
> S. NEWTON

Royal Foundation
in Chr. Hospll.
> *March* 18
> 170$\frac{6}{7}$

NOTE

(1) Bodleian New College MS. IV, fo. 121 v. For Samuel Newton, mathematical master at Christ's Hospital, see vol. IV, p. 104, note (5).

X.715.2 LOWNDES TO THE MINT
18 March 1706/7. Clerical copy. For the answer see Letter X.715.3

Godolphin orders the Mint to make recommendations concerning the continuation of the Edinburgh Mint, taking into consideration the Act of Union and a recent Order of Council respecting its execution (copy enclosed). (P.R.O., Mint/1, 8, p. 135.)

X.715.3 THE MINT TO GODOLPHIN
24 March 1706/7. Clerical copy. Reply to Letter X.715.2

Recommends provision of standard weights and trial plates for the Edinburgh Mint, and that the practice there should be homogeneous with that at the Tower Mint. (P.R.O., Mint/1, 8, p. 136.) Compare Letters 717 and 720, vol. IV, and Letter X.715.4. At Mint Papers, III, fo. 150, is a draft Warrant, undated, in Newton's hand, concerning the preparation of the standard weights. Compare Letter 723, vol. IV.

X.715.4   TAYLOUR TO THE MINT

3 April 1707.   Clerical copy. For the answer see Letter 717, vol. IV

The Mint is to report on the enclosed list of tools and equipment recommended as requisite for the Edinburgh Mint. (P.R.O., Mint/1, 8, p. 137.)

717   THE MINT TO GODOLPHIN

12 April 1707

The Mint Officers attended at the Treasury on 12 April to discuss this report. It was decided that everything absolutely necessary should be provided. Godolphin said he would discuss with Sir David Nairn the items that the Mint Officers had not understood, and also the best way of transporting all the equipment to Scotland. (See *Cal. Treas. Books*, XXI (Part II), 1706–7, p. 26.)

X.722   THE MINT TO GODOLPHIN

2 June 1707.   Clerical copy

Reports (in reply to an order of 'Saturday last') on the resumption of the Edinburgh coinage, weights, Trial of the Pyx, authorizing Warrants, etc. (P.R.O., Mint/1, 8, p. 143.) Compare Letter 722, vol. IV, Newton's cognate, but not identical draft, dated 31 May 1707.

724   NEWTON TO GODOLPHIN

24 June 1707

A holograph draft, dated 23 June 1707, is at Mint Papers, III, fo. 144. A further draft, undated, is at fo. 151.

725   NEWTON TO THE GOLDSMITHS' COMPANY

June 1707

On 1 July Godolphin informed the Mint Officers that the Trial of the Pyx would be held on 28 July, but on 18 July he wrote again, postponing the Pyx to 31 July. (See *Cal. Treas. Books*, XXI (Part II), 1706–7, pp. 343 and 364.

726   THE MINT TO GODOLPHIN

July 1707

A clerical copy of this letter, dated 21 July 1707, is at P.R.O., Mint/1, 8, p. 67. A copy of the Warrant granting the requested £100 is on the same page.

X.726.1   TAYLOUR TO THE MINT

8 July 1707.   Clerical copy. For the answer see Letter X.726.2

Refers Memorial from William Drummond, Warden of the Edinburgh Mint, requesting despatch of officers from the Tower Mint to Scotland. The Tower Mint is to suggest the officers and suitable allowances for them. (P.R.O., Mint/1, 8, p. 145.) Compare Letter 724, vol. IV.

X.726.2   THE MINT TO GODOLPHIN

9 July 1707.   Clerical copy. Reply to Letter X.726.1

Proposes David Gregory as overseer of the Edinburgh Mint and all its officers, and Richard Morgan (Newton's Mint Clerk) as instructor of the Edinburgh clerks. Thomas Saybrook, Henry Halley, and Richard Collard are to go as Moneyers. Allowances are stated. Dyes and puncheons are to be furnished by the Tower Mint. (P.R.O., T/17, 1, pp. 77–8. There is another, undated copy at P.R.O., Mint/1, 8, pp. 145–6.)

Compare Letters 724, 727 and 732, vol. IV. Copies of the Warrant appointing Saybrook, Halley and Collard, dated 12 July 1707, are at Mint Papers, III, fos. 11–12 and 136–7; Saybrook died whilst in office, and Halley and Collard were recalled in March 1709; see Letter X.753.1. Richard Morgan was recalled in October 1707 (see Mint Papers, III, fos. 136–7.) A copy of the Warrant appointing Gregory, dated 12 July 1707, is at Mint Papers, III, fo. 180.

X.726.3   NEWTON TO ?SEAFIELD

SUMMER 1707

From holograph drafts in the Mint Papers[1]

My Lord

I received your Lordps Letter about getting a settlement of the revenues which will from time to time grow due to her Ma[jes]ties Mint at Edinburght [so yt salaries & other expenses of the Mint may be duely paid for ye future,][2] And the same being a thing of consequence: I am humbly of opinion that ye properest method to get it done would be for your Lordp to lay a Memorial before my Lord H. Treasurer[3] about it so soon as this Vacation time is over, & his Lp (who is now in the country) returns to London to do business, wch may be about the end of October, & I believe that it may not be amiss for your Lordp to send the Memorial to ye Secretary of State & desire him to lay it before my Lord Treasurer in your Lordps name. I am apt to think that the method used in the Mint in the Tower will be followed. I am

My Ld

Your

NOTES

(1) III, fo. 164. The letter is probably addressed to the Earl of Seafield, Lord Chancellor of Scotland. The date is uncertain, but it seems most likely that the letter was written during the early stages of the Scottish recoinage, that is in 1707. It may, however, have been composed in 1708, as a result of the difficulties arising out of Stuart's death (see Letter 740, vol. IV, and Letters X.738.1, X.738.2 and X.744.2). In either case, it must have been written during the summer, when Godolphin was out of town.

(2) The square brackets are Newton's, indicating intended deletion.

(3) Godolphin.

⟨726⟩   THE MINT TO GODOLPHIN

21 July 1707

See the correction to Letter 726 above.

X.727.1   TAYLOUR TO THE MINT

28 August 1707.   Clerical copy. For the answer see Letter X.727.2

Encloses an extract of a memorial from the Edinburgh Mint saying that many dyes and puncheons have not yet arrived. These cannot be made in Scotland, as no-one there has the requisite skill. (P.R.O., T/17, 1, p. 115.) Compare Letter X.727.3.

X.727.2   THE MINT TO GODOLPHIN

AUGUST 1707

From a holograph draft in the Mint Papers[1]
Reply to Letter X.727.1

To the R.H. the E[arl] of G[odolphin] Ld H. Tr. of great Br.

May it pl. your L[ordshi]p

In obedience to your L[ordshi]ps Order signified to us by Mr Taylour ye 28th of this I[n]stant,[2] that we should consider ye matter in an abstract of a memorial wch your Lo[rdshi]p has rec[eive]d from Scotland relating to ye Mint & recoynage of ye Moneys there & report to your L[ordshi]p our opinion what is fit to be done therein we humbly represent that all the things wch were desired in a Schedule & letter formerly sent from ye Mint in Scotland[3] were provided & the last of them shipt off five or six weeks ago & that upon a new survey of that Mint made by the Moneyers sent down from hence a new schedule being drawn up of what they further wanted & copies thereof sent hither by Dr Greg[ory][4] the Master & Worker of this Mint gave order immediately for providing the same with all dispatch that no time might be lost. And we further represent that besides the cutters mentioned in the said

450

Abstract there are two flatting Engins & one sixpenny Press set down in the
new Schedule as wanting for the use of ye Moneyers: all wch tools since they
are not yet in that Mint we humbly conceive necessary to be sent from hence,
& they are preparing & will cost about 120 *lb* sterling. And we represent also
that the Puncheons for the crowns & half crowns were made, but failed in the
hardning & new ones are making; & that in ye said Schedule the Assaymaster
desires a new Assay-ballance & the Graver desires Puncheons for the small
arms & letters & they are providing: and that, if your Lordp pleases, all these
things shall be delivered to Sr David Nairn or his Order to be sent by land
carriage for expedition as soon as they can be got ready, wch we reccon will be
before the 10th of the next month, excepting the Puncheons.[5]

All which &c

And we further humbly represent that the Officers of that Mint may coyn
shillings & sixpences as they desire till the Puncheons & Dyes can be got ready
for Crowns & half crowns & by her Ma[jesty']s Warrant are obliged only to
make up the proportions therein expressed at any time before the end of this
coynage.

NOTES

(1) III, fo. 78. A second, rougher draft, also undated, is at fo. 204.

(2) Letter X.727.1.

(3) See Letter X.715.4 and Letter 717, vol. IV.

(4) Gregory's list of further requirements was sent to Newton on 12 August 1707; see
Letter 727, vol. IV.

(5) The complaints of the Edinburgh Mint seem very well justified, since they had received
hardly any usable puncheons, and without these they could not proceed. The crown and half-
crown puncheons, and the small puncheons for arms and inscriptions had not arrived, and—
as we learn from Letter X.727.3—the puncheons for the shillings and sixpences were faulty.

X.727.3   JAMES CLARK TO ?NEWTON

9 SEPTEMBER 1707

From the original in the Mint Papers[1]

Edenbrough Septr 9th 1707

Honoured Sr.

I am much at a stand what to think or say concerning my Office of Engraver
and Sinker to the Mint at Edenbrough for which I have a gift during life:[2]
Formerly when ye Punshions were made by Mr. Rotiers our Master worker
had 250 *lib* allowed to him to get yem made and he bargained with ye Rotiers
for 200 *lib* which he payd: And yen every thing needfull to ye compleeting of

451

ye Coyn came down with ye Punshions. which I my selfe received, and gave receit to ye Privy-Counsel for yem. And when I made ye Punshions here at Edenbrough I had ye 250 *lib* payd to me for making of yem, for my Sallary is but 50 *lib* per Annum as sinker and that betwixt me and my Conjunct or College But now of late I have received Punshions for ye heads & Reverses of ye Shilling & Sixpence with some dyes sunk & finished from yem but our Warden either did not know it or your gravers forgott to send down ye small Punshions for ye Armes of ye Reverses & ye letters for ye inscriptions. the six-pence head Punshion was broken before I did see it and now it is all shaken & split with ye sinking only of two dyes. I have made a new sixpence head and am going to make a new Reverse punshion for ye shilling by reason ye sides of ye shields are some sunk and some broken but upon what account I should do this I know not And ye best expedient I can think on is to refer my selfe to your honour's advise & Counsell what I shall do in ye Matter. So beging pardon for this trouble & your honours Answer I am

Sr your most humble servant

JAMES CLARK

NOTES

(1) III, fo. 183.

(2) It is clear from this letter that Clark, employed as an engraver at the Edinburgh Mint, was in fact able to make dyes and puncheons of some sort, but presumably of low quality (compare Letter X.727.1). He seems to have taken upon himself the task of mending and re-placing the broken puncheons from London.

X.727.4 DAVID GREGORY TO NEWTON

16 SEPTEMBER 1707

From the original in the Public Record Office[1]

Edenborough. 16 *Septr.* 1707

Sir

I wrote to you last post, & gave account of what straitens us, & begged your advice & direction. Although I gave express orders for pieces of the first & last of both pots already melted: yet they have so mingled the barrs, that I can only send you pieces of the two pots distinct; but am not able to tell you whither these pieces are of the first or last of the respective pots. I referr you to my last for their Essays & for the alloy put into these pots.

You need not give your self trouble about the name of the ship in which our things are: for we have a bill of loading sent us.

The Officers & servants of the Mint are bussy about trying the Pix of the

moneys coined before the Union: but in a post or two, I hope to send you some
of our Coin, & further nottices about our Meltings & Pot Essays.

I must beg of you to acquaint by your next, in what forwardness the
Puncheons & Dyes are & when we may expect them.[2] I am

> Sir   Your Most humble and most
>   Oblidged servant
>   D GREGORY

Pray let me have your Essays
masters Opinion of these two
pieces of silver

NOTES

(1) Mint/19, 5, fo. 51. Possibly the reply is Letter 731, vol. IV. Compare also Letters 727
and 728, vol. IV. Clearly a considerable part of the correspondence between Newton and
Gregory is missing.

(2) Compare Letters X.727.1, X.727.2 and X.727.3.

729   NEWTON TO GODOLPHIN
14 November 1707

A copy of the Memorial referred to is at P.R.O., T/17, 1, pp. 172–3. Note (2) is effectively a summary of this Memorial; the Officers of the Edinburgh Mint had asked to be
allowed to continue their ancient method of alloying. Lowndes referred Newton's
letter back to the Officers of the Edinburgh Mint; see T/17, 1, p. 174.

X.729.1   NEWTON TO GREGORY
?15 NOVEMBER 1707[1]
From the holograph draft in the Mint Papers[2]

Sr

I am enquiring for a melter to be sent down to you, But its not practicable
for any man to undertake the meltings with your Pit Coal untill he has had
some experience in working with it, & finds out by that experience how the
fire may be governed so as not to over heat the metal. For no man can undertake to do a thing before he knows how to do it, nor know how to do a thing of
this nature without experience. All that we can do is to send you down a
Melter who may assist you in moderating the fire as well as he can till he has
learnt by experience to do it as well as he should.[3] By your putting in the
allay when you are ready to pour off I know that your metal is much too hot
for otherwise the allay would not be melted so quickly. We put in the allay

soon after the silver begins to run, that it may have time to melt, I heare also that the Barrs are sandy. For if the metal be too hot it will disturbe the sand & make it blow. I Feare your Melter makes too much hast wth the melting & putts in too much fire to make the silver melt quickly, & that your potts are too thin & your furnace not so substantial as ours. For if the silver heat too fast it will grow too hot, & if the pot be too thin it will heat faster then if the Pot be thick, & a thin furnace heats sooner then a thick one, & may grow so hot on the outside as to scorch the Melter when he lades off ye pot. Our Pots are about an inch & a quarter or an inch & a half thick & we melt three times a day in the same furnace & no more & the silver (according to the best of my memory) is about 2 hours or $2\frac{1}{2}$ in melting the first time. But the furnace being once heated the silver runs sooner the second & third time. Pray therefore the next melting see that no more coals be put into the furnace then suffice to make the silver run in two hours time or $2\frac{1}{2}$, & see that the heat be not encreased suddenly to make the silver run, but [be] kept eaven & let ye allay be put [in] soon after the silver begins to run that it may have time to melt & mix well wth the silver, & as soon as the silver is all melted & hot enough to run to the bottom of the mold before it congeales let the melter lade it off wth all dispatch. We lade off a pot of 5 or 600 weight in 20 or 24 minutes. If the furnace holds too many coals, the Melter may put in some brick-bats or stones instead of coales. And if the furnace grow too hot on the outside it may be cased with bricks. The more substantial the Pot & Furnace is the [more] eaven will the heat of ye silver bee. Since the scissel is too fine to be remelted without allay, I doubt it must be run into Ingots & the Ingots rated & standarded & allayed a new but it may be kept a while till the meltings are setled.

As soon as the melting is put into better order, wch I hope will be in two or three meltings more, You may come away.[4] For my Ld Tr[easurer] has given me leave to dismiss you. But before you come away, I desire that you would take a survey of the whole Mint & see particularly how many hours the Mill works in a day & whether there be room for another mill, & how many sizers are at work & how many more there is room for & what is further to be done that they may coine 8 or 9000 lb [troy] weekly & I desire you would leave a correspondence between me & Mr Scot or Mr Drummond or both of them for promoting the service.[5]

Your most h. s.

I. N.

NOTES

(1) The date of the letter is uncertain. The last paragraph indicates that it was written shortly before Gregory's return to England in November 1707, and possibly is the letter of 15 November referred to in the last paragraph of Letter 732 (vol. IV, p. 504).

(2) I, fo. 201. The procedure at the Edinburgh Mint was also discussed in Letter X.727.4, Letters 728 and 729, vol. IV, and Letter 729.2. Compare particularly Letter 729, note (2) (vol. IV, pp. 500–1).

(3) There is no indication that a melter was sent to Edinburgh.

(4) Gregory returned to London on 21 November.

(5) Newton corresponded with both Patrick Scott and William Drummond.

X.729.2 NEWTON FOR ALLARDES TO GODOLPHIN

?NOVEMBER 1707

From the draft in the Mint Papers[1]

May it please your Lo[rdshi]p

It has been found by experience in her Majts Mint at Edinburgh that in melting the silver & lading it out into moulds, the allay fumes away while the silver is lading out so that when the silver first laded out is standard, that wch is last laded out proves about three halfpenny weight finer then standard. And the scissel when remelted proves all of it finer then standard & some of it without the remedy. Whence it has been the practise of the said Mint to add a halfpenny weight of copper to every pound weight of silver for supplying the wast of the allay made by its fuming away & to putt this copper into the pot when the silver is half laded off that the silver wch remains to be laded off may become of the same standard fineness wth that which was laded off before. We use scotch or Pit coale wch causes a greater heat then the coale used in the Tower & makes the copper fume away further, & that coale not being to be had in Scotland & carriage by sea being uncertain & there being at present no time to try experiments for regulating the fire nor any artificer in London who is experienced in melting wth Scotch coale, I humbly pray your Lordp that the officers of her Majts Mint at Edinburgh may be still allowed if her Majty pleases to use their ancient Method of reducing the molten silver to standard untill the present recoinage of the moneys in Scotland shall be finished, it being otherwise impracticable to make the moneys of due standard fineness or to coyn with dispatch, & safety [wthout so great a charge in melting the silver offten as will exceed the profits of my place besides the danger of being without the remedy in some pieces of ye money.]

All wch &c.

Allardes Master of her Majts Mint at Edinb.

NOTE

(1) I, fo. 183, bottom. This draft, in Newton's hand, was clearly intended to act as a guide for a letter from Allardes to Godolphin. It is closely related to Letter 729.

X.734   THOMAS CHAMBERS TO ?NEWTON

30 December 1707.   Original

Should his proposal for supplying the Mint with copper be accepted, he suggests details of how the work will best be organized, and points out that he will be absent from London from 11 January 1708 to 7 February. (Mint Papers, II, fo. 384.) Although the letter is amongst Newton's papers, it is not clear whether it was addressed to Newton himself. Chambers' proposals were later discussed by the Mint Officers; see vol. v, p. 358.

X.735 NEWTON TO GODOLPHIN

21 JANUARY 1707/8

From the clerical copy in the Public Record Office[1]

To the Right Honourable the Lord High Treasurer of Great Britain

May it please your Lordship

For ye speedy setting to work of ye Mint at Edinburgh after ye same way as Her Majesties Mint in ye Tower, your Lordship was pleased to direct that severall Tools Utensills & other materialls thereunto required should be forthwith provided & Conveyed to ye said Mint at Edinburgh,[2] & ye same were accordingly provided & sent, the Account whereof stands as followeth—

	£ s d
For Charcoale, Allum, Argol, flat smooth bastard files, Blistered Steel, A Sweep Mill, Quick Silver brass & Iron Sieves, brushes of several sorts & several other Goods bought of sundry persons as by their severall acquittances amount together to ye sum of	133 – 16 – 10.
For four Cutting Engins, two flatting presses, one coyning presse, two paire of Rollers & other things bought of ye Ingeneer in ye Tower as by his Bill of particulars . .	150 – 00 – 00.
For severall Ballances, Weights, & Counterpoises &c bought of Mr John Smart as by his Bill of particulars .	41 – 02 – 06.
For Puncheons & Dyes for all ye species of Money & for small Puncheons & Letters had of ye Graver & Smith in ye Tower as by theire Bill of particulars	207 – 14 – 00.
Total	532 – 13 – 04

Besides these sums disbursed by me there is ye Sum of Sixty Pounds due to my Clerk Mr. Richard Morgan by Her Majesties Warrant[3] whereby he was sent from here to instruct & assist ye Clerk of ye said Mint at Edinburgh which service he performed duly as appears by ye Certificate of ye Officers of ye said Mint.

There is also due to Dr David Gregory by Her Majesties Warrant[4] the sum of two hundred & fifty Pounds for his Charges & Service in going from here to assist oversee & direct ye Officers of ye said Mint at Edinburgh according to ye methods of ye Mint in ye Tower and attending in this service for ye space of three months dated from ye time of his going from hence besides a month more spent in ye said service for wch your Lordship is empowered by ye said Warrant to make such further allowance as your Lordship shall think reasonable which allowance I humbly propose to be 50£ if your Lordship shall think fitt.

The payment of all which summs I humbly pray your Lordship to order or cause to be paid out of ye proper Fund wch I conceive to be ye Bullion which is appropriated to ye service of Her Majesties Mint at Edinburgh and is at present in ye hands of Daniel Stewart Collector thereof or ye Civil List to be reimbursed out of that Fund hereafter

All which is most humbly submitted to your Lordships great wisdom

ISAAC NEWTON

Mint Office
January 21th: 170⅞

NOTES

(1) Mint/1, 8, p. 149.

(2) Compare Letter 717, vol. IV; there Newton gave an estimate of the cost of equipping the Edinburgh Mint, giving a total of £169. 12s. 8d. This was based on an inventory sent from Scotland. After vising the Edinburgh Mint, Gregory found the requirements to be considerably in excess of those estimated. See also Letters X.715.4, X.727.4 and Letter 727, vol. IV.

(3) It is entered at Mint/1, 8, p. 147, dated 12 July 1707.

(4) It is entered at *ibid.*, p. 146 with the same date.

X.736 NEWTON TO GODOLPHIN

11 FEBRUARY 1707/8

From a holograph draft in the Mint Papers[1]

To the Rt Honble the Lord High Treasurer of Great Britain.

May it please your Lordp

In obedience to your Lordps Order of Reference of ye 22th of December last, we have perused the annexed memorial[2] of Mr Allardes the Master of her Majties Mint at Edinburgh & humbly represent that ye charges allowed to the Moneyers by her Majesties Warrant above 9d per Lwt Troy for coinage of silver are extraordinary & so are the charges of refining the coursest ingots of silver to bring the rest to standard: and therefore both these charges are in our humble opinion to be born by her Majty out of the bullion for coinage belonging to that Mint & to be placed by the Master among the incident charges in his accounts. The extraordinary charges allowed to the Moneyers may amount to about three farthings per Lwt Troy, besides the charges of their journey.

And we are further of opinion that for preventing any stop to the coinage the Rt Honble the Lords Commissioners of her Majties Treasury in Scotland be desired to give orders from time to time to the Collector of the Bullion to pay upon account such summs of money to the said Master as they shall find sufficient for defraying as well the extraordinary as the ordinary charges of the coinage, & to direct the Wardens of that Mint to see that the moneys so paid be applied [by] the Master duly & in just proportion to every man's service without [hindran]ce preference or neglect.

As for the melting of the old moneys into ingots we are humbly of [the opinion th]at a penny per pound weight Troy is a reasonable allowance for the [melters] it being the usual price wch merchants pay the Goldsmiths for melting [their] silver into ingots in London. But this allowance cannot be paid [out of] the bullion belonging to the Mint because this melting is no part of the coinage and silver ought to be in the ingot when imported into the Mint, & if it [is] not in the ingot the owner causes it to be melted into ingots by whom he pleas[es] at his own charge & bears the loss by wast. And this must be done before the Master of the Mint can receive it from him by weight & assay upon his Note in order to coin it, & therefore is always done at the owners charge. According to this method the Importer did beare the whole loss by this melting in the late recoinage of the hammered moneys in England, & the same should be now born by the Importers in Scotland & placed among the losses mentioned in the Act of Union in these words. *It is agreed that in the first place out of the aforesaid summ* (that is out of the Equivalent) *what consideration shall be found necessary to be had for any losses which private persons may sustein by reducing the coin of Scotland to the standard & value of the coin of England may be made good.* But her Majesties most honble Privy Council of Scotland having appointed this melting of the old moneys into ingots, we humbly offer to your Lordps consideration whether it may not be proper for their Lordships if they think fit to appoint also the recompence for performing this melting.

All which is most humbly submitted

NOTES

(1) III, fo. 152. An earlier draft is at fo. 158–9; the wording is considerably different, but the general import remains the same. We date the letter from the clerical copy at Mint/1, 8, p. 152.

(2) A copy of Allardes' Memorial (undated) is at Mint 1/8, p. 151; he asks for payments to the Moneyers (in excess of the 9d per lb. wt coined normally paid) to be considered as incident payments, and therefore not to be found by Allardes himself out of *his* payment per lb wt coined. Similarly he asks for separate payment for melting of bullion, under normal circumstances carried out by the importer. Godolphin agreed with the Mint report, and caused its recommendations to be implemented. (See Mint/1, 8, p. 158, and compare also Letter 736, and subsequent correspondence in vol. IV concerning the payment of George Allardes.)

X.738.1 DRUMMOND TO NEWTON
27 MARCH 1708
From the original in the Mint Papers[1]

Much Honoured Sir

When your Letter of the 4th, with my Lord High Treasurers warrant,[2] to the Lords Commissioners of the Treasurie for payment of £892: 13: 4 Sterline to yourself Dr Gregorie and Mr Morgan came to this place, I hapned to be in the countrie fourty miles from this, about pressing business. Which I was verry much concerned at, But my friend att Edinbrugh, haveing by the order I left him, made open the letter, went immediately about the affair, and procured the Lords of the Treasury their precept upon Mr Daniel Stuart[3] to your order for £532: 13: 4 ster: being the money laid out by your self the copy whereof is inclosed, ane other precept to Dr Gregory for £300:—and a third to Mr Morgan for £60. Mr Stuart himself being adieing I mett this day with his freind Mr Broun about the payment of the money, which on account of the uncertainty of Mr Stuarts Life I pressed, by desiring a bill upon Sir David Nairn[4] payable to you the Doctor and Mr Morgan for your several Summs upon dischargeing the precepts and your giving up the workmens vouchers, and I hade accordingly got it if Sir David Nairns freind hade been in town. But Mr Brown assoores me against nixt post I shall have it,[5] so I have keept the precepts till then when I hope to send you the same with the bill upon Sir David, I gave the Clerk of the Treasury for your precept a guinea and as much for Mr Morgans being what is ordinary and to the under Clerks 9s: 6d, I shall be verry mindefull of the matter and you doe me a great deale of honour in giving me the least opportunity of acknowledgeing my self to be

Much honoured Sir
Your much oblidged and
most obedient humbl servant
W DRUMMOND

Edinbrugh March 27th
1708

459

NOTES

(1) III, fos. 133–4.

(2) Newton's 'letter of the 4th' has not been traced. A Warrant from Godolphin to Lauderdale, dated 26 February 1707/8, authorizing payment of £892. 13s. 4d to Newton, Gregory and Morgan for services at the Edinburgh Mint, is at P.R.O., T/17, 1, p. 320.

(3) Daniel Stuart, Collector of Bullion at the Edinburgh Mint, was lying mortally ill; he was thus unable to perform his duties, and that part of the bullion set aside by him for payment of salaries at the Edinburgh Mint was deposited in his own name, and hence could not be drawn upon. A copy of the precept, dated 26 March 1708, and signed by the Commissioners of Her Majesty's Treasury in Scotland, is at Mint Papers, III, fo. 161.

(4) See vol. IV, p. 498, note (3).

(5) Drummond did not act quickly enough; Stuart was dead by 14 April (see Letter 740), and hence the Warrant became invalid and the bill still was not paid.

X.738.2 NEWTON TO ?DRUMMOND

APRIL 1708[1]

From a draft in private possession

I received yours wth the two Precepts[2] inclosed & if Mr Stuarts father in Law pleases to order the payment of ye money in London upon delivering to his Orders the Precepts signed on the backside & the Vouchers, we are all ready to deliver them. We are satisfied that he may do it with ease & safety & if he refuses once more to do it, I shall not expect that he will pay the money in Scotland after Mr Stuarts death,[3] but apply my self to my Lord Treasurer for a new Order of another kind. Pray let me know his answer by the first Post. I am wth many thanks for the trouble I have given you about this matter

Your most humble &

NOTES

(1) Compare Letter X.738.1 and Letter 740, vol. IV. Clearly the present letter was written sometime between the two, that is, between 27 March and 14 April.

(2) These are mentioned (there were in fact three of them) by Drummond in Letter X.738.1, as not yet sent. They concerned payment to Newton, Gregory and Morgan of money due to them for services concerning the Edinburgh Mint.

(3) Stuart was dead by 14 April; see Letter 740.

X.740.1 NEWTON TO GODOLPHIN
14 APRIL 1708
From a clerical copy in the Public Record Office[1]

To the Rt. Honble The Lord High Treas[urer] of Great Britain

May it please your Lordsp

In obedience to your Lordsps order of Reference of ye 23th March last, I have considered ye Annexed pe[titi]on of Gabriel Le Clerk one of ye Engravers of ye Mint, concerning his salary, & humbly represent that before his going beyond seas I was dissatisfied with his carriage & gave him no leave to work in Foreign Mints, & that his salary was interrupted in my last years account by his absence abroad, & in my most humble opinion wants your Lordsps Order to revive & continue it. But if your Lordsp pleases I will pay it upon your Lordsps verbal order, & if ye Auditor scruples ye Account apply hereafter for a Warrant.

All which is most humbly submitted
to your Lordsps great Wisdome
Is: NEWTON

Mint Office
14*th Apr* 1708

NOTE

(1) Mint/1, 8, p. 69. For Le Clerk see Letter X.684. Le Clerk's petition, dated 23 March 1708, precedes at p. 68. He claims that he went to Germany on personal business in June 1706, having obtained leave from the Mint Officers, work at the Mint being slack at the time. While he was there the Electors of Hanover persuaded him to do occasional work in their Mint. He considers he should be paid his salary of £100 by the Tower Mint for his services abroad.

Godolphin ordered Newton to pay the money (see Mint/1, 8, p. 69).

X.740.2 NEWTON TO GODOLPHIN
14 APRIL 1708
From the holograph original in the Mint Papers[1]

To the Rt Honble the Earl of Godolphin
Ld High Treasurer of great Britain

May it please your Lordship

Your Lordps Warrant & the Precepts[2] of the Lords Commissioners of her Majties Treasury of Scotland thereupon for paying moneys due to Dr Gregory,

461

me & my Clerk,[3] being (after an unnecessary delay of payment) become void by the death[4] of Mr. Stuart, I humbly pray that your Lordship will please to give such new order for the payment of the said moneys as your Lordp in your great wisdome shall think fit

[Is. NEWTON][5]

Mint Office
14 *Apr.* 1708

NOTES

(1) III, fo. 131. Compare Letter 740, vol. IV, of the same date, also concerning the effects of Stuart's death on the financing of the Edinburgh Mint.
(2) See Letter X.738.2, note (2), p. 460.
(3) William Morgan.
(4) Stuart's death must have occurred some time between 27 March and 14 April 1708.
(5) The signature has been torn out.

X.741 NEWTON TO GODOLPHIN

MAY 1708

From a holograph draft in the Bodleian Library, Oxford[1]

To the Rt Honble the Earl of Godolphin
Lord High Treasurer of great Britain

May it please your Lordp

In obedience to your Lordps order of Reference of May 6th upon the annexed Proposal[2] of Mr Wm Morgan & others (brought to us by the Proposer June 30th[)] for taking in the old copper money & coyning a thousand Tunns of better copper money in its stead within the term of seven years, provided the loss wch they may sustein by changing the old money for the new & the interest of forty thousand pounds dead stock may be allowed them over & above the price of the new metal & charges of coynage. We have considered the same & are humbly of opinion that the loss wch would be sustein[ed] by melting down the old copper money & the interest of the dead stock would amount to above eighty thousand pounds, & are an unnecessary charge, & that the coynage of a thousand Tunns would make a very great clamour, six hundred Tunns being sufficient to stock the nation. And we further humbly represent that a constant coynage of about eight or ten Tunns per an may be sufficient to supply the yearly wast of the present copper money, that such a small coynage is safest but not yet wanting, that it may be above ten or twelve years before the coynage of an hundred Tunns shall be wanting & that a greater coinage will not be advisable untill there be a great & general complaint of the want of copper money

All wch is most humbly submitted to your
Lordps great wisdome

NOTES

(1) Bodleian New College MS. 361, II, fo. 24.

(2) A petition from William Morgan and partners that they be allowed to coin 1000 tons of English copper coin over a period of seven years was referred to the Mint for its opinion on 6 May 1708 (*Cal. Treas. Books*, XXII (Part II), 1708, p. 227). For William Morgan see also vol. v, p. 359. Morgan's petition, dated 31 March 1708, is at P.R.O., T/1, 106, no. 38.)

X.742   LOWNDES TO NEWTON AND GREGORY

23 June 1708. Clerical copy

Lowndes refers a Memorial from the Commissioners of the Edinburgh Mint, dated 8 June 1708, to Newton and Gregory. The Commissioners (Robert Rutherford, Archibald Brown and Robert Bruce) claim that the receiving of Scots and Foreign money into the Mint is proving such an arduous task (they are spending two or three days a week there) that they require special remuneration. (P.R.O., T/17, 1, p. 383).

X.744.1   W. BOSWELL TO NEWTON

10 July 1708. Original

The writer, Deputy Comptroller at the Edinburgh Mint, complains that his salary of £60 p.a. is too low. He has complained to Gregory, who suggests that he be given an additional allowance as Weigher and Teller. Similarly the Queen's Clerk, who receives £40, also fills the office of Clerk of the Irons, and should therefore have his salary augmented. A number of new clerks have also been appointed, and these should be paid from 1 August last. James Shiels, Melter, is paid daily, and only when the Mint is working, and Boswell feels he, too, should be given the encouragement of an annual salary of £10 as Purveyor; otherwise he may leave. Drummond is also to write to Newton about all these matters. (Mint Papers, III, fos. 156–7.)

X.744.2 NEWTON TO GODOLPHIN

?JULY 1708

From a holograph draft in the Mint Papers[1]

To the Right Honble the Earl of Godolphin
Lord High Treasurer of Great Britain

May it please your Lordp

In obedience to your Lordps Order of Reference signified to us by Mr Taylor's Letter of July 22th Instant upon the anexed Memorial[2] of ye General & other Officers of her Majties Mint at Edinburgh concerning the Bullion given by Acts of Parliamt. for maintaining a free coynage in that Mint & collected by Mr Daniel Stuart now deceased & left in ye hands of his Executors, we have considered the same, & whereas there is a suspicion that ye said Act of Parliament may be expired by reason of the Act of Union establishing an equality of the duties of South & North Britain we do not yet see how the said Act of Union repeals the Act for raising the Bullion for the maintenance of her Majties Mint in Scotland since it does not repleale [sic] the Act for raising the coynage Duty for the maintaining of her Majties Mint in England. That the two nations may be upon an equal foot, either each nation must maintain its own mint or great Brittain must maintain both Mints with an allowance to each proportional to the taxes raised in each nation. And therefore we humbly offer it to your Lordps Consideration whether the Bullion of her Majties Mint in Scotland should not be still collected & kept apart as formerly untill the opinion of her Majties Cou[n]cil of both nations learned in the law can be had upon this matter if it be needfull

And whereas the Officers of her Majties Mint at Edinburgh represent that they have received no salaries for the last half year nor any money has been lately issued to them for defraying the charges of coynage, [the Commission of the Lds Commissioners of the Treasury expiring before your Lordps late Warrant directed to them for supplying the Mint with moneys could be executed,][3] so that the present coinage is in danger of being retarded or stopt for want of moneys to carry it on, we humbly propose that for the more speedy & effectual supplying the said Mint with moneys, your Lordp will please to authorize and direct Sr George Ushart & others the Executor or Executors of Daniel Stewart Esqr late Collector of the Bullion belonging to her Majties Mint at Edinburgh upon account for ye use & service of that Mint, we being informed that there is a greater summ in their hands, & that [*paragraph ends here*]

464

And we are further most humbly of opinion that her Majesties Warrant be directed to the Generall, Master & Wardens of her Majties said Mint & her Majesties Cashkeeper or Receivers or such Receiver as your Lordp shal please to nominate, directing them to keep all such bullion as shall be impressed to the said Generall or Master, in the Treasury of the said Mint under one key to be kept by the said General or Master, two other keys to be kept by the Warden & Counter Warden & a fourth key to be kept by the said Cashkeeper or Receiver & to issue thence such summs as shall become due quarterly for the payment of salaries & charges of repairs & monthly for defraying the charges of coynage.

And we are further of Opinion that the Master of her Majties said Mint do give security in her Majties Exchequer for making good accounts of such summs of money as shall be imprested to him for the service of the said Mint wch security we humbly propose may be in 2000 *lb* besides his own bond.

And whereas ye Executors of Mr Steward alledge that by the 15th article of ye Treaty of Union, the recoynage of the moneys & the expences thereof should be made good out of the Equivalent, we are humbly of opinion that the charges of melting the money into Ingots are a part of the loss wch the people sustein in recoyning their money & therefore by the Act of Union ought in ye first place to be defrayed out of ye equivalent. But the charges of coyning the Ingots are to be defrayed out of the bullion belonging to ye Mint.

And lastly whereas the Master the Warden & ye Counter-Warden have each of them a new Clerk, but the salaries of these Clerks are not yet appointed: we are humbly of opinion that those said Mr Warden & Counter-Warden may be allowed salaries for these three new Clerks after the rate of 40 *lb* a piece per annum during her Majts pleasure, to commence from ye 1st of August 1707

All wch we most humbly propose to be done by her Mats Warrant directed to the Gen[eral] Master & Wardens of her Majts Mint at Edinburgh & entred in proper Offices

NOTES

(1) III, fos. 29–30. A very rough draft. This is probably Newton's response to the Memorial forwarded to him by William Drummond on 12 July (see Letter 745, vol. IV) and also passed directly to him by Godolphin on 22 July. It is clear from the tone Newton adopts that he is losing patience over the matter of the financing of the Edinburgh Mint. Stuart's death had merely aggravated an already difficult situation. Acts of Parliament had been passed to the effect that the Edinburgh Mint should be run on precisely the same lines as the Mint at the Tower. No account was taken, however, of the fact that circumstances at the two Mints were not identical. George Allardes, Master of the Edinburgh Mint, found himself in great difficulties over the cost of smelting bullion and the payment of the Moneyers. The London Mint itself had not been very quick to provide equipment, expertise or personnel, and the present document shows that the salaries of various employees had not even been properly settled.

(2) Possibly the undated Memorial at Mint Papers, III, fo. 13 from the Mint Officers at Edinburgh to Godolphin, complaining that Mint salaries were still unpaid as a result of Stuart's death.

(3) The brackets are Newton's, as usual indicating his intention to delete the enclosed words.

X.748.1   P. SCOTT TO ?NEWTON

10 AUGUST 1708

From the original in the Mint Papers[1]

Honored Sir

Mr Allardes having been pleased to send me his Letter[2] for you open that I might put in it a Copie of Mr Millars Representation[3] I beg pardon to inform you that it seems Mr Allardes has mistaken the Matter, for tho' Mr Millar be joined with the other three Commissioners to the Receiving the Money from the Bank, His pains is only in assisting at the Tale and Weight and Certifieing the same and the Recept thereof in to the Mint: But he does not concern himself to Witness or observe the melting of it: As you will see by the Copy of his Supplication subjoined. The Original was sent up two posts ago to Sir David Nairn

Our Coinage goes on very well, And we have received the Thesaurers Warands[4] for a supplee of money So that I hope we shall have no stop on that account.

Pray whatever Depursments[5] you are at on our Warands from time to time be pleased to let [me] know them as you have occasion to write and your re-imbursement will be ordered.

I give you my hearty thanks for your kindnes and civilities to me when at London I shall wish for an Opportunity to let you know how much I am

Honored Sir
Your most obedient humble servant

Edinbr 10th Aug: 1708   PAT: SCOTT

NOTES

(1) III, fo. 195v. For the writer see Letter 736, vol. IV; vol. v, p. 57, note (4); and Letter 798, vol. v.

(2) Allardes' letter is now missing.

(3) On the same folded sheet (fo. 194r) is Robert Millar's [or Miller's] supplication (un-dated) requesting payment for his services. Compare also Letter 750, vol. IV.

(4) 'Treasurer's warrants'.

(5) 'Disbursements'.

X.748.2 NEWTON TO SEAFIELD
12 AUGUST 170[8]
From the holograph original in the Scottish Record Office[1]

London Aug. 12. 1707[2]

May it please your Lordp

Upon the first notice of the death of Mr Stewart[3] I laid a Memorial[4] before my Lord Treasurer about the money in the hands of his Executors that it might be paid into ye Mint: but this memorial not coming from the proper Officer, I desired Dr Gregory to signify to one of the Officers of the Mint at Edinburgh that it would be proper for the Officers of that Mint to lay a memorial about that matter before my Lord Treasurer. And accordingly the General of that Mint laid a Memorial before my Lord Treasurer about it in the name of the Officers & we made a report[5] upon it, & two Warrants[6] were sent down to Edinburgh from her Majty the one to the Executors of Mr Steward to pay the money to the General & Master, the other to the General Master & Wardens to lock up the same under their several keys for paying of salaries & other charges as they shall become due whereof the Master is to give an Account annually. If the Executors do scruple to pay the whole at once, the Officers may receive it by parcells. And when they begin to want the money in the hands of the collectors of the Customes it will be proper for them to desire the General to put in another Memorial to my Lord Treasurer about that money. The Executors should also make up their Accounts in the Exchequer: but the method of bringing them to account I do not know. I hope they will do it voluntarily upon paying in the money into the Mint. If there be any thing in wch I can serve your Lordp. or Mr Allardes you may command

> My Lord
> Your Lordps most humble
> & most obedient servant
> Is. NEWTON

Ld Seafield

NOTES

(1) GD 248/571/2/58. The Earl of Seafield was Lord Chancellor of Scotland; see vol. IV, p. 493, note (2).

(2) The letter is clearly misdated; Stuart died about the beginning of April 1708; see Letter 740, vol. IV.

(3) Variously spelt Stuart, Stewart, Steward.

(4) Presumably Letter 740, vol. IV (pp. 515–16).

(5) Probably Letter 744.2.

(6) See Godolphin's letter to Lauderdale of 29 July 1708 with the two Warrants enclosed (copies are at P.R.O., T/17, 1, pp. 409–11).

X.748.3   RECEIVERS AT EDINBURGH TO NEWTON
28 September 1708.   Original

The receivers, Robert Rutherford, Robert Bruce and Archibald Brown, have petitioned Godolphin for a payment of one-third per cent of all monies received in recompense for their service. They hope Newton will also petition Godolphin on their behalf. (Mint Papers, III, fos. 162–3). The petition from the receivers to Godolphin, dated 24 September 1708, is at P.R.O., T/1, 109, no. 19.

X.748.4   THE MINT TO GODOLPHIN
28 DECEMBER 1708
From the holograph original in the Mint Papers[1]

To the Rt Honble the Ld High Treasurer
of great Britain

May it please your Lordps

In obedience to your Lordps Order of Reference of the 27th November last, upon the annexed Memorial[2] of the three Commissioners appointed by her Ma[jes]tys late Privy Council of Scotland to receive all the Scotts & foreign coin see it melted into Ingotts & deliver the Ingotts by weight & assay to be coined & to certify the deficiency; in which Memorial they humbly desire that your Lordsp will be pleased to appoint them a suitable reward for their own attendance & trouble & for paying their Clerk & other servants & defraying the daily charge they are put to, & grownd their Petition upon their Commission wch bears that they shall be sufficiently rewarded: We have considered the said Memorial, & because the business of these Commissioners requires diligent attendance in their own persons at the meltings & their trust is also considerable, We are humbly of opinion that they may deserve for themselves & their Clerk & other servants the summ of seven hundred pounds to be paid out of the bullion or coynage duty of that mint so soon as the present coynage shall be ended.

All which is most humbly submitted to your
Lordships great wisdome.

Mint Office
28th Decemb. 1708

C. PEYTON
IS. NEWTON
J. ELLIS

NOTES

(1) III, fo. 74. The letter is in Newton's hand and bears the autograph signatures of all three Mint Officers. Further drafts are at fos. 129–30 and 154–5.

(2) Presumably Letter X.748.3.

X.749.1 A TESTIMONIAL BY NEWTON AND HALLEY

7 JANUARY 1708/9

From the holograph original in the
Library of the Royal Society[1]

Germin Street St James's
7th Jan 170$\frac{8}{9}$

I have seen some books published by Mr William Jones concerning Algebra, Geometry, Navigation & some other Mathematical Sciences, & find him well skilled in Mathematicks; & believe that in point of knowledge & by his exercise in teaching & experience at Sea he is well qualified for teaching Navigation to the poor boys in the School of the Royall foundation in Christ's Hospital. And I am credibly informed that he is also very industrious & understands Latin & French.

Is. NEWTON

I have long known Mr William Jones, and do humbly subscribe to the opinion of Sr Isaac Newton, that he is very well qualified to teach Navigation in the School aforesaid

EDM: HALLEY

NOTES

(1) MS. 657, separately mounted for exhibition in a wooden frame. Each testimonial appears to be in the autograph of the signatory. The document has been previously published in *Nature*, **150**, no. 3816 (19 December 1942), 731. For William Jones see vol. v, p. 95, note (1) and Index. He was not appointed Mathematics Master at Christ's Hospital upon the resignation of Samuel Newton, the post being filled by the selection of Thomas Hodgson (1672–1735), Flamsteed's assistant (see vol. IV, pp. 104–5 notes and p. 370, note (6)); evidently Flamsteed's influence was greater than that of either Newton or Halley. Both Flamsteed and Newton were governors of the Hospital (see Letter X.714 above). The wording of the testimonial makes it likely that it was Halley who introduced Jones to Newton. Of Jones' early publications, *A New Compendium of the Whole Art of Navigation* (London, 1702) and *Synopsis Palmarium Matheosis* (London, 1706) were in Newton's library.

X.749.2   LOWNDES TO THE MINT

24 January 1708/9.   Clerical copy. For the answer see Letter X.756

Refers a petition from Robert Millar, Clerk of the Meltings at the Edinburgh Mint, for extra remuneration (formerly promised) above his salary of £40 p.a. (P.R.O., Mint/1, 8, p. 156.) Compare Letter 750, vol. IV, and Letter X.748.1.

X.752.1   HENRY HUNT TO NEWTON

February 1708/9.   Original

Has summoned members of the Council of the Royal Society to meet on 23 February [1708/9] according to order. (Jewish National and University Library, Jerusalem. Yahuda Collection, Newton MS. 7(4).)

X.752.2   THE MINT TO GODOLPHIN

10 March 1708/9.   Clerical copy

Report on a Memorial from the Commissioners for Managing her 'Majesty's Dutys on Stampt Vellom Parchment and Paper'. The Mint Officers discuss the preparation of stamps and the prevention of counterfeiting. (P.R.O., Mint/1, 7, p. 55.) For an earlier report see Letter 671, vol. IV.

X.752.3   DRUMMOND TO NEWTON
12 MARCH 1708/9
From the original in the Mint Papers[1]

Honoured Sir

The Recoinage being now over, and the Monyers work at ane end, they are desirous to be discharged and to return home, which the officers are very well satisfied they bee. Ther are two attestations drawn up which the officers of the Mint propose to signe in the Monyers favours, copy whereof they have transmitted this night, to their provost to be shoun yow, which I hope shall please, if any thing furder is requisit for them they will not be denied it.

The Queens warrant appointing them to come doun requires that they continue att Edinbrugh not only till the whole recoinage be over, but likewise untill such time as her Majestie have otherwise ordaind and provided for the Coynage in the Mint here, Or that they shall obtain Leave from my Lord Treasurer. It therefore seems necessar Leave be obtained, for their departure and they defer till they are directed in this from London, mean time the officers of the Mint here to prevent charges ther stay may occasion are willing to do every thing for their dispatch and this is the occasion of giving yow this trouble to signefie this to yow, the freedom of doeing which you will please excuse from

Honoured Sir

Edinbrugh March: 12*th*:
1709

Your most humble and
most obedient servant
W DRUMMOND

NOTE

(1) III, fo. 127. The letter repeats Allardes' requests in the opening of his Memorial to Newton (Letter 751, vol. IV).

X.753.1 NEWTON TO GODOLPHIN
14 MARCH 1708/9
From a clerical copy in the Public Record Office[1]

To the Rt Honble. The Lord High Treasurer
of Great Britaine

May it please Your Lordsp.

Her Majtie haveing sent down to Her Mint in Scotland three of ye Monyers to stay in ye service of that Mint till they should be recall'd by Her Majtie. or your Lordsps Order[2] & ye Recoinage of ye Scottish Money being now at an end so that there is no further occasion for theire staying there; & One of them being dead[3] & ye other two being allow'd by Her Majties Warr[an]t three shillings per diem to each dureing theire stay at that Mint when their is no Coinage: at ye desire of the Mastr. of that Mint[4] I humbly lay the matter before your Lordsps in order to theire being recaled for saving that charge theire Names are Richd. Collard & Henry Haley

All wch is most humbly submitted &c
Is: NEWTON

Mint Office
14 *March* 170⅞

NOTES

(1) Mint/1, 8, p. 154; it is followed by a copy of Godolphin's letter to Hayley and Collard releasing them from their offices.

(2) Newton appears forgetful. Three Moneyers (Collard, Hayley and Saybrook) were originally sent to Edinburgh; after Saybrook's death two more Moneyers (Hopper and Sutton) were sent to replace him (see Letter 751, vol. IV).

(3) Thomas Saybrook or Seabrook; see Letter 751, vol. IV.

(4) Allardes had brought the matter to Newton's notice on 12 February; see Letter 751, vol. IV.

X.753.2 LOWNDES TO NEWTON
25 MARCH 1709
From a clerical copy in the Public Record Office[1]

Sr

It being represented to my Lord Trea[su]rer yt R[ichard] Collard Junr one of ye Monyers at Edenburgh hath been in a languishing Condition for some months past & thereby become useless to ye Company, & it being desired yt he may have leave to returne home for the recovery of his health. His Lordp directs you to permit him to returne & yt You send another person in his roome in case you find it necessary I am Sr &c 25th March 1708[/09]

WM LOWNDES

NOTE

(1) T/27, 18, p. 406. This instruction presumably crossed with Newton's Letter X.753.1.

X.754.1 CRAVEN PEYTON TO NEWTON
14 APRIL 1709
From the original in the Public Record Office[1]

House of Commons 14*th April*. 1709

Sr

The description of what Bullion shall be received [at] is, two pence half penny per Ounce, for every Ounce of Foreign Coynes & foreign or British wrought Plate of the standard of Eleven Ounces two penny weight fine or reduced there unto, I heard you was inquiring for me about this matter, so I have taken the liberty to send it to you, and am

Your most humble sert.

CRAV. PEYTON

To Sr. Isaac Newton
    present

NOTE

(1) Mint/19, 5, fo. 37 v. For the writer, see Letter 743, note (3) (vol. IV, p. 521). A Warrant dated 30 April 1709, and addressed to Newton, authorized the payment of a premium of $2\frac{1}{2}d$ per ounce for foreign coin, and foreign or British plate imported into the Mint, above the normal price. (See copies of the Warrant at P.R.O., Mint/1, 8, p. 70, and in the British Museum, Add. 5751, no. 95, fo. 333.)

X.754.2   MARY PILKINGTON TO NEWTON

26 April 1709.   Original

Written from Nottingham. A brief letter concerning a payment to the writer's cousin, Mary Holden. (Jewish University and National Library, Jerusalem. Yahuda Collection, Newton MS. 7(4).) The writer was Newton's half-niece; see Appendix II, Newton's Genealogy.

X.755   THE MINT TO GODOLPHIN

1 June 1709.   Clerical copy. Reply to Letter X.749.2

Recommends payment of £60 to Millar [or Miller]. (P.R.O., Mint/1, 8, p. 156.) The recommendation was accepted; see Letter 756, vol. IV.

760   NEWTON TO FRANS BURMAN

2 June 1702

The letter is not, of course, addressed to Burman, but is a letter of introduction addressed to Halley. It is printed in *Francisci Burmanni, V. D. M. Viri Clarissimi, itineris Anglicani acta diurna* (Amsterdam, 1828), p. 10. Frans Burman (1671–1719), theologian, brother of the more famous Pieter Burman, was Chaplain to the Dutch embassy which came to England to congratulate Queen Anne on her accession. His visit to England, from May to November 1702, is recorded in detail in his book. (There is an English translation in J. E. B. Mayor, *Cambridge under Queen Anne* (Cambridge, 1911), pp. 312–14.) In 1715 he became Professor of Theology at Utrecht University.

According to his diary, Burman inspected the premises of the Royal Society, at Gresham College, on 28 May 1702; then on 2 June he writes, 'Invisi ad celeberrimum *D. Isaacum Neuwtonum, qui me humanissime excepit vel solius Volderi causa, cujus me discipulum ferebam. Actum præcipue de Systemate Universi.*' ('I visited the most famous Mr Newton, who received me with the greatest kindness, if only for Volder's sake, whose pupil I declared myself to be. The discussion was chiefly about the system of the universe.') Burman then outlines their discussion, which concerned the problem of the resistance presented to the motion of the planets by the æther.

Newton was apparently ready to introduce Burman at a meeting of the Royal Society, but was too involved with Mint business to do so. Instead he wrote the letter of introduction to Halley printed in vol. IV, p. 543. On 3 June Burman visited Halley, but found him absent.

764   THE MINT TO GODOLPHIN

10 August 1709

A clerical copy, with virtually the same wording, but dated 20 July 1709, is at P.R.O., Mint/1, 8, p. 158.

X.767   HENRY HUNT TO NEWTON

October 1709.   Original

Has given notice to Fellows of the Royal Society to resume meetings on Wednesday 26 October 1709. (Jewish National and University Library, Jerusalem. Yahuda Collection, Newton MS. 7(4).)

X.770   THE MINT TO GODOLPHIN

18 January 1709/10.   Clerical copy

Reports (in reply to an order of 10 December 1709) on a bill of John Roos, Engraver of public seals, for payment of £1086. 16s. 4½d. Payment is recommended. (P.R.O., Mint/1, 7, p. 92.) Compare Letter 1340.

X.773   J. BLOW TO NEWTON

4 March 1709/10.   Original

Assessment of Church rates due for Newton's house in Chelsea, for his first six months of residence from Michaelmas [29 September] 1709 to Lady Day [25 March] 1709/10. As the rate was at 3d in the pound per year, and Newton's house was valued at £50, he was charged 6s. 3d for the half year. The writer was one of the churchwardens. (Babson College, Sir Isaac Newton Collection, MS. 435v.)

826   NEWTON TO COTES

24 March 1710/11

Note (2) (vol. v, p. 104): A number of incomplete holograph drafts of the paper are, however, extant in U.L.C., Add. 3965, and in a private collection. The existence of these drafts considerably clarifies Cotes' response to the paper (see Letter 829, vol. v, pp. 107–10, and p. 112, note (10)). The fairest draft seems to be that at U.L.C., Add. 3965(12), fos. 247–59. This gives Propositions 36–40, Book II (1713 edition) and part of the Scholium following. The major differences between the draft and the version finally printed are as follows: (i) The opening paragraphs of Proposition 36 follow the version Cotes transcribes in Letter 829 (see vol. v, p. 111, note (6)). (ii) The whole discussion of the *vena contracta* is absent from Case 1 in the draft; Newton clearly added something on this matter before sending a draft to Cotes with Letter 826. (iii) Corollaries 2 and 6–10 of the printed text are also missing from the draft; probably Newton added these Corollaries after his reconsideration of Proposition 37. (iv) The draft version of Proposition 37 differs radically from the final version, so we print this section in full as Number X.826, below. (v) In the calculation in Proposition 40 the correction term L is omitted from the draft, and the tabulated values differ slightly from those in the printed version. The draft ends at the beginning of Experiment 13 of the Scholium.

X.826 DRAFT OF PROPOSITION 37, BOOK II, FOR THE SECOND EDITION OF THE *PRINCIPIA*

MARCH 1711

From a holograph draft in the University Library, Cambridge[1]

Prop XXXVII.[2] Theor XXVIII[3]

Cylindri in Fluido compresso infinito, secundum longitudinem suam, uniformiter progredientis resistentia est ad vim qua totus ejus motus, quo tempore quadruplum longitudinis suæ describit; vel tolli possit vel generari, ut densitas Medij ad densitatem Cylindri.

Ponamus fundum vasis non in medio perforatum esse sed rimam circularem habere in circuitu inter ipsum et latera vasis et aquam effluere per hanc rimam. Et pondus aquæ totius in vase, ponderis pars quam fundum vasis premitur & ponderis pars altera quæ in defluxum aquæ impenditur erunt inter se ut summa fundi[4] et duplæ rimæ fundum ambientis, fundum rima circumdatum et dupla rima sunt inter se per Corol. 2, 3 et 4 Prop. XXXVI.[5] Hoc autem ita se habebit sive aqua gravitate sua descendat sive alia quacunque vi gravitatem suam æquante impellatur deorsum, sive columna glaciei eadem vi descendat et premat aquam in vase, et in hoc ultimo casu vis tota qua glacies aquam premit vicissim ab aqua premitur erit ad vim qua aqua premit fundum rima circumdatum et vicissim a fundo premitur, ut summa fundi et duplæ rimæ fundum ambientis est ad fundum rima circumdatum & eadem vis tota erit ad vim quæ in aquæ defluxum impenditur ut summa eadem ad duplam aream rimæ.[6]

In systemate autem glaciei cylindricæ & vasis et aquæ in vase motus omnes et omnes pressiones et vires quibus motus generantur eadem sunt inter se sive systema quiescat sive moveatur idem uniformiter quacunque cum velocitate in directum (per Legum Corol. 5): et propterea si glacies cylindrica quiescat & vas cum fundo suo eadem velocitate qua glacies descendebat ascendat in glaciem, motus omnes aquæ in vase respectu vasis & glaciei, et omnes ejus pressiones eadem erunt atque prius. id est vis qua glacies et aqua se mutuo premunt, vis qua aqua et fundum vasis se mutuo premunt, & vis quæ in motum aquæ generandum impenditur, erunt inter se ut sunt summa fundi et duplæ rimæ, fundum rima circumdatum, et duplum rimæ respective. Ideoque cum vis qua glacies & aqua se mutuo premunt sit æqualis ponderi aquæ totius in vase per hypothesin et vis qua aqua et fundum vasis se mutuo premunt sit resistentia fundi ascendentis; hæc resistentia erit ad pondus aquæ totius in vase ut est fundum illud rima circundatum ad summam fundi illius et duplæ rimæ.

475

Sed pondus aquæ totius in vase est ad pondus cylindricæ columnæ cujus basis est fundum vasis et altitudo est eadem cum altitudine aquæ in vase, ut summa rimæ et fundi ad fundum solum et ex æquo perturbate resistentia erit ad pondus cylindricæ illius columnæ ut summa rimæ et fundi vasis est ad summam fundi et duplæ rimæ; hoc est, si rima et latitudo vasis in infinitum augeantur manente magnitudine fundi, ut unum est ad duo. Quare si plano ad axem perpendiculari columna illa cylindrica bisecetur, & dimidium ejus dicatur D, resistentia fundi ascendentis æqualis erit ponderi cylindri D.

Sed hic cylindrus D, pondere suo cadendo in vacuo & casu suo describendo altitudinem aquæ in vase, id est, duplum longitudinis propriæ, velocitatem acquiret qua fundum prædictum ascendit, et eodem tempore ascendo duplam altitudinem aquæ describet.[7] Et cylindrus D cum hac velocitate secundum longitudinem axis sui progrediendo in aqua, resistentiam patietur æqualem resistentiæ fundi ascendentis, per Lem. IV, id est, æqualem ponderi proprio. Et propterea resistentia ejus æqualis est vi qua totus ejus motus interea generari potest dum duplam altitudinem aquæ, seu quadruplum longitudinis propriæ uniformiter progrediendo describit. Et si cylindrus D sit densior vel rarior quam aqua, vis qua totus ejus motus eodem tempore vel generari potest vel tolli, erit major vel minor in eadem ratione. Ideoque resistentia est ad hanc vim ut densitas Medij ad densitatem cylindri. Q.E.D.

Translation

Proposition 37,[2] Theorem 28[3]

> The resistance of a cylinder, moving uniformly in the direction of its length in an infinite, compressed fluid, is to the force by which its whole motion may be either destroyed or generated in the time in which it describes four times its own length, as the density of the medium to the density of the cylinder.

Suppose the base of the vessel not to be pierced in the centre, but let there be a circular aperture going around between [the base] itself and the sides of the vessel, and let the water flow out through this aperture. And the weight of all the water in the vessel, the part of the weight which presses the base of the vessel, and the other part which is used up in the flowing out of the water, will be to one another as the sum of the [area of the][4] base and twice the aperture surrounding the base, the base bounded by the aperture, and twice the aperture are to one another, by Corollories[5] 2, 3, and 4 of Proposition 36. This will be the case, however, whether the water falls by its own weight, or if it is impelled downwards by some other force equalling its weight, or if a column of ice descends with the same force and presses the water in the vessel; and in this last case the whole force with which the ice presses the water, and in turn is pressed by the water, will be to the force with which the water presses the base bounded by the aperture, and is in turn pressed by the base, as the sum of the base and twice the aperture surrounding the base is to the base bounded by the aperture, and the same total force will be to the force which is used up in the flowing of the water as this same sum to twice the area the aperture.[6]

However, in this system of a cylinder of ice, a vessel, and water in the vessel, all the motions and all the pressures and forces by which the same motions are generated, are to one another the same, whether the system is at rest or is moving uniformly in a straight line with any velocity whatever (by Corollary 5 of the Laws [of Motion]): and moreover, if the cylinder of ice is at rest and the vessel, together with its base, rises towards the ice with the same velocity as that with which the ice descended, all the motions of the water in the vessel with respect to the vessel and to the ice, and all their pressures, will be the same as before. That is to say, the force with which the ice and water press each other, the force with which the water and the base of the vessel press one another, and the force which is used up in generating the motion of the water, will be to one another as the sum of the base and twice the aperture, the base bounded by the aperture, and twice the aperture, respectively. And therefore as the force with which the ice and water press one another is equal to the whole weight of the water in the vessel, by hypothesis, and the force with which the water and the base of the vessel press one another is [equal to] the resistance of the ascending base, this resistance will be to the weight of all the water in the vessel as that base bounded by the aperture is to the sum of that base and twice the aperture. But the weight of all the water in the vessel is to the weight of the cylindrical column of water whose base is the [same as the] base of the vessel and whose height is the same as the height of the water in the vessel, as the sum of the aperture and base to the base alone; and by substitution of equal quantities, the resistance will be to the weight of that cylindrical column as the sum of the aperture and the base of the vessel is to the sum of the base and twice the aperture; that is, if the aperture and the width of the vessel are enlarged to infinity, the base remaining the same, as one is to two. Whereby if the cylindrical column is bisected by a plane perpendicular to its axis, and the half of it is called D, the resistance of the ascending base will be equal to the weight of the cylinder D.

But this cylinder D, falling by its own weight in a vacuum and in its fall describing the height of the water in the vessel (that is, twice its own length) acquires the velocity by which the aforesaid base ascends, and in ascending in the same time describes twice the height of the water.[7] And the cylinder D, in moving forward with this velocity in the direction of its length, suffers a resistance equal to the resistance of the ascending base, by Lemma 4; that is, [a resistance] equal to its own weight. And therefore its resistance is equal to the force by which its whole motion can be generated when it describes twice the height of the water or four times its own length, moving uniformly. And if the cylinder D is denser or rarer than the water, the force by which its whole motion may be generated or destroyed in that same time will be greater or less in the same ratio. And therefore the resistance is to this force as the density of the medium to the density of the cylinder. Q.E.D.

477

NOTES

(1) Add. 3965(12), fos. 249–51. This is a rough draft for part of the paper sent by Newton to Cotes with Letter 826; see our correction to Letter 826, above.

(2) Newton considerably rewrote Section VII, Book II of the *Principia* where this Proposition appears. Because of the omission of the old Proposition 34 from the second edition, renumbering of the Propositions also ensued. Hence Proposition 37 in the first edition became (after considerable alteration) Proposition 36 in the second; the matter of Proposition 37 in the second edition was entirely new. The draft given here is considerably different from the final version. For a figure, see vol. v, p. 109.

(3) *Read:* 29. Newton clearly forgot that the deletion of Proposition 34, and hence of Theorem 27, necessitated the renumbering of the Theorems.

(4) Throughout the argument which follows, Newton has left it as understood that he is referring to the areas of the cross-sections he discusses.

(5) The Corollaries referred to seem to be those finally printed. The model set up in the present Proposition is the same as that used in Cases 6–10 of Proposition 36. Newton states the argument very crudely in this draft, making no allowance for the *vena contracta*.

(6) Newton's explanation is exceedingly prolix. He merely repeats here the argument at the beginning of the Proposition. If we consider the area of the aperture to be A, and of the base to be a, then total volume of fluid: volume of cataract: volume of static column $= 2A + a : 2A : a$ (these proportions are derived in Proposition 36 by using arguments based solely on continuity of flow). The cataract and column are considered to act independently, hence the forces they exert are proportional to their masses and hence to their volumes.

(7) This sentence is the one to which Cotes rightly objects in Letter 829 (vol. v, p. 108), on the ground that it is contrary to the principle of hydraulic continuity. His objection caused Newton to rewrite the Proposition completely.

829   COTES TO NEWTON

31 March 1711

For further comments on the draft paper Cotes here discusses, see the correction to Letter 826, above, and also Number X.826.

844   COTES TO NEWTON

4 June 1711

Note (4) (vol. v, p. 153): There is a very rough draft in private possession which indicates that Newton was considering further extensive reorganization of Propositions 36 and 37, with the addition of a number of further Corollaries.

X.880 NEWTON TO ?BENTLEY

c. NOVEMBER 1711

From the holograph draft in the Jewish National and University Library, Jerusalem[1]

Sr

I rem[em]ber very well that about three or four years ago I examined Mr. Hussey[2] for an hour together or above, in the general parts of Mathematicks, in order to know his ability for teaching a Mathematical school, & found him very well qualified. And I believe that in respect of his mathematical abilities he will make an able Professor if the heads of Colleges should think fit to chuse him. But his qualifications in other respects I do not know, being a stranger to him   I am

       Sr   Your most humble servant

                                         Is. NEWTON

Sr

I have made a resolution not to meddle with this election of a Mathematick Professor any further then in answering Letters & have given this answer to some who have desired a certificate from me. In answer to your Letter & another I received from Mr Cotes I send you the two inclosed & give you both & your friends leave to shew them to any of the Electors. If Mr Hussey be as well qualified in temper & manners as in Mathematical skill he will be a grace to ye university.

NOTES

(1) Yahuda Collection, Newton MSS, 7, bundle 4. The letter is clearly related to Cotes' letter to Newton, vol. v, Letter 879, pp. 202–3. There Cotes recommends Christopher Hussey for the Lucasian Professorship in Cambridge, and mentions that Bentley is about to make a similar recommendation.

(2) See Letter 879, note (4) (vol. v, pp. 202–3).

X.887 ARBUTHNOT TO NEWTON

?EARLY 1712

From the original in private possession[1]

Frontispiece	li	s	d[2]	
Monument[3]	102.	2.	2	*Tuesday 5 a'clock*
3 Head-pieces[4]				

Sir

I have agreed with the Graver thad did the head pieces for 20 *lb* for Graving the inscription with the Ornaments.[3] he sayes 29 *lb* but I beleive he will not insist on more than 20. So that ther will be in all 51 *lb* due to Du Guernier[5] & 30 *lb* to the other Graver George Vertue,[6] there must be a present to Catinar,[7] the painter, who has really been at a great deal of trouble if it were my own business I could not offer him less than 20 *lb*. Dr Hally expects as much as is sav'd viz 125 *lb* that should have been given to Mr Flamsted I think if he had 25 *lb* more it will not be Grudg'd by Mr lord Treasurer so that will make in all [150 *lb*.] the Account will stand thus

for printing &c	——	102
To Dr Hally	——	150
To the tuo Gravers	——	80
To Catinar	——	20

These are all the requisites for the acct The sooner it is given in the better. (I forgott indeed some tuo Guineas or so for drawing the Architecture[3] to Mr Gibbs.[8]) I refer all to your own judgement being with the greatest respect

Sir
Your most humble servant
Jo. Arbuthnott

NOTES

(1) The letter concerns the account for printing the plates in Flamsteed's *Historia Cœlestis*, which Newton had been asked to draw up for Oxford (see Letters 887 [misprinted 877] and 892, vol. v, pp. 212, 224–5). Presumably Newton had asked Arbuthnot for estimates of the costs involved, and this is Arbuthnot's reply. The sums of money mentioned do not all agree exactly with those mentioned in Letter 892, but are of the same order. Compare also Letter 1177 *a*, vol. vi.

(2) This is presumably the detailed account for printing the plates. It is possibly not in Arbuthnot's hand.

(3) The 'mausoleum' mentioned in Letter 892; see vol. v, p. 225, note (3).

(4) The finished work in fact has five distinct head-pieces, all but one drawn by Catenaro, and engraved by Du Guernier. One, depicting astronomical instruments, is both drawn and engraved by Du Guernier.

(5) See vol. v, p. 225, note (3).

(6) See vol. v, p. 225, note (4).

(7) Catenaro. See vol. v, p. 225, note (6).

(8) James Gibbs.

X.945 NEWTON TO CAWOOD

?1712

From the holograph draft in the possession of Lord Lymington[1]

Mr Cawood

I have not yet been able to strike such a meridian line as we were speaking of, but can do it at any time by a magnetical needle. I desire that you would examin your needles before you bring them & when you are satisfied that either of them will stand constantly without variation let us know it & we will appoint a day for trial.

NOTE

(1) The draft seems to be clearly associated with Letters 940 and 945, vol. v, its 'we' being Newton and Halley.

X.960 NEWTON TO CASPAR NEUMANN

?1712

From a draft in the University Library, Cambridge[1]

Sr

A copy of your book entituled Clavis Domus Heber[2] was presented to me in your name some months ago & I then desired the person who brought me the present to return my thanks, & told him that I did not understand Hebrew. I have looked over it & find the designe very good. The Hebrew tongue is said to be narrow, & the just signification of several words to be almost lost & few books are extant written in the ancient Hebrew, & the designe of recovering the ancient signification of the words must be very commendable: but for want of skill in that tongue I am unable to make a further judgement of the success in this designe then I am perswaded that a person of your abilities has wanted the success desired. I am

NOTES

(1) Add. 3968(41), fo. 98v.

(2) Dr Isaiah Shachar has kindly informed us that only one work in Hebrew bears this title: Caspar Neumann's מפתח בית עבר *hoc est Clavis Domus Heber*, the first part of which was published at Breslau in 1712; subsequent parts were published at Breslau in 1714 and 1715. It is probable that Neumann sent only the first part to Newton, as only this part is recorded as being in Newton's library (De Villamil), hence we date the letter 1712.

Caspar Neumann (1648–1715), German orientalist, became Professor of Theology at the two gymnasia of Breslau in 1697. He had published earlier *Genesis linguæ sanctæ Veteris Testament: Docens Vulgo sic Dictas Radices non esse Vera Hebræorum Primitiva, sed Voces ab alio quodam Radicibus his priore et simpliciore Principio deductas* (Nuremburg, 1696).

X.1006 FAUQUIER TO NEWTON

?JULY 1713

From the original in the Jewish National and
University Library, Jerusalem[1]

Sr

Mr Cartlisch[2] tells me you will him melt the Gold that was weighed in
some time since. but as you ordered me to keep it till I had heard from you, I
have sent my servant to know your pleasure therein; as likewise to desire you
to lett me know whether you will be at the Mint this day with the rest of the
Officers, and if you will have me provide a dinner accordingly

> Your most obedient
> humb. servant
> FAUQUIER

NOTES

(1) Yahuda Collection, Newton MS. 7(4).
(2) John Cartlich. In 1709 he supplied silver, aqua fortis and fine copper to the Mint. In
1711 he received payment for melting down wrought plate brought into the Mint. In 1713 he
supplied fine gold to the Mint for the manufacture of Peace Medals (see Letter 1006, vol. VI).

1137 NEWTON TO ——

22 March 1715

From Letter X.1235 we see that this letter was in fact addressed to Josiah Burchett, and
sent on 22 March 1717.

X.1226 JOSIAH BURCHETT TO NEWTON

2 November 1716. Clerical copy

Requests Newton to call at the Admiralty tomorrow morning 'to view an Instrument
prepared for finding out the Longitude'. (P.R.O., Adm/2, 450, p. 30.)

⟨654⟩ NEWTON TO ?LOWNDES

?1717

See above, correction to Letter 654.

X.1234.1 BURCHETT TO NEWTON

1 March 1717. Clerical copy

Desires report upon instrument for solving triangles invented by the bearer, Caleb
Bassingwhite, 'which may be of great use in the Practice of Navigation'. (P.R.O.,
Adm/2, 450, p. 313.)

X.1234.2   WILLIAM WATERS TO NEWTON

4 March 1717.   Printed notice

Summons to a meeting of the Commissioners for Building Fifty New Churches on 8 March 1717. (Jewish National and University Library, Jerusalem. Yahuda Collection, Newton MS. 7(3).) Compare Letter 1255, vol. vi.

X.1235   BURCHETT TO NEWTON

21 March 1717.   Clerical copy

For the answer see Letter 1137, vol. vi. Desires report on method of finding the longitude proposed by the bearer, John Vat. (P.R.O., Adm/2, 450, p. 360.) Compare vol. vi, p. 211.

⟨1137⟩   NEWTON TO BURCHETT

22 March 1717. See correction to Letter 1137, above.

X.1248   BURCHETT TO NEWTON

19 July 1717.   Clerical copy

Requests report on a project for the longitude in French sent by the King to Mr Secretary [Joseph] Addison. (P.R.O., Adm/2, 450, p. 578.) Compare Admiralty minute, 17 July 1717 (P.R.O., Adm/1, 4100 (unpaginated)).

X.1257   BURCHETT TO NEWTON

20 August 1717.   Clerical copy

Requests report on method for finding longitude by newly discovered properties of the magnetic needle, devised by the bearer John French 'teacher of Mathematicks, and a Schoolmaster on board some of his Majesty's ships.' (P.R.O., Adm/2, 452, p. 22.) Compare Letter 1121, vol. vi.

X.1287   HENRY NEWMAN TO NEWTON

May 1718   Printed notice

Concerning Newton's contributions to the Commissioners for relieving poor proselytes (Mint Papers, ii, fo. 106v.)

X.1289 NEWTON TO THE TREASURY
10 JUNE 1718
From the holograph original in Goldsmiths' Hall[1]

To the Rt Honble the Lords Commissioners of his Ma[jes]ties Treasury
May it please your Lordps
I most humbly pray that the moneys in his Ma[jesty']s Pix may be tried this summer, it being two years since there was a tryall[2]
Which is most humbly sumitted to your Lo[rdshi]ps great wisdome

ISAAC NEWTON

Mint Office
June 10th
   1718.

NOTES
(1) G.I.2; printed in E. G. V. Newman, 'The Gold Metallurgy of Isaac Newton', *Gold Bulletin*, 8, no. 3 (Johannesburg, 1975), p. 94.
(2) The trial took place on 4 August. See Letter 1292, note (2) (vol. VI, p. 450).

X.1329   —— TO NEWTON
October 1719.   Printed notice

Summons to a meeting of the Commissioners for finishing St Paul's Cathedral on 13 October [1719]. (Bodleian Library, New College MS. 361, II, fo. 77v.) Compare Letter 1255, note (1) (vol. VI, 407).

X.1339   WATERS TO NEWTON
2 July 1720.   Printed notice

Summons to a meeting on 6 July 1720 of the Commissioners for Building Fifty New Churches. (Jewish University and National Library, Jerusalem. Yahuda Collection, Newton MS. 7(3).) Compare Letter 1255, vol. VI.

X.1498   —— TO NEWTON
7 March 1727.   Original (torn)

Summons to a vestry meeting on 8 March, to consider the workhouse and other matters. (U.L.C., Add. 3964(8), fo. 16 *bis*.)

APPENDIX II. NEWTON'S GENEALOGY

We have used three main sources for the construction of the family tree following: Foster, who corrects a number of errors made in earlier accounts, including those of Newton himself; Newton's own rough notes and those made by his contemporaries shortly after his death, now amongst the Keynes MSS. in King's College Library, Cambridge; and information which arises incidentally in the *Correspondence*. The aim of the genealogy we have constructed is to identify for the reader the various relatives of Newton mentioned in the course of the *Correspondence* rather than to give a comprehensive account of his whole family. We also take this opportunity to correct a number of minor errors in earlier volumes of the *Correspondence*.

There are a number of relatives whom we have been unable to identify clearly. In addition to Hannah Tonstall and Ralph Ayscough (see note (14) below), there was a Mr Short of Keal, and perhaps Hannah Clarke was also a relative (see Keynes MSS. 127A and 136 and note (25) below). Others who claimed consanguinity may have done so only in the hope of pecuniary benefit—hence the connections of the Chapman family (see Letters 1314 and 1314A), of William Newton (see Letters 1220 and 1236, vol. VI, and Letter 1311), and of Robert Newton (see Letter 1466), with Newton's own family cannot be traced.

As an adjunct to the genealogy we print the frontispiece from Turnor, which shows a map of the area around Woolsthorpe and Colsterworth. This gives the locations of most of the villages where the various branches of Newton's family lived (see Plate II).

NOTES

(1) The Newton branch of the family is fully discussed in Foster's article. Newton himself in his own draft pedigree (see Keynes MS. 112) and Stukeley, following him (see Keynes MS. 136), made various errors as a result of the inadequacy of the information available to them. In particular William Newton of Skillington and Gonerby is stated to be the second John Newton of Westby's brother, not his son. These errors are perpetuated in vol. IV, p. 461, note (2).

(2) The genealogy of the Ayscough family is roughly drafted by Newton himself in Keynes MS. 112 down to the second generation (Newton's cousins) in greater detail than we give here (see in particular note (4) below). The little we know of the third generation (apart from the Smith branch) is derived incidentally from the *Correspondence*.

(3) See Letter 2 (vol. I, pp. 2–3). In note (2) to that letter his grandmother's name is given as Margaret; Newton, in Keynes MS. 112, gives it as Margery.

(4) Details of Sarah Ayscough's three marriages are given in Keynes MS. 112 by Newton. None of her children is mentioned in the *Correspondence*.

(5) For a letter from Hannah Ayscough to her son, Isaac, see Letter 2, vol. I. Her death is often stated, erroneously, as having occurred in 1689; see vol. II, p. 303, note (2).

(6) See vol. I, p. 3, note (2), where it is stated that James Ayscough was Isaac's guardian, and compare Turnor, p. 158, where it is implied that it was James Ayscough senior Isaac's grandfather, who was his guardian.

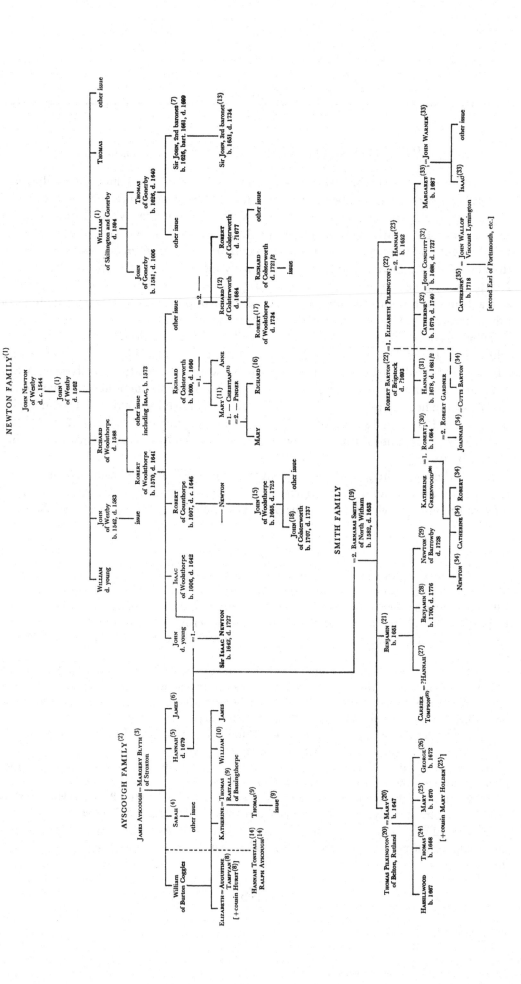

(7) For a discussion of this branch of Newton's family, see Letter 1466, p. 319, note (2), where small errors in biographical information given in earlier volumes are pointed out. Letter 248, vol. II, is from Newton to this Sir John Newton; Letter 583, vol. IV may be from either him or his son, and Letter 1506 was presumably to one of the two.

(8) See Letter 1406, where Tampyan mentions a 'Cous[in] Hurst'.

(9) See Letter 1110, vol. VI.

(10) See Letter 950, vol. V.

(11) Mary Newton and her family are discussed by Foster, p. 23. He gives the surname of her first husband as Christian, whereas Newton, in Keynes MS. 112, gives it as Tompson. This could be a confusion on Newton's part with Benjamin Smith's son-in-law, Carrier Tompson.

(12) The date of Richard's death has been misprinted, confusingly, by Foster in his 'Pedigree IV'.

(13) Letter 719, vol. IV, Letter 1145, vol. VI, and perhaps Letter 1506 are to Sir John Newton, third baronet; Letter 1059, vol. VI, and perhaps Letter 583, vol. IV are from him. He is mentioned in Letter 999, vol. V. See also note (7) above.

(14) Two other Ayscoughs are mentioned in the *Correspondence*: Hannah Tonstall, granddaughter of his 'Unkle Ayscough' (see Letters 1502 and 1549), and Ralph Ayscough (vol. VI, p. xxi), whose relationship to Newton is unknown.

(15) See Letter 1456, note (3), p. 303. Possibly Letter 1507 is addressed to this John Newton. According to a letter from Stukeley to Conduitt, 16 January 1728 (see Keynes MS. 136), Newton gave John Newton land worth £30 p.a. in 1723.

(16) See Letter 1471.

(17) See Letter 1456, note (2), p. 303, and Letter 1488, p. 347. According to a letter from Stukeley to Conduitt, 16 January 1728 (see Keynes MS. 136), Newton purchased a farm for this Robert Newton.

(18) John Newton was Newton's heir, and inherited both the farm Newton had given to his father (see note (15) above), and estates at Sewstern nearby, in all worth £80 p.a. (See Keynes MSS. 136 and 129.) Letter 1507 may have been addressed either to him or to his father.

(19) See Letter 2, vol. I; there the date of Smith's death is given as 1656, whereas Foster gives it as 1653. See also vol. IV, p. 188, note (1), which is incorrect in a number of details (see note (21) below).

(20) For Mary Smith and her husband Thomas Pilkington see vol. V, p. 252, note (1). Mary is also mentioned in Letter 2, vol. I and Letter 906a, vol. V. For their children see *Miscellanea Genealogica et Heraldica*, New Series, 1 (1874), 174.

(21) Benjamin Smith wrote to Newton in 1695; see Letter 540, vol. IV. For his early life see vol. I, p. 3, note (2), and vol. II, p. 303, note (2). Vol. IV, p. 188, note (1) wrongly states that he was the eldest child of Barnabas Smith and Hannah Newton; Foster shows that Mary was in fact the eldest of the three children. Benjamin was baptized in August 1651, hence it is likely that he was born that year, rather than in 1650.

(22) Robert Barton's relationship to Cutts Barton, and his marriage to Elizabeth Pilkington and to Hannah Smith, are discussed in vol. V, p. 200, note (1), and vol. VI, p. xxii. His wife speaks of his illness in Letter 419, vol. III, and Letter 466, vol. III, is possibly addressed to him. Elizabeth Pilkington was Thomas Pilkington's sister; see *Miscellanea Genealogica et Heraldica*, New Series, 1 (1874), 174.

(23) Hannah's marriage and children are discussed in vol. V, p. 200, note (1). She writes to

Newton in Letter 419, vol. III and Letter 1091, vol. VI, and is mentioned in Letter 2, vol. I and Letter 466, vol. III.

(24) Thomas Pilkington was one of Newton's surviving half-nephews and was involved in the administration of Newton's estate after his death. See Keynes MS. 127 A.

(25) For Mary Pilkington's letters to Newton see Letters 906 and 955, vol. V, and Letter X.754.2. In the last of these she mentions a cousin, Mary Holden. Mary Pilkington was one of Newton's eight surviving half-nephews and half-nieces.

In a genealogy given in Nichols, p. 38, the daughter of Thomas and Mary Pilkington is named Hannah, not Mary, and is described as having married J. Clarke. Hannah Clarke's signature also appears on one of the documents relating to the administration of Newton's estate. (See Keynes MS. 136 and 127 A).

(26) George Pilkington is mentioned in Letter 955, vol. V. He was also involved in the administration of Newton's estate after his death. See Keynes MS. 127 A.

(27) In Letter 1515 Newton mentions Carrier Tompson as marrying his niece Smith. The name also appears in the documents relating to the winding up of Newton's estate (see Keynes MS. 127 A). Possibly his niece's christian name was Hannah, as the name Hannah Tompson appears on similar documents in P.R.O., PROB/3/26/66, p. 57. Also this is the name mentioned in Nichols, p. 38.

(28) In Keynes MS. 127 A a Benjamin Smith is mentioned as one of the administrators of Sir Isaac Newton's estate. In Nichols, pp. 32–4, an anecdote concerning Benjamin Smith, Newton's nephew, is reported, derived from a William Sheepshanks. Benjamin is described as having moved to Newton's house in 1718, and having 'chiefly lived' there thereafter. Newton left him £500 p.a., in the form of the income from estates in Nottinghamshire and Rutland. We are also told that 'Among Mr. Smith's papers were several letters from Sir Isaac Newton. In these he addressed his nephew by the familiar name of Ben, and pressed him to chuse a profession. There was some vulgar phraseology in them which induced me to burn them, when I arranged his papers after his death.' Benjamin is described as Rector of Linton in Craven.

(29) In Keynes MS. 127 A a Newton Smith is stated as a relative with a right to administrate Sir Isaac Newton's estate; according to Stukeley (Keynes MS. 136) he was 'in a languishing condition' shortly after Newton's death.

(30) See vol. V, pp. 199–201 and p. 345, and vol. VI, p. xii. Robert Barton is also referred to in Letter 878a, vol. V, but the 'Katherine' mentioned there is in fact Catherine Barton, not Robert's wife Katherine.

(31) According to Turnor, this Hannah Barton was baptized at Colsterworth in 1678. Turnbull (vol. III, p. 279, note (1)) states that she died aged 8½ years, but we have not traced the source of this statement. According to *Miscellanea Genealogica et Heraldica*, New Series, 1 (1874), 174, she died in 1681/2.

(32) References in the *Correspondence* to Catherine Barton and John Conduitt are too numerous to list here; see the indexes to the separate volumes. Foster (p. 18) introduces a confusion by spelling *this* Catherine's name as Katherine. It is stated in vol. V, p. 200, note (1) that Catherine Conduitt had five children; in fact she had one child only, also named Catherine (see note (35) below), who had five children. John Conduitt was instrumental in the winding up of Newton's estate (see Keynes MS. 127 A) and succeeded him as Master of the Mint.

(33) For an undated letter from Margaret Warner to Newton, see Letter 1551. The annuity given by Newton to Margaret is mentioned in Letter 878a, vol. V. Her husband, John Warner, helped in the administration of Newton's estates (see Keynes MS. 127 A and also De Villamil, pp. 49–61, and P.R.O., PROB/3/26/66). In 1726 Newton granted Isaac Warner,

their son, the rents from land in his possession worth £100 p.a. (see British Museum, Add. 5017*(1), fo. 73).

(34) For Katherine Greenwood's children, and the marriage of her daughter Joannah to Cutts Barton, grandson of Robert Barton's first marriage, see vol. VI, p. xxii, and compare Letter 949, vol. V.

(35) Newton bought the child Catherine Conduitt an estate in Kensington shortly before his death; see Keynes MS. 129. She married John Wallop, Viscount Lymington, eldest son of the first Earl of Portsmouth in 1740, by whom she had five children, her eldest son becoming the second Earl of Portsmouth. Through her Newton's papers passed into the hands of the Portsmouth family.

NOTE ADDED IN PROOF

Since this volume was prepared, D. T. Whiteside has in *Mathematical Papers*, VII, p. xxii, note (52) called attention to a group of unpublished drafts (in U.L.C. Add. 4005) of letters addressed to Nathaniel Hawes and Edward Paget in May–July 1694, concerning the Mathematical School at Christ's Hospital (compare vol. III, Letters 452–453, 455). In the final volume of the *Mathematical Papers* will be found a letter from Newton to David Gregory of 11 February 1696/7 concerning the brachistochrone (Christ Church College, Oxford, MS. 346).

Another possible draft letter (Add. 3964 (8), fo. 12r) to an unknown person gives Newton's opinion (*c.* 1707?) of 'the Pump proposed by Dr. Papin'.

We have already noted the existence of other Letters which we have not been able to print in this *Correspondence*, and there are indubitably others of whose existence we remain as yet unaware.

INDEX

BERNOULLI, JOHANN I (*cont.*)
'Extrait de la Reponse à M. Herman':
79 n. 6
Hartsoeker, criticism of: 218–22
and the inverse problem of central forces:
68 n. 13, 76, 78, 79 n. 6
and Keill: xxxii–xxxiii, 12 n. 3, 23 n. 1,
47 n. 6, 48, 81 n. 1, 83, 91 n. 4, 123
n. 10, 196, 197
and Keill's challenge: 11–14 nn. 2 & 3
Kruse's defence of: *see under* KRUSE
and Leibniz: xxxiv, 75, 77, 161–4
and 'Leibniz's' problem: 82
on luminescence: 222 n. 5
and Mariotte's experiment: 67 n. 7
his mathematical studies: 76, 78
Mencke his champion: xxxiii, 12 n. 3,
255 n. 1
and Monmort, correspondence with: 12–13
n. 3, 76, 78, 82
Monmort's critique of: 14 n. 3, 21, 22
Naudé on: 225, 233–4
and Newton: xxiv, xliv, 91, 161–5, 178,
179, 183–5
correspondence with: xxxiii, xxxiv, xxxv,
82, 178, 179, 200, 201
criticized by: xxxiv, 156, 167–8
and Newton's *Opticks*: xxxi–xxxii, 160, 163,
169
English edition (1717): 15–16, 43–7,
51–2
Latin edition (1719): 43–7, 51–2, 62, 64
and Newton's portrait: 163, 165, 178, 179
and Newton's *Principia*
errors in: 54, 76, 78
errors in Book II, Prop. 10: xxxiii, 13 n. 3,
48, 54, 68 n. 1, 69 n. 5
third edition: xxxviii
and Newton's *Traité d'Optique* (1722): 178,
179, 218–20
'nouveau phosphore': 222 n. 5
Opera Omnia: xxv
and Raphson: 82–4
Appendix to the *History of Fluxions*:
xxxiii, 44, 46, 76, 78, 79 n. 3
'Responsio' (1719): xxxiii, 12 n. 3, 54,
56 n. 13
Royal Society
his election to: 44, 46, 47 n. 11, 81 n. 1
his 'expulsion' from: xxxiii, 76, 78,
79 nn. 7 & 8, 80, 82–3, 91 n. 4

and second differences: 48
and Brook Taylor: xxxiii, 11–14, 23 n. 1,
37, 39–40, 76, 77, 79 n. 5, 80 n. 1,
91 n. 4, 151
and Taylor's challenge problem: 30
n. 6
and Varignon, correspondence with: xxxiv,
xxxv, 3 n. 1, 66 n. 1, 69, 70, 80 n. 1,
91 n. 4, 106 n. 2, 122 n. 1, 178, 179,
183–5, 196–8
BERNOULLI, NIKOLAUS I
'Addition': 56 n. 12
and error in *Principia*, Book II, Prop. 10: 54,
56 nn. 11 & 12, 63, 65, 68 n. 11
and Keill's pecuniary challenge: 37, 38
n. 3
and the motion of a pendulum in a resisting
medium: 53
Newton, correspondence with: 54, 56 n. 11
and Stirling: 53, 55 n. 3
Traité d'Optique received by: 160, 163, 169
BERNOULLI, NIKOLAUS II
and the calculus dispute: 23 n. 1
and Keill's 'challenge' problem: xxxii,
xxxiii, 11, 14 n. 4
Monmort's letter to: 12 n. 3
Taylor answered by: xxxiii, 38 n. 4
Traité d'Optique received by: 160, 163, 169
BIANCHINI, FRANCESCO: 255, 357
bibles: 368
biblical chronology: 358; *see also* Newton,
Abrégé de Chronologie
BIESTER, JOHANN PETER: 344 n. 1, 357
LETTER to Newton: 1 February 1726, **1485**,
342–4
BIGNON, JEAN-PAUL
and Des Maizeaux's *Recueil*: 135
and Newton's *Opticks* (1717): 16, 213,
214 n. 4
and Newton's portrait: 106 n. 2
BILLERS, Mr: 34
BILLINGSLEY, CASE: 331
binomial expansion: 240 n. 16
BIRDIKIN, THOMAS: 415–16, 418–19
bishops: *see under relevant see*
BLACKBORN or BLACKBOURN, Mr: 304
BLACKWELL, JOHN: 404, 407
LETTERS to Newton: 8 October 1698,
x.593, 404; 24 October 1698, **x.595**,
404; 12 November 1698, **x.596.1**, 404;
28 November 1698, **x.596.3**, 405

medals: *see under* Mint
men-of-war: 442
MENCKE, JOHANN BURCHARD: *255 n. 1*
LETTER from Newton: 1724, **1419**, 254–5
Johann I Bernoulli championed by: xxxiii
'Epistola ad Broock Taylor': 12 n. 3, 38
n. 4, 150 n. 7, 152 n. 4, 255 n. 1
and Monmort's letters: 12 n. 3
omission of name from R.S. list: 254–5
Mercator's series: 394, 396 n. 8
mercury, transit of, 1723: 269, 270 n. 2, 271,
272
metallurgical assays: xxix
METCALF, EDMUND: 289
Military Knights of Windsor: 359 n. 3
MILLAR *or* MILLER, ROBERT: 466, 470, 473
MILLS, CHARLES
LETTER to the Mint: 23 February 1720,
1335, 86–7
The Mine Adventurers: 420
Mint, the
LETTERS
to the Chester Mint: ?August 1697,
x.571, 400–1
to Sidney Godolphin: 5 August 1702,
x.650.2, 427–8; 13 January 1702/3,
x.657, 429; 20 January 1702/3, **x.
658**, 429–30; 1 September 1703, **x.666**,
432; 4 November 1703, **x.667.1**, 432;
9 December 1703, **x.677.2**, 432; 19
April 1704, **x.672**, 433; 15 November
1704, **x.678**, 436; 10 January 1704/5,
x.684, 436; 2 January 1705/6, **x.705**,
442; ?1705/6, **x.707**, 442; 10 May
1706, **x.708.2**, 444; 20 June 1706,
x.708.5, 446; 5 September 1706, **x.
709.1**, 446; 25 September 1706, **x.
709.2**, 446; 24 March 1706/7, **x.715.3**,
447; 2 June 1707, **x.722**, 448; 9 July
1707, **x.726.2**, 449; August 1707,
x.727.2, 450–1; 28 December 1708,
x.748.4, 468–9; 10 March 1708/9, **x.
752.2**, 470; 1 June 1709, **x.755**, 473;
18 January 1709/10, **x.770**, 474
to the Treasury: 18 August 1698, **x.591.1**,
403; 20 August 1698, **x.591.2**, 403;
8 April 1699, **x.610.1**, 409; May 1699,
x.610.3, 409–10; January 1700/1,
x.630.1, 414; 2 April 1701,
x.633.1, 417; *c.* April 1701, **x.633.3**,
420; 7 May 1701, **x.633.7**, 421; *c.*

End of 1701, **x.643.2**, 423; 10 December 1701, **x.643.3**, 423; 18 March
1701/2, **x.645.1**, 423; 26 March 1702,
x.645.2, 423; ?October 1702, **x.651**,
428; 24 April 1703, **x.663.2**, 430; 5
July 1720, **1340**, 93–4
from Baillie: 21 April 1724, **1433**, 275–6
from Dodington: 21 April 1724, **1433**,
275–6
from Fowle: 27 April 1699, **x.610.2**, 409
from Lowndes: 20 May 1704, **x.673.1**,
434; 8 August 1704, **x.673.3**, 434;
18 March 1706/7, **x.715.2**, 447; 24
January 1708/9, **x.749.2**, 470
from the Ordnance: 23 February 1720,
1335, 86–7
from Scrope: 23 September 1725, **1478**,
333–4; 16 August 1726, **1494**, 350–1
from Taylour: 3 April 1707, **x.715.4**, 448;
8 July 1707, **x.726.1**, 449
from the Treasury: 21 April 1724, **1433**,
275–6
from R. Walpole: 8 February 1722, **1338**,
191–2; 21 April 1724, **1433**, 275–6
from William Yonge: 21 April 1724, **1433**,
275–6
for further correspondence concerning the Mint, see
LETTERS from John Conduitt, Fauquier
and Pinckney, *and also* LETTERS to and
from the Treasury
PETITION TO THE QUEEN: 25 July 1705,
x.694.1, 440
REPRESENTATION TO THE TREASURY:
December 1698, **x.597**, 405
addenda and corrigenda to
Number 550, vol. IV: 398
Letter 607, vol. IV, 407
Letter 650, vol. IV: 425
Letters 717 and 726, vol. IV: 448, 450
Letter 764, vol. IV: 473
assay of ore: 355, 366–7
bimetallism: xli
buildings, repair of: 86–7, 94, 384, 410–11,
414
Charters of: 370
clerks at the Mint: 202–3
coin design and marks: 420, 442
coinage Acts: 6–8
coinage of gold and silver: 8–10
common dollar of the Empire: 89
Comptroller's house: 414

INDEX

Mint (*cont.*)
conformability with Edinburgh Mint: 6
copper coinage: xli–xliii, 32, 33 n. 1, 42, 56–9, 217
assay of: 57–8, 276–8
cutting of copper coin: 36
methods used in coining: 35–6, 58
supply of copper for: 456, 462–3
and William Wood: 272–8
counterfeiters: *see under* counterfeiters
the Country Mints: *see under* Country Mints *and individually under town concerned*
defence of: 410–11.
ducats: 9
East Indian coin: 413–14
engravers: xli, 93–4, 350–1, 433, 435–6
accommodation of: 436
appointment of: 203, 210–12
the engraving of seals: 31–2
exchange rates: xlii, 8–10
of bullion: 472
in Her Majesty's Plantations: 432, 434
with Sweden: 89
exemption of Mint Officers from public duties: 370, 440
export of gold and silver: 8–10
farthings: 33 n.1
foreign exchange: *see* Mint: exchange rates, *above*
French money: 9
gold: xlii, 482
assay of: 340–1
coinage of: 57
importation into the Mint: xli, 349, 433, 482
Newton's comments on: 5, 8–10
guinea, reduction in value of: xlii, 5, 8–10
gulde *or* guilder: 89
half-guineas: 28
halfpence: 33 n.1
hammered coin: 399
Imperial dollars, value of: 87–9
Italian coins: 377
Marien grosch: 89
medals
commemorating great actions: 446
engraving of: 436
for Quenn Anne's coronation: 424–5
and the Vigo Booty: 431, 433
moneyers: 385, 446
and the copper coinage: 57–8

corporation of: 217
petition for payment: 409–10, 421
and the Provost's house: 414
Newton's responsibilities at: xxix, xli–xliii, 384
pistoles: 416–17
porter at the Mint: 202
Portuguese gold coin: 333–4, 340–1
Pyx, trial of the
of 1706: 444
of 1707: 448
of 1718: 484
of 1722: 212
of 1724: 287
of 1725: 334
of Wood's coin: 273 n. 3, 274–8
quarrels with the Ordnance: 94–5, 310 n.1
quarter-guineas: 28
refining furnace: 35
reichthalers: 87–9
Rix dollars of Sweden: 89
salaries at: xlii, 202–3, 210–2, 385
schellings Lubs of Hamburgh: 89
seals: 93–4, 350–1, 423
for the Bahamas: 292
public seals: 191–2
Royal seals: 31–2
silver: xlii
coinage of: 57
value of: 5, 8–10
from wrecks: 181–2
smith at the Mint
his forge: 94–5
his house: 86–7, 409–12
joint employee of Mint and Ordnance: 95, 410
Spanish silver: 181
styvers: 89
Swedish coin: 87–9
tin coin: 217
tin stocks: 428–9
export of: 439–40, 442
receiver of: 432–3
weights, standardization of: 278–9
MOIVRE, *see* DE MOIVRE
MOLYNEUX, THOMAS
LETTERS
to the Chester Mint: ?August 1697, **x.571**, 400–1
to the Treasury: 8 April 1699, **x.610.1**, 409

508

NICOLE, FRANÇOIS: 105, 106, 107 n. 9
a nobleman
   LETTER from Newton: n.d., **1516**, 369
Northampton Mercury: 187–8
Norwich Mint: 427–8
Nouvelles Litteraires: 47 nn. 8 & 9, 86 nn. 20 &
   24, 99, 120, 121, 134, 219, 221

OLDENBURG, HENRY
   LETTERS from Newton: n.d., **1508**, 364;
      January 1674/5, **x.132**, 387–9
   addendum to Letter 166, vol. I: 388
   Leibniz's correspondence with: 18, 161, 163
   Newton's correspondence with: xxxviii,
      107, 161, 163
   and the Royal Society: 364, 387–8
Old Testament, chronology of: xl–xli
OMERIQUE, ANTONIUS HUGO DE: *see* DE
   OMERIQUE
optics
   French interest in: xxxvii, 111–17
   experiments: 51, 52, 111–18
Ordnance, the
   LETTERS
      to the Mint: 23 February 1720, **1335**,
      86–7
      to ?Newton: 2 December 1699, **x.617**, 412
   quarrels with the Mint: xlii, 86–7, 94–6,
      310 n. 1, 409–12
ORFFYRAEUS, JOHANN ERNST ELIAS: *146 n. 3*
   perpetual motion machine of: xxxix, 143–7,
      253–4
orthogonals problem: 42, 82, 139 n. 6
OXFORD, possible letter to: 422 n. 1
OUGHTON, Colonel, M.P. for Coventry: 188
OUGHTRED, WILLIAM: 397–8
OVERTON, BENJAMIN: 411, 420
OVERTON, HENRY: 186, 187 n. 3
OXENDEN, GEORGE: 396 n. 13
OXENDEN, Sir HENRY: 396 n. 14

PAGET, EDWARD: 488
PALMER, SAMUEL: 287–8
PAPIAS OF HIERAPOLIS: 431
Papin's pump: 488
PARKINS, WILLIAM, Rector of Colsterworth:
   365, 371, 373
Parliamentary elections: 385
PARSONS, WILLIAM: 435
PASCAL, ETIENNE: 20 n. 11, 386
PELHAM, H.

WARRANT signed by: 1 November 1721,
   1380, 176–7
PELLET, THOMAS: 128 n. 6
PEMBERTON, HENRY: 248 n. 1
   LETTERS to Newton: ?October 1723, **1413**,
      248–9; ?November 1723, **1414**, 249–
      50; ?November 1723, **1415**, 250–1;
      ?December 1723, **1416**, 251–2; January
      1724, **1420**, 255–62; 11 February 1724,
      1421, 262–3; 18 February 1724, **1422**,
      263; ?March 1724, **1426**, 268–9;
      ?May 1724, **1439**, 282–3; ?August
      1724, **1443**, 288–9; ?September 1724,
      1445, 289–92; ?October 1724, **1447**,
      292–3; ?November 1724, **1448**, 293;
      ?December 1724, **1451**, 297–8; ?Jan-
      uary 1725, **1454**, 301; ?February
      1725, **1457**, 304; ?February 1725,
      1458, 304–5; ?April 1725, **1462**,
      312–13; ?May 1725, **1467**, 320–1;
      17 May 1725, **1468**, 321–2; 31
      May 1725, **1470**, 323; 22 June 1725,
      1472, 326–7; 17 July 1725, **1473**, 327–
      8; 9 February 1726, **1486**, 334–6
   QUERIES ON *PRINCIPIA*, 2nd Edition, pp.
      321–60: **1451a**, 298–300
   QUERIES ON *PRINCIPIA*, 2nd Edition, pp.
      364–8: **1458a**, 306–7
   ENCLOSED PAPER: **1458b**, 308–9
   QUERIES ON *PRINCIPIA*, 2nd Edition, pp.
      389–426: **1462a**, 313–15
   QUERIES ON *PRINCIPIA*, 2nd Edition, pp.
      464–74: **1470a**, 323–5
   on cometary motion: 326–7
   Epistola ad amicum de Cotesii: 321, 322 n. 4
   and Newton
      correspondence with: xxix, xxx, xxxvii–
      xxxix
      Principia, 3rd edition: xxxvii–xxxix, 248–
      52, 255–63, 264–9, 282–3, 288–93, 297–
      301, 304–10, 312–15, 320–8, 344–6
      Principia, English translation of: xxxvii
   A view of Sir Isaac Newton's Philosophy: xxxvii
   and Wilson: 110 n. 1
PEMBERTON, PETER: 400 n. 1
pendulum in a resisting medium: 53–5, 250
PERCIVAL, ?THOMAS: 317 n. 1
   LETTER from Newton: 12 May 1725, **1464**,
      317–18
perpetual motion machines: xxxix, xli,
   143–7, 253–4